AIRPLANE FLYING HANDBOOK

AIRPLANE FLYING HANDBOOK

FAA-H-8083-3A

U.S. DEPARTMENT OF TRANSPORTATION
FEDERAL AVIATION ADMINISTRATION
Flight Standards Service

Foreword by David Soucie

SKYHORSE PUBLISHING

All inquiries should be addressed to Skyhorse Publishing, 307 West 36th Street, 11th Floor, New York, NY 10018.

Skyhorse Publishing books may be purchased in bulk at special discounts for sales promotion, corporate gifts, fund-raising, or educational purposes. Special editions can also be created to specifications. For details, contact the Special Sales Department, Skyhorse Publishing, 307 West 36th Street, 11th Floor, New York, NY 10018 or Skyhorse@skyhorsepublishing.com.

Visit our website at www.skyhorsepublishing.com.

10 9 8 7 6 5 4 3 2

Library of Congress Cataloging-in-Publication Data is available on file.

Print ISBN: 978-1-62914-590-7
Ebook ISBN: 978-1-62914-854-0

Printed in China

THE ART OF FLYING

Freedom from the bounds of gravity, reserved for the Gods in eras past, has somehow evaded the fate of other wondrous artful discoveries that eventually became commonplace. Photography, railways, telephones, automobiles, television, Internet, and countless other modern discoveries born of art and invention seem to have fallen victim to the ordinary. Ancient history also provides lessons of art relegated to science.

Greek artists had visions of glorious structures to honor and shelter the gods. Magnificent coliseums, housing hundreds of thousands of people, were built to provide a view of gladiators ruling over man and beast. These structures, while beautiful to behold, ultimately failed, structurally crumbling under their own weight and taking hundreds of lives with them. Mourning citizens demanded that Caesar take action to assure their safety. This compelled the advancement of scientific design, architecture, and government regulation. Workers with only rote skills replaced the sculptor's talent. Repetitive and routine designs replaced creative invention. Today, architecture and structural engineering has come into balance with art and beauty. Magnificent, modern architectural achievements around the world, safely sheltering hundreds of thousands, stand in beautiful testament to the balance of art, function, and safety.

Flight was also born of visionaries. They conceived of glorious flying machines suspended by lighter-than-air balloons that would someday grace the skies, transporting passengers over impassable land routes to new and exciting places.

These early visionaries gave way to inventors with grand ideas of leaving the earth behind in heavier-than-air flying machines. From within the depths of Devinci's artistic mind, innovation, and creativity, these visions materialized. He discovered the principle of flight was balance. Lift must balance with weight, thrust must balance with drag. His works of art then evolved into practical and viable transportation. Like the marvelous ancient structures, flying machines also crumbled under their own weight, claiming the lives of those inside. Before commercial air travel was viable, citizens demanded the government take action to assure their safety. This compelled the advancement of aeronautical design and systems engineering. Similar to the evolution of architecture, aeronautics and engineering have come into balance with innovation and creativity. Magnificent flying machines with incredible complexity gracefully fly over our heads, carrying hundreds of people, proudly displaying their balance of beauty, function, and safety. With every balance, however, there is a new imbalance.

As modern technology encroaches on the controls in the cockpit, the art and intrigue of flying persists. Flying has not yet relegated itself to rote repetitive skills as have other arts. Technology is exceeding its role of making flight safer and creating a new imbalance. Will flying continue to transcend the commonplace or will technology push it into routine?

Perhaps the wonder and fascination of technology and automated flight started to crumble under its own weight? Hundreds of people have recently lost their lives as technology was allowed to take the controls. Will citizens call upon government regulation to be more restrictive, to take action to protect the passengers, crew, and their families from more technological failures? Or is it in the hands and minds of the flyers—the artists of aviation—to find a balance?

I am hopeful the art of flying will find the perfect balance with technology and automated flight. Only time will tell. Only the minds of flyers whose artful hands are on the controls will be able to find this balance—the hands of flyers like you, who find beauty and art in being free from the bounds of earth. Flyers who won't succumb to the lure of technology, automation, and rote skills that would otherwise commandeer the art of flight.

The joy of piercing clouds, soaring over majestic mountains and vast oceans like gods, can never be replaced with technology. Flying will continue to defy its presumed destiny of becoming routine. It will attain yet another natural balance: a perfect balance of technology and art, function, and safety.

—David Soucie

PREFACE

The *Airplane Flying Handbook* is designed as a technical manual to introduce basic pilot skills and knowledge that are essential for piloting airplanes. It provides information on transition to other airplanes and the operation of various airplane systems. It is developed by the Flight Standards Service, Airman Testing Standards Branch, in cooperation with various aviation educators and industry.

This handbook is developed to assist student pilots learning to fly airplanes. It is also beneficial to pilots who wish to improve their flying proficiency and aeronautical knowledge, those pilots preparing for additional certificates or ratings, and flight instructors engaged in the instruction of both student and certificated pilots. It introduces the future pilot to the realm of flight and provides information and guidance in the performance of procedures and maneuvers required for pilot certification. Topics such as navigation and communication, meteorology, use of flight information publications, regulations, and aeronautical decision making are available in other Federal Aviation Administration (FAA) publications.

This handbook conforms to pilot training and certification concepts established by the FAA. There are different ways of teaching, as well as performing flight procedures and maneuvers, and many variations in the explanations of aerodynamic theories and principles. This handbook adopts a selective method and concept of flying airplanes. The discussion and explanations reflect the most commonly used practices and principles. Occasionally the word "must" or similar language is used where the desired action is deemed critical. The use of such language is not intended to add to, interpret, or relieve a duty imposed by Title 14 of the Code of Federal Regulations (14 CFR).

It is essential for persons using this handbook to also become familiar with and apply the pertinent parts of 14 CFR and the *Aeronautical Information Manual (AIM)*. The AIM is available online at **http://www.faa.gov/atpubs**. Performance standards for demonstrating competence required for pilot certification are prescribed in the appropriate airplane practical test standard.

The current Flight Standards Service airman training and testing material and subject matter knowledge codes for all airman certificates and ratings can be obtained from the Flight Standards Service Web site at **http://av-info.faa.gov**.

The FAA greatly acknowledges the valuable assistance provided by many individuals and organizations throughout the aviation community whose expertise contributed to the preparation of this handbook.

This handbook supersedes FAA-H-8083-3, *Airplane Flying Handbook*, dated 1999. This handbook also supersedes AC 61-9B, *Pilot Transition Courses for Complex Single-Engine and Light Twin-Engine Airplanes*, dated 1974; and related portions of AC 61-10A, *Private and Commercial Pilots Refresher Courses*, dated 1972. This revision expands all technical subject areas from the previous edition, FAA-H-8083-3. It also incorporates new areas of safety concerns and technical information not previously covered. The chapters covering transition to seaplanes and skiplanes have been removed. They will be incorporated into a new handbook (under development), FAA-H-8083-23, *Seaplane, Skiplane and Float/Ski Equipped Helicopter Operations Handbook.*

This handbook is available for download from the Flight Standards Service Web site at **http://av-info.faa.gov**. This web site also provides information about availablity of printed copies.

This handbook is published by the U.S. Department of Transportation, Federal Aviation Administration, Airman Testing Standards Branch, AFS-630, P.O. Box 25082, Oklahoma City, OK 73125. Comments regarding this handbook should be sent in e-mail form to **AFS630comments@faa.gov**.

AC 00-2, Advisory Circular Checklist, transmits the current status of FAA advisory circulars and other flight information publications. This checklist is available via the Internet at **http://www.faa.gov/aba/html_policies/ac00_2.html**.

CONTENTS

Chapter 6—Ground Reference Maneuvers

Chapter 7—Airport Traffic Patterns

Chapter 8—Approaches and Landings

Chapter 9—Performance Maneuvers

Chapter 10—Night Operations

Chapter 11—Transition to Complex Airplanes

Chapter 12—Transition to Multiengine Airplanes

Chapter 13—Transition to Tailwheel Airplanes

Chapter 14—Transition to Turbopropeller Powered Airplanes

Chapter 15—Transition to Jet Powered Airplanes

Chapter 16—Emergency Procedures

Chapter 1

Introduction to Flight Training

PURPOSE OF FLIGHT TRAINING

The overall purpose of primary and intermediate flight training, as outlined in this handbook, is the acquisition and honing of **basic airmanship skills. Airmanship** can be defined as:

- A sound acquaintance with the principles of flight,
- The ability to operate an airplane with competence and precision both on the ground and in the air, and
- The exercise of sound judgment that results in optimal operational safety and efficiency.

Learning to fly an airplane has often been likened to learning to drive an automobile. This analogy is misleading. Since an airplane operates in a different environment, three dimensional, it requires a type of motor skill development that is more sensitive to this situation such as:

- **Coordination**—The ability to use the hands and feet together subconsciously and in the proper relationship to produce desired results in the airplane.
- **Timing**—The application of muscular coordination at the proper instant to make flight, and all maneuvers incident thereto, a constant smooth process.
- **Control touch**—The ability to sense the action of the airplane and its probable actions in the immediate future, with regard to attitude and speed variations, by the sensing and evaluation of varying pressures and resistance of the control surfaces transmitted through the cockpit flight controls.
- **Speed sense**—The ability to sense instantly and react to any reasonable variation of airspeed.

An airman becomes one with the airplane rather than a machine operator. An accomplished airman demonstrates the ability to assess a situation quickly and accurately and deduce the correct procedure to be followed under the circumstance; to analyze accurately the probable results of a given set of circumstances or of a proposed procedure; to exercise care and due regard for safety; to gauge accurately the performance of the airplane; and to recognize personal limitations and limitations of the airplane and avoid approaching the critical points of each. The development of airmanship skills requires effort and dedication on the part of both the student pilot and the flight instructor, beginning with the very first training flight where proper habit formation begins with the student being introduced to good operating practices.

Every airplane has its own particular flight characteristics. The purpose of primary and intermediate flight training, however, is not to learn how to fly a particular make and model airplane. The underlying purpose of flight training is to develop skills and safe habits that are transferable to any airplane. Basic airmanship skills serve as a firm foundation for this. The pilot who has acquired necessary airmanship skills during training, and demonstrates these skills by flying training-type airplanes with precision and safe flying habits, will be able to easily transition to more complex and higher performance airplanes. It should also be remembered that the goal of flight training is a safe and competent pilot, and that passing required practical tests for pilot certification is only incidental to this goal.

ROLE OF THE FAA

The Federal Aviation Administration (FAA) is empowered by the U.S. Congress to promote aviation safety by prescribing safety standards for civil aviation. This is accomplished through the Code of Federal Regulations (CFRs) formerly referred to as Federal Aviation Regulations (FARs).

Title 14 of the Code of Federal Regulations (14 CFR) part 61 pertains to the certification of pilots, flight instructors, and ground instructors. 14 CFR part 61 prescribes the eligibility, aeronautical knowledge, flight proficiency, and training and testing requirements for each type of pilot certificate issued.

14 CFR part 67 prescribes the medical standards and certification procedures for issuing medical certificates for airmen and for remaining eligible for a medical certificate.

14 CFR part 91 contains general operating and flight rules. The section is broad in scope and provides general guidance in the areas of general flight rules, visual flight rules (VFR), instrument flight rules (IFR), aircraft maintenance, and preventive maintenance and alterations.

Within the FAA, the Flight Standards Service sets the aviation standards for airmen and aircraft operations in the United States and for American airmen and aircraft around the world. The FAA Flight Standards Service is headquartered in Washington, D.C., and is broadly organized into divisions based on work function (Air Transportation, Aircraft Maintenance, Technical Programs, a Regulatory Support Division based in Oklahoma City, OK, and a General Aviation and Commercial Division). Regional Flight Standards division managers, one at each of the FAA's nine regional offices, coordinate Flight Standards activities within their respective regions.

The interface between the FAA Flight Standards Service and the aviation community/general public is the local Flight Standards District Office (FSDO). [Figure 1-1] The approximately 90 FSDOs are strategically located across the United States, each office having jurisdiction over a specific geographic area. The individual FSDO is responsible for all air activity occurring within its geographic boundaries. In addition to accident investigation and the enforcement of aviation regulations, the individual FSDO is responsible for the certification and surveillance of air carriers, air operators, flight schools/training centers, and airmen including pilots and flight instructors.

Each FSDO is staffed by aviation safety inspectors whose specialties include operations, maintenance, and avionics. General aviation operations inspectors are highly qualified and experienced aviators. Once accepted for the position, an inspector must satisfactorily complete a course of indoctrination training conducted at the FAA Academy, which includes airman evaluation and pilot testing techniques and procedures. Thereafter, the inspector must complete recurrent training on a regular basis. Among other duties, the FSDO inspector is responsible for administering FAA practical tests for pilot and flight instructor certificates and associated ratings. All questions concerning pilot certification (and/or requests for other aviation information or services) should be directed to the FSDO having jurisdiction in the particular geographic area. FSDO telephone numbers are listed in the blue pages of the telephone directory under United States Government offices, Department of Transportation, Federal Aviation Administration.

Figure 1-1. FAA FSDO.

ROLE OF THE PILOT EXAMINER

Pilot and flight instructor certificates are issued by the FAA upon satisfactory completion of required knowledge and practical tests. The administration of these tests is an FAA responsibility normally carried out at the FSDO level by FSDO inspectors. The FAA, however, being a U.S. government agency, has limited resources and must prioritize its responsibilities. The agency's highest priority is the surveillance of certificated air carriers, with the certification of airmen (including pilots and flight instructors) having a lower priority.

In order to satisfy the public need for pilot testing and certification services, the FAA delegates certain of these responsibilities, as the need arises, to private individuals who are not FAA employees. A designated pilot examiner (DPE) is a private citizen who is designated as a representative of the FAA Administrator to perform specific (but limited) pilot certification tasks on behalf of the FAA, and may charge a reasonable fee for doing so. Generally, a DPE's authority is limited to accepting applications and conducting practical tests leading to the issuance of specific pilot certificates and/or ratings. A DPE operates under the direct supervision of the FSDO that holds the examiner's designation file. A FSDO inspector is assigned to monitor the DPE's certification activities. Normally, the DPE is authorized to conduct these activities only within the designating FSDO's jurisdictional area.

The FAA selects only highly qualified individuals to be designated pilot examiners. These individuals must have good industry reputations for professionalism, high integrity, a demonstrated willingness to serve the public, and adhere to FAA policies and procedures in certification matters. A designated pilot examiner is expected to administer practical tests with the same degree of professionalism, using the same methods, procedures, and standards as an FAA aviation safety inspector. It should be remembered, however, that a DPE is not an FAA aviation safety inspector. A DPE cannot initiate enforcement action, investigate accidents, or perform surveillance activities on behalf of the FAA. However, the majority of FAA practical tests at the recreational, private, and commercial pilot level are administered by FAA designated pilot examiners.

ROLE OF THE FLIGHT INSTRUCTOR

The flight instructor is the cornerstone of aviation safety. The FAA has adopted an operational training concept that places the full responsibility for student training on the authorized flight instructor. In this role, the instructor assumes the total responsibility for training the student pilot in all the knowledge areas and skills necessary to operate safely and competently as a certificated pilot in the National Airspace System. This training will include airmanship skills, pilot judgment and decision making, and accepted good operating practices.

An FAA certificated flight instructor has to meet broad flying experience requirements, pass rigid knowledge and practical tests, and demonstrate the ability to apply recommended teaching techniques before being certificated. In addition, the flight instructor's certificate must be renewed every 24 months by showing continued success in training pilots, or by satisfactorily completing a flight instructor's refresher course or a practical test designed to upgrade aeronautical knowledge, pilot proficiency, and teaching techniques.

A pilot training program is dependent on the quality of the ground and flight instruction the student pilot receives. A good flight instructor will have a thorough understanding of the learning process, knowledge of the fundamentals of teaching, and the ability to communicate effectively with the student pilot.

A good flight instructor will use a syllabus and insist on correct techniques and procedures from the beginning of training so that the student will develop proper habit patterns. The syllabus should embody the "building block" method of instruction, in which the student progresses from the known to the unknown. The course of instruction should be laid out so that each new maneuver embodies the principles involved in the performance of those previously undertaken. Consequently, through each new subject introduced, the student not only learns a new principle or technique, but broadens his/her application of those previously learned and has his/her deficiencies in the previous maneuvers emphasized and made obvious.

The flying habits of the flight instructor, both during flight instruction and as observed by students when conducting other pilot operations, have a vital effect on safety. Students consider their flight instructor to be a paragon of flying proficiency whose flying habits they, consciously or unconsciously, attempt to imitate. For this reason, a good flight instructor will meticulously observe the safety practices taught the students. Additionally, a good flight instructor will carefully observe all regulations and recognized safety practices during all flight operations.

Generally, the student pilot who enrolls in a pilot training program is prepared to commit considerable time, effort, and expense in pursuit of a pilot certificate. The student may tend to judge the effectiveness of the flight instructor, and the overall success of the pilot training program, solely in terms of being able to pass the requisite FAA practical test. A good flight instructor, however, will be able to communicate to the student that evaluation through practical tests is a mere sampling of pilot ability that is compressed into a short period of time. The flight instructor's role, however, is to train the "total" pilot.

SOURCES OF FLIGHT TRAINING

The major sources of flight training in the United States include FAA-approved pilot schools and training centers, non-certificated (14 CFR part 61) flying schools, and independent flight instructors. FAA "approved" schools are those flight schools certificated by the FAA as pilot schools under 14 CFR part 141. [Figure 1-2] Application for certification is voluntary, and the school must meet stringent requirements for personnel, equipment, maintenance, and facilities. The school must operate in accordance with an established curriculum, which includes a training course outline (TCO)

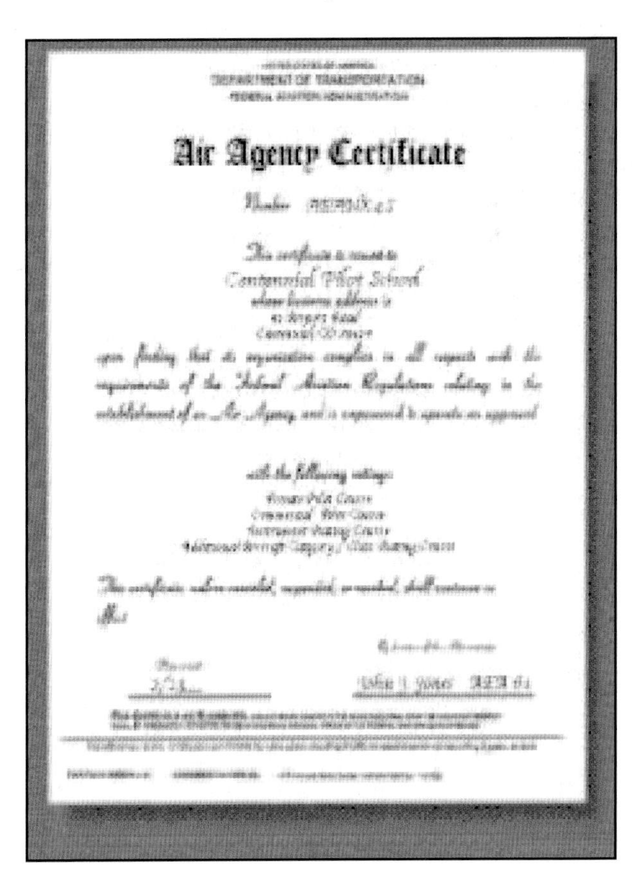
Air Agency Certificate

Figure 1-2. FAA-approved pilot school certificate.

approved by the FAA. The TCO must contain student enrollment prerequisites, detailed description of each lesson including standards and objectives, expected accomplishments and standards for each stage of training, and a description of the checks and tests used to measure a student's accomplishments. FAA-approved pilot school certificates must be renewed every 2 years. Renewal is contingent upon proof of continued high quality instruction and a minimum level of instructional activity. Training at an FAA certificated pilot school is structured. Because of this structured environment, the CFRs allow graduates of these pilot schools to meet the certification experience requirements of 14 CFR part 61 with less flight time. Many FAA certificated pilot schools have designated pilot examiners (DPEs) on their staff to administer FAA practical tests. Some schools have been granted examining authority by the FAA. A school with examining authority for a particular course or courses has the authority to recommend its graduates for pilot certificates or ratings without further testing by the FAA. A list of FAA certificated pilot schools and their training courses can be found in Advisory Circular (AC) 140-2, *FAA Certificated Pilot School Directory*.

FAA-approved training centers are certificated under 14 CFR part 142. Training centers, like certificated pilot schools, operate in a structured environment with approved courses and curricula, and stringent standards for personnel, equipment, facilities, operating procedures and record keeping. Training centers certificated under 14 CFR part 142, however, specialize in the use of flight simulation (flight simulators and flight training devices) in their training courses.

The overwhelming majority of flying schools in the United States are not certificated by the FAA. These schools operate under the provisions of 14 CFR part 61. Many of these non-certificated flying schools offer excellent training, and meet or exceed the standards required of FAA-approved pilot schools. Flight instructors employed by non-certificated flying schools, as well as independent flight instructors, must meet the same basic 14 CFR part 61 flight instructor requirements for certification and renewal as those flight instructors employed by FAA certificated pilot schools. In the end, any training program is dependent upon the quality of the ground and flight instruction a student pilot receives.

PRACTICAL TEST STANDARDS

Practical tests for FAA pilot certificates and associated ratings are administered by FAA inspectors and designated pilot examiners in accordance with FAA-developed practical test standards (PTS). [Figure 1-3] 14 CFR part 61 specifies the areas of operation in which knowledge and skill must be demonstrated by the applicant. The CFRs provide the flexibility to permit the FAA to publish practical test standards containing the areas of operation and specific tasks in which competence must be demonstrated. The FAA requires that all practical tests be conducted in accordance with the appropriate practical test standards and the policies set forth in the Introduction section of the practical test standard book.

It must be emphasized that the practical test standards book is a testing document rather than a teaching document. An appropriately rated flight instructor is responsible for training a pilot applicant to acceptable standards in all subject matter areas, procedures, and maneuvers included in the tasks within each area of operation in the appropriate practical test standard. The pilot applicant should be familiar with this book and refer to the standards it contains during training. However, the practical test standard book is not intended to be used as a training syllabus. It contains the standards to which maneuvers/procedures on FAA practical tests must be performed and the FAA policies governing the administration of practical tests. Descriptions of tasks, and information on how to perform maneuvers and procedures are contained in reference and teaching documents such as this handbook. A list of reference documents is contained in the Introduction section of each practical test standard book.

Practical test standards may be downloaded from the Regulatory Support Division's, AFS-600, Web site at **http://afs600.faa.gov**. Printed copies of practical test standards can be purchased from the Superintendent of Documents, U.S. Government Printing Office, Washington, DC 20402. The official online bookstore Web site for the U.S. Government Printing Office is **www.access.gpo.gov**.

FLIGHT SAFETY PRACTICES

In the interest of safety and good habit pattern formation, there are certain basic flight safety practices and procedures that must be emphasized by the flight instructor, and adhered to by both instructor and student, beginning with the very first dual instruction flight. These include, but are not limited to, collision avoidance procedures including proper scanning techniques and clearing procedures, runway incursion avoidance, stall awareness, positive transfer of controls, and cockpit workload management.

COLLISION AVOIDANCE

All pilots must be alert to the potential for midair collision and near midair collisions. The general operating and flight rules in 14 CFR part 91 set forth the concept of "See and Avoid." This concept requires that vigilance shall be maintained at all times, by each person operating an aircraft regardless of whether the operation is conducted under instrument

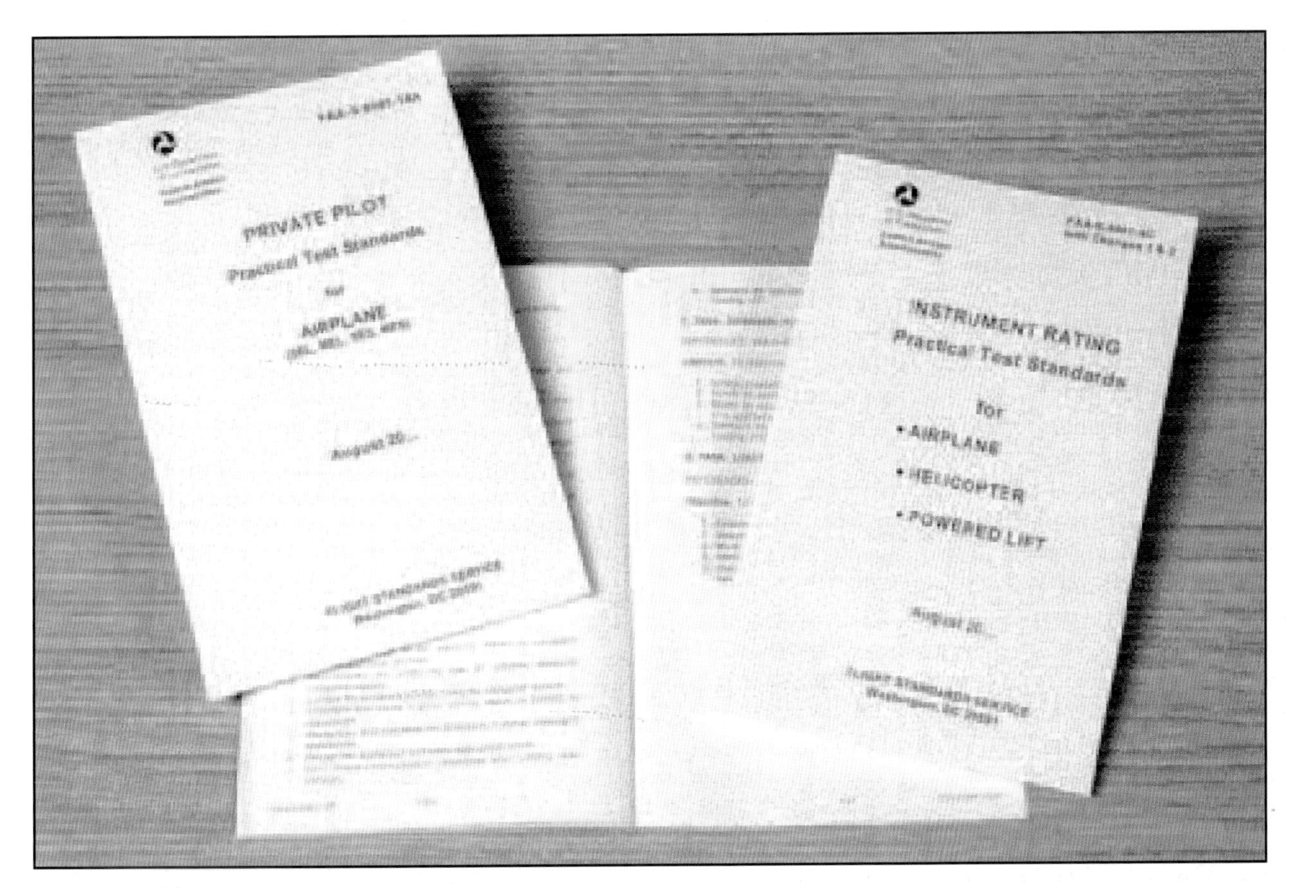

Figure 1-3. PTS books.

flight rules (IFR) or visual flight rules (VFR). Pilots should also keep in mind their responsibility for continuously maintaining a vigilant lookout regardless of the type of aircraft being flown and the purpose of the flight. Most midair collision accidents and reported near midair collision incidents occur in good VFR weather conditions and during the hours of daylight. Most of these accident/incidents occur within 5 miles of an airport and/or near navigation aids.

The "See and Avoid" concept relies on knowledge of the limitations of the human eye, and the use of proper visual scanning techniques to help compensate for these limitations. The importance of, and the proper techniques for, visual scanning should be taught to a student pilot at the very beginning of flight training. The competent flight instructor should be familiar with the visual scanning and collision avoidance information contained in Advisory Circular (AC) 90-48, *Pilots' Role in Collision Avoidance*, and the *Aeronautical Information Manual* (AIM).

There are many different types of clearing procedures. Most are centered around the use of clearing turns. The essential idea of the clearing turn is to be certain that the next maneuver is not going to proceed into another airplane's flightpath. Some pilot training programs have hard and fast rules, such as requiring two 90° turns in opposite directions before executing any training maneuver. Other types of clearing procedures may be developed by individual flight instructors. Whatever the preferred method, the flight instructor should teach the beginning student an effective clearing procedure and insist on its use. The student pilot should execute the appropriate clearing procedure before all turns and before executing any training maneuver. Proper clearing procedures, combined with proper visual scanning techniques, are the most effective strategy for collision avoidance.

RUNWAY INCURSION AVOIDANCE

A runway incursion is any occurrence at an airport involving an aircraft, vehicle, person, or object on the ground that creates a collision hazard or results in a loss of separation with an aircraft taking off, landing, or intending to land. The three major areas contributing to runway incursions are:

- Communications,
- Airport knowledge, and
- Cockpit procedures for maintaining orientation.

Taxi operations require constant vigilance by the entire flight crew, not just the pilot taxiing the airplane. This is especially true during flight training operations. Both the student pilot and the flight instructor need to be continually aware of the movement and location of other aircraft and ground vehicles on the airport movement area. Many flight training activities are conducted at non-tower controlled airports. The absence of an operating airport control tower creates a need for increased vigilance on the part of pilots operating at those airports.

Planning, clear communications, and enhanced situational awareness during airport surface operations will reduce the potential for surface incidents. Safe aircraft operations can be accomplished and incidents eliminated if the pilot is properly trained early on and, throughout his/her flying career, accomplishes standard taxi operating procedures and practices. This requires the development of the formalized teaching of safe operating practices during taxi operations. The flight instructor is the key to this teaching. The flight instructor should instill in the student an awareness of the potential for runway incursion, and should emphasize the runway incursion avoidance procedures contained in Advisory Circular (AC) 91-73, Part 91 *Pilot and Flightcrew Procedures During Taxi Operations and Part 135 Single-Pilot Operations.*

STALL AWARENESS

14 CFR part 61 requires that a student pilot receive and log flight training in stalls and stall recoveries prior to solo flight. During this training, the flight instructor should emphasize that the direct cause of every stall is an excessive angle of attack. The student pilot should fully understand that there are any number of flight maneuvers which may produce an increase in the wing's angle of attack, but the stall does not occur until the angle of attack becomes excessive. This "critical" angle of attack varies from 16 to 20° depending on the airplane design.

The flight instructor must emphasize that low speed is not necessary to produce a stall. The wing can be brought to an excessive angle of attack at any speed. High pitch attitude is not an absolute indication of proximity to a stall. Some airplanes are capable of vertical flight with a corresponding low angle of attack. Most airplanes are quite capable of stalling at a level or near level pitch attitude.

The key to stall awareness is the pilot's ability to visualize the wing's angle of attack in any particular circumstance, and thereby be able to estimate his/her margin of safety above stall. This is a learned skill that must be acquired early in flight training and carried through the pilot's entire flying career. The pilot must understand and appreciate factors such as airspeed, pitch attitude, load factor, relative wind, power setting, and aircraft configuration in order to develop a reasonably accurate mental picture of the wing's angle of attack at any particular time. It is essential to flight safety that a pilot take into consideration this visualization of the wing's angle of attack prior to entering any flight maneuver.

USE OF CHECKLISTS

Checklists have been the foundation of pilot standardization and cockpit safety for years. The checklist is an aid to the memory and helps to ensure that critical items necessary for the safe operation of aircraft are not overlooked or forgotten. However, checklists are of no value if the pilot is not committed to its use. Without discipline and dedication to using the checklist at the appropriate times, the odds are on the side of error. Pilots who fail to take the checklist seriously become complacent and the only thing they can rely on is memory.

The importance of consistent use of checklists cannot be overstated in pilot training. A major objective in primary flight training is to establish habit patterns that will serve pilots well throughout their entire flying career. The flight instructor must promote a positive attitude toward the use of checklists, and the student pilot must realize its importance. At a minimum, prepared checklists should be used for the following phases of flight.

- Preflight Inspection.
- Before Engine Start.
- Engine Starting.
- Before Taxiing.
- Before Takeoff.
- After Takeoff.
- Cruise.
- Descent.
- Before Landing.
- After Landing.
- Engine Shutdown and Securing.

POSITIVE TRANSFER OF CONTROLS

During flight training, there must always be a clear understanding between the student and flight instructor of who has control of the aircraft. Prior to any dual training flight, a briefing should be conducted that includes the procedure for the exchange of flight controls. The following three-step process for the exchange of flight controls is highly recommended.

When a flight instructor wishes the student to take control of the aircraft, he/she should say to the student, "You have the flight controls." The student should acknowledge immediately by saying, "I have the flight controls." The flight instructor confirms by again saying, "You have the flight controls." Part of the procedure should be a visual check to ensure that the other person actually has the flight controls. When returning the controls to the flight instructor, the student should follow the same procedure the instructor used when giving control to the student. The student should stay on the controls until the instructor says: "I have the flight controls." There should never be any doubt as to who is flying the airplane at any one time. Numerous accidents have occurred due to a lack of communication or misunderstanding as to who actually had control of the aircraft, particularly between students and flight instructors. Establishing the above procedure during initial training will ensure the formation of a very beneficial habit pattern.

Chapter 2

Ground Operations

VISUAL INSPECTION

The accomplishment of a safe flight begins with a careful visual inspection of the airplane. The purpose of the preflight visual inspection is twofold: to determine that the airplane is legally airworthy, and that it is in condition for safe flight. The airworthiness of the airplane is determined, in part, by the following certificates and documents, which must be on board the airplane when operated. [Figure 2-1]

- Airworthiness certificate.
- Registration certificate.
- FCC radio station license, if required by the type of operation.
- Airplane operating limitations, which may be in the form of an FAA-approved Airplane Flight Manual and/or Pilot's Operating Handbook (AFM/POH), placards, instrument markings, or any combination thereof.

Airplane logbooks are not required to be kept in the airplane when it is operated. However, they should be inspected prior to flight to show that the airplane has had required tests and inspections. Maintenance records for the airframe and engine are required to be kept. There may also be additional propeller records.

At a minimum, there should be an annual inspection within the preceding 12-calendar months. In addition, the airplane may also be required to have a 100-hour inspection in accordance with Title14 of the Code of Federal Regulations (14 CFR) part 91, section 91.409(b).

If a transponder is to be used, it is required to be inspected within the preceding 24-calendar months. If the airplane is operated under instrument flight rules (IFR) in controlled airspace, the pitot-static system is also required to be inspected within the preceding 24-calendar months.

The emergency locator transmitter (ELT) should also be checked. The ELT is battery powered, and the battery replacement or recharge date should not be exceeded.

Airworthiness Directives (ADs) have varying compliance intervals and are usually tracked in a separate area of the appropriate airframe, engine, or propeller record.

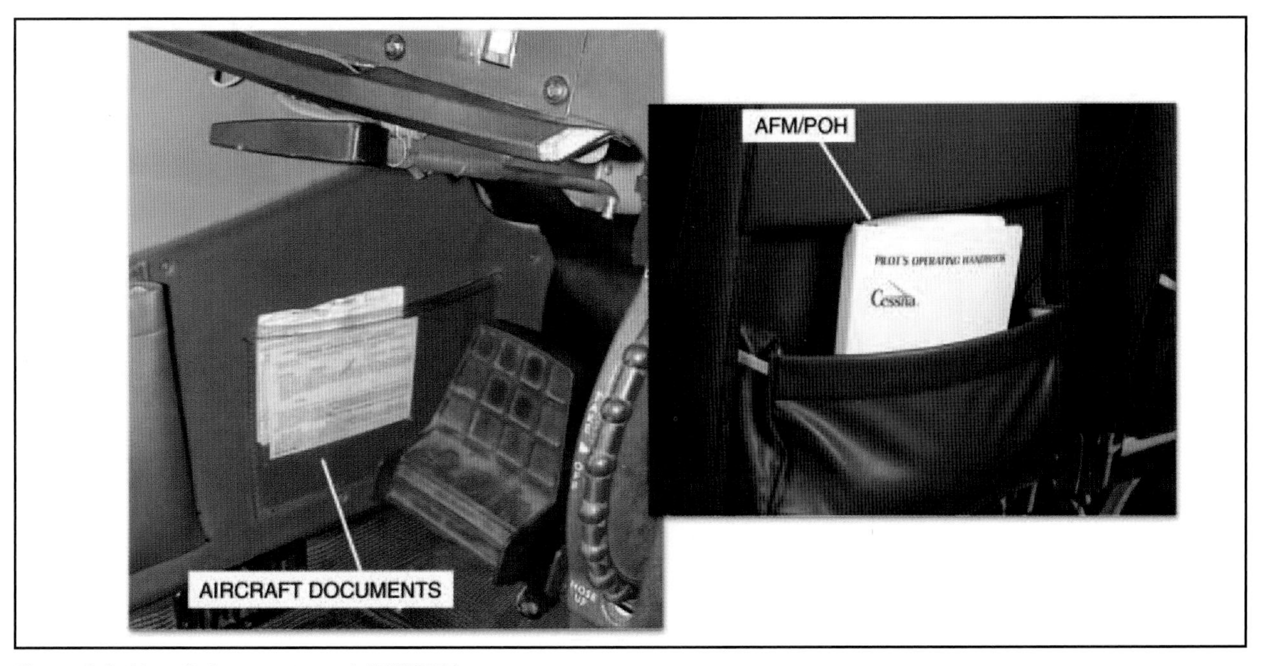

Figure 2-1. Aircraft documents and AFM/POH.

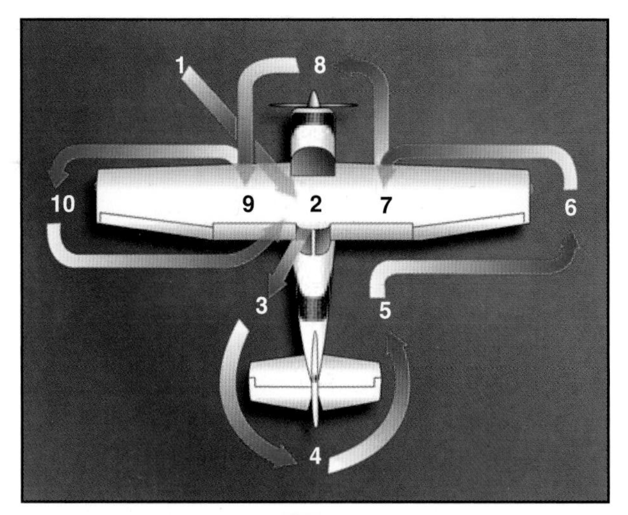

Figure 2-2. Preflight inspection.

The determination of whether the airplane is in a condition for safe flight is made by a preflight inspection of the airplane and its components. [Figure 2-2] The preflight inspection should be performed in accordance with a printed checklist provided by the airplane manufacturer for the specific make and model airplane. However, the following general areas are applicable to all airplanes.

The preflight inspection of the airplane should begin while approaching the airplane on the ramp. The pilot should make note of the general appearance of the airplane, looking for obvious discrepancies such as a landing gear out of alignment, structural distortion, skin damage, and dripping fuel or oil leaks. Upon reaching the airplane, all tiedowns, control locks, and chocks should be removed.

INSIDE THE COCKPIT

The inspection should start with the cabin door. If the door is hard to open or close, or if the carpeting or seats are wet from a recent rain, there is a good chance that the door, fuselage, or both are misaligned. This may be a sign of structural damage.

The windshield and side windows should be examined for cracks and/or crazing. Crazing is the first stage of delamination of the plastic. Crazing decreases visibility, and a severely crazed window can result in near zero visibility due to light refraction at certain angles to the sun.

The pilot should check the seats, seat rails, and seat belt attach points for wear, cracks, and serviceability. The seat rail holes where the seat lock pins fit should

Figure 2-3. Inside the cockpit.

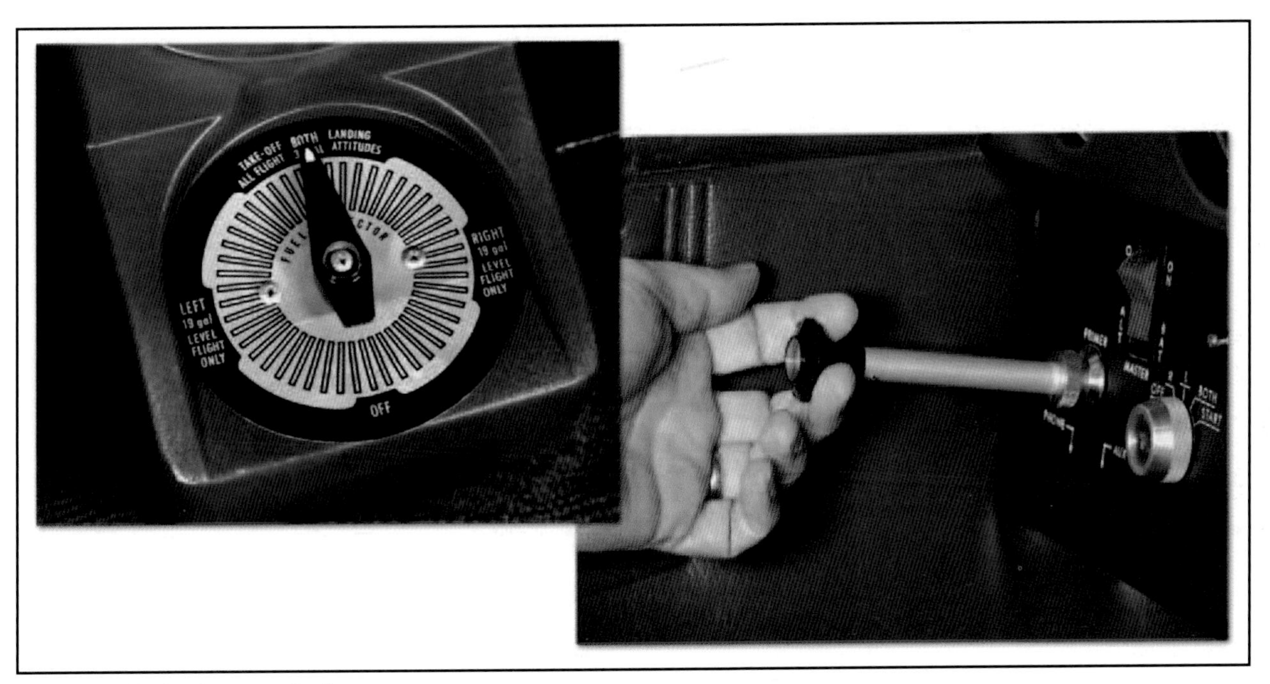

Figure 2-4. Fuel selector and primer.

also be inspected. The holes should be round and not oval. The pin and seat rail grips should also be checked for wear and serviceability.

Inside the cockpit, three key items to be checked are: (1) battery and ignition switches—off, (2) control column locks—*removed*, (3) landing gear control—*down and locked*. [Figure 2-3]

The fuel selectors should be checked for proper operation in all positions—including the OFF position. Stiff selectors, or ones where the tank position is hard to find, are unacceptable. The primer should also be exercised. The pilot should feel resistance when the primer is both pulled out and pushed in. The primer should also lock securely. Faulty primers can interfere with proper engine operation. [Figure 2-4] The engine controls should also be manipulated by slowly moving each through its full range to check for binding or stiffness.

The airspeed indicator should be properly marked, and the indicator needle should read zero. If it does not, the instrument may not be calibrated correctly. Similarly, the vertical speed indicator (VSI) should also read zero when the airplane is on the ground. If it does not, a small screwdriver can be used to zero the instrument. The VSI is the only flight instrument that a pilot has the prerogative to adjust. All others must be adjusted by an FAA certificated repairman or mechanic.

The magnetic compass is a required instrument for both VFR and IFR flight. It must be securely mounted, with a correction card in place. The instrument face must be clear and the instrument case full of fluid. A cloudy instrument face, bubbles in the fluid, or a partially filled case renders the instrument unusable. [Figure 2-5]

The gyro driven attitude indicator should be checked before being powered. A white haze on the inside of

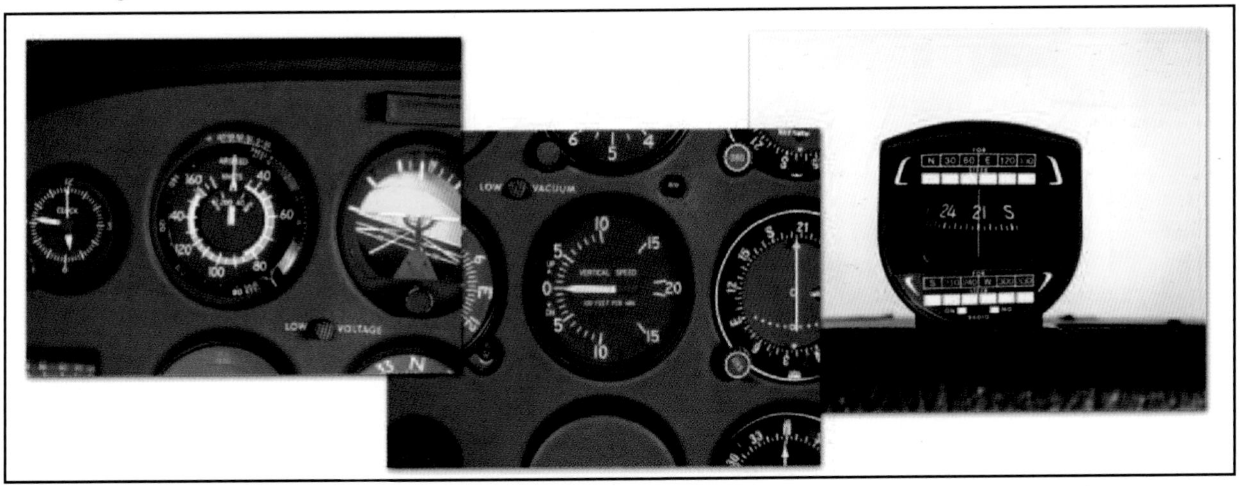

Figure 2-5. Airspeed indicator, VSI, and magnetic compass.

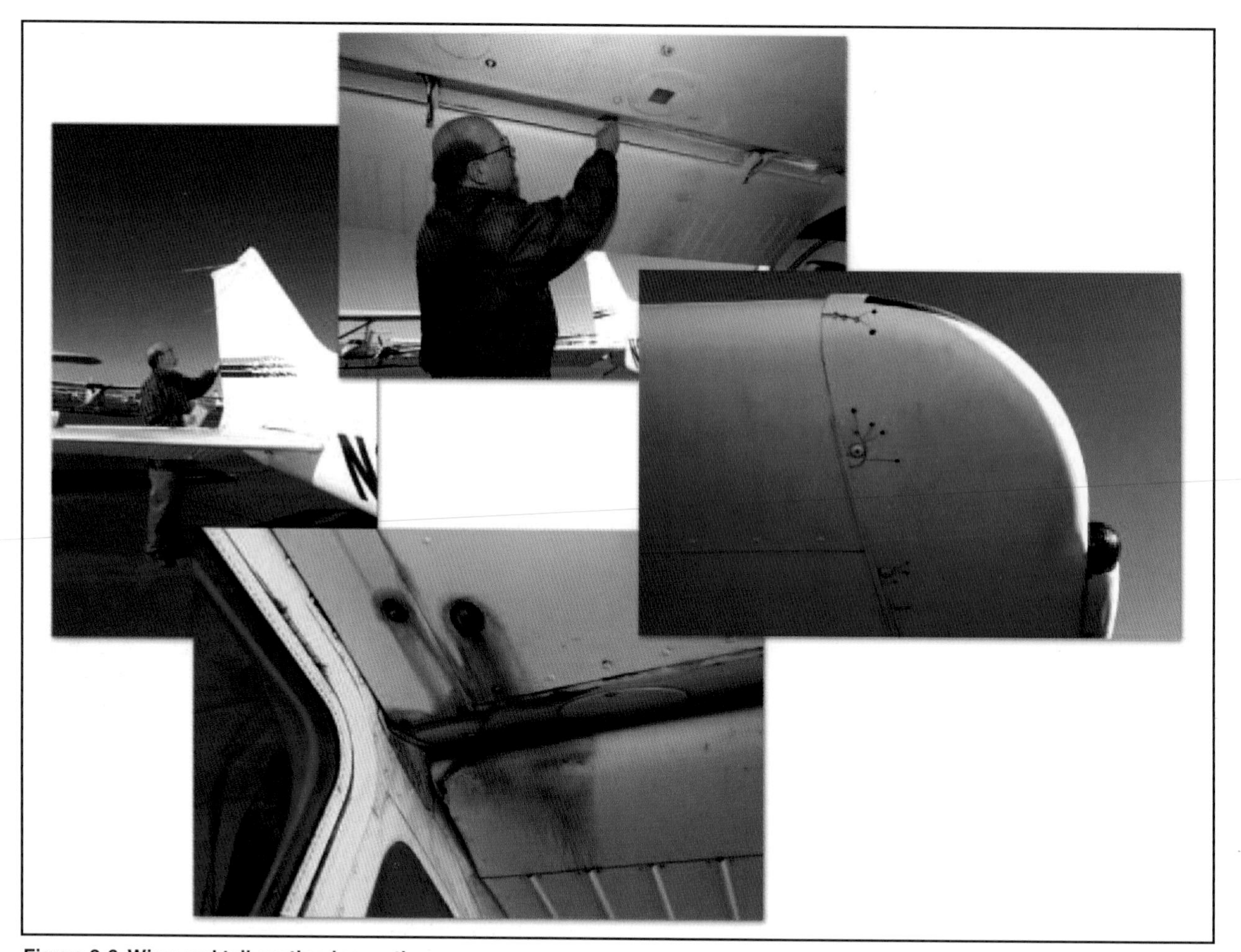

Figure 2-6. Wing and tail section inspection.

the glass face may be a sign that the seal has been breached, allowing moisture and dirt to be sucked into the instrument.

The altimeter should be checked against the ramp or field elevation after setting in the barometric pressure. If the variation between the known field elevation and the altimeter indication is more than 75 feet, its accuracy is questionable.

The pilot should turn on the battery master switch and make note of the fuel quantity gauge indications for comparison with an actual visual inspection of the fuel tanks during the exterior inspection.

OUTER WING SURFACES AND TAIL SECTION

The pilot should inspect for any signs of deterioration, distortion, and loose or missing rivets or screws, especially in the area where the outer skin attaches to the airplane structure. [Figure 2-6] The pilot should look along the wing spar rivet line—from the wingtip to the fuselage—for skin distortion. Any ripples and/or waves may be an indication of internal damage or failure.

Loose or sheared aluminum rivets may be identified by the presence of black oxide which forms rapidly when the rivet works free in its hole. Pressure applied to the skin adjacent to the rivet head will help verify the loosened condition of the rivet.

When examining the outer wing surface, it should be remembered that any damage, distortion, or malformation of the wing leading edge renders the airplane unairworthy. Serious dents in the leading edge, and disrepair of items such as stall strips, and deicer boots can cause the airplane to be aerodynamically unsound. Also, special care should be taken when examining the wingtips. Airplane wingtips are usually fiberglass. They are easily damaged and subject to cracking. The pilot should look at stop drilled cracks for evidence of crack progression, which can, under some circumstances, lead to in-flight failure of the wingtip.

The pilot should remember that fuel stains anywhere on the wing warrant further investigation—no matter how old the stains appear to be. Fuel stains are a sign of probable fuel leakage. On airplanes equipped with integral fuel tanks, evidence of fuel leakage can be found along rivet lines along the underside of the wing.

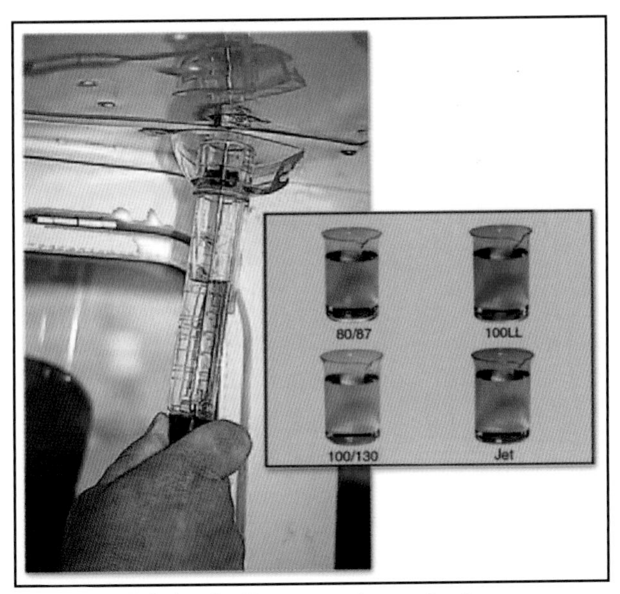

Figure 2-7. Aviation fuel types, grades, and colors.

FUEL AND OIL

Particular attention should be paid to the fuel quantity, type and grade, and quality. [Figure 2-7] Many fuel tanks are very sensitive to airplane attitude when attempting to fuel for maximum capacity. Nosewheel strut extension, both high as well as low, can significantly alter the attitude, and therefore the fuel capacity. The airplane attitude can also be affected laterally by a ramp that slopes, leaving one wing slightly higher than another. Always confirm the fuel quantity indicated on the fuel gauges by visually inspecting the level of each tank.

The type, grade, and color of fuel are critical to safe operation. The only widely available aviation gasoline (AVGAS) grade in the United States is low-lead 100-octane, or 100LL. AVGAS is dyed for easy recognition of its grade and has a familiar gasoline scent. Jet-A, or jet fuel, is a kerosene-based fuel for turbine powered airplanes. It has disastrous consequences when inadvertently introduced into reciprocating airplane engines. The piston engine operating on jet fuel may start, run, and power the airplane, but will fail because the engine has been destroyed from detonation.

Jet fuel has a distinctive kerosene scent and is oily to the touch when rubbed between fingers. Jet fuel is clear or straw colored, although it may appear dyed when mixed in a tank containing AVGAS. When a few drops of AVGAS are placed upon white paper, they evaporate quickly and leave just a trace of dye. In comparison, jet fuel is slower to evaporate and leaves an oily smudge. Jet fuel refueling trucks and dispensing equipment are marked with JET-A placards in white letters on a black background. Prudent pilots will supervise fueling to ensure that the correct tanks are filled with the right quantity, type, and grade of fuel. The pilot should always ensure that the fuel caps have been securely replaced following each fueling.

Engines certificated for grades 80/87 or 91/96 AVGAS will run satisfactorily on 100LL. The reverse is not true. Fuel of a lower grade/octane, if found, should never be substituted for a required higher grade. Detonation will severely damage the engine in a very short period of time.

Automotive gasoline is sometimes used as a substitute fuel in certain airplanes. Its use is acceptable only when the particular airplane has been issued a supplemental type certificate (STC) to both the airframe and engine allowing its use.

Checking for water and other sediment contamination is a key preflight element. Water tends to accumulate in fuel tanks from condensation, particularly in partially filled tanks. Because water is heavier than fuel, it tends to collect in the low points of the fuel system. Water can also be introduced into the fuel system from deteriorated gas cap seals exposed to rain, or from the supplier's storage tanks and delivery vehicles. Sediment contamination can arise from dust and dirt entering the tanks during refueling, or from deteriorating rubber fuel tanks or tank sealant.

The best preventive measure is to minimize the opportunity for water to condense in the tanks. If possible, the fuel tanks should be completely filled with the proper grade of fuel after each flight, or at least filled after the last flight of the day. The more fuel there is in the tanks, the less opportunity for condensation to occur. Keeping fuel tanks filled is also the best way to slow the aging of rubber fuel tanks and tank sealant.

Sufficient fuel should be drained from the fuel strainer quick drain and from each fuel tank sump to check for fuel grade/color, water, dirt, and smell. If water is present, it will usually be in bead-like droplets, different in color (usually clear, sometimes muddy), in the bottom of the sample. In extreme cases, do not overlook the possibility that the entire sample, particularly a small sample, is water. If water is found in the first fuel sample, further samples should be taken until no water appears. Significant and/or consistent water or sediment contamination are grounds for further investigation by qualified maintenance personnel. Each fuel tank sump should be drained during preflight and after refueling.

The fuel tank vent is an important part of a preflight inspection. Unless outside air is able to enter the tank as fuel is drawn out, the eventual result will be fuel gauge malfunction and/or fuel starvation. During the preflight inspection, the pilot should be alert for any signs of vent tubing damage, as well as vent blockage. A functional check of the fuel vent system can be done simply by opening the fuel cap. If there is a rush of air when the fuel tank cap is cracked, there could be a serious problem with the vent system.

The oil level should be checked during each preflight and rechecked with each refueling. Reciprocating airplane engines can be expected to consume a small amount of oil during normal operation. If the consumption grows or suddenly changes, qualified maintenance personnel should investigate. If line service personnel add oil to the engine, the pilot should ensure that the oil cap has been securely replaced.

LANDING GEAR, TIRES, AND BRAKES

Tires should be inspected for proper inflation, as well as cuts, bruises, wear, bulges, imbedded foreign object, and deterioration. As a general rule, tires with cord showing, and those with cracked sidewalls are considered unairworthy.

Brakes and brake systems should be checked for rust and corrosion, loose nuts/bolts, alignment, brake pad wear/cracks, signs of hydraulic fluid leakage, and hydraulic line security/abrasion.

An examination of the nose gear should include the shimmy damper, which is painted white, and the torque link, which is painted red, for proper servicing and general condition. All landing gear shock struts should also be checked for proper inflation.

ENGINE AND PROPELLER

The pilot should make note of the condition of the engine cowling. [Figure 2-8] If the cowling rivet heads reveal aluminum oxide residue, and chipped paint surrounding and radiating away from the cowling rivet heads, it is a sign that the rivets have been rotating until the holes have been elongated. If allowed to continue, the cowling may eventually separate from the airplane in flight.

Certain engine/propeller combinations require installation of a prop spinner for proper engine cooling. In these cases, the engine should not be operated unless the spinner is present and properly installed. The pilot should inspect the propeller spinner and spinner mounting plate for security of attachment, any signs of chafing of propeller blades, and defects such as cracking. A cracked spinner is unairworthy.

The propeller should be checked for nicks, cracks, pitting, corrosion, and security. The propeller hub should be checked for oil leaks, and the alternator/generator drive belt should be checked for proper tension and signs of wear.

When inspecting inside the cowling, the pilot should look for signs of fuel dye which may indicate a fuel leak. The pilot should check for oil leaks, deterioration of oil lines, and to make certain that the oil cap, filter, oil cooler and drain plug are secure. The exhaust system should be checked for white stains caused by exhaust leaks at the cylinder head or cracks in the stacks. The heat muffs should also be checked for general condition and signs of cracks or leaks.

The air filter should be checked for condition and secure fit, as well as hydraulic lines for deterioration and/or leaks. The pilot should also check for loose or foreign objects inside the cowling such as bird nests, shop rags, and/or tools. All visible wires and lines should be checked for security and condition. And lastly, when the cowling is closed, the cowling fasteners should be checked for security.

Figure 2-8. Check the propeller and inside the cowling.

COCKPIT MANAGEMENT

After entering the airplane, the pilot should first ensure that all necessary equipment, documents, checklists, and navigation charts appropriate for the flight are on board. If a portable intercom, headsets, or a hand-held global positioning system (GPS) is used, the pilot is responsible for ensuring that the routing of wires and cables does not interfere with the motion or the operation of any control.

Regardless of what materials are to be used, they should be neatly arranged and organized in a manner that makes them readily available. The cockpit and cabin should be checked for articles that might be tossed about if turbulence is encountered. Loose items should be properly secured. All pilots should form the habit of good housekeeping.

The pilot must be able to see inside and outside references. If the range of motion of an adjustable seat is inadequate, cushions should be used to provide the proper seating position.

When the pilot is comfortably seated, the safety belt and shoulder harness (if installed) should be fastened and adjusted to a comfortably snug fit. The shoulder harness must be worn at least for the takeoff and landing, unless the pilot cannot reach or operate the controls with it fastened. The safety belt must be worn at all times when the pilot is seated at the controls.

If the seats are adjustable, it is important to ensure that the seat is locked in position. Accidents have occurred as the result of seat movement during acceleration or pitch attitude changes during takeoffs or landings. When the seat suddenly moves too close or too far away from the controls, the pilot may be unable to maintain control of the airplane.

14 CFR part 91 requires the pilot to ensure that each person on board is briefed on how to fasten and unfasten his/her safety belt and, if installed, shoulder harness. This should be accomplished before starting the engine, along with a passenger briefing on the proper use of safety equipment and exit information. Airplane manufacturers have printed briefing cards available, similar to those used by airlines, to supplement the pilot's briefing.

GROUND OPERATIONS

It is important that a pilot operates an airplane safely on the ground. This includes being familiar with standard hand signals that are used by ramp personnel. [Figure 2-9]

ENGINE STARTING

The specific procedures for engine starting will not be discussed here since there are as many different methods as there are different engines, fuel systems, and starting conditions. The before engine starting and engine starting checklist procedures should be followed. There are, however, certain precautions that apply to all airplanes.

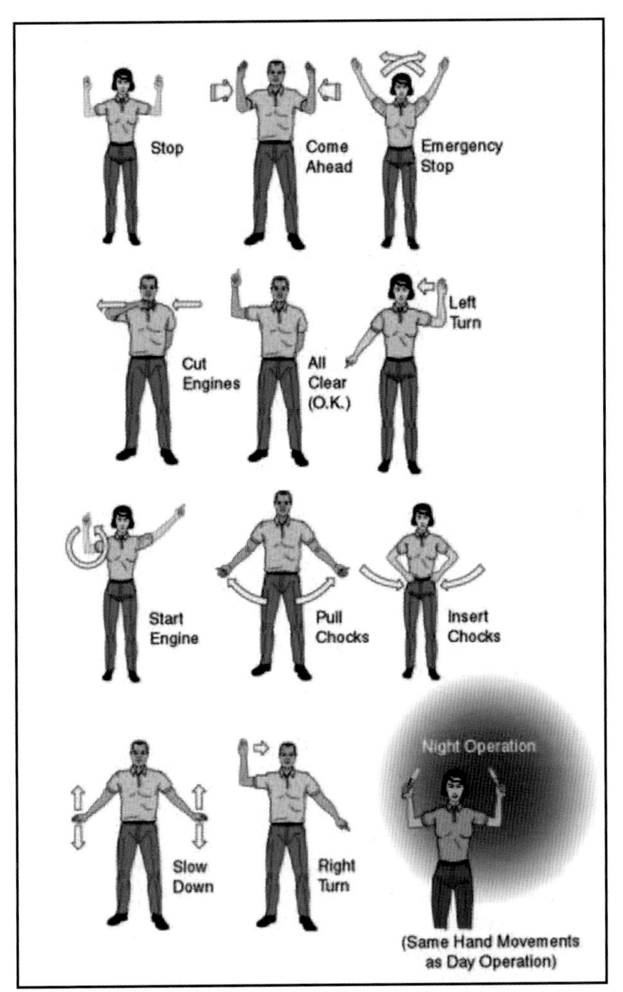

Figure 2-9. Standard hand signals.

Some pilots have started the engine with the tail of the airplane pointed toward an open hangar door, parked automobiles, or a group of bystanders. This is not only discourteous, but may result in personal injury and damage to the property of others. Propeller blast can be surprisingly powerful.

When ready to start the engine, the pilot should look in all directions to be sure that nothing is or will be in the vicinity of the propeller. This includes nearby persons and aircraft that could be struck by the propeller blast or the debris it might pick up from the ground. The anticollision light should be turned on prior to engine start, even during daytime operations. At night, the position (navigation) lights should also be on.

The pilot should always call "CLEAR" out of the side window and wait for a response from persons who may be nearby before activating the starter.

When activating the starter, one hand should be kept on the throttle. This allows prompt response if the engine falters during starting, and allows the pilot to rapidly retard the throttle if revolutions per minute (r.p.m.) are excessive after starting. A low r.p.m. setting (800 to 1,000) is recommended immediately following engine start. It is highly undesirable to allow the r.p.m. to race immediately after start, as there will be insufficient lubrication until the oil pressure rises. In freezing temperatures, the engine will also be exposed to potential mechanical distress until it warms and normal internal operating clearances are assumed.

As soon as the engine is operating smoothly, the oil pressure should be checked. If it does not rise to the manufacturer's specified value, the engine may not be receiving proper lubrication and should be shut down immediately to prevent serious damage.

Although quite rare, the starter motor may remain on and engaged after the engine starts. This can be detected by a continuous very high current draw on the ammeter. Some airplanes also have a starter engaged warning light specifically for this purpose. The engine should be shut down immediately should this occur.

Starters are small electric motors designed to draw large amounts of current for short periods of cranking. Should the engine fail to start readily, avoid continuous starter operation for periods longer than 30 seconds without a cool down period of at least 30 seconds to a minute (some AFM/POH specify even longer). Their service life is drastically shortened from high heat through overuse.

HAND PROPPING

Even though most airplanes are equipped with electric starters, it is helpful if a pilot is familiar with the procedures and dangers involved in starting an engine by turning the propeller by hand (hand propping). Due to the associated hazards, this method of starting should be used only when absolutely necessary and when proper precautions have been taken.

An engine should not be hand propped unless two people, both familiar with the airplane and hand propping techniques, are available to perform the procedure. The person pulling the propeller blades through directs all activity and is in charge of the procedure. The other person, thoroughly familiar with the controls, must be seated in the airplane with the brakes set. As an additional precaution, chocks may be placed in front of the main wheels. If this is not feasible, the airplane's tail may be securely tied. Never allow a person unfamiliar with the controls to occupy the pilot's seat when hand propping. The procedure should never be attempted alone.

When hand propping is necessary, the ground surface near the propeller should be stable and free of debris. Unless a firm footing is available, consider relocating the airplane. Loose gravel, wet grass, mud, oil, ice, or snow might cause the person pulling the propeller through to slip into the rotating blades as the engine starts.

Both participants should discuss the procedure and agree on voice commands and expected action. To begin the procedure, the fuel system and engine controls (tank selector, primer, pump, throttle, and mixture) are set for a normal start. The ignition/magneto switch should be checked to be sure that it is OFF. Then the descending propeller blade should be rotated so that it assumes a position slightly above the horizontal. The person doing the hand propping should face the descending blade squarely and stand slightly less than one arm's length from the blade. If a stance too far away were assumed, it would be necessary to lean forward in an unbalanced condition to reach the blade. This may cause the person to fall forward into the rotating blades when the engine starts.

The procedure and commands for hand propping are:

- Person out front says, "GAS ON, SWITCH OFF, THROTTLE CLOSED, BRAKES SET."
- Pilot seat occupant, after making sure the fuel is ON, mixture is RICH, ignition/magneto switch is OFF, throttle is CLOSED, and brakes SET, says, "GAS ON, SWITCH OFF, THROTTLE CLOSED, BRAKES SET."
- Person out front, after pulling the propeller through to prime the engine says, "BRAKES AND CONTACT."
- Pilot seat occupant checks the brakes SET and turns the ignition switch ON, then says, "BRAKES AND CONTACT."

The propeller is swung by forcing the blade downward rapidly, pushing with the palms of both hands. If the blade is gripped tightly with the fingers, the person's body may be drawn into the propeller blades should the engine misfire and rotate momentarily in the opposite direction. As the blade is pushed down, the person should step backward, away from the propeller. If the engine does not start, the propeller should not be repositioned for another attempt until it is certain the ignition/magneto switch is turned OFF.

The words CONTACT (mags ON) and SWITCH OFF (mags OFF) are used because they are significantly different from each other. Under noisy conditions or high winds, the words CONTACT and SWITCH OFF are less likely to be misunderstood than SWITCH ON and SWITCH OFF.

When removing the wheel chocks after the engine starts, it is essential that the pilot remember that the propeller is almost invisible. Incredible as it may seem, serious injuries and fatalities occur when people who have just started an engine walk or reach into the propeller arc to remove the chocks. Before the chocks are removed, the throttle should be set to idle and the chocks approached from the rear of the propeller. Never approach the chocks from the front or the side.

The procedures for hand propping should always be in accordance with the manufacturer's recommendations and checklist. Special starting procedures are used when the engine is already warm, very cold, or when flooded or vapor locked. There will also be a different starting procedure when an external power source is used.

TAXIING

The following basic taxi information is applicable to both nosewheel and tailwheel airplanes.

Taxiing is the controlled movement of the airplane under its own power while on the ground. Since an airplane is moved under its own power between the parking area and the runway, the pilot must thoroughly understand and be proficient in taxi procedures.

An awareness of other aircraft that are taking off, landing, or taxiing, and consideration for the right-of-way of others is essential to safety. When taxiing, the pilot's eyes should be looking outside the airplane, to the sides, as well as the front. The pilot must be aware of the entire area around the airplane to ensure that the airplane will clear all obstructions and other aircraft. If at any time there is doubt about the clearance from an object, the pilot should stop the airplane and have someone check the clearance. It may be necessary to have the airplane towed or physically moved by a ground crew.

It is difficult to set any rule for a single, safe taxiing speed. What is reasonable and prudent under some conditions may be imprudent or hazardous under others. The primary requirements for safe taxiing are positive control, the ability to recognize potential hazards in time to avoid them, and the ability to stop or turn where and when desired, without undue reliance on the brakes. Pilots should proceed at a cautious speed on congested or busy ramps. Normally, the speed should be at the rate where movement of the airplane is dependent on the throttle. That is, slow enough so when the throttle is closed, the airplane can be stopped promptly. When yellow taxiway centerline stripes are provided, they should be observed unless necessary to clear airplanes or obstructions.

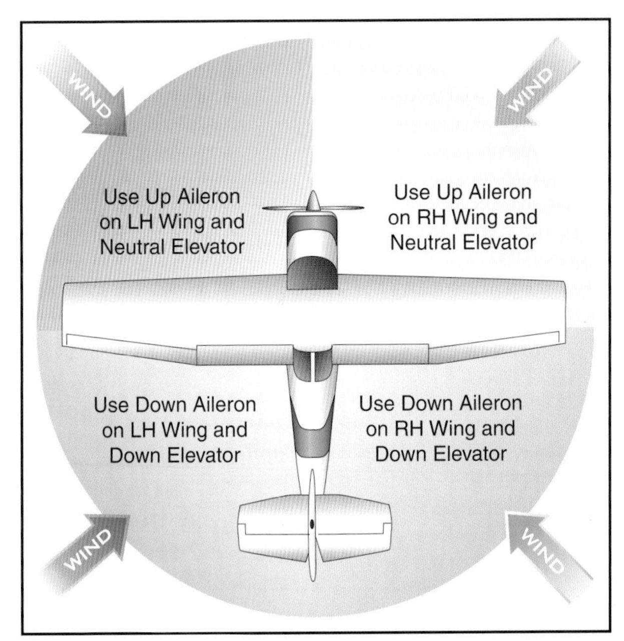

Figure 2-10. Flight control positions during taxi.

When taxiing, it is best to slow down before attempting a turn. Sharp, high-speed turns place undesirable side loads on the landing gear and may result in an uncontrollable swerve or a ground loop. This swerve is most likely to occur when turning from a downwind heading toward an upwind heading. In moderate to high-wind conditions, pilots will note the airplane's tendency to weathervane, or turn into the wind when the airplane is proceeding crosswind.

When taxiing at appropriate speeds in no-wind conditions, the aileron and elevator control surfaces have little or no effect on directional control of the airplane. The controls should not be considered steering devices and should be held in a neutral position. Their proper use while taxiing in windy conditions will be discussed later. [Figure 2-10]

Steering is accomplished with rudder pedals and brakes. To turn the airplane on the ground, the pilot should apply rudder in the desired direction of turn and use whatever power or brake that is necessary to control the taxi speed. The rudder pedal should be held in the direction of the turn until just short of the point where the turn is to be stopped. Rudder pressure is then released or opposite pressure is applied as needed.

More engine power may be required to start the airplane moving forward, or to start a turn, than is required to keep it moving in any given direction. When using additional power, the throttle should immediately be retarded once the airplane begins moving, to prevent excessive acceleration.

When first beginning to taxi, the brakes should be tested for proper operation as soon as the airplane is put in motion. Applying power to start the airplane moving forward slowly, then retarding the throttle and simultaneously applying pressure smoothly to both brakes does this. If braking action is unsatisfactory, the engine should be shut down immediately.

The presence of moderate to strong headwinds and/or a strong propeller slipstream makes the use of the elevator necessary to maintain control of the pitch attitude while taxiing. This becomes apparent when considering the lifting action that may be created on the horizontal tail surfaces by either of those two factors. The elevator control in nosewheel-type airplanes should be held in the neutral position, while in tailwheel-type airplanes it should be held in the aft position to hold the tail down.

Downwind taxiing will usually require less engine power after the initial ground roll is begun, since the wind will be pushing the airplane forward. [Figure 2-11] To avoid overheating the brakes when taxiing downwind, keep engine power to a minimum. Rather than continuously riding the brakes to control speed, it is better to apply brakes only occasionally. Other than sharp turns at low speed, the throttle should always be at idle before the brakes are applied. It is a common student error to taxi with a power setting that requires controlling taxi speed with the brakes. This is the aeronautical equivalent of driving an automobile with both the accelerator and brake pedals depressed.

When taxiing with a quartering headwind, the wing on the upwind side will tend to be lifted by the wind unless the aileron control is held in that direction (upwind aileron UP). [Figure 2-12] Moving the aileron

Figure 2-11. Downwind taxi.

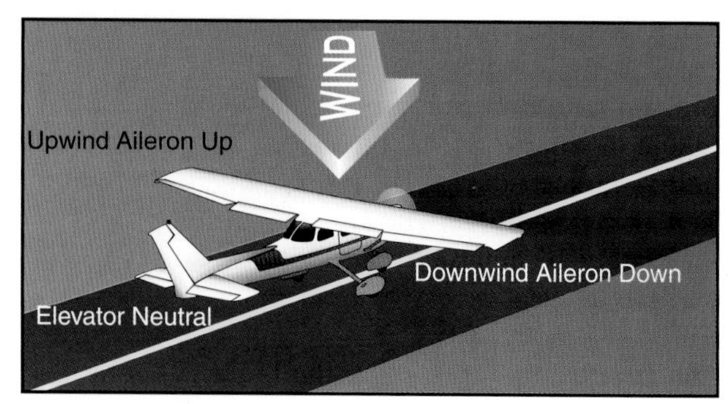

Figure 2-12. Quartering headwind.

into the UP position reduces the effect of the wind striking that wing, thus reducing the lifting action. This control movement will also cause the downwind aileron to be placed in the DOWN position, thus a small amount of lift and drag on the downwind wing, further reducing the tendency of the upwind wing to rise.

When taxiing with a quartering tailwind, the elevator should be held in the DOWN position, and the upwind aileron, DOWN. [Figure 2-13] Since the wind is striking the airplane from behind, these control positions reduce the tendency of the wind to get under the tail and the wing and to nose the airplane over.

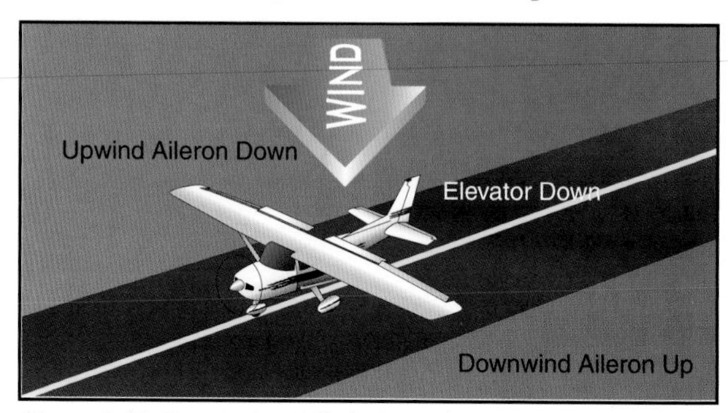

Figure 2-13. Quartering tailwind.

The application of these crosswind taxi corrections helps to minimize the weathervaning tendency and ultimately results in making the airplane easier to steer.

Normally, all turns should be started using the rudder pedal to steer the nosewheel. To tighten the turn after full pedal deflection is reached, the brake may be applied as needed. When stopping the airplane, it is advisable to always stop with the nosewheel straight ahead to relieve any side load on the nosewheel and to make it easier to start moving ahead.

During crosswind taxiing, even the nosewheel-type airplane has some tendency to weathervane. However,

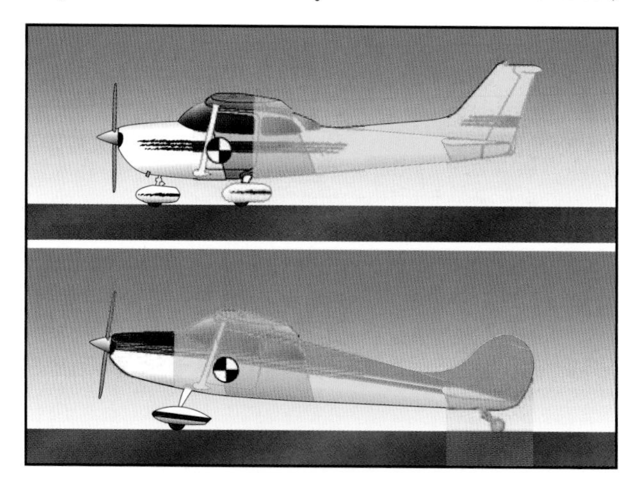

Figure 2-14. Surface area most affected by wind.

the weathervaning tendency is less than in tailwheel-type airplanes because the main wheels are located farther aft, and the nosewheel's ground friction helps to resist the tendency. [Figure 2-14] The nosewheel linkage from the rudder pedals provides adequate steering control for safe and efficient ground handling, and normally, only rudder pressure is necessary to correct for a crosswind.

BEFORE TAKEOFF CHECK

The before takeoff check is the systematic procedure for making a check of the engine, controls, systems, instruments, and avionics prior to flight. Normally, it is performed after taxiing to a position near the takeoff end of the runway. Taxiing to that position usually allows sufficient time for the engine to warm up to at least minimum operating temperatures. This ensures adequate lubrication and internal engine clearances before being operated at high power settings. Many engines require that the oil temperature reach a minimum value as stated in the AFM/POH before high power is applied.

Air-cooled engines generally are closely cowled and equipped with pressure baffles that direct the flow of air to the engine in sufficient quantities for cooling in flight. On the ground, however, much less air is forced through the cowling and around the baffling. Prolonged ground operations may cause cylinder overheating long before there is an indication of rising oil temperature. Cowl flaps, if available, should be set according to the AFM/POH.

Before beginning the before takeoff check, the airplane should be positioned clear of other aircraft. There should not be anything behind the airplane that might be damaged by the prop blast. To minimize overheating during engine runup, it is recommended that the airplane be headed as nearly as possible into the wind. After the airplane is properly positioned for the runup, it should be allowed to roll forward slightly so that the nosewheel or tailwheel will be aligned fore and aft.

During the engine runup, the surface under the airplane should be firm (a smooth, paved, or turf surface if possible) and free of debris. Otherwise, the propeller may pick up pebbles, dirt, mud, sand, or other loose objects and hurl them backwards. This damages the propeller and may damage the tail of the airplane. Small chips in the leading edge of the propeller form stress risers, or lines of concentrated high stress. These are highly undesirable and may lead to cracks and possible propeller blade failure.

While performing the engine runup, the pilot must divide attention inside and outside the airplane. If the parking brake slips, or if application of the toe brakes is inadequate for the amount of power applied, the airplane could move forward unnoticed if attention is fixed inside the airplane.

Each airplane has different features and equipment, and the before takeoff checklist provided by the airplane manufacturer or operator should be used to perform the runup.

AFTER LANDING

During the after-landing roll, the airplane should be gradually slowed to normal taxi speed before turning off the landing runway. Any significant degree of turn at faster speeds could result in ground looping and subsequent damage to the airplane.

To give full attention to controlling the airplane during the landing roll, the after-landing check should be performed only after the airplane is brought to a complete stop clear of the active runway. There have been many cases of the pilot mistakenly grasping the wrong handle and retracting the landing gear, instead of the flaps, due to improper division of attention while the airplane was moving. However, this procedure may be modified if the manufacturer recommends that specific after-landing items be accomplished during landing rollout. For example, when performing a short-field landing, the manufacturer may recommend retracting the flaps on rollout to improve braking. In this situation, the pilot should make a positive identification of the flap control and retract the flaps.

CLEAR OF RUNWAY

Because of different features and equipment in various airplanes, the after-landing checklist provided by the manufacturer should be used. Some of the items may include:

- Flaps . . . . Identify and retract
- Cowl flaps . . . . Open
- Propeller control . . . . Full increase
- Trim tabs . . . . Set

PARKING

Unless parking in a designated, supervised area, the pilot should select a location and heading which will prevent the propeller or jet blast of other airplanes from striking the airplane broadside. Whenever possible, the airplane should be parked headed into the existing or forecast wind. After stopping on the desired heading, the airplane should be allowed to roll straight ahead enough to straighten the nosewheel or tailwheel.

Engine Shutdown

Finally, the pilot should always use the procedures in the manufacturer's checklist for shutting down the engine and securing the airplane. Some of the important items include:

- Set the parking brakes ON.
- Set throttle to IDLE or 1,000 r.p.m. If turbocharged, observe the manufacturer's spool down procedure.
- Turn ignition switch OFF then ON at idle to check for proper operation of switch in the OFF position.
- Set propeller control (if equipped) to FULL INCREASE.
- Turn electrical units and radios OFF.
- Set mixture control to IDLE CUTOFF.
- Turn ignition switch to OFF when engine stops.
- Turn master electrical switch to OFF.
- Install control lock.

Postflight

A flight is never complete until the engine is shut down and the airplane is secured. A pilot should consider this an essential part of any flight.

Securing and Servicing

After engine shutdown and deplaning passengers, the pilot should accomplish a postflight inspection. This includes checking the general condition of the aircraft. For a departure, the oil should be checked and fuel added if required. If the aircraft is going to be inactive, it is a good operating practice to fill the tanks to the top to prevent water condensation from forming. When the flight is completed for the day, the aircraft should be hangared or tied down and the flight controls secured.

THE FOUR FUNDAMENTALS

There are four fundamental basic flight maneuvers upon which all flying tasks are based: straight-and-level flight, turns, climbs, and descents. All controlled flight consists of either one, or a combination or more than one, of these basic maneuvers. If a student pilot is able to perform these maneuvers well, and the student's proficiency is based on accurate "feel" and control analysis rather than mechanical movements, the ability to perform any assigned maneuver will only be a matter of obtaining a clear visual and mental conception of it. The flight instructor must impart a good knowledge of these basic elements to the student, and must combine them and plan their practice so that perfect performance of each is instinctive without conscious effort. The importance of this to the success of flight training cannot be overemphasized. As the student progresses to more complex maneuvers, discounting any difficulties in visualizing the maneuvers, most student difficulties will be caused by a lack of training, practice, or understanding of the principles of one or more of these fundamentals.

EFFECTS AND USE OF THE CONTROLS

In explaining the functions of the controls, the instructor should emphasize that the controls never change in the results produced in relation to the pilot. The pilot should always be considered the center of movement of the airplane, or the reference point from which the movements of the airplane are judged and described. The following will always be true, regardless of the airplane's attitude in relation to the Earth.

- When back pressure is applied to the elevator control, the airplane's nose rises in relation to the pilot.
- When forward pressure is applied to the elevator control, the airplane's nose lowers in relation to the pilot.
- When right pressure is applied to the aileron control, the airplane's right wing lowers in relation to the pilot.
- When left pressure is applied to the aileron control, the airplane's left wing lowers in relation to the pilot.
- When pressure is applied to the right rudder pedal, the airplane's nose moves (yaws) to the right in relation to the pilot.
- When pressure is applied to the left rudder pedal, the airplane's nose moves (yaws) to the left in relation to the pilot.

The preceding explanations should prevent the beginning pilot from thinking in terms of "up" or "down" in respect to the Earth, which is only a relative state to the pilot. It will also make understanding of the functions of the controls much easier, particularly when performing steep banked turns and the more advanced maneuvers. Consequently, the pilot must be able to properly determine the control application required to place the airplane in any attitude or flight condition that is desired.

The flight instructor should explain that the controls will have a natural "live pressure" while in flight and that they will remain in neutral position of their own accord, if the airplane is trimmed properly.

With this in mind, the pilot should be cautioned never to think of movement of the controls, but of exerting a force on them against this live pressure or resistance. Movement of the controls should not be emphasized; it is the duration and amount of the force exerted on them that effects the displacement of the control surfaces and maneuvers the airplane.

The amount of force the airflow exerts on a control surface is governed by the airspeed and the degree that the surface is moved out of its neutral or streamlined position. Since the airspeed will not be the same in all maneuvers, the actual amount the control surfaces are moved is of little importance; but it is important that the pilot maneuver the airplane by applying sufficient control *pressure* to obtain a desired result, regardless of how far the control surfaces are actually moved.

The controls should be held lightly, with the fingers, not grabbed and squeezed. Pressure should be exerted on the control yoke with the fingers. A common error in beginning pilots is a tendency to "choke the stick." This tendency should be avoided as it prevents the development of "feel," which is an important part of aircraft control.

The pilot's feet should rest comfortably against the rudder pedals. Both heels should support the weight of the feet on the cockpit floor with the ball of each foot touching the individual rudder pedals. The legs and feet should not be tense; they must be relaxed just as when driving an automobile.

When using the rudder pedals, pressure should be applied smoothly and evenly by pressing with the ball of one foot. Since the rudder pedals are interconnected, and act in opposite directions, when pressure is applied to one pedal, pressure on the other must be relaxed proportionately. When the rudder pedal must be moved significantly, heavy pressure changes should be made by applying the pressure with the ball of the foot while the heels slide along the cockpit floor. Remember, the ball of each foot must rest comfortably on the rudder pedals so that even slight pressure changes can be felt.

In summary, during flight, it is the *pressure* the pilot exerts on the control yoke and rudder pedals that causes the airplane to move about its axes. When a control surface is moved out of its streamlined position (even slightly), the air flowing past it will exert a force against it and will try to return it to its streamlined position. It is this force that the pilot feels as pressure on the control yoke and the rudder pedals.

FEEL OF THE AIRPLANE

The ability to sense a flight condition, without relying on cockpit instrumentation, is often called "feel of the airplane," but senses in addition to "feel" are involved.

Sounds inherent to flight are an important sense in developing "feel." The air that rushes past the modern light plane cockpit/cabin is often masked by soundproofing, but it can still be heard. When the level of sound increases, it indicates that airspeed is increasing. Also, the powerplant emits distinctive sound patterns in different conditions of flight. The sound of the engine in cruise flight may be different from that in a climb, and different again from that in a dive. When power is used in fixed-pitch propeller airplanes, the loss of r.p.m. is particularly noticeable. The amount of noise that can be heard will depend on how much the slipstream masks it out. But the relationship between slipstream noise and powerplant noise aids the pilot in estimating not only the present airspeed but the trend of the airspeed.

There are three sources of actual "feel" that are very important to the pilot. One is the pilot's own body as it responds to forces of acceleration. The "G" loads imposed on the airframe are also felt by the pilot. Centripetal accelerations force the pilot down into the seat or raise the pilot against the seat belt. Radial accelerations, as they produce slips or skids of the airframe, shift the pilot from side to side in the seat. These forces need not be strong, only perceptible by the pilot to be useful. An accomplished pilot who has excellent "feel" for the airplane will be able to detect even the minutest change.

The response of the aileron and rudder controls to the pilot's touch is another element of "feel," and is one that provides direct information concerning airspeed. As previously stated, control surfaces move in the airstream and meet resistance proportional to the speed of the airstream. When the airstream is fast, the controls are stiff and hard to move. When the airstream is slow, the controls move easily, but must be deflected a greater distance. The pressure that must be exerted on the controls to effect a desired result, and the lag between their movement and the response of the airplane, becomes greater as airspeed decreases.

Another type of "feel" comes to the pilot through the airframe. It consists mainly of vibration. An example is the aerodynamic buffeting and shaking that precedes a stall.

Kinesthesia, or the sensing of changes in direction or speed of motion, is one of the most important senses a pilot can develop. When properly developed, kinesthesia can warn the pilot of changes in speed and/or the beginning of a settling or **mushing** of the airplane.

The senses that contribute to "feel" of the airplane are inherent in every person. However, "feel" must be developed. The flight instructor should direct the beginning pilot to be attuned to these senses and teach an awareness of their meaning as it relates to various conditions of flight. To do this effectively, the flight instructor must fully understand the difference between perceiving something and merely noticing it. It is a well established fact that the pilot who develops a "feel" for the airplane early in flight training will have little difficulty with advanced flight maneuvers.

ATTITUDE FLYING

In contact (VFR) flying, flying by attitude means visually establishing the airplane's attitude with reference to the natural horizon. [Figure 3-1] Attitude is the angular difference measured between an airplane's axis and the line of the Earth's horizon. Pitch attitude is the angle formed by the longitudinal axis, and bank attitude is the angle formed by the lateral axis. Rotation about the airplane's vertical axis (yaw) is termed an attitude relative to the airplane's flightpath, but not relative to the natural horizon.

In attitude flying, airplane control is composed of four components: pitch control, bank control, power control, and trim.

- Pitch control is the control of the airplane about the lateral axis by using the elevator to raise and lower the nose in relation to the natural horizon.
- Bank control is control of the airplane about the longitudinal axis by use of the ailerons to attain a desired bank angle in relation to the natural horizon.

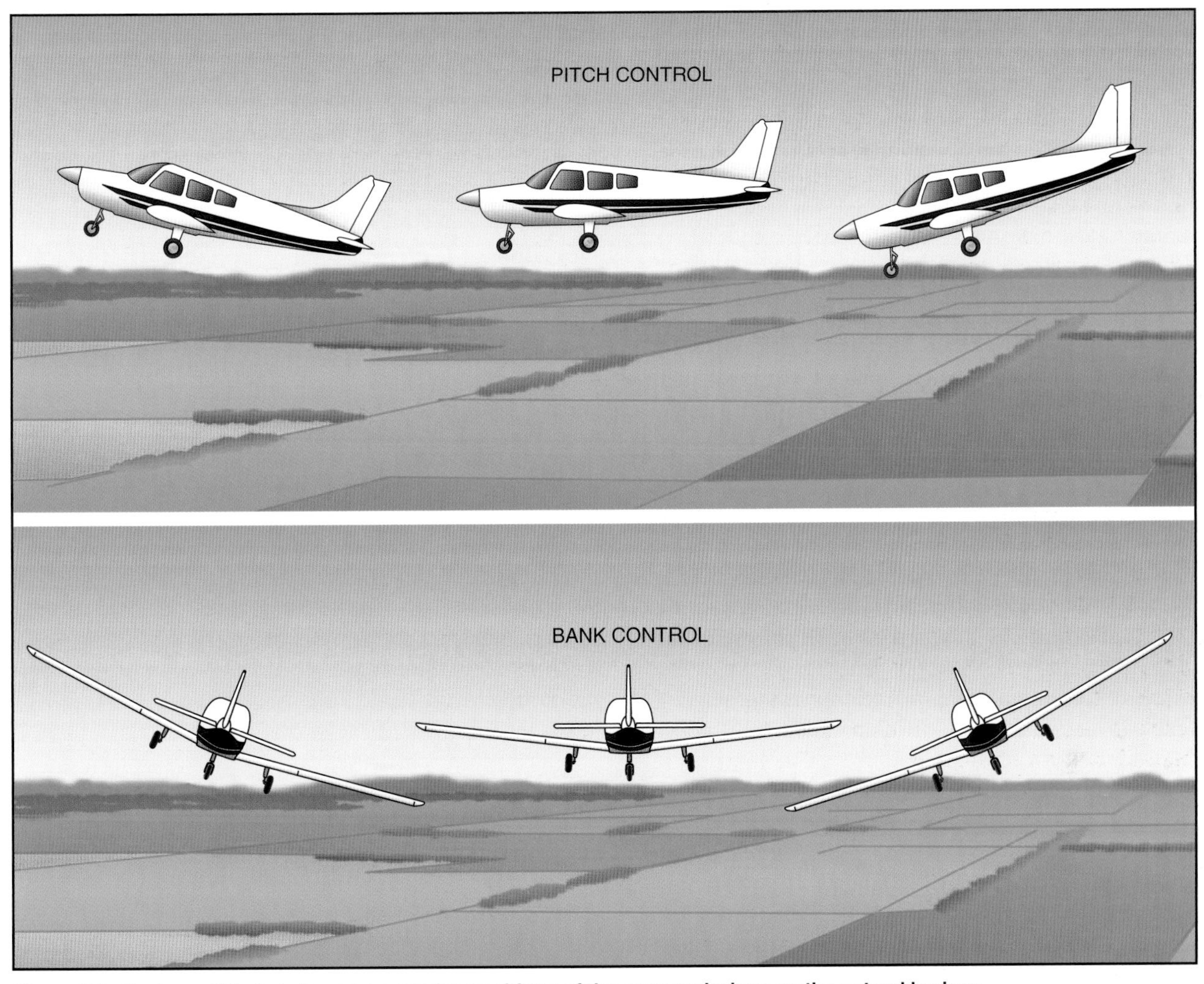

Figure 3-1. Airplane attitude is based on relative positions of the nose and wings on the natural horizon.

- Power control is used when the flight situation indicates a need for a change in thrust.
- Trim is used to relieve all possible control pressures held after a desired attitude has been attained.

The primary rule of attitude flying is:

ATTITUDE + POWER = PERFORMANCE

INTEGRATED FLIGHT INSTRUCTION

When introducing basic flight maneuvers to a beginning pilot, it is recommended that the "Integrated" or "Composite" method of flight instruction be used. This means the use of outside references and flight instruments to establish and maintain desired flight attitudes and airplane performance. [Figure 3-2] When beginning pilots use this technique, they achieve a more precise and competent overall piloting ability. Although this method of airplane control may become second nature with experience, the beginning pilot must make a determined effort to master the technique. The basic elements of which are as follows.

- The airplane's attitude is established and maintained by positioning the airplane in relation to the natural horizon. At least 90 percent of the pilot's attention should be devoted to this end, along with

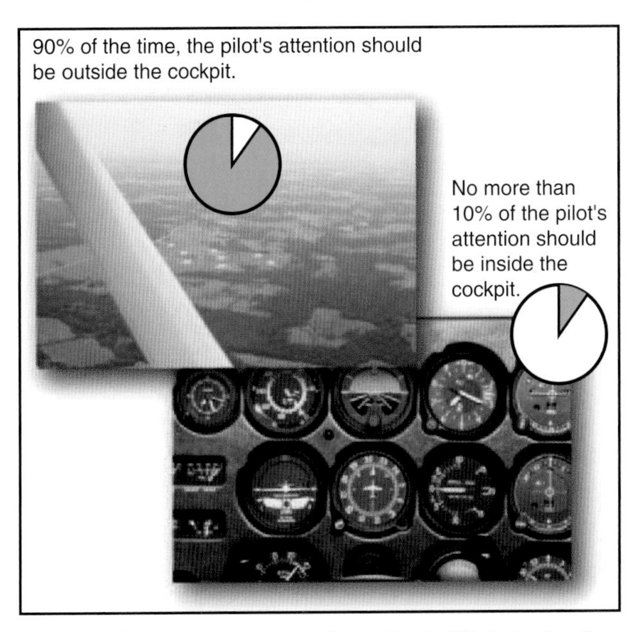

Figure 3-2. Integrated or composite method of flight instruction.

scanning for other airplanes. If, during a recheck of the pitch and/or bank, either or both are found to be other than desired, an immediate correction is made to return the airplane to the proper attitude. Continuous checks and immediate corrections will allow little chance for the airplane to deviate from the desired heading, altitude, and flightpath.

- The airplane's attitude is confirmed by referring to flight instruments, and its performance checked. If airplane performance, as indicated by flight instruments, indicates a need for correction, a specific amount of correction must be determined, then applied with reference to the natural horizon. The airplane's attitude and performance are then rechecked by referring to flight instruments. The pilot then maintains the corrected attitude by reference to the natural horizon.
- The pilot should monitor the airplane's performance by making numerous quick glances at the flight instruments. No more than 10 percent of the pilot's attention should be inside the cockpit. The pilot must develop the skill to instantly focus on the appropriate flight instrument, and then immediately return to outside reference to control the airplane's attitude.

The pilot should become familiar with the relationship between outside references to the natural horizon and the corresponding indications on flight instruments inside the cockpit. For example, a pitch attitude adjustment may require a movement of the pilot's reference point on the airplane of several inches in relation to the natural horizon, but correspond to a small fraction of an inch movement of the reference bar on the airplane's attitude indicator. Similarly, a deviation from desired bank, which is very obvious when referencing the wingtip's position relative to the natural horizon, may be nearly imperceptible on the airplane's attitude indicator to the beginning pilot.

The use of integrated flight instruction does not, and is not intended to prepare pilots for flight in instrument weather conditions. The most common error made by the beginning student is to make pitch or bank corrections while still looking inside the cockpit. Control pressure is applied, but the beginning pilot, not being familiar with the intricacies of flight by references to instruments, including such things as instrument lag and gyroscopic precession, will invariably make excessive attitude corrections and end up "chasing the instruments." Airplane attitude by reference to the natural horizon, however, is immediate in its indications, accurate, and presented many times larger than any instrument could be. Also, the beginning pilot must be made aware that anytime, for whatever reason, airplane attitude by reference to the natural horizon cannot be established and/or maintained, the situation should be considered a bona fide emergency.

STRAIGHT-AND-LEVEL FLIGHT

It is impossible to emphasize too strongly the necessity for forming correct habits in flying straight and level. All other flight maneuvers are in essence a deviation from this fundamental flight maneuver. Many flight instructors and students are prone to believe that perfection in straight-and-level flight will come of itself, but such is not the case. It is not uncommon to find a pilot whose basic flying ability consistently falls just short of minimum expected standards, and upon analyzing the reasons for the shortcomings to discover that the cause is the inability to fly straight and level properly.

Straight-and-level flight is flight in which a constant heading and altitude are maintained. It is accomplished by making immediate and measured corrections for deviations in direction and altitude from unintentional slight turns, descents, and climbs. *Level flight*, at first, is a matter of consciously fixing the relationship of the position of some portion of the airplane, used as a reference point, with the horizon. In establishing the reference points, the instructor should place the airplane in the desired position and aid the student in selecting reference points. The instructor should be aware that no two pilots see this relationship exactly the same. The references will depend on where the pilot is sitting, the pilot's height (whether short or tall), and the pilot's manner of sitting. It is, therefore, important that during the fixing of this relationship, the pilot sit in a normal manner; otherwise the points will not be the same when the normal position is resumed.

In learning to control the airplane in level flight, it is important that the student be taught to maintain a light grip on the flight controls, and that the control forces desired be exerted lightly and just enough to produce the desired result. The student should learn to associate the apparent movement of the references with the forces which produce it. In this way, the student can develop the ability to regulate the change desired in the airplane's attitude by the amount and direction of forces applied to the controls without the necessity of referring to instrument or outside references for each minor correction.

The pitch attitude for *level flight* (constant altitude) is usually obtained by selecting some portion of the airplane's nose as a reference point, and then keeping that point in a fixed position relative to the horizon. [Figure 3-3] Using the principles of attitude flying, that position should be cross-checked occasionally against the altimeter to determine whether or not the pitch attitude is correct. If altitude is being gained or lost, the pitch attitude should be readjusted in relation to the horizon and then the altimeter rechecked to determine if altitude is now being maintained. The

Figure 3-3. Nose reference for straight-and-level flight.

application of forward or back-elevator pressure is used to control this attitude.

The pitch information obtained from the attitude indicator also will show the position of the nose relative to the horizon and will indicate whether elevator pressure is necessary to change the pitch attitude to return to level flight. However, the primary reference source is the natural horizon.

In all normal maneuvers, the term "increase the pitch attitude" implies raising the nose in relation to the horizon; the term "decreasing the pitch attitude" means lowering the nose.

Straight flight (laterally level flight) is accomplished by visually checking the relationship of the airplane's wingtips with the horizon. Both wingtips should be equidistant above or below the horizon (depending on whether the airplane is a high-wing or low-wing type), and any necessary adjustments should be made with the ailerons, noting the relationship of control pressure and the airplane's attitude. [Figure 3-4] The student should understand that anytime the wings are banked, even though very slightly, the airplane will turn. The objective of straight-and-level flight is to detect small deviations from laterally level flight as soon as they occur, necessitating only small corrections. Reference to the heading indicator should be made to note any change in direction.

Figure 3-4. Wingtip reference for straight-and-level flight.

Continually observing the wingtips has advantages other than being the only positive check for leveling the wings. It also helps divert the pilot's attention from the airplane's nose, prevents a fixed stare, and automatically expands the pilot's area of vision by increasing the range necessary for the pilot's vision to cover. In practicing straight-and-level-flight, the wingtips can be used not only for establishing the airplane's laterally level attitude or bank, but to a lesser degree, its pitch attitude. This is noted only for assistance in learning straight-and-level flight, and is not a recommended practice in normal operations.

The scope of a student's vision is also very important, for if it is obscured the student will tend to look out to one side continually (usually the left) and consequently lean that way. This not only gives the student a biased angle from which to judge, but also causes the student to exert unconscious pressure on the controls in that direction, which results in dragging a wing.

With the wings approximately level, it is possible to maintain straight flight by simply exerting the necessary forces on the rudder in the desired direction. However, the instructor should point out that the practice of using rudder alone is not correct and may make precise control of the airplane difficult. Straight–and-level flight requires almost no application of control pressures if the airplane is properly trimmed and the air is smooth. For that reason, the student must not form the habit of constantly moving the controls unnecessarily. The student must learn to recognize when corrections are necessary, and then to make a measured response easily and naturally.

To obtain the proper conception of the forces required on the rudder during straight-and-level-flight, the airplane must be held level. One of the most common faults of beginning students is the tendency to concentrate on the nose of the airplane and attempting to hold the wings level by observing the curvature of the nose cowling. With this method, the reference line is very short and the deviation, particularly if very slight, can go unnoticed. Also, a very small deviation from level, by this short reference line, becomes considerable at the wingtips and results in an appreciable dragging of one wing. This attitude requires the use of additional rudder to maintain straight flight, giving a false conception of neutral control forces. The habit of dragging one wing, and compensating with rudder pressure, if allowed to develop is particularly hard to break, and if not corrected will result in considerable difficulty in mastering other flight maneuvers.

For all practical purposes, the airspeed will remain constant in straight-and-level flight with a constant power setting. Practice of intentional airspeed changes, by increasing or decreasing the power, will provide an excellent means of developing proficiency in maintaining straight-and-level flight at various speeds. Significant changes in airspeed will, of course, require considerable changes in pitch attitude and pitch trim to maintain altitude. Pronounced changes in pitch attitude and trim will also be necessary as the flaps and landing gear are operated.

Common errors in the performance of straight-and-level flight are:

- Attempting to use improper reference points on the airplane to establish attitude.
- Forgetting the location of preselected reference points on subsequent flights.
- Attempting to establish or correct airplane attitude using flight instruments rather than outside visual reference.
- Attempting to maintain direction using only rudder control.
- Habitually flying with one wing low.
- "Chasing" the flight instruments rather than adhering to the principles of attitude flying.
- Too tight a grip on the flight controls resulting in overcontrol and lack of feel.
- Pushing or pulling on the flight controls rather than exerting pressure against the airstream.
- Improper scanning and/or devoting insufficient time to outside visual reference. (Head in the cockpit.)
- Fixation on the nose (pitch attitude) reference point.
- Unnecessary or inappropriate control inputs.
- Failure to make timely and measured control inputs when deviations from straight-and-level flight are detected.
- Inadequate attention to sensory inputs in developing feel for the airplane.

TRIM CONTROL

The airplane is designed so that the primary flight controls (rudder, aileron, and elevator) are streamlined with the nonmovable airplane surfaces when the airplane is cruising straight-and-level at normal weight and loading. If the airplane is flying out of that basic balanced condition, one or more of the control surfaces is going to have to be held out of its streamlined position by continuous control input. The use of trim tabs relieves the pilot of this requirement. Proper trim technique is a very important and often overlooked basic flying skill. An improperly trimmed airplane requires constant control pressures, produces pilot tension and fatigue, distracts the pilot from scanning, and contributes to abrupt and erratic airplane attitude control.

Because of their relatively low power and speed, not all light airplanes have a complete set of trim tabs that are adjustable from the cockpit. In airplanes where rudder, aileron, and elevator trim are available, a definite sequence of trim application should be used. Elevator/stabilator should be trimmed first to relieve the need for control pressure to maintain constant airspeed/pitch attitude. Attempts to trim the rudder at varying airspeed are impractical in propeller driven airplanes because of the change in the torque correcting offset of the vertical fin. Once a constant airspeed/pitch attitude has been established, the pilot should hold the wings level with aileron pressure while rudder pressure is trimmed out. Aileron trim should then be adjusted to relieve any lateral control yoke pressure.

A common trim control error is the tendency to overcontrol the airplane with trim adjustments. To avoid this the pilot must learn to establish and hold the airplane in the desired attitude using the primary flight controls. The proper attitude should be established with reference to the horizon and then verified by reference to performance indications on the flight instruments. The pilot should then apply trim in the above sequence to relieve whatever hand and foot pressure had been required. The pilot must avoid using the trim to establish or correct airplane attitude. The airplane attitude must be established and held first, then control pressures trimmed out so that the airplane will maintain the desired attitude in "hands off" flight. Attempting to "fly the airplane with the trim tabs" is a common fault in basic flying technique even among experienced pilots.

A properly trimmed airplane is an indication of good piloting skills. Any control pressures the pilot feels should be a result of deliberate pilot control input during a planned change in airplane attitude, not a result of pressures being applied by the airplane because the pilot is allowing it to assume control.

LEVEL TURNS

A turn is made by banking the wings in the direction of the desired turn. A specific angle of bank is selected by the pilot, control pressures applied to achieve the desired bank angle, and appropriate control pressures exerted to maintain the desired bank angle once it is established. [Figure 3-5]

Figure 3-5. Level turn to the left.

All four primary controls are used in close coordination when making turns. Their functions are as follows.

- The ailerons bank the wings and so determine the rate of turn at any given airspeed.
- The elevator moves the nose of the airplane up or down in relation to the pilot, and perpendicular to the wings. Doing that, it both sets the pitch attitude in the turn and "pulls" the nose of the airplane around the turn.
- The throttle provides thrust which may be used for airspeed to tighten the turn.
- The rudder offsets any yaw effects developed by the other controls. The rudder does not turn the airplane.

For purposes of this discussion, turns are divided into three classes: shallow turns, medium turns, and steep turns.

- Shallow turns are those in which the bank (less than approximately 20°) is so shallow that the inherent lateral stability of the airplane is acting to level the wings unless some aileron is applied to maintain the bank.
- Medium turns are those resulting from a degree of bank (approximately 20° to 45°) at which the airplane remains at a constant bank.

Steep turns are those resulting from a degree of bank (45° or more) at which the "overbanking tendency" of an airplane overcomes stability, and the bank increases unless aileron is applied to prevent it.

Changing the direction of the wing's lift toward one side or the other causes the airplane to be pulled in that direction. [Figure 3-6] Applying coordinated aileron and rudder to bank the airplane in the direction of the desired turn does this.

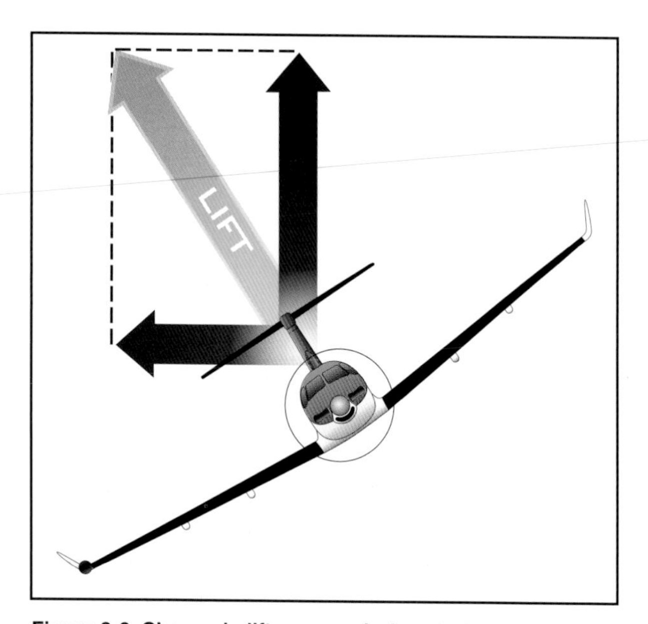

Figure 3-6. Change in lift causes airplane to turn.

When an airplane is flying straight and level, the total lift is acting perpendicular to the wings and to the Earth. As the airplane is banked into a turn, the lift then becomes the resultant of two components. One, the vertical lift component, continues to act perpendicular to the Earth and opposes gravity. Second, the horizontal lift component (centripetal) acts parallel to the Earth's surface and opposes inertia (apparent centrifugal force). These two lift components act at right angles to each other, causing the resultant total lifting force to act perpendicular to the banked wing of the airplane. It is the horizontal lift component that actually turns the airplane—not the rudder. When applying aileron to bank the airplane, the lowered aileron (on the rising wing) produces a greater drag than the raised aileron (on the lowering wing). [Figure 3-7] This increased aileron yaws the airplane toward the rising wing, or opposite to the direction of turn. To counteract this adverse yawing moment, rudder pressure must be applied simultaneously with aileron in the desired direction of turn. This action is required to produce a coordinated turn.

After the bank has been established in a medium banked turn, all pressure applied to the aileron may be relaxed. The airplane will remain at the selected bank

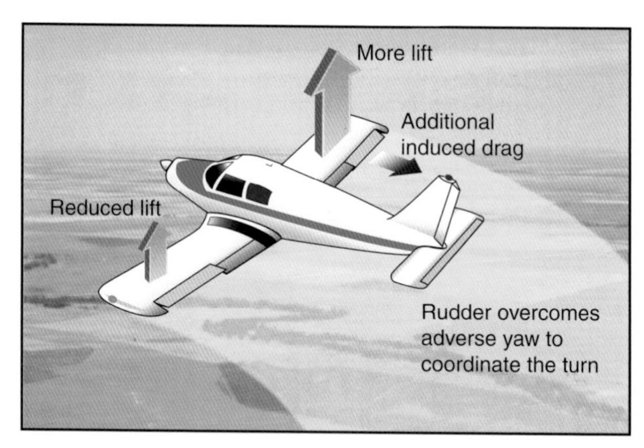

Figure 3-7. Forces during a turn.

with no further tendency to yaw since there is no longer a deflection of the ailerons. As a result, pressure may also be relaxed on the rudder pedals, and the rudder allowed to streamline itself with the direction of the slipstream. Rudder pressure maintained after the turn is established will cause the airplane to skid to the outside of the turn. If a definite effort is made to center the rudder rather than let it streamline itself to the turn, it is probable that some opposite rudder pressure will be exerted inadvertently. This will force the airplane to yaw opposite its turning path, causing the airplane to slip to the inside of the turn. The ball in the turn-and-slip indicator will be displaced off-center whenever the airplane is skidding or slipping sideways. [Figure 3-8] In proper coordinated flight, there is no skidding or slipping. An essential basic airmanship skill is the ability of the pilot to sense or "feel" any uncoordinated condition (slip or skid) without referring to instrument reference. During this stage of training, the flight instructor should stress the development of this ability and insist on its use to attain perfect coordination in all subsequent training.

In all constant altitude, constant airspeed turns, it is necessary to increase the angle of attack of the wing when rolling into the turn by applying up elevator. This is required because part of the vertical lift has been diverted to horizontal lift. Thus, the total lift must be increased to compensate for this loss.

To stop the turn, the wings are returned to level flight by the coordinated use of the ailerons and rudder applied in the opposite direction. To understand the relationship between airspeed, bank, and radius of turn, it should be noted that the rate of turn at any given true airspeed depends on the horizontal lift component. The horizontal lift component varies in proportion to the amount of bank. Therefore, the rate of turn at a given true airspeed increases as the angle of bank is increased. On the other hand, when a turn is made at a higher true airspeed at a given bank angle, the inertia is greater and the horizontal lift component required for the turn is greater, causing the turning rate

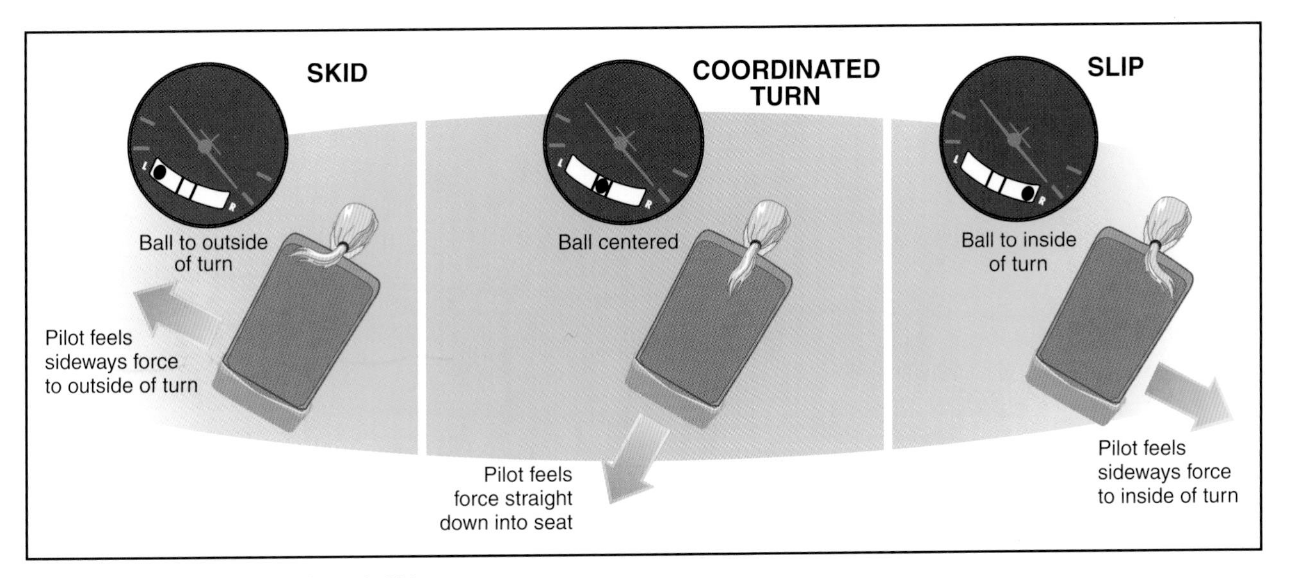

Figure 3-8. Indications of a slip and skid.

to become slower. [Figure 3-9 on next page] Therefore, at a given angle of bank, a higher true airspeed will make the radius of turn larger because the airplane will be turning at a slower rate.

When changing from a shallow bank to a medium bank, the airspeed of the wing on the outside of the turn increases in relation to the inside wing as the radius of turn decreases. The additional lift developed because of this increase in speed of the wing balances the inherent lateral stability of the airplane. At any given airspeed, aileron pressure is not required to maintain the bank. If the bank is allowed to increase from a medium to a steep bank, the radius of turn decreases further. The lift of the outside wing causes the bank to steepen and opposite aileron is necessary to keep the bank constant.

As the radius of the turn becomes smaller, a significant difference develops between the speed of the inside wing and the speed of the outside wing. The wing on the outside of the turn travels a longer circuit than the inside wing, yet both complete their respective circuits in the same length of time. Therefore, the outside wing travels faster than the inside wing, and as a result, it develops more lift. This creates an overbanking tendency that must be controlled by the use of the ailerons. [Figure 3-10] Because the outboard wing is developing more lift, it also has more induced drag. This causes a slight slip during steep turns that must be corrected by use of the rudder.

Sometimes during early training in steep turns, the nose may be allowed to get excessively low resulting in a significant loss in altitude. To recover, the pilot should first reduce the angle of bank with coordinated use of the rudder and aileron, then raise the nose of the airplane to level flight with the elevator. If recovery from an excessively nose-low steep bank condition is attempted by use of the elevator only, it will cause a steepening of the bank and could result in overstressing the airplane. Normally, small corrections for pitch during steep turns are accomplished with the elevator, and the bank is held constant with the ailerons.

To establish the desired angle of bank, the pilot should use outside visual reference points, as well as the bank indicator on the attitude indicator.

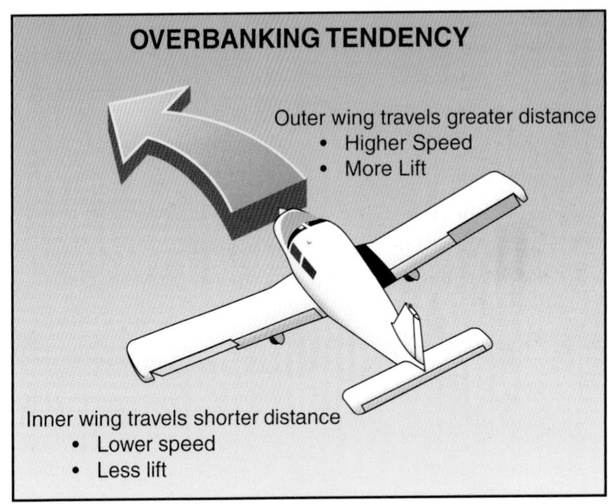

Figure 3-10. Overbanking tendency during a steep turn.

The best outside reference for establishing the degree of bank is the angle formed by the raised wing of low-wing airplanes (the lowered wing of high-wing airplanes) and the horizon, or the angle made by the top of the engine cowling and the horizon. [Figure 3-11 on page 3-11] Since on most light airplanes the engine cowling is fairly flat, its horizontal angle to the horizon will give some indication of the approximate degree of bank. Also, information obtained from the attitude indicator will show the angle of the wing in relation to the horizon. Information from the turn coordinator, however, will not.

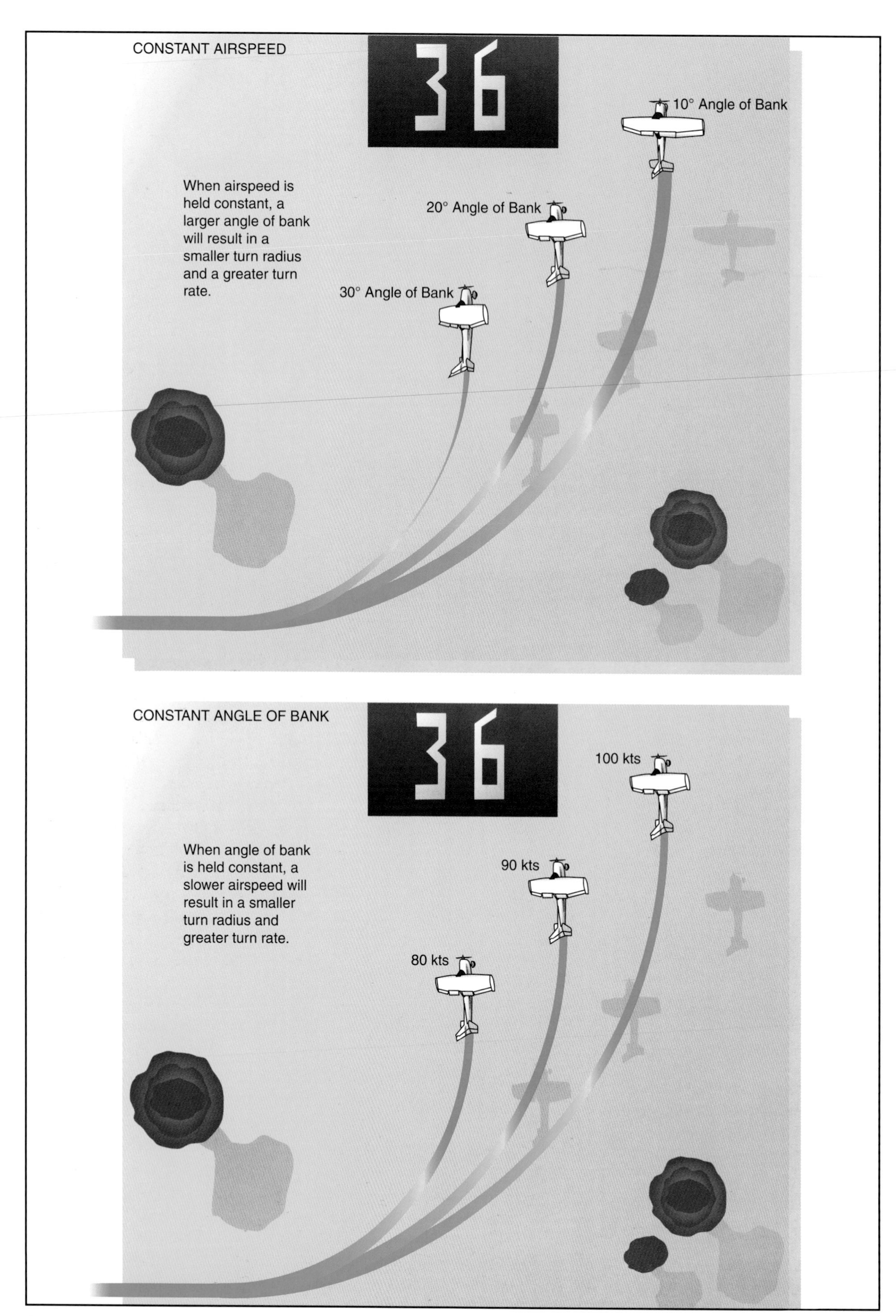

Figure 3-9. Angle of bank and airspeed regulate rate and radius of turn.

Figure 3-11. Visual reference for angle of bank.

Figure 3-12. Right and wrong posture while seated in the airplane.

The pilot's posture while seated in the airplane is very important, particularly during turns. It will affect the interpretation of outside visual references. At the beginning, the student may lean away from the turn in an attempt to remain upright in relation to the ground rather than ride with the airplane. This should be corrected immediately if the student is to properly learn to use visual references. [Figure 3-12]

Parallax error is common among students and experienced pilots. This error is a characteristic of airplanes that have side-by-side seats because the pilot is seated to one side of the longitudinal axis about which the airplane rolls. This makes the nose *appear* to rise when making a left turn and to descend when making right turns. [Figure 3-13]

Beginning students should not use large aileron and rudder applications because this produces a rapid roll rate and allows little time for corrections before the desired bank is reached. Slower (small control displacement) roll rates provide more time to make necessary pitch and bank corrections. As soon as the airplane rolls from the wings-level attitude, the nose should also start to move along the horizon, increasing its rate of travel proportionately as the bank is increased.

The following variations provide excellent guides.

- If the nose starts to move before the bank starts, rudder is being applied too soon.
- If the bank starts before the nose starts turning, or the nose moves in the opposite direction, the rudder is being applied too late.
- If the nose moves up or down when entering a bank, excessive or insufficient up elevator is being applied.

As the desired angle of bank is established, aileron and rudder pressures should be relaxed. This will stop the bank from increasing because the aileron and rudder control surfaces will be neutral in their streamlined position. The up-elevator pressure should not be relaxed, but should be held constant to maintain a constant altitude. Throughout the turn, the pilot should cross-check the airspeed indicator, and if the airspeed has decreased more than 5 knots, additional power should be used. The cross-check should also include outside references, altimeter, and vertical speed indicator (VSI), which can help determine whether or not the pitch attitude is correct. If gaining or losing altitude, the pitch attitude should be adjusted in relation to the horizon, and then the altimeter and VSI rechecked to determine if altitude is being maintained.

Figure 3-13. Parallax view.

During all turns, the ailerons, rudder, and elevator are used to correct minor variations in pitch and bank just as they are in straight-and-level flight.

The rollout from a turn is similar to the roll-in except the flight controls are applied in the opposite direction. Aileron and rudder are applied in the direction of the rollout or toward the high wing. As the angle of bank decreases, the elevator pressure should be relaxed as necessary to maintain altitude.

Since the airplane will continue turning as long as there is any bank, the rollout must be started before reaching the desired heading. The amount of lead required to roll out on the desired heading will depend on the degree of bank used in the turn. Normally, the lead is one-half the degrees of bank. For example, if the bank is 30°, lead the rollout by 15°. As the wings become level, the control pressures should be smoothly relaxed so that the controls are neutralized as the airplane returns to straight-and-level flight. As the rollout is being completed, attention should be given to outside visual references, as well as the attitude and heading indicators to determine that the wings are being leveled and the turn stopped.

Instruction in level turns should begin with medium turns, so that the student has an opportunity to grasp the fundamentals of turning flight without having to deal with overbanking tendency, or the inherent stability of the airplane attempting to level the wings. The instructor should not ask the student to roll the airplane from bank to bank, but to change its attitude from level to bank, bank to level, and so on with a slight pause at the termination of each phase. This pause allows the airplane to free itself from the effects of any misuse of the controls and assures a correct start for the next turn. During these exercises, the idea of control forces, rather than movement, should be emphasized by pointing out the resistance of the controls to varying forces applied to them. The beginning student should be encouraged to use the rudder freely. Skidding in this phase indicates positive control use, and may be easily corrected later. The use of too little rudder, or rudder use in the wrong direction at this stage of training, on the other hand, indicates a lack of proper conception of coordination.

In practicing turns, the action of the airplane's nose will show any error in coordination of the controls. Often, during the entry or recovery from a bank, the nose will describe a vertical arc above or below the horizon, and then remain in proper position after the bank is established. This is the result of lack of timing and coordination of forces on the elevator and rudder controls during the entry and recovery. It indicates that the student has a knowledge of correct turns, but that entry and recovery techniques are in error.

Because the elevator and ailerons are on one control, and pressures on both are executed simultaneously, the beginning pilot is often apt to continue pressure on one of these unintentionally when force on the other only is intended. This is particularly true in left-hand turns, because the position of the hands makes correct movements slightly awkward at first. This is sometimes responsible for the habit of climbing slightly in right-hand turns and diving slightly in left-hand turns. This results from many factors, including the unequal rudder pressures required to the right and to the left when turning, due to the torque effect.

The tendency to climb in right-hand turns and descend in left-hand turns is also prevalent in airplanes having side-by-side cockpit seating. In this case, it is due to the pilot's being seated to one side of the longitudinal axis about which the airplane rolls. This makes the nose appear to rise during a correctly executed left turn and to descend during a correctly executed right turn. An attempt to keep the nose on the same apparent level will cause climbing in right turns and diving in left turns.

Excellent coordination and timing of all the controls in turning requires much practice. It is essential that this coordination be developed, because it is the very basis of this fundamental flight maneuver.

If the body is properly relaxed, it will act as a pendulum and may be swayed by any force acting on it. During a skid, it will be swayed away from the turn, and during a slip, toward the inside of the turn. The same effects will be noted in tendencies to slide on the seat. As the "feel" of flying develops, the properly directed student will become highly sensitive to this last tendency and will be able to detect the presence of, or even the approach of, a slip or skid long before any other indication is present.

Common errors in the performance of level turns are:

- Failure to adequately clear the area before beginning the turn.
- Attempting to execute the turn solely by instrument reference.
- Attempting to sit up straight, in relation to the ground, during a turn, rather than riding with the airplane.
- Insufficient feel for the airplane as evidenced by the inability to detect slips/skids without reference to flight instruments.
- Attempting to maintain a constant bank angle by referencing the "cant" of the airplane's nose.

- Fixating on the nose reference while excluding wingtip reference.

- "Ground shyness"—making "flat turns" (skidding) while operating at low altitudes in a conscious or subconscious effort to avoid banking close to the ground.

- Holding rudder in the turn.

- Gaining proficiency in turns in only one direction (usually the left).

- Failure to coordinate the use of throttle with other controls.

- Altitude gain/loss during the turn.

CLIMBS AND CLIMBING TURNS

When an airplane enters a climb, it changes its flightpath from level flight to an inclined plane or climb attitude. In a climb, weight no longer acts in a direction perpendicular to the flightpath. It acts in a rearward direction. This causes an increase in total drag requiring an increase in thrust (power) to balance the forces. An airplane can only sustain a climb angle when there is sufficient thrust to offset increased drag; therefore, climb is limited by the thrust available.

Like other maneuvers, climbs should be performed using outside visual references and flight instruments. It is important that the pilot know the engine power settings and pitch attitudes that will produce the following conditions of climb.

NORMAL CLIMB—Normal climb is performed at an airspeed recommended by the airplane manufacturer. Normal climb speed is generally somewhat higher than the airplane's best rate of climb. The additional airspeed provides better engine cooling, easier control, and better visibility over the nose. Normal climb is sometimes referred to as "cruise climb." Complex or high performance airplanes may have a specified cruise climb in addition to normal climb.

BEST RATE OF CLIMB—Best rate of climb (V_Y) is performed at an airspeed where the most excess *power* is available over that required for level flight. This condition of climb will produce the most gain in altitude in the least amount of time (maximum rate of climb in feet per minute). The best rate of climb made at full allowable power is a maximum climb. It must be fully understood that attempts to obtain more climb performance than the airplane is capable of by increasing pitch attitude will result in a *decrease* in the rate of altitude gain.

BEST ANGLE OF CLIMB—Best angle of climb (V_X) is performed at an airspeed that will produce the most altitude gain in a given *distance*. Best angle-of-climb airspeed (V_X) is considerably lower than best rate of climb (V_Y), and is the airspeed where the most excess *thrust* is available over that required for level flight. The best angle of climb will result in a steeper climb *path,* although the airplane will take longer to reach the same altitude than it would at best rate of climb. The best angle of climb, therefore, is used in clearing obstacles after takeoff. [Figure 3-14]

It should be noted that, as altitude increases, the speed for best angle of climb increases, and the speed for best rate of climb decreases. The point at which these two speeds meet is the absolute ceiling of the airplane. [Figure 3-15 on next page]

A straight climb is entered by gently increasing pitch attitude to a predetermined level using back-elevator pressure, and simultaneously increasing engine power to the climb power setting. Due to an increase in downwash over the horizontal stabilizer as power is applied, the airplane's nose will tend to immediately begin to rise of its own accord to an attitude higher than

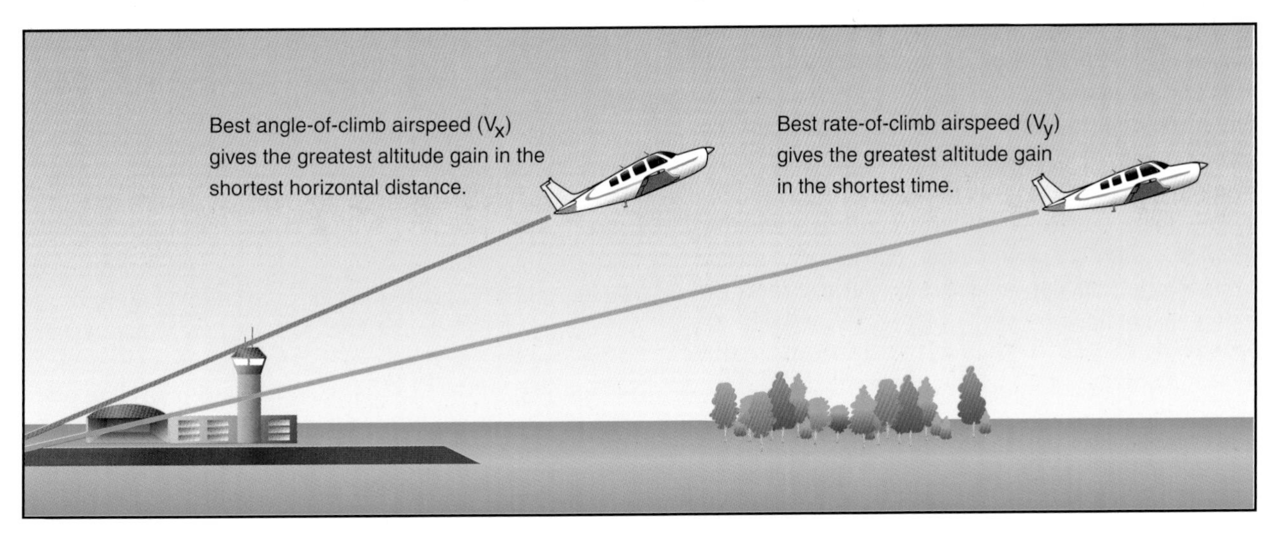

Figure 3-14. Best angle of climb vs. best rate of climb.

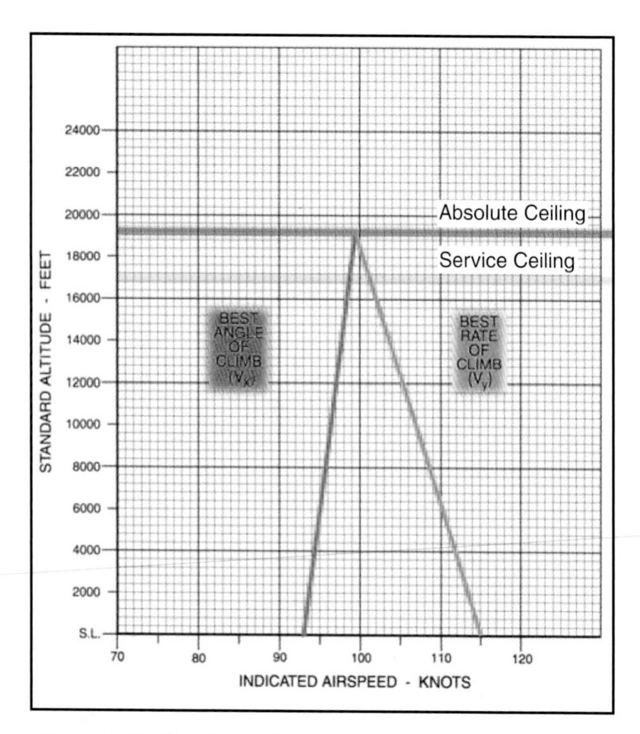

Figure 3-15. Absolute ceiling.

that at which it would stabilize. The pilot must be prepared for this.

As a climb is started, the airspeed will gradually diminish. This reduction in airspeed is gradual because of the initial momentum of the airplane. The thrust required to maintain straight-and-level flight at a given airspeed is not sufficient to maintain the same airspeed in a climb. Climbing flight requires more power than flying level because of the increased drag caused by gravity acting rearward. Therefore, power must be advanced to a higher power setting to offset the increased drag.

The propeller effects at climb power are a primary factor. This is because airspeed is significantly slower than at cruising speed, and the airplane's angle of attack is significantly greater. Under these conditions, torque and asymmetrical loading of the propeller will cause the airplane to roll and yaw to the left. To counteract this, the right rudder must be used.

During the early practice of climbs and climbing turns, this may make coordination of the controls seem awkward (left climbing turn holding right rudder), but after a little practice this correction for propeller effects will become instinctive.

Trim is also a very important consideration during a climb. After the climb has been established, the airplane should be trimmed to relieve all pressures from the flight controls. If changes are made in the pitch attitude, power, or airspeed, the airplane should be retrimmed in order to relieve control pressures.

When performing a climb, the power should be advanced to the climb power recommended by the manufacturer. If the airplane is equipped with a controllable-pitch propeller, it will have not only an engine tachometer, but also a manifold pressure gauge. Normally, the flaps and landing gear (if retractable) should be in the retracted position to reduce drag.

As the airplane gains altitude during a climb, the manifold pressure gauge (if equipped) will indicate a loss in manifold pressure (power). This is because the same volume of air going into the engine's induction system gradually decreases in density as altitude increases. When the volume of air in the manifold decreases, it causes a loss of power. This will occur at the rate of approximately 1-inch of manifold pressure for each 1,000-foot gain in altitude. During prolonged climbs, the throttle must be continually advanced, if constant power is to be maintained.

To enter the climb, simultaneously advance the throttle and apply back-elevator pressure to raise the nose of the airplane to the proper position in relation to the horizon. As power is increased, the airplane's nose will rise due to increased download on the stabilizer. This is caused by increased slipstream. As the pitch attitude increases and the airspeed decreases, progressively more right rudder must be applied to compensate for propeller effects and to hold a constant heading.

After the climb is established, back-elevator pressure must be maintained to keep the pitch attitude constant. As the airspeed decreases, the elevators will try to return to their neutral or streamlined position, and the airplane's nose will tend to lower. Nose-up elevator trim should be used to compensate for this so that the pitch attitude can be maintained without holding back-elevator pressure. Throughout the climb, since the power is fixed at the climb power setting, the airspeed is controlled by the use of elevator.

A cross-check of the airspeed indicator, attitude indicator, and the position of the airplane's nose in relation to the horizon will determine if the pitch attitude is correct. At the same time, a constant heading should be held with the wings level if a straight climb is being performed, or a constant angle of bank and rate of turn if a climbing turn is being performed. [Figure 3-16]

To return to straight-and-level flight from a climb, it is necessary to initiate the level-off at approximately 10 percent of the rate of climb. For example, if the airplane is climbing at 500 feet per minute (f.p.m.), leveling off should start 50 feet below the desired altitude. The nose must be lowered gradually because a loss of altitude will result if the pitch attitude is changed to the level flight position without allowing the airspeed to increase proportionately.

Figure 3-16. Climb indications.

After the airplane is established in level flight at a constant altitude, climb power should be retained temporarily so that the airplane will accelerate to the cruise airspeed more rapidly. When the speed reaches the desired cruise speed, the throttle setting and the propeller control (if equipped) should be set to the cruise power setting and the airplane trimmed. After allowing time for engine temperatures to stabilize, adjust the mixture control as required.

In the performance of climbing turns, the following factors should be considered.

- With a constant power setting, the same pitch attitude and airspeed cannot be maintained in a bank as in a straight climb due to the increase in the total lift required.
- The degree of bank should not be too steep. A steep bank significantly decreases the rate of climb. The bank should always remain constant.
- It is necessary to maintain a constant airspeed and constant rate of turn in both right and left turns. The coordination of all flight controls is a primary factor.
- At a constant power setting, the airplane will climb at a slightly shallower climb angle because some of the lift is being used to turn the airplane.
- Attention should be diverted from fixation on the airplane's nose and divided equally among inside and outside references.

There are two ways to establish a climbing turn. Either establish a straight climb and then turn, or enter the climb and turn simultaneously. Climbing turns should be used when climbing to the local practice area. Climbing turns allow better visual scanning, and it is easier for other pilots to see a turning aircraft.

In any turn, the loss of vertical lift and increased induced drag, due to increased angle of attack, becomes greater as the angle of bank is increased. So shallow turns should be used to maintain an efficient rate of climb.

All the factors that affect the airplane during level (constant altitude) turns will affect it during climbing turns or any other training maneuver. It will be noted that because of the low airspeed, aileron drag (adverse yaw) will have a more prominent effect than it did in straight-and-level flight and more rudder pressure will have to be blended with aileron pressure to keep the airplane in coordinated flight during changes in bank angle. Additional elevator back pressure and trim will also have to be used to compensate for centrifugal force, for the loss of vertical lift, and to keep pitch attitude constant.

During climbing turns, as in any turn, the loss of vertical lift and induced drag due to increased angle of attack becomes greater as the angle of bank is increased, so shallow turns should be used to maintain an efficient rate of climb. If a medium or steep banked turn is used, climb performance will be degraded.

Common errors in the performance of climbs and climbing turns are:

- Attempting to establish climb pitch attitude by referencing the airspeed indicator, resulting in "chasing" the airspeed.
- Applying elevator pressure too aggressively, resulting in an excessive climb angle.
- Applying elevator pressure too aggressively during level-off resulting in negative "G" forces.
- Inadequate or inappropriate rudder pressure during climbing turns.
- Allowing the airplane to yaw in straight climbs, usually due to inadequate right rudder pressure.
- Fixation on the nose during straight climbs, resulting in climbing with one wing low.
- Failure to initiate a climbing turn properly with use of rudder and elevators, resulting in little turn, but rather a climb with one wing low.
- Improper coordination resulting in a slip which counteracts the effect of the climb, resulting in little or no altitude gain.
- Inability to keep pitch and bank attitude constant during climbing turns.
- Attempting to exceed the airplane's climb capability.

DESCENTS AND DESCENDING TURNS

When an airplane enters a descent, it changes its flightpath from level to an inclined plane. It is important that the pilot know the power settings and pitch attitudes that will produce the following conditions of descent.

PARTIAL POWER DESCENT—The normal method of losing altitude is to descend with partial power. This is often termed "cruise" or "enroute" descent. The airspeed and power setting recommended by the airplane manufacturer for prolonged descent should be used. The target descent rate should be 400 – 500 f.p.m. The airspeed may vary from cruise airspeed to that used on the downwind leg of the landing pattern. But the wide range of possible airspeeds should not be interpreted to permit erratic pitch changes. The desired airspeed, pitch attitude, and power combination should be preselected and kept constant.

DESCENT AT MINIMUM SAFE AIRSPEED—A minimum safe airspeed descent is a nose-high, power assisted descent condition principally used for clearing obstacles during a landing approach to a short runway. The airspeed used for this descent condition is recommended by the airplane manufacturer and normally is no greater than 1.3 V_{SO}. Some characteristics of the minimum safe airspeed descent are a steeper than normal descent angle, and the excessive power that may be required to produce acceleration at low airspeed should "mushing" and/or an excessive rate of descent be allowed to develop.

GLIDES—A glide is a basic maneuver in which the airplane loses altitude in a controlled descent with little or no engine power; forward motion is maintained by gravity pulling the airplane along an inclined path and the descent rate is controlled by the pilot balancing the forces of gravity and lift.

Although glides are directly related to the practice of power-off accuracy landings, they have a specific operational purpose in normal landing approaches, and forced landings after engine failure. Therefore, it is necessary that they be performed more subconsciously than other maneuvers because most of the time during their execution, the pilot will be giving full attention to details other than the mechanics of performing the maneuver. Since glides are usually performed relatively close to the ground, accuracy of their execution and the formation of proper technique and habits are of special importance.

Because the application of controls is somewhat different in glides than in power-on descents, gliding maneuvers require the perfection of a technique somewhat different from that required for ordinary power-on maneuvers. This control difference is caused primarily by two factors—the absence of the usual propeller slipstream, and the difference in the relative effectiveness of the various control surfaces at slow speeds.

The glide ratio of an airplane is the distance the airplane will, with power off, travel forward in relation to the altitude it loses. For instance, if an airplane travels 10,000 feet forward while descending 1,000 feet, its glide ratio is said to be 10 to 1.

The glide ratio is affected by all four fundamental forces that act on an airplane (weight, lift, drag, and thrust). If all factors affecting the airplane are constant, the glide ratio will be constant. Although the effect of wind will not be covered in this section, it is a very prominent force acting on the gliding distance of the airplane in relationship to its movement over the ground. With a tailwind, the airplane will glide farther because of the higher groundspeed. Conversely, with a headwind the airplane will not glide as far because of the slower groundspeed.

Variations in weight do not affect the glide angle provided the pilot uses the correct airspeed. Since it is the lift over drag (L/D) ratio that determines the distance the airplane can glide, weight will not affect the distance. The glide ratio is based only on the relationship of the aerodynamic forces acting on the airplane. The only effect weight has is to vary the time the airplane will glide. The heavier the airplane the higher the airspeed must be to obtain the same glide ratio. For example, if two airplanes having the same L/D ratio, but different weights, start a glide from the same altitude, the heavier airplane gliding at a higher airspeed will arrive at the same touchdown point in a shorter time. Both airplanes will cover the same distance, only the lighter airplane will take a longer time.

Under various flight conditions, the drag factor may change through the operation of the landing gear and/or flaps. When the landing gear or the flaps are extended, drag increases and the airspeed will decrease unless the pitch attitude is lowered. As the pitch is lowered, the glidepath steepens and reduces the distance traveled. With the power off, a windmilling propeller also creates considerable drag, thereby retarding the airplane's forward movement.

Although the propeller thrust of the airplane is normally dependent on the power output of the engine, the throttle is in the closed position during a glide so the thrust is constant. Since power is not used during a glide or power-off approach, the pitch attitude must be adjusted as necessary to maintain a constant airspeed.

The best speed for the glide is one at which the airplane will travel the greatest forward distance for a given loss of altitude in still air. This *best glide speed* corresponds to an angle of attack resulting in the least drag on the airplane and giving the best lift-to-drag ratio (L/D_{MAX}). [Figure 3-17]

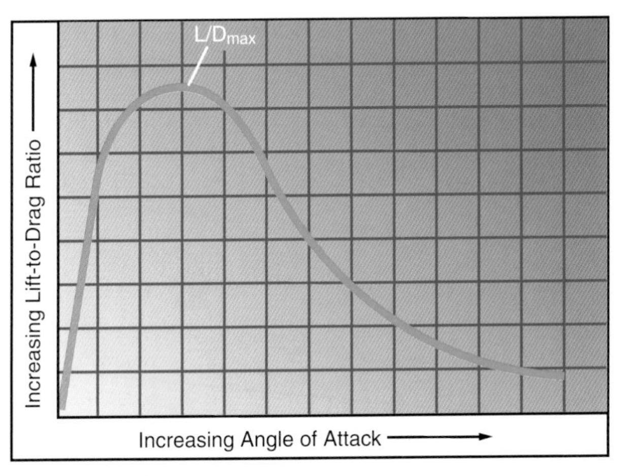

Figure 3-17. L/D_{MAX}.

Any change in the gliding airspeed will result in a proportionate change in glide ratio. Any speed, other than the best glide speed, results in more drag. Therefore, as the glide airspeed is reduced or increased from the optimum or best glide speed, the glide ratio is also changed. When descending at a speed below the best glide speed, induced drag increases. When descending at a speed above best glide speed, parasite drag increases. In either case, the rate of descent will *increase.* [Figure 3-18]

This leads to a cardinal rule of airplane flying that a student pilot must understand and appreciate: The pilot must never attempt to "stretch" a glide by applying back-elevator pressure and reducing the airspeed below the airplane's recommended best glide speed. Attempts to stretch a glide will invariably result in an increase in the rate and angle of descent and may precipitate an inadvertent stall.

To enter a glide, the pilot should close the throttle and advance the propeller (if so equipped) to low pitch (high r.p.m.). A constant altitude should be held with back pressure on the elevator control until the airspeed decreases to the recommended glide speed. Due to a decrease in downwash over the horizontal stabilizer as power is reduced, the airplane's nose will tend to immediately begin to lower of its own accord to an attitude lower than that at which it would stabilize. The pilot must be prepared for this. To keep pitch attitude constant after a power change, the pilot must counteract the immediate trim change. If the pitch attitude is allowed to decrease during glide entry, excess speed will be carried into the glide and retard the attainment of the correct glide angle and airspeed. Speed should be allowed to dissipate before the pitch attitude is decreased. This point is particularly important in so-called clean airplanes as they are very slow to lose their speed and any slight deviation of the nose downwards results in an immediate increase in airspeed. Once the airspeed has dissipated to normal or best glide speed, the pitch attitude should be allowed to decrease to maintain that speed. This should be done with reference to the horizon. When the speed has stabilized, the airplane should be retrimmed for "hands off" flight.

When the approximate gliding pitch attitude is established, the airspeed indicator should be checked. If the airspeed is higher than the recommended speed, the pitch attitude is too low, and if the airspeed is less than recommended, the pitch attitude is too high; therefore, the pitch attitude should be readjusted accordingly referencing the horizon. After the adjustment has been made, the airplane should be retrimmed so that it will maintain this attitude without the need to hold pressure on the elevator control. *The principles of attitude flying require that the proper flight attitude be established using outside visual references first, then using the flight instruments as a secondary check.* It is a good practice to always retrim the airplane after each pitch adjustment.

A stabilized power-off descent at the best glide speed is often referred to as a *normal* glide. The flight instructor should demonstrate a normal glide, and direct the student pilot to memorize the airplane's angle and speed by visually checking the airplane's attitude with reference to the horizon, and noting the pitch of the sound made by the air passing over the structure, the pressure on the controls, and the feel of

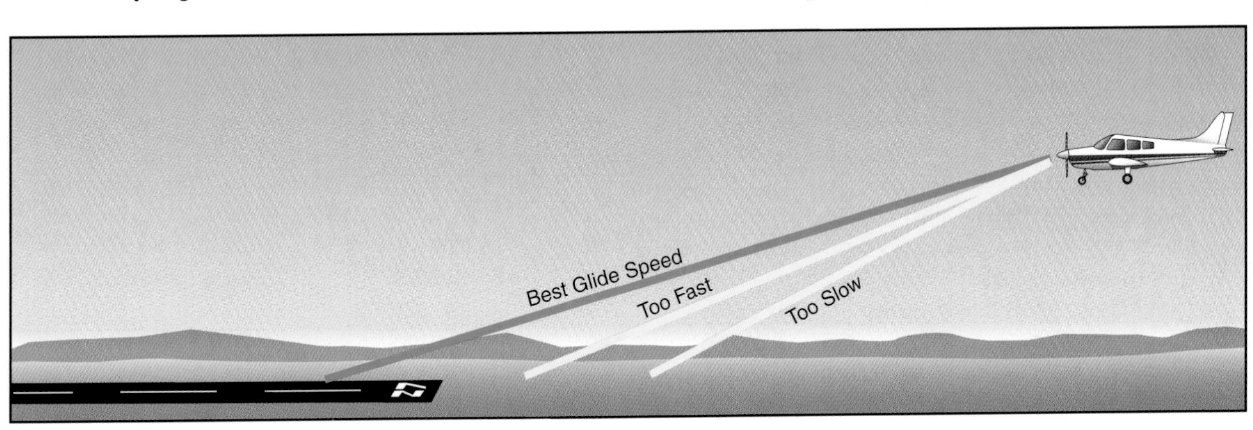

Figure 3-18. Best glide speed provides the greatest forward distance for a given loss of altitude.

the airplane. Due to lack of experience, the beginning student may be unable to recognize slight variations of speed and angle of bank immediately by vision or by the pressure required on the controls. Hearing will probably be the indicator that will be the most easily used at first. The instructor should, therefore, be certain that the student understands that an increase in the pitch of sound denotes increasing speed, while a decrease in pitch denotes less speed. When such an indication is received, the student should consciously apply the other two means of perception so as to establish the proper relationship. The student pilot must use all three elements consciously until they become habits, and must be alert when attention is diverted from the attitude of the airplane and be responsive to any warning given by a variation in the feel of the airplane or controls, or by a change in the pitch of the sound.

After a good comprehension of the normal glide is attained, the student pilot should be instructed in the differences in the results of normal and "abnormal" glides. Abnormal glides being those conducted at speeds other than the normal best glide speed. Pilots who do not acquire an understanding and appreciation of these differences will experience difficulties with accuracy landings, which are comparatively simple if the fundamentals of the glide are thoroughly understood.

Too fast a glide during the approach for landing invariably results in floating over the ground for varying distances, or even overshooting, while too slow a glide causes undershooting, flat approaches, and hard touchdowns. A pilot without the ability to recognize a normal glide will not be able to judge where the airplane will go, or can be made to go, in an emergency. Whereas, in a normal glide, the flightpath may be sighted to the spot on the ground on which the airplane will land. This cannot be done in any abnormal glide.

GLIDING TURNS—The action of the control system is somewhat different in a glide than with power, making gliding maneuvers stand in a class by themselves and require the perfection of a technique different from that required for ordinary power maneuvers. The control difference is caused mainly by two factors—the absence of the usual slipstream, and the difference or relative effectiveness of the various control surfaces at various speeds and particularly at reduced speed. The latter factor has its effect exaggerated by the first, and makes the task of coordination even more difficult for the inexperienced pilot. These principles should be thoroughly explained in order that the student may be alert to the necessary differences in coordination.

After a feel for the airplane and control touch have been developed, the necessary compensation will be automatic; but while any mechanical tendency exists, the student will have difficulty executing gliding turns, particularly when making a practical application of them in attempting accuracy landings.

Three elements in gliding turns which tend to force the nose down and increase glide speed are:

- Decrease in effective lift due to the direction of the lifting force being at an angle to the pull of gravity.
- The use of the rudder acting as it does in the entry to a power turn.
- The normal stability and inherent characteristics of the airplane to nose down with the power off.

These three factors make it necessary to use more back pressure on the elevator than is required for a straight glide or a power turn and, therefore, have a greater effect on the relationship of control coordination.

When recovery is being made from a gliding turn, the force on the elevator control which was applied during the turn must be decreased or the nose will come up too high and considerable speed will be lost. This error will require considerable attention and conscious control adjustment before the normal glide can again be resumed.

In order to maintain the most efficient or normal glide in a turn, more altitude must be sacrificed than in a straight glide since this is the only way speed can be maintained without power. Turning in a glide decreases the performance of the airplane to an even greater extent than a normal turn with power.

Still another factor is the difference in rudder action in turns with and without power. In power turns it is required that the desired recovery point be anticipated in the use of controls and that considerably more pressure than usual be exerted on the rudder. In the recovery from a gliding turn, the same rudder action takes place but without as much pressure being necessary. The actual displacement of the rudder is approximately the same, but it seems to be less in a glide because the resistance to pressure is so much less due to the absence of the propeller slipstream. This often results in a much greater application of rudder through a greater range than is realized, resulting in an abrupt stoppage of the turn when the rudder is applied for recovery. This factor is particularly important during landing practice since the student almost invariably recovers from the last turn too soon and may enter a cross-control condition trying to correct the landing with the rudder alone. This results in landing from a skid that is too easily mistaken for drift.

There is another danger in excessive rudder use during gliding turns. As the airplane skids, the bank will increase. This often alarms the beginning pilot when it occurs close to the ground, and the pilot may respond by applying aileron pressure toward the outside of the turn to stop the bank. At the same time, the rudder forces the nose down and the pilot may apply back-elevator pressure to hold it up. If allowed to progress, this situation may result in a fully developed cross-control condition. A stall in this situation will almost certainly result in a spin.

The level-off from a glide must be started before reaching the desired altitude because of the airplane's downward inertia. The amount of lead depends on the rate of descent and the pilot's control technique. With too little lead, there will be a tendency to descend below the selected altitude. For example, assuming a 500-foot per minute rate of descent, the altitude must be led by 100 – 150 feet to level off at an airspeed higher than the glide speed. At the lead point, power should be increased to the appropriate level flight cruise setting so the desired airspeed will be attained at the desired altitude. The nose tends to rise as both airspeed and downwash on the tail section increase. The pilot must be prepared for this and smoothly control the pitch attitude to attain level flight attitude so that the level-off is completed at the desired altitude.

Particular attention should be paid to the action of the airplane's nose when recovering (and entering) gliding turns. The nose must not be allowed to describe an arc with relation to the horizon, and particularly it must not be allowed to come up during recovery from turns, which require a constant variation of the relative pressures on the different controls.

Common errors in the performance of descents and descending turns are:

- Failure to adequately clear the area.
- Inadequate back-elevator control during glide entry resulting in too steep a glide.
- Failure to slow the airplane to approximate glide speed prior to lowering pitch attitude.
- Attempting to establish/maintain a normal glide solely by reference to flight instruments.
- Inability to sense changes in airspeed through sound and feel.
- Inability to stabilize the glide (chasing the airspeed indicator).
- Attempting to "stretch" the glide by applying back-elevator pressure.
- Skidding or slipping during gliding turns due to inadequate appreciation of the difference in rudder action as opposed to turns with power.
- Failure to lower pitch attitude during gliding turn entry resulting in a decrease in airspeed.
- Excessive rudder pressure during recovery from gliding turns.
- Inadequate pitch control during recovery from straight glides.
- "Ground shyness"—resulting in cross-controlling during gliding turns near the ground.
- Failure to maintain constant bank angle during gliding turns.

PITCH AND POWER

No discussion of climbs and descents would be complete without touching on the question of what controls altitude and what controls airspeed. The pilot must understand the effects of both power and elevator control, working together, during different conditions of flight. The closest one can come to a formula for determining airspeed/altitude control that is valid under all circumstances is a basic principle of attitude flying which states:

> "At any pitch attitude, the amount of power used will determine whether the airplane will climb, descend, or remain level at that attitude."

Through a wide range of nose-low attitudes, a descent is the only possible condition of flight. The addition of power at these attitudes will only result in a greater rate of descent at a faster airspeed.

Through a range of attitudes from very slightly nose-low to about 30° nose-up, a typical light airplane can be made to climb, descend, or maintain altitude depending on the power used. In about the lower third of this range, the airplane will descend at idle power without stalling. As pitch attitude is increased, however, engine power will be required to prevent a stall. Even more power will be required to maintain altitude, and even more for a climb. At a pitch attitude approaching 30° nose-up, all available power will provide only enough thrust to maintain altitude. A slight increase in the steepness of climb or a slight decrease in power will produce a descent. From that point, the least inducement will result in a stall.

Chapter 4

Slow Flight, Stalls, and Spins

INTRODUCTION

The maintenance of lift and control of an airplane in flight requires a certain minimum airspeed. This critical airspeed depends on certain factors, such as gross weight, load factors, and existing density altitude. The minimum speed below which further controlled flight is impossible is called the stalling speed. An important feature of pilot training is the development of the ability to estimate the margin of safety above the stalling speed. Also, the ability to determine the characteristic responses of any airplane at different airspeeds is of great importance to the pilot. The student pilot, therefore, must develop this awareness in order to safely avoid stalls and to operate an airplane correctly and safely at slow airspeeds.

SLOW FLIGHT

Slow flight could be thought of, by some, as a speed that is less than cruise. In pilot training and testing, however, slow flight is broken down into two distinct elements: (1) the establishment, maintenance of, and maneuvering of the airplane at airspeeds and in configurations appropriate to takeoffs, climbs, descents, landing approaches and go-arounds, and, (2) maneuvering at the slowest airspeed at which the airplane is capable of maintaining controlled flight without indications of a stall—usually 3 to 5 knots above stalling speed.

FLIGHT AT LESS THAN CRUISE AIRSPEEDS

Maneuvering during slow flight demonstrates the flight characteristics and degree of controllability of an airplane at less than cruise speeds. The ability to determine the characteristic control responses at the lower airspeeds appropriate to takeoffs, departures, and landing approaches is a critical factor in stall awareness.

As airspeed decreases, control effectiveness decreases disproportionately. For instance, there may be a certain loss of effectiveness when the airspeed is reduced from 30 to 20 m.p.h. above the stalling speed, but there will normally be a much greater loss as the airspeed is further reduced to 10 m.p.h. above stalling. The objective of maneuvering during slow flight is to develop the pilot's sense of feel and ability to use the controls correctly, and to improve proficiency in performing maneuvers that require slow airspeeds.

Maneuvering during slow flight should be performed using both instrument indications and outside visual reference. Slow flight should be practiced from straight glides, straight-and-level flight, and from medium banked gliding and level flight turns. Slow flight at approach speeds should include slowing the airplane smoothly and promptly from cruising to approach speeds without changes in altitude or heading, and determining and using appropriate power and trim settings. Slow flight at approach speed should also include configuration changes, such as landing gear and flaps, while maintaining heading and altitude.

FLIGHT AT MINIMUM CONTROLLABLE AIRSPEED

This maneuver demonstrates the flight characteristics and degree of controllability of the airplane at its *minimum* flying speed. By definition, the term "flight at minimum controllable airspeed" means a speed at which any further increase in angle of attack or load factor, or reduction in power will cause an immediate stall. Instruction in flight at minimum controllable airspeed should be introduced at reduced power settings, with the airspeed sufficiently above the stall to permit maneuvering, but close enough to the stall to sense the characteristics of flight at very low airspeed—which are sloppy controls, ragged response to control inputs, and difficulty maintaining altitude. Maneuvering at minimum controllable airspeed should be performed using both instrument indications and outside visual reference. It is important that pilots form the habit of frequent reference to the flight instruments, especially the airspeed indicator, while flying at very low airspeeds. However, a "feel" for the airplane at very low airspeeds must be developed to avoid inadvertent stalls and to operate the airplane with precision.

To begin the maneuver, the throttle is gradually reduced from cruising position. While the airspeed is decreasing, the position of the nose in relation to the horizon should be noted and should be raised as necessary to maintain altitude.

When the airspeed reaches the maximum allowable for landing gear operation, the landing gear (if equipped with retractable gear) should be extended and all gear down checks performed. As the airspeed reaches the maximum allowable for flap operation, full flaps

Figure 4-1. Slow flight—Low airspeed, high angle of attack, high power, and constant altitude.

should be lowered and the pitch attitude adjusted to maintain altitude. [Figure 4-1] Additional power will be required as the speed further decreases to maintain the airspeed just above a stall. As the speed decreases further, the pilot should note the feel of the flight controls, especially the elevator. The pilot should also note the sound of the airflow as it falls off in tone level.

As airspeed is reduced, the flight controls become less effective and the normal nosedown tendency is reduced. The elevators become less responsive and coarse control movements become necessary to retain control of the airplane. The slipstream effect produces a strong yaw so the application of rudder is required to maintain coordinated flight. The secondary effect of applied rudder is to induce a roll, so aileron is required to keep the wings level. This can result in flying with crossed controls.

During these changing flight conditions, it is important to retrim the airplane as often as necessary to compensate for changes in control pressures. If the airplane has been trimmed for cruising speed, heavy aft control pressure will be needed on the elevators, making precise control impossible. If too much speed is lost, or too little power is used, further back pressure on the elevator control may result in a loss of altitude or a stall. When the desired pitch attitude and minimum control airspeed have been established, it is important to continually cross-check the attitude indicator, altimeter, and airspeed indicator, as well as outside references to ensure that accurate control is being maintained.

The pilot should understand that when flying more slowly than **minimum drag speed ($LD/_{MAX}$)** the airplane will exhibit a characteristic known as "**speed instability**." If the airplane is disturbed by even the slightest turbulence, the airspeed will decrease. As airspeed decreases, the total drag also increases resulting in a further loss in airspeed. The total drag continues to rise and the speed continues to fall. Unless more power is applied and/or the nose is lowered, the speed will continue to decay right down to the stall. This is an extremely important factor in the performance of slow flight. The pilot must understand that, at speed less than minimum drag speed, the airspeed is unstable and will continue to decay if allowed to do so.

When the attitude, airspeed, and power have been stabilized in straight flight, turns should be practiced to determine the airplane's controllability characteristics at this minimum speed. During the turns, power and pitch attitude may need to be increased to maintain the airspeed and altitude. The objective is to acquaint the pilot with the lack of maneuverability at minimum speeds, the danger of incipient stalls, and the tendency of the airplane to stall as the bank is increased. A stall may also occur as a result of abrupt or rough control movements when flying at this critical airspeed.

Abruptly raising the flaps while at minimum controllable airspeed will result in lift suddenly being lost, causing the airplane to lose altitude or perhaps stall.

Once flight at minimum controllable airspeed is set up properly for level flight, a descent or climb at minimum controllable airspeed can be established by adjusting the power as necessary to establish the desired rate of descent or climb. The beginning pilot should note the increased yawing tendency at minimum control airspeed at high power settings with flaps fully extended. In some airplanes, an attempt to climb at such a slow airspeed may result in a *loss* of altitude, even with maximum power applied.

Common errors in the performance of slow flight are:

- Failure to adequately clear the area.
- Inadequate back-elevator pressure as power is reduced, resulting in altitude loss.
- Excessive back-elevator pressure as power is reduced, resulting in a climb, followed by a rapid reduction in airspeed and "mushing."
- Inadequate compensation for adverse yaw during turns.
- Fixation on the airspeed indicator.
- Failure to anticipate changes in lift as flaps are extended or retracted.
- Inadequate power management.
- Inability to adequately divide attention between airplane control and orientation.

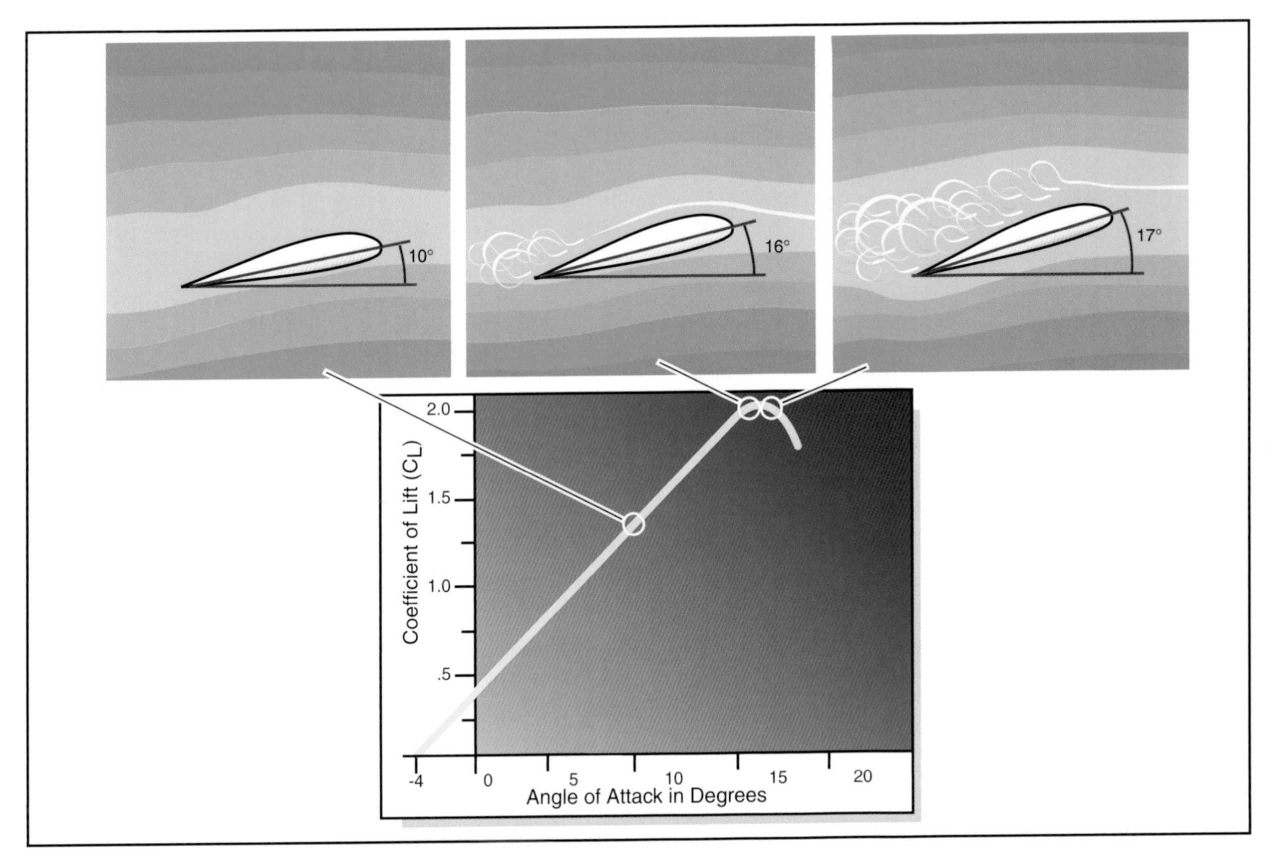

Figure 4-2. Critical angle of attack and stall.

STALLS

A stall occurs when the smooth airflow over the airplane's wing is disrupted, and the lift degenerates rapidly. This is caused when the wing exceeds its critical angle of attack. This can occur at any airspeed, in any attitude, with any power setting. [Figure 4-2]

The practice of stall recovery and the development of awareness of stalls are of primary importance in pilot training. The objectives in performing intentional stalls are to familiarize the pilot with the conditions that produce stalls, to assist in recognizing an approaching stall, and to develop the habit of taking prompt preventive or corrective action.

Intentional stalls should be performed at an altitude that will provide adequate height above the ground for recovery and return to normal level flight. Though it depends on the degree to which a stall has progressed, most stalls require some loss of altitude during recovery. The longer it takes to recognize the approaching stall, the more complete the stall is likely to become, and the greater the loss of altitude to be expected.

RECOGNITION OF STALLS

Pilots must recognize the flight conditions that are conducive to stalls and know how to apply the necessary corrective action. They should learn to recognize an approaching stall by sight, sound, and feel. The following cues may be useful in recognizing the approaching stall.

- Vision is useful in detecting a stall condition by noting the attitude of the airplane. This sense can only be relied on when the stall is the result of an unusual attitude of the airplane. Since the airplane can also be stalled from a normal attitude, vision in this instance would be of little help in detecting the approaching stall.

- Hearing is also helpful in sensing a stall condition. In the case of fixed-pitch propeller airplanes in a power-on condition, a change in sound due to loss of revolutions per minute (r.p.m.) is particularly noticeable. The lessening of the noise made by the air flowing along the airplane structure as airspeed decreases is also quite noticeable, and when the stall is almost complete, vibration and incident noises often increase greatly.

- Kinesthesia, or the sensing of changes in direction or speed of motion, is probably the most important and the best indicator to the trained and experienced pilot. If this sensitivity is properly developed, it will warn of a decrease in speed or the beginning of a settling or mushing of the airplane.

- Feel is an important sense in recognizing the onset of a stall. The feeling of control pressures is very important. As speed is reduced, the resistance to pressures on the controls becomes progressively less. Pressures exerted on the controls tend to become movements of the control surfaces. The lag between these movements and the response of the airplane becomes greater, until in a complete stall all controls can be moved with almost no resistance, and with little immediate effect on the airplane. Just before the stall occurs, buffeting, uncontrollable pitching, or vibrations may begin.

Several types of stall warning indicators have been developed to warn pilots of an approaching stall. The use of such indicators is valuable and desirable, but the reason for practicing stalls is to learn to recognize stalls without the benefit of warning devices.

FUNDAMENTALS OF STALL RECOVERY

During the practice of intentional stalls, the real objective is not to learn how to stall an airplane, but to learn how to recognize an approaching stall and take prompt corrective action. [Figure 4-3] Though the recovery actions must be taken in a coordinated manner, they are broken down into three actions here for explanation purposes.

First, at the indication of a stall, the pitch attitude and angle of attack must be decreased positively and immediately. Since the basic cause of a stall is always an excessive angle of attack, the cause must first be eliminated by releasing the back-elevator pressure that was necessary to attain that angle of attack or by moving the elevator control forward. This lowers the nose and returns the wing to an effective angle of attack. The amount of elevator control pressure or movement used depends on the design of the airplane, the severity of the stall, and the proximity of the ground. In some airplanes, a moderate movement of the elevator control—perhaps slightly forward of neutral—is enough, while in others a forcible push to the full forward position may be required. An excessive negative load on the wings caused by excessive forward movement of the elevator may impede, rather than hasten, the stall recovery. The object is to reduce the angle of attack but only enough to allow the wing to regain lift.

Second, the maximum allowable power should be applied to increase the airplane's airspeed and assist in reducing the wing's angle of attack. The throttle should be promptly, but smoothly, advanced to the maximum allowable power. The flight instructor

Figure 4-3. Stall recognition and recovery.

should emphasize, however, that power is not essential for a safe stall recovery if sufficient altitude is available. Reducing the angle of attack is the only way of recovering from a stall regardless of the amount of power used.

Although stall recoveries should be practiced without, as well as with the use of power, in most actual stalls the application of more power, if available, is an integral part of the stall recovery. Usually, the greater the power applied, the less the loss of altitude.

Maximum allowable power applied at the instant of a stall will usually not cause overspeeding of an engine equipped with a fixed-pitch propeller, due to the heavy air load imposed on the propeller at slow airspeeds. However, it will be necessary to reduce the power as airspeed is gained after the stall recovery so the airspeed will not become excessive. When performing intentional stalls, the tachometer indication should never be allowed to exceed the red line (maximum allowable r.p.m.) marked on the instrument.

Third, straight-and-level flight should be regained with coordinated use of all controls.

Practice in both power-on and power-off stalls is important because it simulates stall conditions that could occur during normal flight maneuvers. For example, the power-on stalls are practiced to show what could happen if the airplane were climbing at an excessively nose-high attitude immediately after takeoff or during a climbing turn. The power-off turning stalls are practiced to show what could happen if the controls are improperly used during a turn from the base leg to the final approach. The power-off straight-ahead stall simulates the attitude and flight characteristics of a particular airplane during the final approach and landing.

Usually, the first few practices should include only approaches to stalls, with recovery initiated as soon as the first buffeting or partial loss of control is noted. In this way, the pilot can become familiar with the indications of an approaching stall without actually stalling the airplane. Once the pilot becomes comfortable with this procedure, the airplane should be slowed in such a manner that it stalls in as near a level pitch attitude as is possible. The student pilot must not be allowed to form the impression that in all circumstances, a high pitch attitude is necessary to exceed the critical angle of attack, or that in all circumstances, a level or near level pitch attitude is indicative of a low angle of attack. Recovery should be practiced first *without* the addition of power, by merely relieving enough back-elevator pressure that the stall is broken and the airplane assumes a normal glide attitude. The instructor should also introduce the student to a secondary stall at this point. Stall recoveries should then be practiced with the addition of power to determine how effective power will be in executing a safe recovery and minimizing altitude loss.

Stall accidents usually result from an inadvertent stall at a low altitude in which a recovery was not accomplished prior to contact with the surface. As a preventive measure, stalls should be practiced at an altitude which will allow recovery no lower than 1,500 feet AGL. To recover with a minimum loss of altitude requires a reduction in the angle of attack (lowering the airplane's pitch attitude), application of power, and termination of the descent without entering another (secondary) stall.

USE OF AILERONS/RUDDER IN STALL RECOVERY

Different types of airplanes have different stall characteristics. Most airplanes are designed so that the wings will stall progressively outward from the wing roots (where the wing attaches to the fuselage) to the wingtips. This is the result of designing the wings in a manner that the wingtips have less angle of incidence than the wing roots. [Figure 4-4] Such a design feature causes the wingtips to have a smaller angle of attack than the wing roots during flight.

Figure 4-4. Wingtip washout.

Exceeding the critical angle of attack causes a stall; the wing roots of an airplane will exceed the critical angle before the wingtips, and the wing roots will stall first. The wings are designed in this manner so that aileron control will be available at high angles of attack (slow airspeed) and give the airplane more stable stalling characteristics.

When the airplane is in a stalled condition, the wingtips continue to provide some degree of lift, and the ailerons still have some control effect. During recovery from a stall, the return of lift begins at the tips and progresses toward the roots. Thus, the ailerons can be used to level the wings.

Using the ailerons requires finesse to avoid an aggravated stall condition. For example, if the right wing dropped during the stall and excessive aileron control were applied to the left to raise the wing, the aileron deflected downward (right wing) would produce a greater angle of attack (and drag), and possibly a more complete stall at the tip as the critical angle of attack is exceeded. The increase in drag created by the high angle of attack on that wing might cause the airplane to yaw in that direction. This adverse yaw could result in a spin unless directional control was maintained by rudder, and/or the aileron control sufficiently reduced.

Even though excessive aileron pressure may have been applied, a spin will not occur if directional (yaw) control is maintained by timely application of coordinated rudder pressure. Therefore, it is important that the rudder be used properly during both the entry and the recovery from a stall. The primary use of the rudder in stall recoveries is to counteract any tendency of the airplane to yaw or slip. The correct recovery technique would be to decrease the pitch attitude by applying forward-elevator pressure to break the stall, advancing the throttle to increase airspeed, and simultaneously maintaining directional control with coordinated use of the aileron and rudder.

STALL CHARACTERISTICS

Because of engineering design variations, the stall characteristics for all airplanes cannot be specifically described; however, the similarities found in small general aviation training-type airplanes are noteworthy enough to be considered. It will be noted that the power-on and power-off stall warning indications will be different. The power-off stall will have less noticeable clues (buffeting, shaking) than the power-on stall. In the power-off stall, the predominant clue can be the elevator control position (full up-elevator against the stops) and a high descent rate. When performing the power-on stall, the buffeting will likely be the predominant clue that provides a positive indication of the stall. For the purpose of airplane certification, the stall warning may be furnished either through the inherent aerodynamic qualities of the airplane, or by a stall warning device that will give a clear distinguishable indication of the stall. Most airplanes are equipped with a stall warning device.

The factors that affect the stalling characteristics of the airplane are balance, bank, pitch attitude, coordination, drag, and power. The pilot should learn the effect of the stall characteristics of the airplane being flown and the proper correction. It should be reemphasized that a stall can occur at any airspeed, in any attitude, or at any power setting, depending on the total number of factors affecting the particular airplane.

A number of factors may be induced as the result of other factors. For example, when the airplane is in a nose-high turning attitude, the angle of bank has a tendency to increase. This occurs because with the airspeed decreasing, the airplane begins flying in a smaller and smaller arc. Since the outer wing is moving in a larger radius and traveling faster than the inner wing, it has more lift and causes an overbanking tendency. At the same time, because of the decreasing airspeed and lift on both wings, the pitch attitude tends to lower. In addition, since the airspeed is decreasing while the power setting remains constant, the effect of torque becomes more prominent, causing the airplane to yaw.

During the practice of power-on turning stalls, to compensate for these factors and to maintain a constant flight attitude until the stall occurs, aileron pressure must be continually adjusted to keep the bank attitude constant. At the same time, back-elevator pressure must be continually increased to maintain the pitch attitude, as well as right rudder pressure increased to keep the ball centered and to prevent adverse yaw from changing the turn rate. If the bank is allowed to become too steep, the vertical component of lift decreases and makes it even more difficult to maintain a constant pitch attitude.

Whenever practicing turning stalls, a constant pitch and bank attitude should be maintained until the stall occurs. Whatever control pressures are necessary should be applied even though the controls appear to be crossed (aileron pressure in one direction, rudder pressure in the opposite direction). During the entry to a power-on turning stall to the right, in particular, the controls will be crossed to some extent. This is due to right rudder pressure being used to overcome torque and left aileron pressure being used to prevent the bank from increasing.

APPROACHES TO STALLS (IMMINENT STALLS)—POWER-ON OR POWER-OFF

An imminent stall is one in which the airplane is approaching a stall but is not allowed to completely stall. This stall maneuver is primarily for practice in retaining (or regaining) full control of the airplane immediately upon recognizing that it is almost in a stall or that a stall is likely to occur if timely preventive action is not taken.

The practice of these stalls is of particular value in developing the pilot's sense of feel for executing maneuvers in which maximum airplane performance is required. These maneuvers require flight with the airplane approaching a stall, and recovery initiated before a stall occurs. As in all maneuvers that involve significant changes in altitude or direction, the pilot must ensure that the area is clear of other air traffic before executing the maneuver.

These stalls may be entered and performed in the attitudes and with the same configuration of the basic full stalls or other maneuvers described in this chapter. However, instead of allowing a complete stall, when the first buffeting or decay of control effectiveness is noted, the angle of attack must be reduced immediately by releasing the back-elevator pressure and applying whatever additional power is necessary. Since the airplane will not be completely stalled, the pitch attitude needs to be decreased only to a point where minimum controllable airspeed is attained or until adequate control effectiveness is regained.

The pilot must promptly recognize the indication of a stall and take timely, positive control action to prevent a full stall. Performance is unsatisfactory if a full stall occurs, if an excessively low pitch attitude is attained, or if the pilot fails to take timely action to avoid excessive airspeed, excessive loss of altitude, or a spin.

FULL STALLS POWER-OFF

The practice of power-off stalls is usually performed with normal landing approach conditions in simulation of an accidental stall occurring during landing approaches. Airplanes equipped with flaps and/or retractable landing gear should be in the landing configuration. Airspeed in excess of the normal approach speed should not be carried into a stall entry since it could result in an abnormally nose-high attitude. Before executing these practice stalls, the pilot must be sure the area is clear of other air traffic.

After extending the landing gear, applying carburetor heat (if applicable), and retarding the throttle to idle (or normal approach power), the airplane should be held at a constant altitude in level flight until the airspeed decelerates to that of a normal approach. The airplane should then be smoothly nosed down into the normal approach attitude to maintain that airspeed. Wing flaps should be extended and pitch attitude adjusted to maintain the airspeed.

When the approach attitude and airspeed have stabilized, the airplane's nose should be smoothly raised to an attitude that will induce a stall. Directional control should be maintained with the rudder, the wings held level by use of the ailerons, and a constant-pitch attitude maintained with the elevator until the stall occurs. The stall will be recognized by clues, such as full up-elevator, high descent rate, uncontrollable nosedown pitching, and possible buffeting.

Recovering from the stall should be accomplished by reducing the angle of attack, releasing back-elevator pressure, and advancing the throttle to maximum allowable power. Right rudder pressure is necessary to overcome the engine torque effects as power is advanced and the nose is being lowered. [Figure 4-5]

The nose should be lowered as necessary to regain flying speed and returned to straight-and-level flight

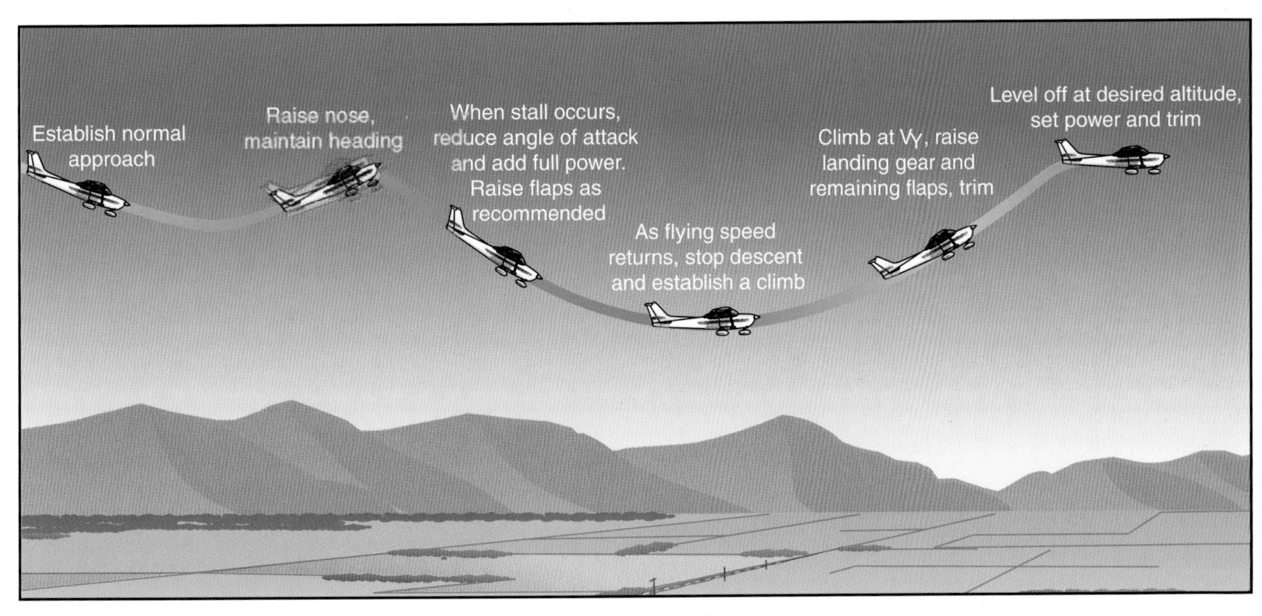

Figure 4-5. Power-off stall and recovery.

attitude. After establishing a positive rate of climb, the flaps and landing gear are retracted, as necessary, and when in level flight, the throttle should be returned to cruise power setting. After recovery is complete, a climb or go-around procedure should be initiated, as the situation dictates, to assure a minimum loss of altitude.

Recovery from power-off stalls should also be practiced from shallow banked turns to simulate an inadvertent stall during a turn from base leg to final approach. During the practice of these stalls, care should be taken that the turn continues at a uniform rate until the complete stall occurs. If the power-off turn is not properly coordinated while approaching the stall, wallowing may result when the stall occurs. If the airplane is in a slip, the outer wing may stall first and whip downward abruptly. This does not affect the recovery procedure in any way; the angle of attack must be reduced, the heading maintained, and the wings leveled by coordinated use of the controls. In the practice of turning stalls, no attempt should be made to stall the airplane on a predetermined heading. However, to simulate a turn from base to final approach, the stall normally should be made to occur within a heading change of approximately 90°.

After the stall occurs, the recovery should be made straight ahead with minimum loss of altitude, and accomplished in accordance with the recovery procedure discussed earlier.

Recoveries from power-off stalls should be accomplished both with, and without, the addition of power, and may be initiated either just after the stall occurs, or after the nose has pitched down through the level flight attitude.

FULL STALLS POWER-ON

Power-on stall recoveries are practiced from straight climbs, and climbing turns with 15 to 20° banks, to simulate an accidental stall occurring during takeoffs and climbs. Airplanes equipped with flaps and/or retractable landing gear should normally be in the takeoff configuration; however, power-on stalls should also be practiced with the airplane in a clean configuration (flaps and/or gear retracted) as in departure and normal climbs.

After establishing the takeoff or climb configuration, the airplane should be slowed to the normal lift-off speed while clearing the area for other air traffic. When the desired speed is attained, the power should be set at takeoff power for the takeoff stall or the recommended climb power for the departure stall while establishing a climb attitude. The purpose of reducing the airspeed to lift-off airspeed before the throttle is advanced to the recommended setting is to avoid an excessively steep nose-up attitude for a long period before the airplane stalls.

After the climb attitude is established, the nose is then brought smoothly upward to an attitude obviously impossible for the airplane to maintain and is held at that attitude until the full stall occurs. In most airplanes, after attaining the stalling attitude, the elevator control must be moved progressively further back as the airspeed decreases until, at the full stall, it will have reached its limit and cannot be moved back any farther.

Recovery from the stall should be accomplished by immediately reducing the angle of attack by positively

Figure 4-6. Power-on stall.

releasing back-elevator pressure and, in the case of a departure stall, smoothly advancing the throttle to maximum allowable power. In this case, since the throttle is already at the climb power setting, the addition of power will be relatively slight. [Figure 4-6]

The nose should be lowered as necessary to regain flying speed with the minimum loss of altitude and then raised to climb attitude. Then, the airplane should be returned to the normal straight-and-level flight attitude, and when in normal level flight, the throttle should be returned to cruise power setting. The pilot must recognize instantly when the stall has occurred and take prompt action to prevent a prolonged stalled condition.

SECONDARY STALL

This stall is called a secondary stall since it may occur after a recovery from a preceding stall. It is caused by attempting to hasten the completion of a stall recovery before the airplane has regained sufficient flying speed. [Figure 4-7] When this stall occurs, the back-elevator pressure should again be released just as in a normal stall recovery. When sufficient airspeed has been regained, the airplane can then be returned to straight-and-level flight.

This stall usually occurs when the pilot uses abrupt control input to return to straight-and-level flight after a stall or spin recovery. It also occurs when the pilot fails to reduce the angle of attack sufficiently during stall recovery by not lowering pitch attitude sufficiently, or by attempting to break the stall by using power only.

ACCELERATED STALLS

Though the stalls just discussed normally occur at a specific airspeed, the pilot must thoroughly understand that all stalls result solely from attempts to fly at excessively high angles of attack. During flight, the angle of attack of an airplane wing is determined by a number of factors, the most important of which are the airspeed, the gross weight of the airplane, and the load factors imposed by maneuvering.

At the same gross weight, airplane configuration, and power setting, a given airplane will consistently stall at the same indicated airspeed if no acceleration is involved. The airplane will, however, stall at a higher indicated airspeed when excessive maneuvering loads are imposed by steep turns, pull-ups, or other abrupt changes in its flightpath. Stalls entered from such flight situations are called "accelerated maneuver stalls," a term, which has no reference to the airspeeds involved.

Stalls which result from abrupt maneuvers tend to be more rapid, or severe, than the unaccelerated stalls, and because they occur at higher-than-normal airspeeds, and/or may occur at lower than anticipated pitch attitudes, they may be unexpected by an inexperienced pilot. Failure to take immediate steps toward recovery when an accelerated stall occurs may result in a complete loss of flight control, notably, power-on spins.

This stall should never be practiced with wing flaps in the extended position due to the lower "G" load limitations in that configuration.

Accelerated maneuver stalls should not be performed in any airplane, which is prohibited from such maneuvers by its type certification restrictions or Airplane Flight Manual (AFM) and/or Pilot's Operating Handbook (POH). If they are permitted, they should be performed with a bank of approximately 45°, and in no case at a speed greater

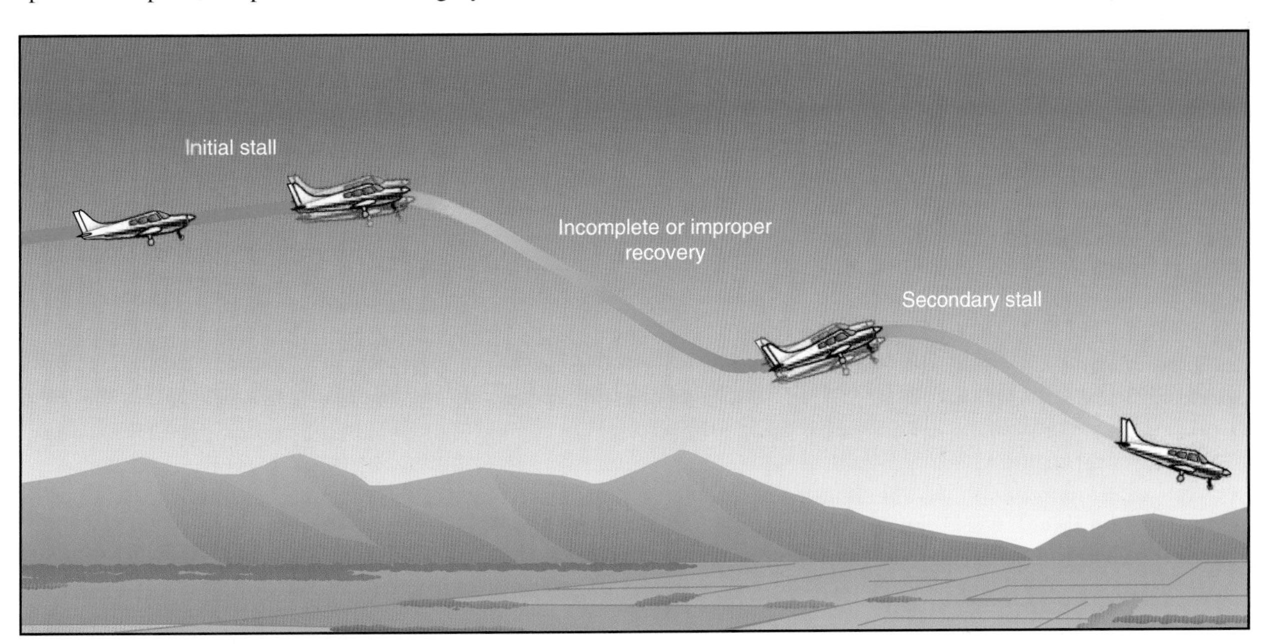

Figure 4-7. Secondary stall.

than the airplane manufacturer's recommended airspeeds or the design maneuvering speed specified for the airplane. The design maneuvering speed is the maximum speed at which the airplane can be stalled or full available aerodynamic control will not exceed the airplane's limit load factor. At or below this speed, the airplane will usually stall before the limit load factor can be exceeded. Those speeds must not be exceeded because of the extremely high structural loads that are imposed on the airplane, especially if there is turbulence. In most cases, these stalls should be performed at no more than 1.2 times the normal stall speed.

The objective of demonstrating accelerated stalls is not to develop competency in setting up the stall, but rather to learn how they may occur and to develop the ability to recognize such stalls immediately, and to take prompt, effective recovery action. It is important that recoveries are made at the first indication of a stall, or immediately after the stall has fully developed; a prolonged stall condition should never be allowed.

An airplane will stall during a coordinated steep turn exactly as it does from straight flight, except that the pitching and rolling actions tend to be more sudden. If the airplane is slipping toward the inside of the turn at the time the stall occurs, it tends to roll rapidly toward the outside of the turn as the nose pitches down because the outside wing stalls before the inside wing. If the airplane is skidding toward the outside of the turn, it will have a tendency to roll to the inside of the turn because the inside wing stalls first. If the coordination of the turn at the time of the stall is accurate, the airplane's nose will pitch away from the pilot just as it does in a straight flight stall, since both wings stall simultaneously.

An accelerated stall demonstration is entered by establishing the desired flight attitude, then smoothly, firmly, and progressively increasing the angle of attack until a stall occurs. Because of the rapidly changing flight attitude, sudden stall entry, and possible loss of altitude, it is extremely vital that the area be clear of other aircraft and the entry altitude be adequate for safe recovery.

This demonstration stall, as in all stalls, is accomplished by exerting excessive back-elevator pressure. Most frequently it would occur during improperly executed steep turns, stall and spin recoveries, and pullouts from steep dives. The objectives are to determine the stall characteristics of the airplane and develop the ability to instinctively recover at the onset of a stall at other-than-normal stall speed or flight attitudes. An accelerated stall, although usually demonstrated in steep turns, may actually be encountered any time excessive back-elevator pressure is applied and/or the angle of attack is increased too rapidly.

From straight-and-level flight at maneuvering speed or less, the airplane should be rolled into a steep level flight turn and back-elevator pressure gradually applied. After the turn and bank are established, back-elevator pressure should be smoothly and steadily increased. The resulting apparent centrifugal force will push the pilot's body down in the seat, increase the wing loading, and decrease the airspeed. After the airspeed reaches the design maneuvering speed or within 20 knots above the unaccelerated stall speed, back-elevator pressure should be firmly increased until a definite stall occurs. These speed restrictions must be observed to prevent exceeding the load limit of the airplane.

When the airplane stalls, recovery should be made promptly, by releasing sufficient back-elevator pressure and increasing power to reduce the angle of attack. If an uncoordinated turn is made, one wing may tend to drop suddenly, causing the airplane to roll in that direction. If this occurs, the excessive back-elevator pressure must be released, power added, and the airplane returned to straight-and-level flight with coordinated control pressure.

The pilot should recognize when the stall is imminent and take prompt action to prevent a completely stalled condition. It is imperative that a prolonged stall, excessive airspeed, excessive loss of altitude, or spin be avoided.

CROSS-CONTROL STALL

The objective of a cross-control stall demonstration maneuver is to show the effect of improper control technique and to emphasize the importance of using coordinated control pressures whenever making turns. This type of stall occurs with the controls crossed—aileron pressure applied in one direction and rudder pressure in the opposite direction.

In addition, when excessive back-elevator pressure is applied, a cross-control stall may result. This is a stall that is most apt to occur during a poorly planned and executed base-to-final approach turn, and often is the result of overshooting the centerline of the runway during that turn. Normally, the proper action to correct for overshooting the runway is to increase the rate of turn by using coordinated aileron and rudder. At the relatively low altitude of a base-to-final approach turn, improperly trained pilots may be apprehensive of steepening the bank to increase the rate of turn, and rather than steepening the bank, they hold the bank constant and attempt to increase the rate of turn by adding more rudder pressure in an effort to align it with the runway.

The addition of inside rudder pressure will cause the speed of the outer wing to increase, therefore, creating greater lift on that wing. To keep that wing from rising and to maintain a constant angle of bank, opposite aileron pressure needs to be applied. The added inside rudder pressure will also cause the nose to lower in relation to the horizon. Consequently, additional back-elevator pressure would be required to maintain a constant-pitch attitude. The resulting condition is a turn with rudder applied in one direction, aileron in the opposite direction, and excessive back-elevator pressure—a pronounced cross-control condition.

Since the airplane is in a skidding turn during the cross-control condition, the wing on the outside of the turn speeds up and produces more lift than the inside wing; thus, the airplane starts to increase its bank. The down aileron on the inside of the turn helps drag that wing back, slowing it up and decreasing its lift, which requires more aileron application. This further causes the airplane to roll. The roll may be so fast that it is possible the bank will be vertical or past vertical before it can be stopped.

For the demonstration of the maneuver, it is important that it be entered at a safe altitude because of the possible extreme nosedown attitude and loss of altitude that may result.

Before demonstrating this stall, the pilot should clear the area for other air traffic while slowly retarding the throttle. Then the landing gear (if retractable gear) should be lowered, the throttle closed, and the altitude maintained until the airspeed approaches the normal glide speed. Because of the possibility of exceeding the airplane's limitations, flaps should not be extended. While the gliding attitude and airspeed are being established, the airplane should be retrimmed. When the glide is stabilized, the airplane should be rolled into a medium-banked turn to simulate a final approach turn that would overshoot the centerline of the runway.

During the turn, excessive rudder pressure should be applied in the direction of the turn but the bank held constant by applying opposite aileron pressure. At the same time, increased back-elevator pressure is required to keep the nose from lowering.

All of these control pressures should be increased until the airplane stalls. When the stall occurs, recovery is made by releasing the control pressures and increasing power as necessary to recover.

In a cross-control stall, the airplane often stalls with little warning. The nose may pitch down, the inside wing may suddenly drop, and the airplane may continue to roll to an inverted position. This is usually the beginning of a spin. It is obvious that close to the ground is no place to allow this to happen.

Recovery must be made before the airplane enters an abnormal attitude (vertical spiral or spin); it is a simple matter to return to straight-and-level flight by coordinated use of the controls. The pilot must be able to recognize when this stall is imminent and must take immediate action to prevent a completely stalled condition. It is imperative that this type of stall not occur during an actual approach to a landing, since recovery may be impossible prior to ground contact due to the low altitude.

The flight instructor should be aware that during traffic pattern operations, any conditions that result in overshooting the turn from base leg to final approach, dramatically increases the possibility of an unintentional accelerated stall while the airplane is in a cross-control condition.

ELEVATOR TRIM STALL

The elevator trim stall maneuver shows what can happen when full power is applied for a go-around and positive control of the airplane is not maintained. [Figure 4-8] Such a situation may occur during a go-around procedure from a normal landing approach

Figure 4-8. Elevator trim stall.

or a simulated forced landing approach, or immediately after a takeoff. The objective of the demonstration is to show the importance of making smooth power applications, overcoming strong trim forces and maintaining positive control of the airplane to hold safe flight attitudes, and using proper and timely trim techniques.

At a safe altitude and after ensuring that the area is clear of other air traffic, the pilot should slowly retard the throttle and extend the landing gear (if retractable gear). One-half to full flaps should be lowered, the throttle closed, and altitude maintained until the airspeed approaches the normal glide speed. When the normal glide is established, the airplane should be trimmed for the glide just as would be done during a landing approach (nose-up trim).

During this simulated final approach glide, the throttle is then advanced smoothly to maximum allowable power as would be done in a go-around procedure. The combined forces of thrust, torque, and back-elevator trim will tend to make the nose rise sharply and turn to the left.

When the throttle is fully advanced and the pitch attitude increases above the normal climbing attitude and it is apparent that a stall is approaching, adequate forward pressure must be applied to return the airplane to the normal climbing attitude. While holding the airplane in this attitude, the trim should then be adjusted to relieve the heavy control pressures and the normal go-around and level-off procedures completed.

The pilot should recognize when a stall is approaching, and take prompt action to prevent a completely stalled condition. It is imperative that a stall not occur during an actual go-around from a landing approach.

Common errors in the performance of intentional stalls are:

- Failure to adequately clear the area.
- Inability to recognize an approaching stall condition through feel for the airplane.
- Premature recovery.
- Over-reliance on the airspeed indicator while excluding other cues.
- Inadequate scanning resulting in an unintentional wing-low condition during entry.
- Excessive back-elevator pressure resulting in an exaggerated nose-up attitude during entry.
- Inadequate rudder control.
- Inadvertent secondary stall during recovery.
- Failure to maintain a constant bank angle during turning stalls.
- Excessive forward-elevator pressure during recovery resulting in negative load on the wings.
- Excessive airspeed buildup during recovery.
- Failure to take timely action to prevent a full stall during the conduct of imminent stalls.

Spins

A spin may be defined as an aggravated stall that results in what is termed "autorotation" wherein the airplane follows a downward corkscrew path. As the airplane rotates around a vertical axis, the rising wing is less stalled than the descending wing creating a rolling, yawing, and pitching motion. The airplane is basically being forced downward by gravity, rolling, yawing, and pitching in a spiral path. [Figure 4-9]

The autorotation results from an unequal angle of attack on the airplane's wings. The rising wing has a decreasing angle of attack, where the relative lift increases and the drag decreases. In effect, this wing is less stalled. Meanwhile, the descending wing has an

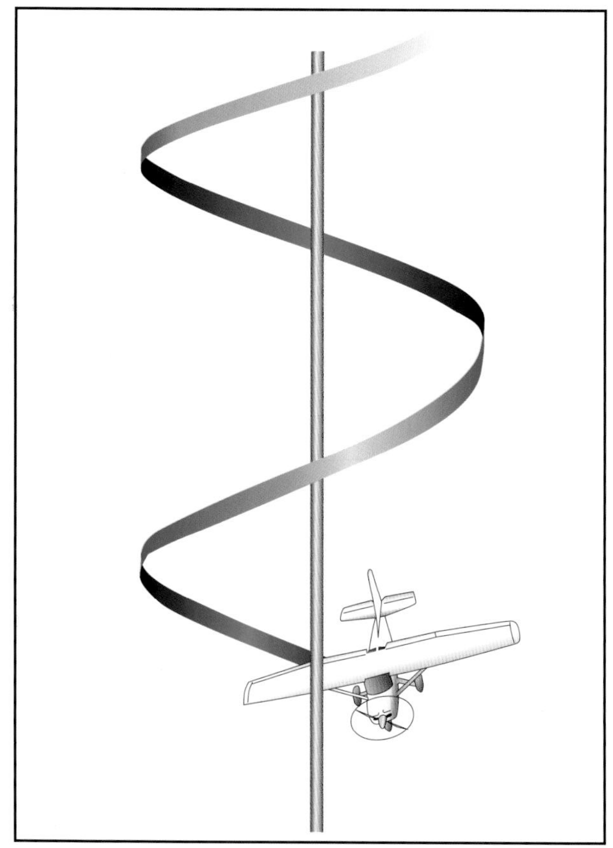

Figure 4-9. Spin—an aggravated stall and autorotation.

increasing angle of attack, past the wing's critical angle of attack (stall) where the relative lift decreases and drag increases.

A spin is caused when the airplane's wing exceeds its critical angle of attack (stall) with a sideslip or yaw acting on the airplane at, or beyond, the actual stall. During this uncoordinated maneuver, a pilot may not be aware that a critical angle of attack has been exceeded until the airplane yaws out of control toward the lowering wing. If stall recovery is not initiated immediately, the airplane may enter a spin.

If this stall occurs while the airplane is in a slipping or skidding turn, this can result in a spin entry and rotation in the direction that the rudder is being applied, regardless of which wingtip is raised.

Many airplanes have to be forced to spin and require considerable judgment and technique to get the spin started. These same airplanes that have to be forced to spin, may be accidentally put into a spin by mishandling the controls in turns, stalls, and flight at minimum controllable airspeeds. This fact is additional evidence of the necessity for the practice of stalls until the ability to recognize and recover from them is developed.

Often a wing will drop at the beginning of a stall. When this happens, the nose will attempt to move (yaw) in the direction of the low wing. This is where use of the rudder is important during a stall. The correct amount of opposite rudder must be applied to keep the nose from yawing toward the low wing. By maintaining directional control and not allowing the nose to yaw toward the low wing, before stall recovery is initiated, a spin will be averted. If the nose is allowed to yaw during the stall, the airplane will begin to slip in the direction of the lowered wing, and will enter a spin. An airplane must be stalled in order to enter a spin; therefore, continued practice in stalls will help the pilot develop a more instinctive and prompt reaction in recognizing an approaching spin. It is essential to learn to apply immediate corrective action any time it is apparent that the airplane is nearing spin conditions. If it is impossible to avoid a spin, the pilot should immediately execute spin recovery procedures.

SPIN PROCEDURES

The flight instructor should demonstrate spins in those airplanes that are approved for spins. Special spin procedures or techniques required for a particular airplane are not presented here. Before beginning any spin operations, the following items should be reviewed.

- The airplane's AFM/POH limitations section, placards, or type certification data, to determine if the airplane is approved for spins.
- Weight and balance limitations.
- Recommended entry and recovery procedures.
- The requirements for parachutes. It would be appropriate to review a current Title 14 of the Code of Federal Regulations (14 CFR) part 91 for the latest parachute requirements.

A thorough airplane preflight should be accomplished with special emphasis on excess or loose items that may affect the weight, center of gravity, and controllability of the airplane. Slack or loose control cables (particularly rudder and elevator) could prevent full anti-spin control deflections and delay or preclude recovery in some airplanes.

Prior to beginning spin training, the flight area, above and below the airplane, must be clear of other air traffic. This may be accomplished while slowing the airplane for the spin entry. All spin training should be initiated at an altitude high enough for a completed recovery at or above 1,500 feet AGL.

It may be appropriate to introduce spin training by first practicing both power-on and power-off stalls, in a clean configuration. This practice would be used to familiarize the student with the airplane's specific stall and recovery characteristics. Care should be taken with the handling of the power (throttle) in entries and during spins. Carburetor heat should be applied according to the manufacturer's recommendations.

There are four phases of a spin: **entry**, **incipient**, **developed**, and **recovery**. [Figure 4-10 on next page]

ENTRY PHASE

The entry phase is where the pilot provides the necessary elements for the spin, either accidentally or intentionally. The entry procedure for demonstrating a spin is similar to a power-off stall. During the entry, the power should be reduced slowly to idle, while simultaneously raising the nose to a pitch attitude that will ensure a stall. As the airplane approaches a stall, smoothly apply full rudder in the direction of the desired spin rotation while applying full back (up) elevator to the limit of travel. Always maintain the ailerons in the neutral position during the spin procedure unless AFM/POH specifies otherwise.

INCIPIENT PHASE

The incipient phase is from the time the airplane stalls and rotation starts until the spin has fully developed. This change may take up to two turns for most airplanes. Incipient spins that are not allowed to develop into a steady-state spin are the most commonly used in the introduction to spin training and recovery techniques. In

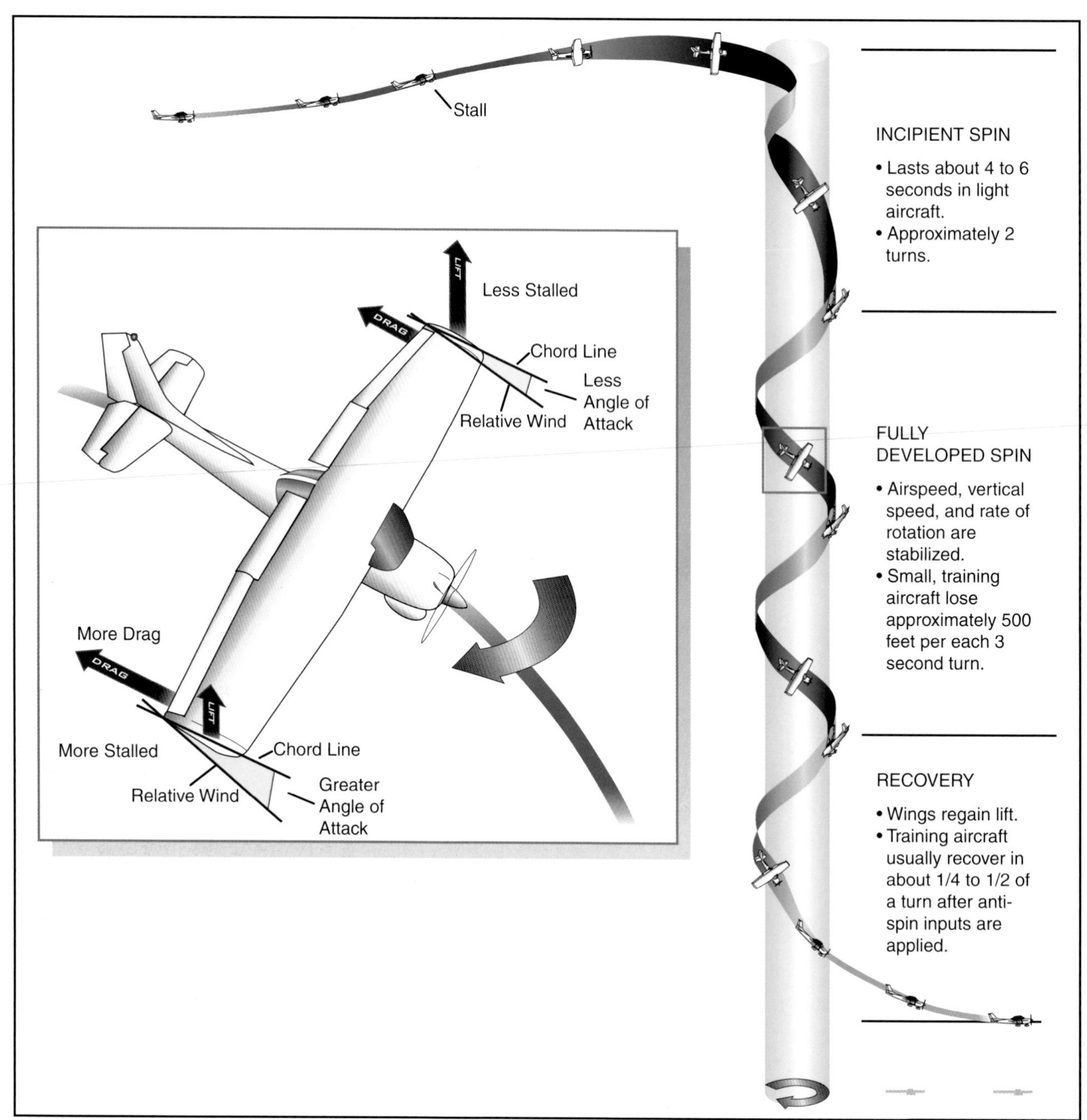

Figure 4-10. Spin entry and recovery.

this phase, the aerodynamic and inertial forces have not achieved a balance. As the incipient spin develops, the indicated airspeed should be near or below stall airspeed, and the turn-and-slip indicator should indicate the direction of the spin.

The incipient spin recovery procedure should be commenced prior to the completion of 360° of rotation. The pilot should apply full rudder opposite the direction of rotation. If the pilot is not sure of the direction of the spin, check the turn-and-slip indicator; it will show a deflection in the direction of rotation.

DEVELOPED PHASE

The developed phase occurs when the airplane's angular rotation rate, airspeed, and vertical speed are stabilized while in a flightpath that is nearly vertical. This is where airplane aerodynamic forces and inertial forces are in balance, and the attitude, angles, and self-sustaining motions about the vertical axis are constant or repetitive. The spin is in equilibrium.

RECOVERY PHASE

The recovery phase occurs when the angle of attack of the wings decreases below the critical angle of attack and autorotation slows. Then the nose steepens and rotation stops. This phase may last for a quarter turn to several turns.

To recover, control inputs are initiated to disrupt the spin equilibrium by stopping the rotation and stall. To accomplish spin recovery, the manufacturer's recommended procedures should be followed. In the absence of the manufacturer's recommended spin recovery procedures and techniques, the following spin recovery procedures are recommended.

Step 1—REDUCE THE POWER (THROTTLE) TO IDLE. Power aggravates the spin characteristics. It usually results in a flatter spin attitude and increased rotation rates.

Step 2—POSITION THE AILERONS TO NEUTRAL. Ailerons may have an adverse effect on spin recovery. Aileron control in the direction of the spin may speed up the rate of rotation and delay the recovery. Aileron control opposite the direction of the spin may cause the down aileron to move the wing deeper into the stall and aggravate the situation. The best procedure is to ensure that the ailerons are neutral.

Step 3—APPLY FULL OPPOSITE RUDDER AGAINST THE ROTATION. Make sure that full (against the stop) opposite rudder has been applied.

Step 4—APPLY A POSITIVE AND BRISK, STRAIGHT FORWARD MOVEMENT OF THE ELEVATOR CONTROL FORWARD OF THE NEUTRAL TO BREAK THE STALL. This should be done immediately after full rudder application. The forceful movement of the elevator will decrease the excessive angle of attack and break the stall. The controls should be held firmly in this position. When the stall is "broken," the spinning will stop.

Step 5—AFTER SPIN ROTATION STOPS, NEUTRALIZE THE RUDDER. If the rudder is not neutralized at this time, the ensuing increased airspeed acting upon a deflected rudder will cause a yawing or skidding effect.

Slow and overly cautious control movements during spin recovery must be avoided. In certain cases it has been found that such movements result in the airplane continuing to spin indefinitely, even with anti-spin inputs. A brisk and positive technique, on the other hand, results in a more positive spin recovery.

Step 6—BEGIN APPLYING BACK-ELEVATOR PRESSURE TO RAISE THE NOSE TO LEVEL FLIGHT. Caution must be used not to apply excessive back-elevator pressure after the rotation stops. Excessive back-elevator pressure can cause a secondary stall and result in another spin. Care should be taken not to exceed the "G" load limits and airspeed limitations during recovery. If the flaps and/or retractable landing gear are extended prior to the spin, they should be retracted as soon as possible after spin entry.

It is important to remember that the above spin recovery procedures and techniques are recommended for use only in the absence of the manufacturer's procedures. Before any pilot attempts to begin spin training, that pilot must be familiar with the procedures provided by the manufacturer for spin recovery.

The most common problems in spin recovery include pilot confusion as to the direction of spin rotation and whether the maneuver is a spin versus spiral. If the airspeed is increasing, the airplane is no longer in a spin but in a spiral. In a spin, the airplane is stalled. The indicated airspeed, therefore, should reflect stall speed.

INTENTIONAL SPINS

The *intentional spinning* of an airplane, for which the spin maneuver is not specifically approved, is NOT authorized by this handbook or by the Code of Federal Regulations. The official sources for determining if the spin maneuver IS APPROVED or NOT APPROVED for a specific airplane are:

- Type Certificate Data Sheets or the Aircraft Specifications.
- The limitation section of the FAA-approved AFM/POH. The limitation sections may provide additional specific requirements for spin authorization, such as limiting gross weight, CG range, and amount of fuel.
- On a placard located in clear view of the pilot in the airplane, NO ACROBATIC MANEUVERS INCLUDING SPINS APPROVED. In airplanes placarded against spins, there is no assurance that recovery from a fully developed spin is possible.

There are occurrences involving airplanes wherein spin restrictions are *intentionally* ignored by some pilots. Despite the installation of placards prohibiting intentional spins in these airplanes, a number of pilots, and some flight instructors, attempt to justify the maneuver, rationalizing that the spin restriction results merely because of a "technicality" in the airworthiness standards.

Some pilots reason that the airplane was spin tested during its certification process and, therefore, no problem should result from demonstrating or practicing spins. However, those pilots overlook the fact that a normal category airplane certification only requires the airplane recover from a one-turn spin in not more than one additional turn or 3 seconds, whichever takes longer. This same test of controllability can also be used in certificating an airplane in the Utility category (14 CFR section 23.221 (b)).

The point is that 360° of rotation (one-turn spin) does not provide a stabilized spin. If the airplane's controllability has not been explored by the engineering test pilot beyond the certification requirements, prolonged spins (inadvertent or intentional) in that airplane place an operating pilot in an unexplored flight situation. Recovery may be difficult or impossible.

In 14 CFR part 23, "Airworthiness Standards: Normal, Utility, Acrobatic, and Commuter Category Airplanes," there are no requirements for investigation of *controllability* in a true spinning condition for the Normal category airplanes. The one-turn "margin of safety" is essentially a check of the airplane's controllability in a delayed recovery from a *stall*. Therefore, *in airplanes placarded against spins there is absolutely no assurance whatever that recovery from a fully developed spin is possible under any circumstances.* The pilot of an airplane placarded against intentional spins should assume that the airplane may well become uncontrollable in a spin.

WEIGHT AND BALANCE REQUIREMENTS

With each airplane that is approved for spinning, the weight and balance requirements are important for safe performance and recovery from the spin maneuver. Pilots must be aware that just minor weight or balance changes can affect the airplane's spin recovery characteristics. Such changes can either alter or enhance the spin maneuver and/or recovery characteristics. For example, the addition of weight in the aft baggage compartment, or additional fuel, may still permit the airplane to be operated within CG, but could seriously affect the spin and recovery characteristics.

An airplane that may be difficult to spin intentionally in the Utility Category (restricted aft CG and reduced weight) could have less resistance to spin entry in the Normal Category (less restricted aft CG and increased weight). This situation is due to the airplane being able to generate a higher angle of attack and load factor. Furthermore, an airplane that is approved for spins in the Utility Category, but loaded in the Normal Category, may not recover from a spin that is allowed to progress beyond the incipient phase.

Common errors in the performance of *intentional* spins are:

- Failure to apply full rudder pressure in the desired spin direction during spin entry.
- Failure to apply and maintain full up-elevator pressure during spin entry, resulting in a spiral.
- Failure to achieve a fully stalled condition prior to spin entry.
- Failure to apply full rudder against the spin during recovery.
- Failure to apply sufficient forward-elevator pressure during recovery.
- Failure to neutralize the rudder during recovery after rotation stops, resulting in a possible secondary spin.
- Slow and overly cautious control movements during recovery.
- Excessive back-elevator pressure after rotation stops, resulting in possible secondary stall.
- Insufficient back-elevator pressure during recovery resulting in excessive airspeed.

General

This chapter discusses takeoffs and departure climbs in tricycle landing gear (nosewheel-type) airplanes under normal conditions, and under conditions which require maximum performance. A thorough knowledge of takeoff principles, both in theory and practice, will often prove of extreme value throughout a pilot's career. It will often prevent an attempted takeoff that would result in an accident, or during an emergency, make possible a takeoff under critical conditions when a pilot with a less well rounded knowledge and technique would fail.

The takeoff, though relatively simple, often presents the most hazards of any part of a flight. The importance of thorough knowledge and faultless technique and judgment cannot be overemphasized.

It must be remembered that the manufacturer's recommended procedures, including airplane configuration and airspeeds, and other information relevant to takeoffs and departure climbs in a specific make and model airplane are contained in the FAA-approved Airplane Flight Manual and/or Pilot's Operating Handbook (AFM/POH) for that airplane. If any of the information in this chapter differs from the airplane manufacturer's recommendations as contained in the AFM/POH, the airplane manufacturer's recommendations take precedence.

Terms and Definitions

Although the takeoff and climb is one continuous maneuver, it will be divided into three separate steps for purposes of explanation: (1) the takeoff roll, (2) the lift-off, and (3) the initial climb after becoming airborne. [Figure 5-1]

- **Takeoff Roll (ground roll)**—the portion of the takeoff procedure during which the airplane is accelerated from a standstill to an airspeed that provides sufficient lift for it to become airborne.
- **Lift-off (rotation)**—the act of becoming airborne as a result of the wings lifting the airplane off the ground or the pilot rotating the nose up, increasing the angle of attack to start a climb.
- **Initial Climb**—begins when the airplane leaves the ground and a pitch attitude has been established to climb away from the takeoff area. Normally, it is considered complete when the airplane has reached a safe maneuvering altitude, or an en route climb has been established.

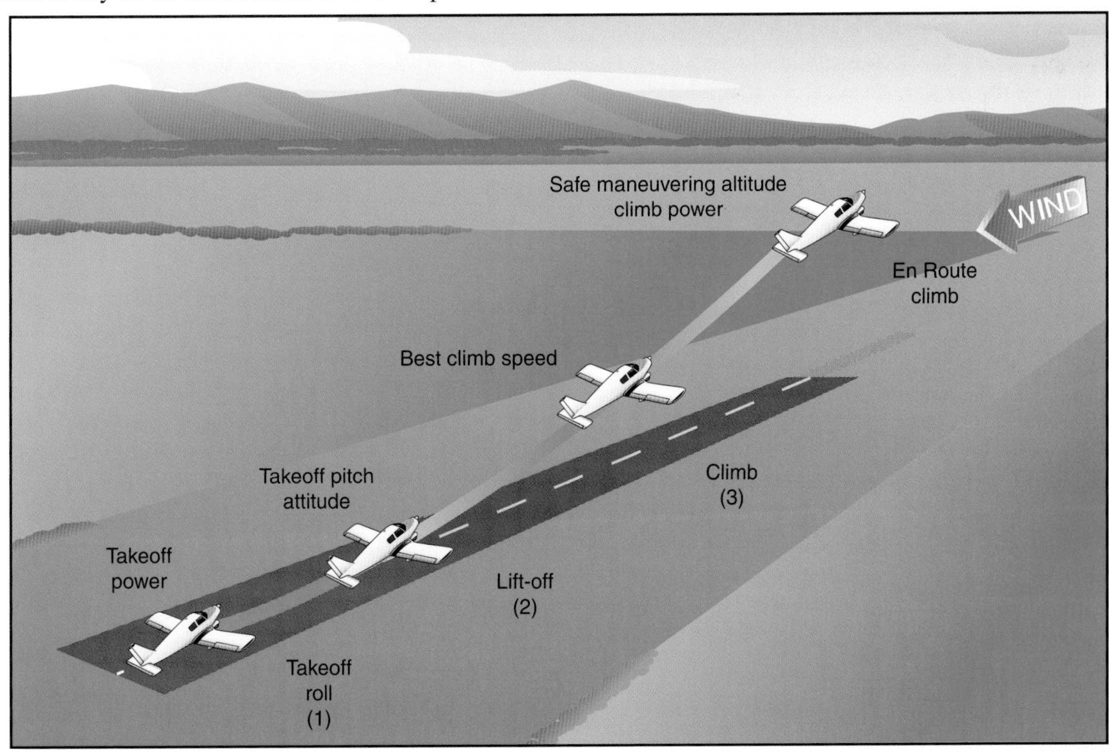

Figure 5-1. Takeoff and climb.

PRIOR TO TAKEOFF

Before taxiing onto the runway or takeoff area, the pilot should ensure that the engine is operating properly and that all controls, including flaps and trim tabs, are set in accordance with the before takeoff checklist. In addition, the pilot must make certain that the approach and takeoff paths are clear of other aircraft. At uncontrolled airports, pilots should announce their intentions on the common traffic advisory frequency (CTAF) assigned to that airport. When operating from an airport with an operating control tower, pilots must contact the tower operator and receive a takeoff clearance before taxiing onto the active runway.

It is not recommended to take off immediately behind another aircraft, particularly large, heavily loaded transport airplanes, because of the wake turbulence that is generated.

While taxiing onto the runway, the pilot can select ground reference points that are aligned with the runway direction as aids to maintaining directional control during the takeoff. These may be runway centerline markings, runway lighting, distant trees, towers, buildings, or mountain peaks.

NORMAL TAKEOFF

A normal takeoff is one in which the airplane is headed into the wind, or the wind is very light. Also, the takeoff surface is firm and of sufficient length to permit the airplane to gradually accelerate to normal lift-off and climb-out speed, and there are no obstructions along the takeoff path.

There are two reasons for making a takeoff as nearly into the wind as possible. First, the airplane's speed while on the ground is much less than if the takeoff were made downwind, thus reducing wear and stress on the landing gear. Second, a shorter ground roll and therefore much less runway length is required to develop the minimum lift necessary for takeoff and climb. Since the airplane depends on airspeed in order to fly, a headwind provides some of that airspeed, even with the airplane motionless, from the wind flowing over the wings.

TAKEOFF ROLL

After taxiing onto the runway, the airplane should be carefully aligned with the intended takeoff direction, and the nosewheel positioned straight, or centered. After releasing the brakes, the throttle should be advanced smoothly and continuously to takeoff power. An abrupt application of power may cause the airplane to yaw sharply to the left because of the torque effects of the engine and propeller. This will be most apparent in high horsepower engines. As the airplane starts to roll forward, the pilot should assure both feet are on the rudder pedals so that the toes or balls of the feet are on the rudder portions, not on the brake portions. Engine instruments should be monitored during the takeoff roll for any malfunctions.

In nosewheel-type airplanes, pressures on the elevator control are not necessary beyond those needed to steady it. Applying unnecessary pressure will only aggravate the takeoff and prevent the pilot from recognizing when elevator control pressure is actually needed to establish the takeoff attitude.

As speed is gained, the elevator control will tend to assume a neutral position if the airplane is correctly trimmed. At the same time, directional control should be maintained with smooth, prompt, positive rudder corrections throughout the takeoff roll. The effects of engine torque and P-factor at the initial speeds tend to pull the nose to the left. The pilot must use whatever rudder pressure and aileron needed to correct for these effects or for existing wind conditions to keep the nose of the airplane headed straight down the runway. The use of brakes for steering purposes should be avoided, since this will cause slower acceleration of the airplane's speed, lengthen the takeoff distance, and possibly result in severe swerving.

While the speed of the takeoff roll increases, more and more pressure will be felt on the flight controls, particularly the elevators and rudder. If the tail surfaces are affected by the propeller slipstream, they become effective first. As the speed continues to increase, all of the flight controls will gradually become effective enough to maneuver the airplane about its three axes. It is at this point, in the taxi to flight transition, that the airplane is being flown more than taxied. As this occurs, progressively smaller rudder deflections are needed to maintain direction.

The feel of resistance to the movement of the controls and the airplane's reaction to such movements are the only real indicators of the degree of control attained. This feel of resistance is not a measure of the airplane's speed, but rather of its controllability. To determine the degree of controllability, the pilot must be conscious of the reaction of the airplane to the control pressures and immediately adjust the pressures as needed to control the airplane. The pilot must wait for the reaction of the airplane to the applied control pressures and attempt to sense the control resistance to pressure rather than attempt to control the airplane by movement of the controls. Balanced control surfaces increase the importance of this point, because they materially reduce the intensity of the resistance offered to pressures exerted by the pilot.

At this stage of training, beginning takeoff practice, a student pilot will normally not have a full appreciation of the variations of control pressures with the speed of the airplane. The student, therefore, may tend to move the controls through wide ranges seeking the pressures that are familiar and expected, and as a consequence over-control the airplane. The situation may be aggravated by the sluggish reaction of the airplane to these movements. The flight instructor should take measures to check these tendencies and stress the importance of the development of feel. The student pilot should be required to feel lightly for resistance and accomplish the desired results by applying pressure against it. This practice will enable the student pilot, as experience is gained, to achieve a sense of the point when sufficient speed has been acquired for the takeoff, instead of merely guessing, fixating on the airspeed indicator, or trying to force performance from the airplane.

LIFT-OFF

Since a good takeoff depends on the proper takeoff attitude, it is important to know how this attitude appears and how it is attained. The ideal takeoff attitude requires only minimum pitch adjustments shortly after the airplane lifts off to attain the speed for the best rate of climb (V_Y). [Figure 5-2] The pitch attitude necessary for the airplane to accelerate to V_Y speed should be demonstrated by the instructor and memorized by the student. Initially, the student pilot may have a tendency to hold excessive back-elevator pressure just after lift-off, resulting in an abrupt pitch-up. The flight instructor should be prepared for this.

Each type of airplane has a best pitch attitude for normal lift-off; however, varying conditions may make a difference in the required takeoff technique. A rough field, a smooth field, a hard surface runway, or a short or soft, muddy field, all call for a slightly different technique, as will smooth air in contrast to a strong, gusty wind. The different techniques for those other-than-normal conditions are discussed later in this chapter.

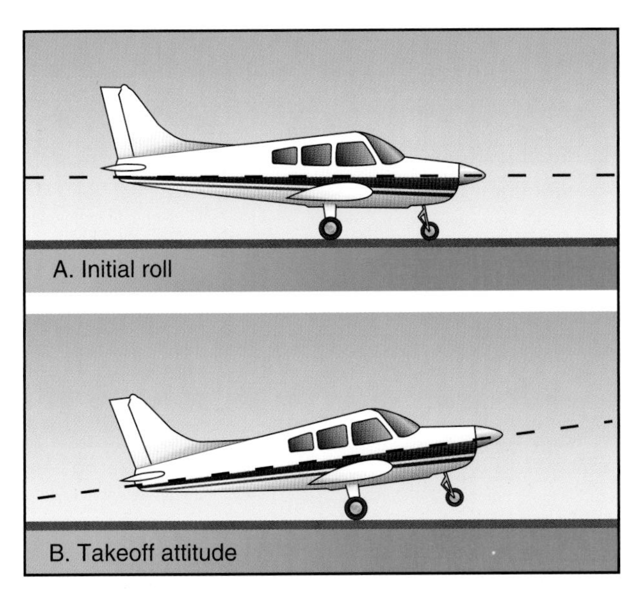

Figure 5-2. Initial roll and takeoff attitude.

When all the flight controls become effective during the takeoff roll in a nosewheel-type airplane, back-elevator pressure should be gradually applied to raise the nosewheel slightly off the runway, thus establishing the takeoff or lift-off attitude. This is often referred to as "rotating." At this point, the position of the nose in relation to the horizon should be noted, then back-elevator pressure applied as necessary to hold this attitude. The wings must be kept level by applying aileron pressure as necessary.

The airplane is allowed to fly off the ground while in the normal takeoff attitude. Forcing it into the air by applying excessive back-elevator pressure would only result in an excessively high pitch attitude and may delay the takeoff. As discussed earlier, excessive and rapid changes in pitch attitude result in proportionate changes in the effects of torque, thus making the airplane more difficult to control.

Although the airplane can be forced into the air, this is considered an unsafe practice and should be avoided under normal circumstances. If the airplane is forced to leave the ground by using too much back-elevator pressure before adequate flying speed is attained, the wing's angle of attack may be excessive, causing the airplane to settle back to the runway or even to stall. On the other hand, if sufficient back-elevator pressure is not held to maintain the correct takeoff attitude after becoming airborne, or the nose is allowed to lower excessively, the airplane may also settle back to the runway. This would occur because the angle of attack is decreased and lift diminished to the degree where it will not support the airplane. It is important, then, to hold the correct attitude constant after rotation or lift-off.

As the airplane leaves the ground, the pilot must continue to be concerned with maintaining the wings in a level attitude, as well as holding the proper pitch attitude. Outside visual scan to attain/maintain proper airplane pitch and bank attitude must be intensified at this critical point. The flight controls have not yet become fully effective, and the beginning pilot will often have a tendency to fixate on the airplane's pitch attitude and/or the airspeed indicator and neglect the natural tendency of the airplane to roll just after breaking ground.

During takeoffs in a strong, gusty wind, it is advisable that an extra margin of speed be obtained before the airplane is allowed to leave the ground. A takeoff at the normal takeoff speed may result in a lack of positive control, or a stall, when the airplane encounters a sudden lull in strong, gusty wind, or other turbulent air currents. In this case, the pilot should allow the airplane to stay on the ground longer to attain more speed; then make a smooth, positive rotation to leave the ground.

INITIAL CLIMB

Upon lift-off, the airplane should be flying at approximately the pitch attitude that will allow it to accelerate to V_Y. This is the speed at which the airplane will gain the most altitude in the shortest period of time.

If the airplane has been properly trimmed, some back-elevator pressure may be required to hold this attitude until the proper climb speed is established. On the other hand, relaxation of any back-elevator pressure before this time may result in the airplane settling, even to the extent that it contacts the runway.

The airplane will pick up speed rapidly after it becomes airborne. Once a positive rate of climb is established, the flaps and landing gear can be retracted (if equipped).

It is recommended that takeoff power be maintained until reaching an altitude of at least 500 feet above the surrounding terrain or obstacles. The combination of V_Y and takeoff power assures the maximum altitude gained in a minimum amount of time. This gives the pilot more altitude from which the airplane can be safely maneuvered in case of an engine failure or other emergency.

Since the power on the initial climb is fixed at the takeoff power setting, the airspeed must be controlled by making slight pitch adjustments using the elevators. However, the pilot should not fixate on the airspeed indicator when making these pitch changes, but should, instead, continue to scan outside to adjust the airplane's attitude in relation to the horizon. In accordance with the principles of attitude flying, the pilot should first make the necessary pitch change with reference to the natural horizon and hold the new attitude momentarily, and then glance at the airspeed indicator as a check to see if the new attitude is correct. Due to inertia, the airplane will not accelerate or decelerate immediately as the pitch is changed. It takes a little time for the airspeed to change. If the pitch attitude has been over or under corrected, the airspeed indicator will show a speed that is more or less than that desired. When this occurs, the cross-checking and appropriate pitch-changing process must be repeated until the desired climbing attitude is established.

When the correct pitch attitude has been attained, it should be held constant while cross-checking it against the horizon and other outside visual references. The airspeed indicator should be used only as a check to determine if the attitude is correct.

After the recommended climb airspeed has been established, and a safe maneuvering altitude has been reached, the power should be adjusted to the recommended climb setting and the airplane trimmed to relieve the control pressures. This will make it easier to hold a constant attitude and airspeed.

During initial climb, it is important that the takeoff path remain aligned with the runway to avoid drifting into obstructions, or the path of another aircraft that may be taking off from a parallel runway. Proper scanning techniques are essential to a safe takeoff and climb, not only for maintaining attitude and direction, but also for collision avoidance in the airport area.

When the student pilot nears the solo stage of flight training, it should be explained that the airplane's takeoff performance will be much different when the instructor is out of the airplane. Due to decreased load, the airplane will become airborne sooner and will climb more rapidly. The pitch attitude that the student has learned to associate with initial climb may also differ due to decreased weight, and the flight controls may seem more sensitive. If the situation is unexpected, it may result in increased tension that may remain until after the landing. Frequently, the existence of this tension and the uncertainty that develops due to the perception of an "abnormal" takeoff results in poor performance on the subsequent landing.

Common errors in the performance of normal takeoffs and departure climbs are:

- Failure to adequately clear the area prior to taxiing into position on the active runway.
- Abrupt use of the throttle.
- Failure to check engine instruments for signs of malfunction after applying takeoff power.
- Failure to anticipate the airplane's left turning tendency on initial acceleration.
- Overcorrecting for left turning tendency.
- Relying solely on the airspeed indicator rather than developed feel for indications of speed and airplane controllability during acceleration and lift-off.
- Failure to attain proper lift-off attitude.
- Inadequate compensation for torque/P-factor during initial climb resulting in a sideslip.
- Over-control of elevators during initial climb-out.

- Limiting scan to areas directly ahead of the airplane (pitch attitude and direction), resulting in allowing a wing (usually the left) to drop immediately after lift-off.
- Failure to attain/maintain best rate-of-climb airspeed (V_Y).
- Failure to employ the principles of attitude flying during climb-out, resulting in "chasing" the airspeed indicator.

CROSSWIND TAKEOFF

While it is usually preferable to take off directly into the wind whenever possible or practical, there will be many instances when circumstances or judgment will indicate otherwise. Therefore, the pilot must be familiar with the principles and techniques involved in crosswind takeoffs, as well as those for normal takeoffs. A crosswind will affect the airplane during takeoff much as it does in taxiing. With this in mind, it can be seen that the technique for crosswind correction during takeoffs closely parallels the crosswind correction techniques used in taxiing.

TAKEOFF ROLL

The technique used during the initial takeoff roll in a crosswind is generally the same as used in a normal takeoff, except that aileron control must be held INTO the crosswind. This raises the aileron on the upwind wing to impose a downward force on the wing to counteract the lifting force of the crosswind and prevents the wing from rising.

As the airplane is taxied into takeoff position, it is essential that the windsock and other wind direction indicators be checked so that the presence of a crosswind may be recognized and anticipated. If a crosswind is indicated, FULL aileron should be held into the wind as the takeoff roll is started. This control position should be maintained while the airplane is accelerating and until the ailerons start becoming sufficiently effective for maneuvering the airplane about its longitudinal axis.

With the aileron held into the wind, the takeoff path must be held straight with the rudder. [Figure 5-3]

Normally, this will require applying downwind rudder pressure, since on the ground the airplane will tend to **weathervane** into the wind. When takeoff power is applied, torque or P-factor that yaws the airplane to the left may be sufficient to counteract the weathervaning tendency caused by a crosswind from the right. On the other hand, it may also aggravate the tendency to

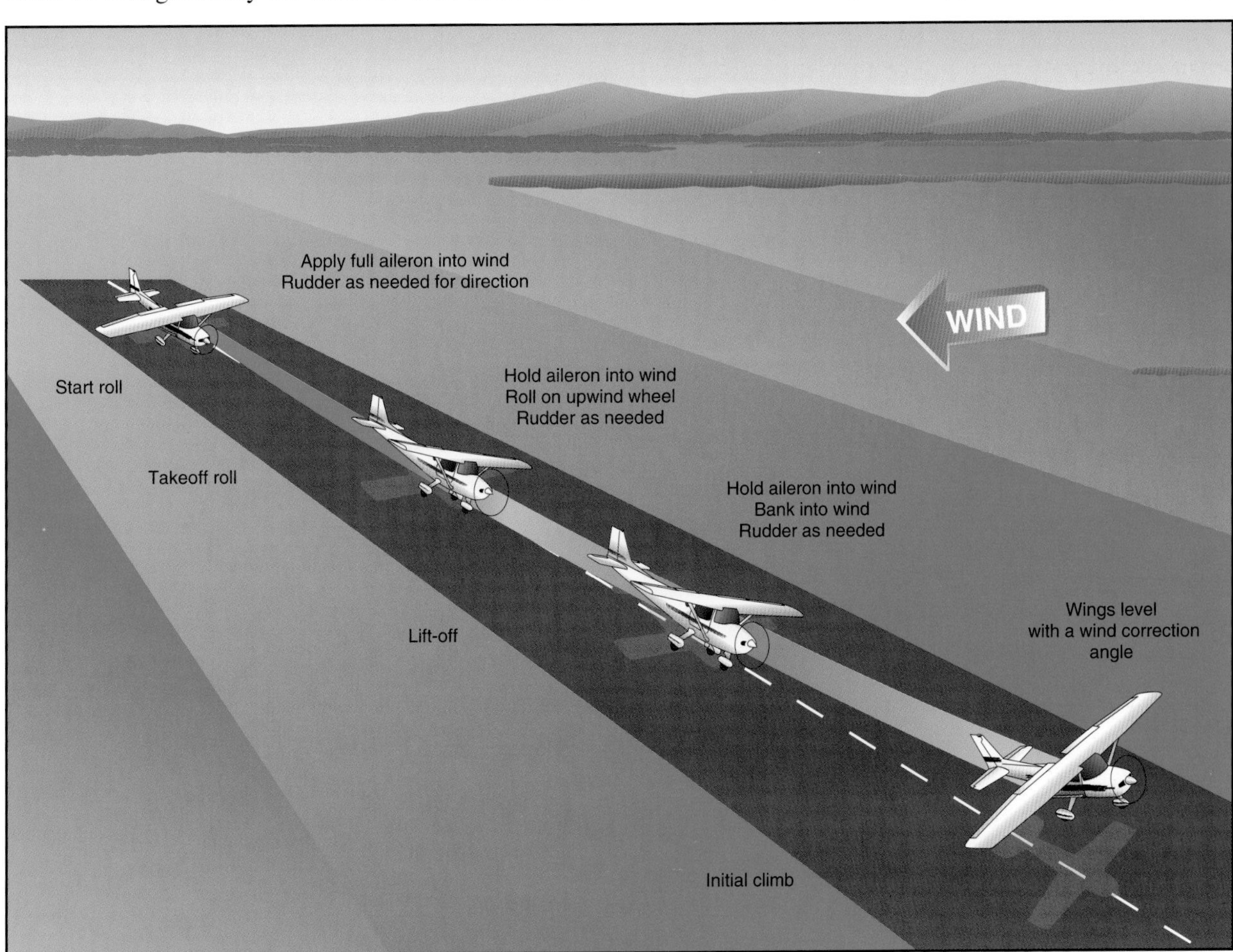

Figure 5-3. Crosswind takeoff roll and initial climb.

swerve left when the wind is from the left. In any case, whatever rudder pressure is required to keep the airplane rolling straight down the runway should be applied.

As the forward speed of the airplane increases and the crosswind becomes more of a relative headwind, the mechanical holding of full aileron into the wind should be reduced. It is when increasing pressure is being felt on the aileron control that the ailerons are becoming more effective. As the aileron's effectiveness increases and the **crosswind component** of the relative wind becomes less effective, it will be necessary to gradually reduce the aileron pressure. The crosswind component effect does not completely vanish, so some aileron pressure will have to be maintained throughout the takeoff roll to keep the crosswind from raising the upwind wing. If the upwind wing rises, thus exposing more surface to the crosswind, a "skipping" action may result. [Figure 5-4]

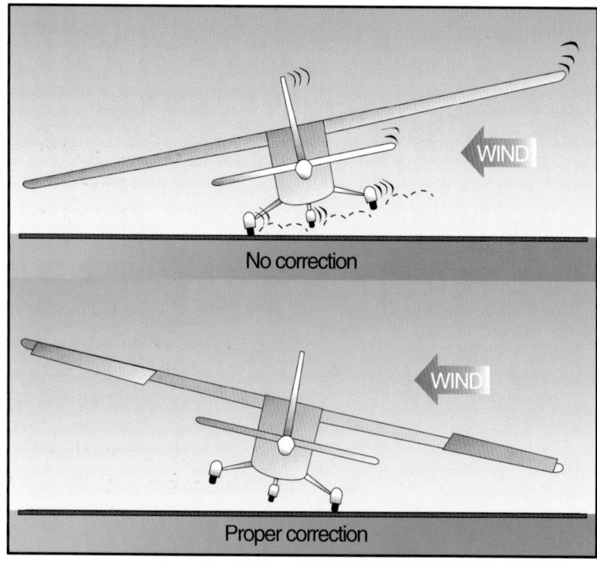

Figure 5-4. Crosswind effect.

This is usually indicated by a series of very small bounces, caused by the airplane attempting to fly and then settling back onto the runway. During these bounces, the crosswind also tends to move the airplane sideways, and these bounces will develop into side-skipping. This side-skipping imposes severe side stresses on the landing gear and could result in structural failure.

It is important, during a crosswind takeoff roll, to hold sufficient aileron into the wind not only to keep the upwind wing from rising but to hold that wing down so that the airplane will, immediately after lift-off, be **sideslipping** into the wind enough to counteract drift.

LIFT-OFF

As the nosewheel is being raised off the runway, the holding of aileron control into the wind may result in the downwind wing rising and the downwind main wheel lifting off the runway first, with the remainder of the takeoff roll being made on that one main wheel. This is acceptable and is preferable to side-skipping.

If a significant crosswind exists, the main wheels should be held on the ground slightly longer than in a normal takeoff so that a smooth but very definite lift-off can be made. This procedure will allow the airplane to leave the ground under more positive control so that it will definitely remain airborne while the proper amount of wind correction is being established. More importantly, this procedure will avoid imposing excessive side-loads on the landing gear and prevent possible damage that would result from the airplane settling back to the runway while drifting.

As both main wheels leave the runway and ground friction no longer resists drifting, the airplane will be slowly carried sideways with the wind unless adequate drift correction is maintained by the pilot. Therefore, it is important to establish and maintain the proper amount of crosswind correction prior to lift-off by applying aileron pressure toward the wind to keep the upwind wing from rising and applying rudder pressure as needed to prevent weathervaning.

INITIAL CLIMB

If proper crosswind correction is being applied, as soon as the airplane is airborne, it will be sideslipping into the wind sufficiently to counteract the drifting effect of the wind. [Figure 5-5] This sideslipping should be continued until the airplane has a positive rate of climb. At that time, the airplane should be turned into the wind to establish just enough wind correction angle to counteract the wind and then the wings rolled level. Firm and aggressive use of the rudders will be required to keep the airplane headed straight down the runway. The climb with a wind correction angle should be continued to follow a ground track aligned with the runway direction. However, because the force of a crosswind may vary markedly within a few hundred feet of the ground, frequent checks of actual ground track should be made, and the wind correction adjusted as necessary. The remainder of the climb technique is the same used for normal takeoffs and climbs.

Common errors in the performance of crosswind takeoffs are:

- Failure to adequately clear the area prior to taxiing onto the active runway.
- Using less than full aileron pressure into the wind initially on the takeoff roll.
- Mechanical use of aileron control rather than sensing the need for varying aileron control input through feel for the airplane.

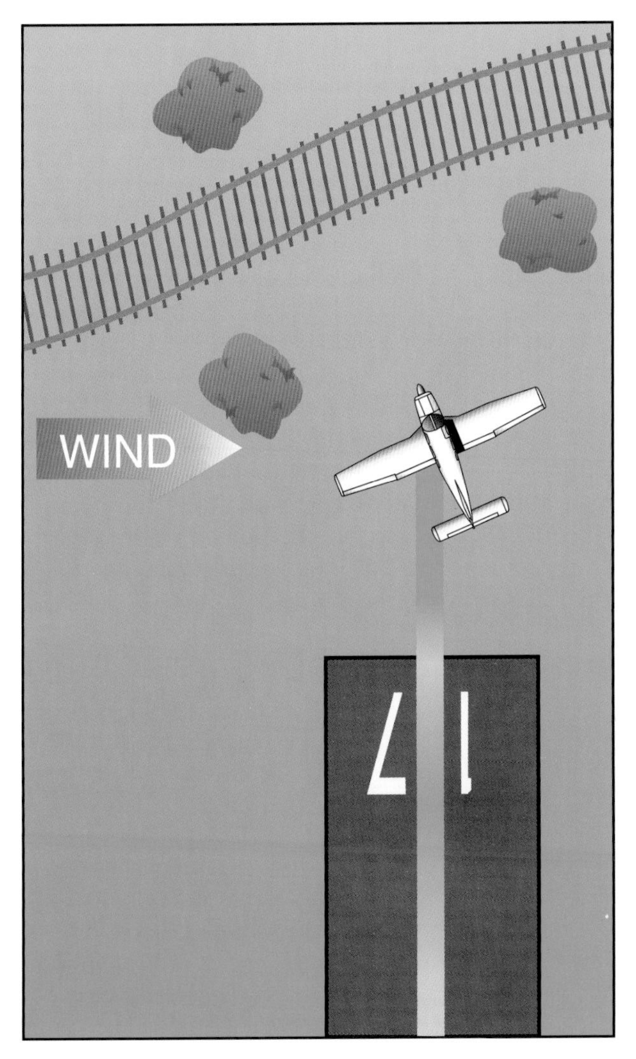

Figure 5-5. Crosswind climb flightpath.

- Premature lift-off resulting in side-skipping.
- Excessive aileron input in the latter stage of the takeoff roll resulting in a steep bank into the wind at lift-off.
- Inadequate drift correction after lift-off.

GROUND EFFECT ON TAKEOFF

Ground effect is a condition of improved performance encountered when the airplane is operating very close to the ground. Ground effect can be detected and measured up to an altitude equal to one wingspan above the surface. [Figure 5-6] However, ground effect is most significant when the airplane (especially a low-wing airplane) is maintaining a constant attitude at low airspeed at low altitude (for example, during takeoff when the airplane lifts off and accelerates to climb speed, and during the landing flare before touchdown).

When the wing is under the influence of ground effect, there is a reduction in upwash, downwash, and wingtip vortices. As a result of the reduced wingtip vortices, induced drag is reduced. When the wing is at a height equal to one-fourth the span, the reduction in induced drag is about 25 percent, and when the wing is at a height equal to one-tenth the span, the reduction in induced drag is about 50 percent. At high speeds where parasite drag dominates, induced drag is a small part of the total drag. Consequently, the effects of ground effect are of greater concern during takeoff and landing.

On takeoff, the takeoff roll, lift-off, and the beginning of the initial climb are accomplished in the ground effect area. The ground effect causes local increases in static pressure, which cause the airspeed indicator and altimeter to indicate slightly less than they should, and usually results in the vertical speed indicator indicating a descent. As the airplane lifts off and climbs out of the ground effect area, however, the following will occur.

- The airplane will require an increase in angle of attack to maintain the same lift coefficient.
- The airplane will experience an increase in induced drag and thrust required.
- The airplane will experience a pitch-up tendency and will require less elevator travel because of an increase in downwash at the horizontal tail.

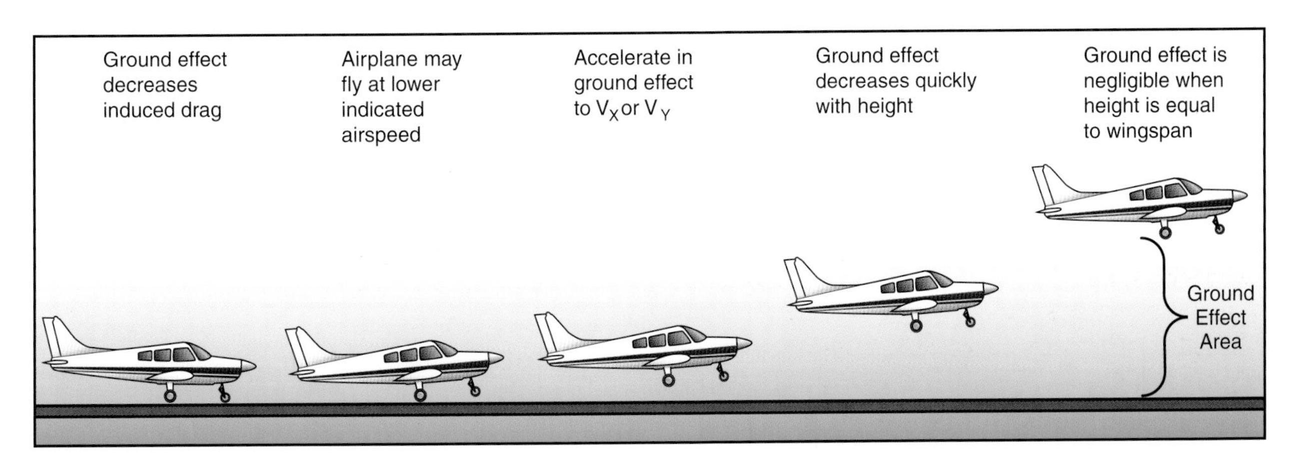

Figure 5-6. Takeoff in ground effect area.

- The airplane will experience a reduction in static source pressure as it leaves the ground effect area and a corresponding increase in indicated airspeed.

Due to the reduced drag in ground effect, the airplane may seem to be able to take off below the recommended airspeed. However, as the airplane rises out of ground effect with an insufficient airspeed, initial climb performance may prove to be marginal because of the increased drag. Under conditions of high-density altitude, high temperature, and/or maximum gross weight, the airplane may be able to become airborne at an insufficient airspeed, but unable to climb out of ground effect. Consequently, the airplane may not be able to clear obstructions, or may settle back on the runway. The point to remember is that additional power is required to compensate for increases in drag that occur as an airplane leaves ground effect. But during an initial climb, the engine is already developing maximum power. The only alternative is to lower pitch attitude to gain additional airspeed, which will result in inevitable altitude loss. Therefore, under marginal conditions, it is important that the airplane takes off at the recommended speed that will provide adequate initial climb performance.

Ground effect is important to normal flight operations. If the runway is long enough, or if no obstacles exist, ground effect can be used to an advantage by using the reduced drag to improve initial acceleration. Additionally, the procedure for takeoff from unsatisfactory surfaces is to take as much weight on the wings as possible during the ground run, and to lift off with the aid of ground effect before true flying speed is attained. It is then necessary to reduce the angle of attack to attain normal airspeed before attempting to fly away from the ground effect area.

SHORT-FIELD TAKEOFF AND MAXIMUM PERFORMANCE CLIMB

Takeoffs and climbs from fields where the takeoff area is short or the available takeoff area is restricted by obstructions require that the pilot operate the airplane at the limit of its takeoff performance capabilities. To depart from such an area safely, the pilot must exercise positive and precise control of airplane attitude and airspeed so that takeoff and climb performance results in the shortest ground roll and the steepest angle of climb. [Figure 5-7]

The achieved result should be consistent with the performance section of the FAA-approved Airplane Flight Manual and/or Pilot's Operating Handbook (AFM/POH). In all cases, the power setting, flap setting, airspeed, and procedures prescribed by the airplane's manufacturer should be followed.

In order to accomplish a maximum performance takeoff safely, the pilot must have adequate knowledge in the use and effectiveness of the best angle-of-climb speed (V_X) and the best rate-of-climb speed (V_Y) for the specific make and model of airplane being flown.

The speed for V_X is that which will result in the greatest gain in altitude for a given distance over the ground. It is usually slightly less than V_Y which provides the greatest gain in altitude per unit of time. The specific speeds to be used for a given airplane are stated in the FAA-approved AFM/POH. It should be emphasized that in some airplanes, a deviation of 5 knots from the recommended speed will result in a significant reduction in climb performance. Therefore, precise control of airspeed has an important bearing on the successful execution as well as the safety of the maneuver.

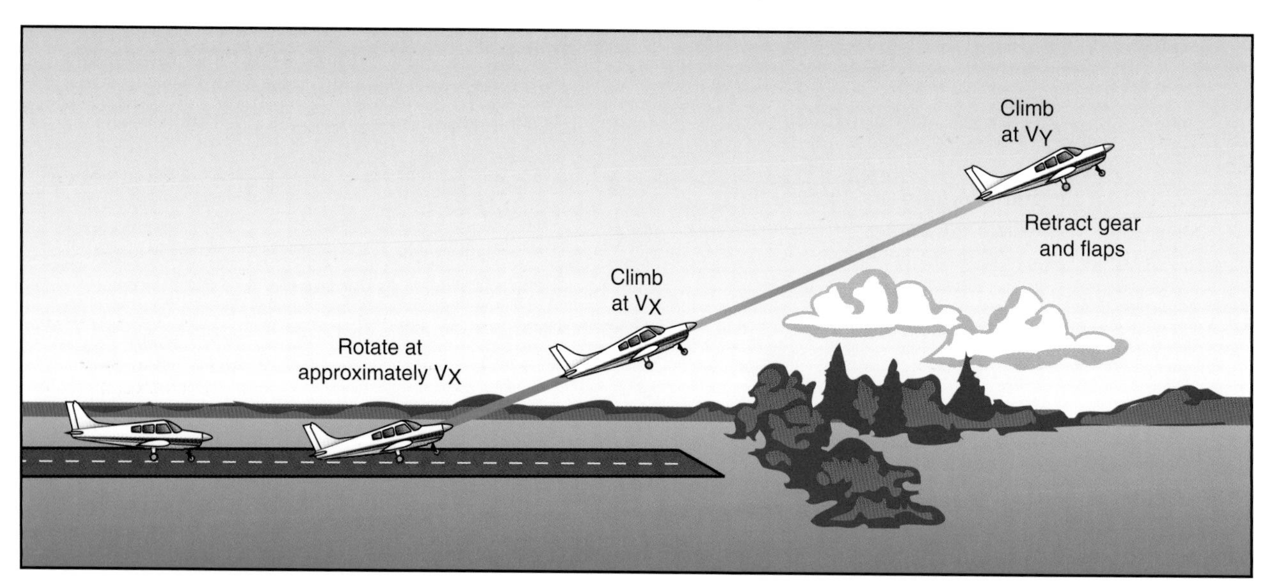

Figure 5-7. Short-field takeoff.

TAKEOFF ROLL

Taking off from a short field requires the takeoff to be started from the very beginning of the takeoff area. At this point, the airplane is aligned with the intended takeoff path. If the airplane manufacturer recommends the use of flaps, they should be extended the proper amount before starting the takeoff roll. This permits the pilot to give full attention to the proper technique and the airplane's performance throughout the takeoff.

Some authorities prefer to hold the brakes until the maximum obtainable engine r.p.m. is achieved before allowing the airplane to begin its takeoff run. However, it has not been established that this procedure will result in a shorter takeoff run in all light single-engine airplanes. Takeoff power should be applied smoothly and continuously—without hesitation—to accelerate the airplane as rapidly as possible. The airplane should be allowed to roll with its full weight on the main wheels and accelerated to the lift-off speed. As the takeoff roll progresses, the airplane's pitch attitude and angle of attack should be adjusted to that which results in the minimum amount of drag and the quickest acceleration. In nosewheel-type airplanes, this will involve little use of the elevator control, since the airplane is already in a low drag attitude.

LIFT-OFF

Approaching best angle-of-climb speed (V_X), the airplane should be smoothly and firmly lifted off, or rotated, by applying back-elevator pressure to an attitude that will result in the best angle-of-climb airspeed (V_X). Since the airplane will accelerate more rapidly after lift-off, additional back-elevator pressure becomes necessary to hold a constant airspeed. After becoming airborne, a wings level climb should be maintained at V_X until obstacles have been cleared or, if no obstacles are involved, until an altitude of at least 50 feet above the takeoff surface is attained. Thereafter, the pitch attitude may be lowered slightly, and the climb continued at best rate-of-climb speed (V_Y) until reaching a safe maneuvering altitude. Remember that an attempt to pull the airplane off the ground prematurely, or to climb too steeply, may cause the airplane to settle back to the runway or into the obstacles. Even if the airplane remains airborne, the initial climb will remain flat and climb performance/obstacle clearance ability seriously degraded until best angle-of-climb airspeed (V_X) is achieved. [Figure 5-8]

The objective is to rotate to the appropriate pitch attitude at (or near) best angle-of-climb airspeed. It should be remembered, however, that some airplanes will have a natural tendency to lift off well before reaching V_X. In these airplanes, it may be necessary to allow the airplane to lift off in ground effect and then reduce pitch attitude to level until the airplane accelerates to best angle-of-climb airspeed with the wheels just clear of the runway surface. This method is preferable to forcing the airplane to remain on the ground with forward-elevator pressure until best angle-of-climb speed is attained. Holding the airplane on the ground unnecessarily puts excessive pressure on the nosewheel, may result in "**wheelbarrowing**," and will hinder both acceleration and overall airplane performance.

INITIAL CLIMB

On short-field takeoffs, the landing gear and flaps should remain in takeoff position until clear of obstacles (or as recommended by the manufacturer) and V_Y has been established. It is generally unwise for the pilot to be looking in the cockpit or reaching for landing gear and flap controls until obstacle clearance is assured. When the airplane is stabilized at V_Y, the gear (if equipped) and then the flaps should be retracted. It is usually advisable to raise the flaps in increments to avoid sudden loss of lift and settling of the airplane. Next, reduce the power to the normal climb setting or as recommended by the airplane manufacturer.

Common errors in the performance of short-field takeoffs and maximum performance climbs are:

- Failure to adequately clear the area.
- Failure to utilize all available runway/takeoff area.
- Failure to have the airplane properly trimmed prior to takeoff.
- Premature lift-off resulting in high drag.
- Holding the airplane on the ground unnecessarily with excessive forward-elevator pressure.
- Inadequate rotation resulting in excessive speed after lift-off.
- Inability to attain/maintain best angle-of-climb airspeed.

Figure 5-8. Effect of premature lift-off.

- Fixation on the airspeed indicator during initial climb.
- Premature retraction of landing gear and/or wing flaps.

SOFT/ROUGH-FIELD TAKEOFF AND CLIMB

Takeoffs and climbs from soft fields require the use of operational techniques for getting the airplane airborne as quickly as possible to eliminate the drag caused by tall grass, soft sand, mud, and snow, and may or may not require climbing over an obstacle. The technique makes judicious use of ground effect and requires a feel for the airplane and fine control touch. These same techniques are also useful on a rough field where it is advisable to get the airplane off the ground as soon as possible to avoid damaging the landing gear.

Soft surfaces or long, wet grass usually reduces the airplane's acceleration during the takeoff roll so much that adequate takeoff speed might not be attained if normal takeoff techniques were employed.

It should be emphasized that the correct takeoff procedure for soft fields is quite different from that appropriate for short fields with firm, smooth surfaces. To minimize the hazards associated with takeoffs from soft or rough fields, support of the airplane's weight must be transferred as rapidly as possible from the wheels to the wings as the takeoff roll proceeds. Establishing and maintaining a relatively high angle of attack or nose-high pitch attitude as early as possible does this. Wing flaps may be lowered prior to starting the takeoff (if recommended by the manufacturer) to provide additional lift and to transfer the airplane's weight from the wheels to the wings as early as possible.

Stopping on a soft surface, such as mud or snow, might bog the airplane down; therefore, it should be kept in continuous motion with sufficient power while lining up for the takeoff roll.

TAKEOFF ROLL

As the airplane is aligned with the takeoff path, takeoff power is applied smoothly and as rapidly as the powerplant will accept it without faltering. As the airplane accelerates, enough back-elevator pressure should be applied to establish a positive angle of attack and to reduce the weight supported by the nosewheel.

When the airplane is held at a nose-high attitude throughout the takeoff run, the wings will, as speed increases and lift develops, progressively relieve the wheels of more and more of the airplane's weight, thereby minimizing the drag caused by surface irregularities or adhesion. If this attitude is accurately maintained, the airplane will virtually fly itself off the ground, becoming airborne at airspeed slower than a safe climb speed because of ground effect. [Figure 5-9]

LIFT-OFF

After becoming airborne, the nose should be lowered very gently with the wheels clear of the surface to allow the airplane to accelerate to V_Y, or V_X if obstacles must be cleared. Extreme care must be exercised immediately after the airplane becomes airborne and while it accelerates, to avoid settling back onto the surface. An attempt to climb prematurely or too steeply may cause the airplane to settle back to the surface as a result of losing the benefit of ground effect. An attempt to climb out of ground effect before sufficient climb airspeed is attained may result in the airplane being unable to climb further as the ground effect area is transited, even with full power. Therefore, it is essential that the airplane remain in ground effect until at least V_X is reached. This requires feel for the airplane, and a very fine control touch, in order to avoid over-controlling the elevator as required control pressures change with airplane acceleration.

INITIAL CLIMB

After a positive rate of climb is established, and the airplane has accelerated to V_Y, retract the landing gear and flaps, if equipped. If departing from an airstrip with wet snow or slush on the takeoff surface, the gear should not be retracted immediately. This allows for any wet snow or slush to be air-dried. In the event an obstacle must be cleared after a soft-field takeoff, the climb-out is performed at V_X until the obstacle has been cleared. After reaching this point, the pitch attitude is adjusted to V_Y and the gear and flaps are retracted. The power may then be reduced to the normal climb setting.

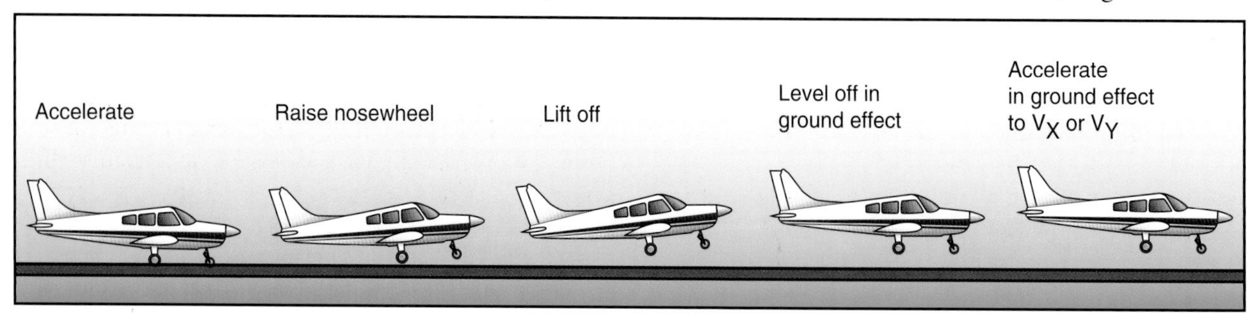

Figure 5-9. Soft-field takeoff.

Common errors in the performance of soft/rough field takeoff and climbs are:

- Failure to adequately clear the area.
- Insufficient back-elevator pressure during initial takeoff roll resulting in inadequate angle of attack.
- Failure to cross-check engine instruments for indications of proper operation after applying power.
- Poor directional control.
- Climbing too steeply after lift-off.
- Abrupt and/or excessive elevator control while attempting to level off and accelerate after lift-off.
- Allowing the airplane to "mush" or settle resulting in an inadvertent touchdown after lift-off.
- Attempting to climb out of ground effect area before attaining sufficient climb speed.
- Failure to anticipate an increase in pitch attitude as the airplane climbs out of ground effect.

REJECTED TAKEOFF/ENGINE FAILURE

Emergency or abnormal situations can occur during a takeoff that will require a pilot to reject the takeoff while still on the runway. Circumstances such as a malfunctioning powerplant, inadequate acceleration, runway incursion, or air traffic conflict may be reasons for a rejected takeoff.

Prior to takeoff, the pilot should have in mind a point along the runway at which the airplane should be airborne. If that point is reached and the airplane is not airborne, immediate action should be taken to discontinue the takeoff. Properly planned and executed, chances are excellent the airplane can be stopped on the remaining runway without using extraordinary measures, such as excessive braking that may result in loss of directional control, airplane damage, and/or personal injury.

In the event a takeoff is rejected, the power should be reduced to idle and maximum braking applied while maintaining directional control. If it is necessary to shut down the engine due to a fire, the mixture control should be brought to the idle cutoff position and the magnetos turned off. In all cases, the manufacturer's emergency procedure should be followed.

What characterizes all power loss or engine failure occurrences after lift-off is urgency. In most instances, the pilot has only a few seconds after an engine failure to decide what course of action to take and to execute it. Unless prepared in advance to make the proper decision, there is an excellent chance the pilot will make a poor decision, or make no decision at all and allow events to rule.

In the event of an engine failure on initial climb-out, the pilot's first responsibility is to maintain aircraft control. At a climb pitch attitude without power, the airplane will be at or near a stalling angle of attack. At the same time, the pilot may still be holding right rudder. It is essential the pilot immediately lower the pitch attitude to prevent a stall and possible spin. The pilot should establish a controlled glide toward a plausible landing area (preferably straight ahead on the remaining runway).

NOISE ABATEMENT

Aircraft noise problems have become a major concern at many airports throughout the country. Many local communities have pressured airports into developing specific operational procedures that will help limit aircraft noise while operating over nearby areas. For years now, the FAA, airport managers, aircraft operators, pilots, and special interest groups have been working together to minimize aircraft noise for nearby sensitive areas. As a result, noise abatement procedures have been developed for many of these airports that include standardized profiles and procedures to achieve these lower noise goals.

Airports that have noise abatement procedures provide information to pilots, operators, air carriers, air traffic facilities, and other special groups that are applicable to their airport. These procedures are available to the aviation community by various means. Most of this information comes from the *Airport/Facility Directory*, local and regional publications, printed handouts, operator bulletin boards, safety briefings, and local air traffic facilities.

At airports that use noise abatement procedures, reminder signs may be installed at the taxiway hold positions for applicable runways. These are to remind pilots to use and comply with noise abatement procedures on departure. Pilots who are not familiar with these procedures should ask the tower or air traffic facility for the recommended procedures. In any case, pilots should be considerate of the surrounding community while operating their airplane to and from such an airport. This includes operating as quietly, yet safely as possible.

Chapter 6

Ground Reference Maneuvers

PURPOSE AND SCOPE

Ground reference maneuvers and their related factors are used in developing a high degree of pilot skill. Although most of these maneuvers are not performed as such in normal everyday flying, the elements and principles involved in each are applicable to performance of the customary pilot operations. They aid the pilot in analyzing the effect of wind and other forces acting on the airplane and in developing a fine control touch, coordination, and the division of attention necessary for accurate and safe maneuvering of the airplane.

All of the early part of the pilot's training has been conducted at relatively high altitudes, and for the purpose of developing technique, knowledge of maneuvers, coordination, feel, and the handling of the airplane in general. This training will have required that most of the pilot's attention be given to the actual handling of the airplane, and the results of control pressures on the action and attitude of the airplane.

If permitted to continue beyond the appropriate training stage, however, the student pilot's concentration of attention will become a fixed habit, one that will seriously detract from the student's ease and safety as a pilot, and will be very difficult to eliminate. Therefore, it is necessary, as soon as the pilot shows proficiency in the fundamental maneuvers, that the pilot be introduced to maneuvers requiring outside attention on a practical application of these maneuvers and the knowledge gained.

It should be stressed that, during ground reference maneuvers, it is equally important that basic flying technique previously learned be maintained. The flight instructor should not allow any relaxation of the student's previous standard of technique simply because a new factor is added. This requirement should be maintained throughout the student's progress from maneuver to maneuver. Each new maneuver should embody some advance and include the principles of the preceding one in order that continuity be maintained. Each new factor introduced should be merely a step-up of one already learned so that orderly, consistent progress can be made.

MANEUVERING BY REFERENCE TO GROUND OBJECTS

Ground track or ground reference maneuvers are performed at a relatively low altitude while applying wind drift correction as needed to follow a predetermined track or path over the ground. They are designed to develop the ability to control the airplane, and to recognize and correct for the effect of wind while dividing attention among other matters. This requires planning ahead of the airplane, maintaining orientation in relation to ground objects, flying appropriate headings to follow a desired ground track, and being cognizant of other air traffic in the immediate vicinity.

Ground reference maneuvers should be flown at an altitude of approximately 600 to 1,000 feet AGL. The actual altitude will depend on the speed and type of airplane to a large extent, and the following factors should be considered.

- The speed with relation to the ground should not be so apparent that events happen too rapidly.
- The radius of the turn and the path of the airplane over the ground should be easily noted and changes planned and effected as circumstances require.
- Drift should be easily discernable, but not tax the student too much in making corrections.
- Objects on the ground should appear in their proportion and size.
- The altitude should be low enough to render any gain or loss apparent to the student, but in no case lower than 500 feet above the highest obstruction.

During these maneuvers, both the instructor and the student should be alert for available forced-landing fields. The area chosen should be away from communities, livestock, or groups of people to prevent possible annoyance or hazards to others. Due to the altitudes at which these maneuvers are performed, there is little time available to search for a suitable field for landing in the event the need arises.

DRIFT AND GROUND TRACK CONTROL

Whenever any object is free from the ground, it is affected by the medium with which it is surrounded. This means that a free object will move in whatever direction and speed that the medium moves.

For example, if a powerboat is crossing a river and the river is still, the boat could head directly to a point on the opposite shore and travel on a straight course to that point without drifting. However, if the river were flowing swiftly, the water current would have to be considered. That is, as the boat progresses forward with its own power, it must also move upstream at the same rate the river is moving it downstream. This is accomplished by angling the boat upstream sufficiently to counteract the downstream flow. If this is done, the boat will follow the desired **track** across the river from the departure point directly to the intended destination point. Should the boat not be headed sufficiently upstream, it would drift with the current and run aground at some point downstream on the opposite bank. [Figure 6-1]

As soon as an airplane becomes airborne, it is free of ground friction. Its path is then affected by the air mass in which it is flying; therefore, the airplane (like the boat) will not always track along the ground in the exact direction that it is headed. When flying with the longitudinal axis of the airplane aligned with a road, it may be noted that the airplane gets closer to or farther from the road without any turn having been made. This would indicate that the air mass is moving sideward in relation to the airplane. Since the airplane is flying within this moving body of air (wind), it moves or drifts with the air in the same direction and speed, just like the boat moved with the river current. [Figure 6-1]

When flying straight and level and following a selected ground track, the preferred method of correcting for wind drift is to head the airplane (wind correction angle) sufficiently into the wind to cause the airplane to move forward into the wind at the same rate the wind is moving it sideways. Depending on the wind velocity, this may require a large wind correction angle or one of only a few degrees. When the drift has been neutralized, the airplane will follow the desired ground track.

To understand the need for drift correction during flight, consider a flight with a wind velocity of 30 knots from the left and 90° to the direction the airplane is headed. After 1 hour, the body of air in which the airplane is flying will have moved 30 nautical miles (NM) to the right. Since the airplane is moving with this body of air, it too will have drifted 30 NM to the right. In relation to the air, the airplane moved forward, but in relation to the ground, it moved forward as well as 30 NM to the right.

There are times when the pilot needs to correct for drift while in a turn. [Figure 6-2] Throughout the turn the wind will be acting on the airplane from constantly changing angles. The relative wind angle and speed

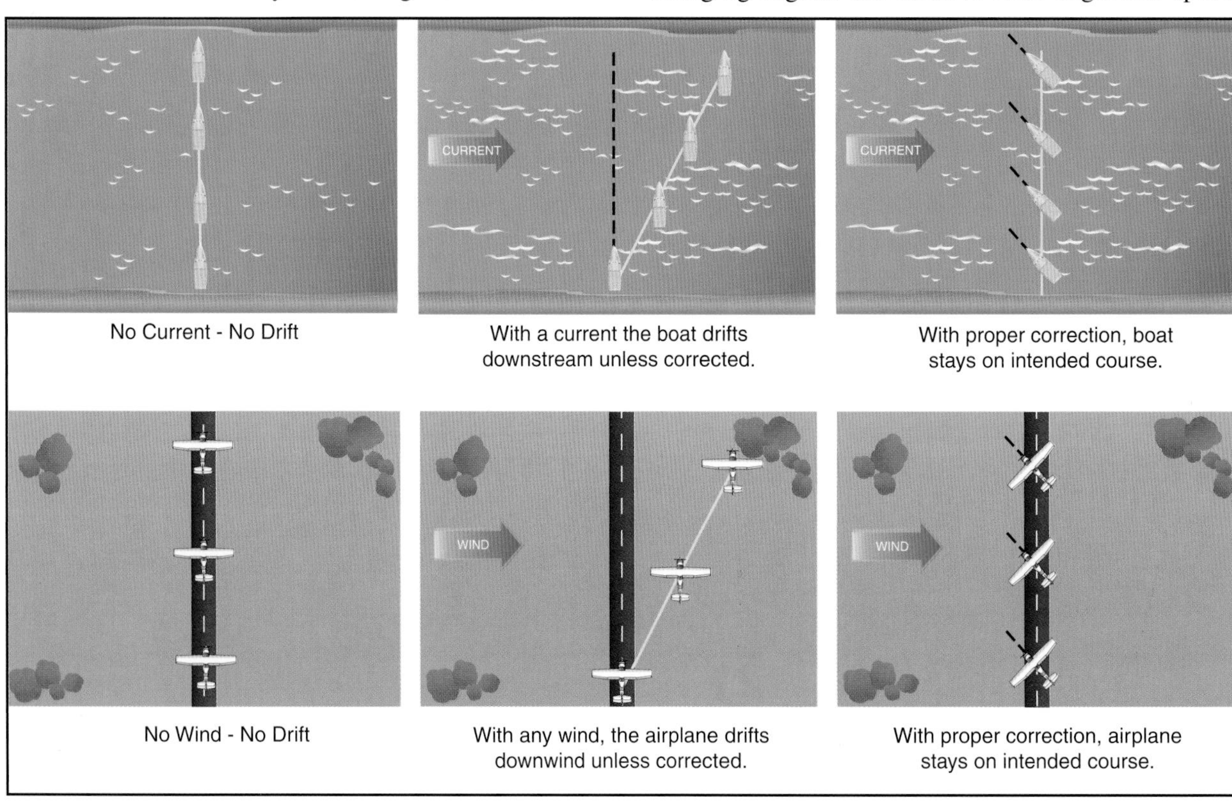

Figure 6-1. Wind drift.

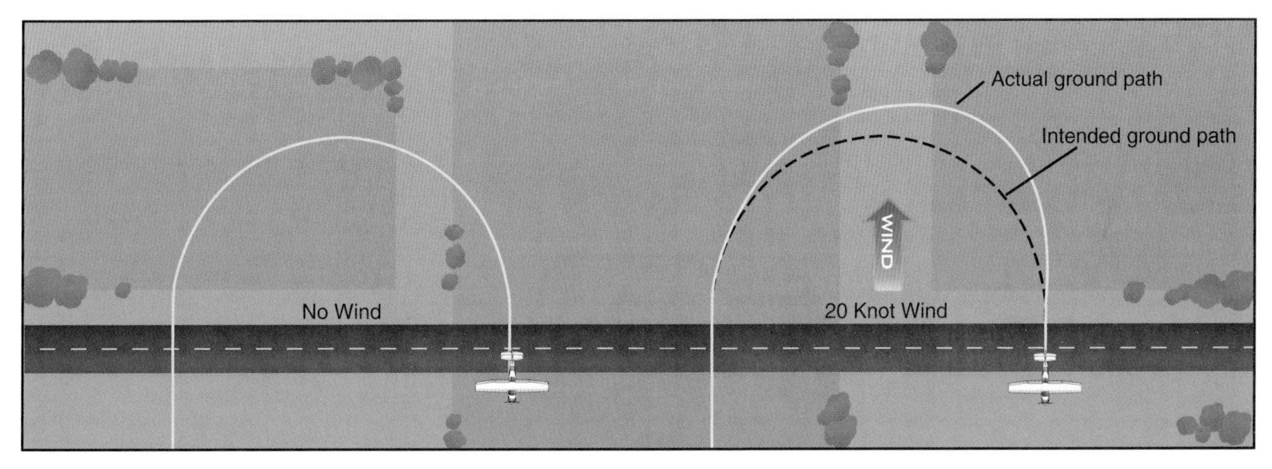

Figure 6-2. Effect of wind during a turn.

govern the time it takes for the airplane to progress through any part of a turn. This is due to the constantly changing **groundspeed**. When the airplane is headed into the wind, the groundspeed is decreased; when headed downwind, the groundspeed is increased. Through the crosswind portion of a turn, the airplane must be turned sufficiently into the wind to counteract drift.

To follow a desired circular ground track, the wind correction angle must be varied in a timely manner because of the varying groundspeed as the turn progresses. The faster the groundspeed, the faster the wind correction angle must be established; the slower the groundspeed, the slower the wind correction angle may be established. It can be seen then that the steepest bank and fastest rate of turn should be made on the downwind portion of the turn and the shallowest bank and slowest rate of turn on the upwind portion.

The principles and techniques of varying the angle of bank to change the rate of turn and wind correction angle for controlling wind drift during a turn are the same for all ground track maneuvers involving changes in direction of flight.

When there is no wind, it should be simple to fly along a ground track with an arc of exactly 180° and a constant radius because the flightpath and ground track would be identical. This can be demonstrated by approaching a road at a 90° angle and, when directly over the road, rolling into a medium-banked turn, then maintaining the same angle of bank throughout the 180° of turn. [Figure 6-2]

To complete the turn, the rollout should be started at a point where the wings will become level as the airplane again reaches the road at a 90° angle and will be directly over the road just as the turn is completed. This would be possible only if there were absolutely no wind and if the angle of bank and the rate of turn remained constant throughout the entire maneuver.

If the turn were made with a constant angle of bank and a wind blowing directly across the road, it would result in a constant radius turn through the air. However, the wind effects would cause the ground track to be distorted from a constant radius turn or semicircular path. The greater the wind velocity, the greater would be the difference between the desired ground track and the flightpath. To counteract this drift, the flightpath can be controlled by the pilot in such a manner as to neutralize the effect of the wind, and cause the ground track to be a constant radius semicircle.

The effects of wind during turns can be demonstrated after selecting a road, railroad, or other ground reference that forms a straight line parallel to the wind. Fly into the wind directly over and along the line and then make a turn with a constant medium angle of bank for 360° of turn. [Figure 6-3] The airplane will return to a point directly over the line but slightly downwind from the starting point, the amount depending on the wind velocity and the time required to complete the turn. The path over the ground will be an elongated circle, although in reference to the air it is a perfect circle. Straight flight during the upwind segment after completion of the turn is necessary to bring the airplane back to the starting position.

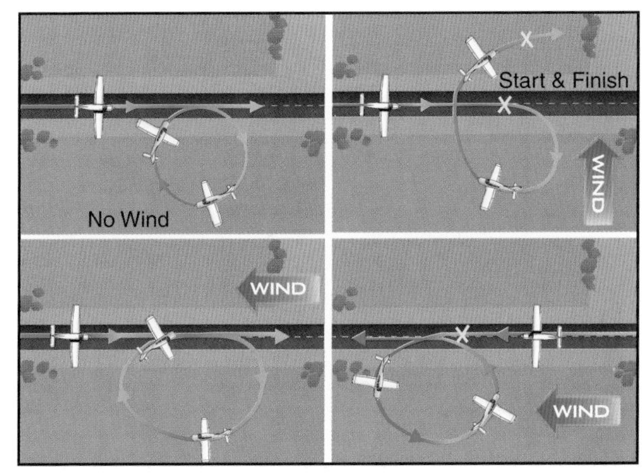

Figure 6-3. Effect of wind during turns.

A similar 360° turn may be started at a specific point over the reference line, with the airplane headed directly downwind. In this demonstration, the effect of wind during the constant banked turn will drift the airplane to a point where the line is reintercepted, but the 360° turn will be completed at a point downwind from the starting point.

Another reference line which lies directly crosswind may be selected and the same procedure repeated, showing that if wind drift is not corrected the airplane will, at the completion of the 360° turn, be headed in the original direction but will have drifted away from the line a distance dependent on the amount of wind.

From these demonstrations, it can be seen where and why it is necessary to increase or decrease the angle of bank and the rate of turn to achieve a desired track over the ground. The principles and techniques involved can be practiced and evaluated by the performance of the ground track maneuvers discussed in this chapter.

RECTANGULAR COURSE

Normally, the first ground reference maneuver the pilot is introduced to is the rectangular course. [Figure 6-4] The rectangular course is a training maneuver in which the ground track of the airplane is equidistant from all sides of a selected rectangular area on the ground. The maneuver simulates the conditions encountered in an airport traffic pattern. While performing the maneuver, the altitude and airspeed should be held constant. The maneuver assists the student pilot in perfecting:

- Practical application of the turn.
- The division of attention between the flightpath, ground objects, and the handling of the airplane.
- The timing of the start of a turn so that the turn will be fully established at a definite point over the ground.
- The timing of the recovery from a turn so that a definite ground track will be maintained.
- The establishing of a ground track and the determination of the appropriate "crab" angle.

Like those of other ground track maneuvers, one of the objectives is to develop division of attention between the flightpath and ground references, while controlling the airplane and watching for other aircraft in the

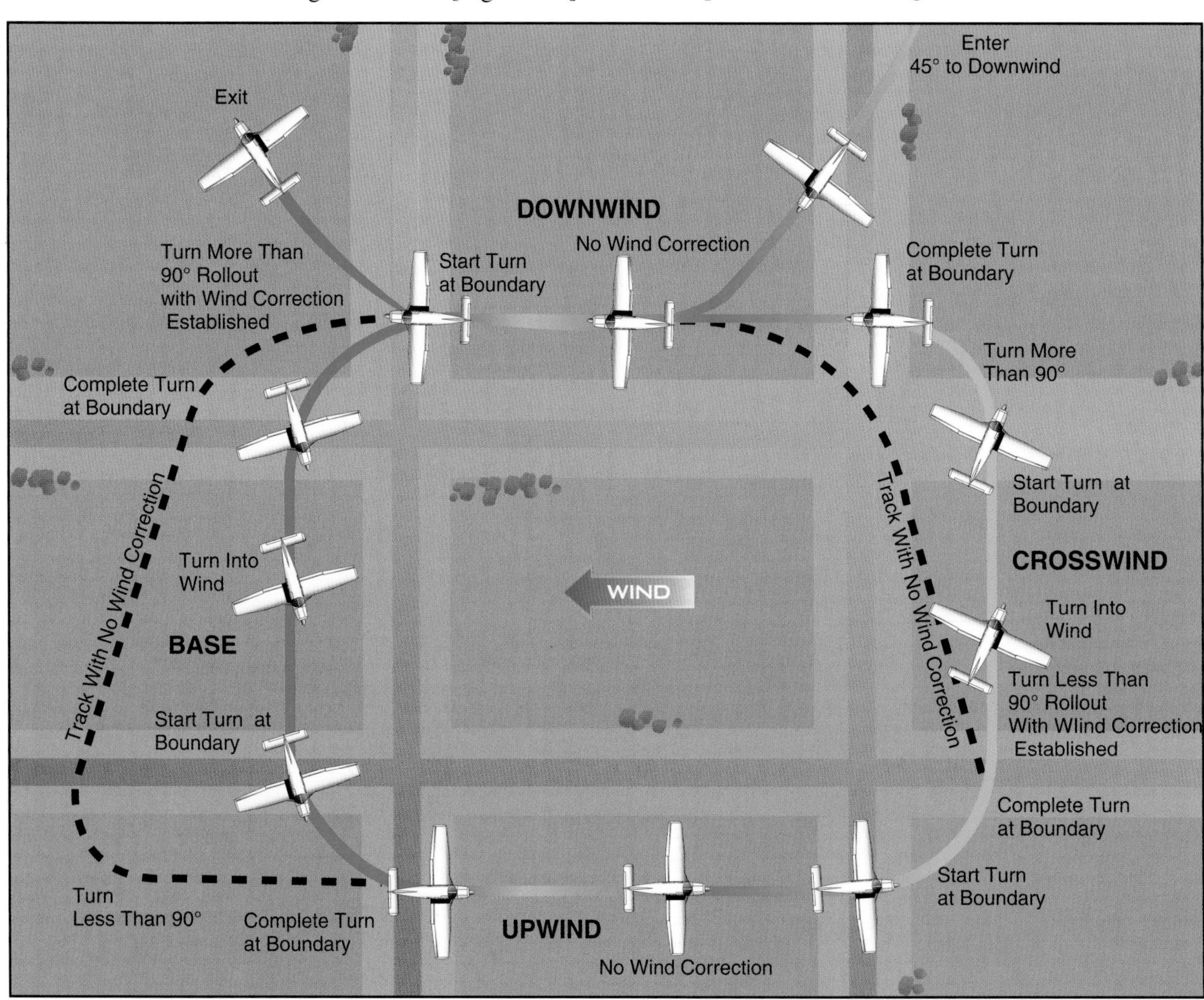

Figure 6-4. Rectangular course.

vicinity. Another objective is to develop recognition of drift toward or away from a line parallel to the intended ground track. This will be helpful in recognizing drift toward or from an airport runway during the various legs of the airport traffic pattern.

For this maneuver, a square or rectangular field, or an area bounded on four sides by section lines or roads (the sides of which are approximately a mile in length), should be selected well away from other air traffic. The airplane should be flown parallel to and at a uniform distance about one-fourth to one-half mile away from the field boundaries, not above the boundaries. For best results, the flightpath should be positioned outside the field boundaries just far enough that they may be easily observed from either pilot seat by looking out the side of the airplane. If an attempt is made to fly directly above the edges of the field, the pilot will have no usable reference points to start and complete the turns. The closer the track of the airplane is to the field boundaries, the steeper the bank necessary at the turning points. Also, the pilot should be able to see the edges of the selected field while seated in a normal position and looking out the side of the airplane during either a left-hand or right-hand course. The distance of the ground track from the edges of the field should be the same regardless of whether the course is flown to the left or right. All turns should be started when the airplane is abeam the corner of the field boundaries, and the bank normally should not exceed 45°. These should be the determining factors in establishing the distance from the boundaries for performing the maneuver.

Although the rectangular course may be entered from any direction, this discussion assumes entry on a downwind.

On the downwind leg, the wind is a tailwind and results in an increased groundspeed. Consequently, the turn onto the next leg is entered with a fairly fast rate of roll-in with relatively steep bank. As the turn progresses, the bank angle is reduced gradually because the tailwind component is diminishing, resulting in a decreasing groundspeed.

During and after the turn onto this leg (the equivalent of the base leg in a traffic pattern), the wind will tend to drift the airplane away from the field boundary. To compensate for the drift, the amount of turn will be more than 90°.

The rollout from this turn must be such that as the wings become level, the airplane is turned slightly toward the field and into the wind to correct for drift. The airplane should again be the same distance from the field boundary and at the same altitude, as on other legs. The base leg should be continued until the upwind leg boundary is being approached. Once more the pilot should anticipate drift and turning radius. Since drift correction was held on the base leg, it is necessary to turn less than 90° to align the airplane parallel to the upwind leg boundary. This turn should be started with a medium bank angle with a gradual reduction to a shallow bank as the turn progresses. The rollout should be timed to assure paralleling the boundary of the field as the wings become level.

While the airplane is on the upwind leg, the next field boundary should be observed as it is being approached, to plan the turn onto the crosswind leg. Since the wind is a headwind on this leg, it is reducing the airplane's groundspeed and during the turn onto the crosswind leg will try to drift the airplane toward the field. For this reason, the roll-in to the turn must be slow and the bank relatively shallow to counteract this effect. As the turn progresses, the headwind component decreases, allowing the groundspeed to increase. Consequently, the bank angle and rate of turn are increased gradually to assure that upon completion of the turn the crosswind ground track will continue the same distance from the edge of the field. Completion of the turn with the wings level should be accomplished at a point aligned with the upwind corner of the field.

Simultaneously, as the wings are rolled level, the proper drift correction is established with the airplane turned into the wind. This requires that the turn be less than a 90° change in heading. If the turn has been made properly, the field boundary will again appear to be one-fourth to one-half mile away. While on the crosswind leg, the wind correction angle should be adjusted as necessary to maintain a uniform distance from the field boundary.

As the next field boundary is being approached, the pilot should plan the turn onto the downwind leg. Since a wind correction angle is being held into the wind and away from the field while on the crosswind leg, this next turn will require a turn of more than 90°. Since the crosswind will become a tailwind, causing the groundspeed to increase during this turn, the bank initially should be medium and progressively increased as the turn proceeds. To complete the turn, the rollout must be timed so that the wings become level at a point aligned with the crosswind corner of the field just as the longitudinal axis of the airplane again becomes parallel to the field boundary. The distance from the field boundary should be the same as from the other sides of the field.

Usually, drift should not be encountered on the upwind or the downwind leg, but it may be difficult to find a situation where the wind is blowing exactly parallel to the field boundaries. This would make it necessary to use a slight wind correction angle on all the legs. It is important to anticipate the turns to correct for groundspeed, drift, and turning radius. When the wind is behind the airplane, the turn must be faster and steeper; when it is ahead of the airplane, the turn must be slower and shallower. These same techniques apply while flying in airport traffic patterns.

Common errors in the performance of rectangular courses are:

- Failure to adequately clear the area.
- Failure to establish proper altitude prior to entry. (Typically entering the maneuver while descending.)
- Failure to establish appropriate wind correction angle resulting in drift.
- Gaining or losing altitude.
- Poor coordination. (Typically skidding in turns from a downwind heading and slipping in turns from an upwind heading.)
- Abrupt control usage.
- Inability to adequately divide attention between airplane control and maintaining ground track.
- Improper timing in beginning and recovering from turns.
- Inadequate visual lookout for other aircraft.

S-TURNS ACROSS A ROAD

An S-turn across a road is a practice maneuver in which the airplane's ground track describes semicircles of equal radii on each side of a selected straight line on the ground. [Figure 6-5] The straight line may be a road, fence, railroad, or section line that lies perpendicular to the wind, and should be of sufficient length for making a series of turns. A constant altitude should be maintained throughout the maneuver.

S-turns across a road present one of the most elementary problems in the practical application of the turn and in the correction for wind drift in turns. While the application of this maneuver is considerably less advanced in some respects than the rectangular course, it is taught after the student has been introduced to that maneuver in order that the student may have a knowledge of the correction for wind drift in straight flight along a reference line before the student attempt to correct for drift by playing a turn.

The objectives of S-turns across a road are to develop the ability to compensate for drift during turns, orient the flightpath with ground references, follow an assigned ground track, arrive at specified points on assigned headings, and divide the pilot's attention. The

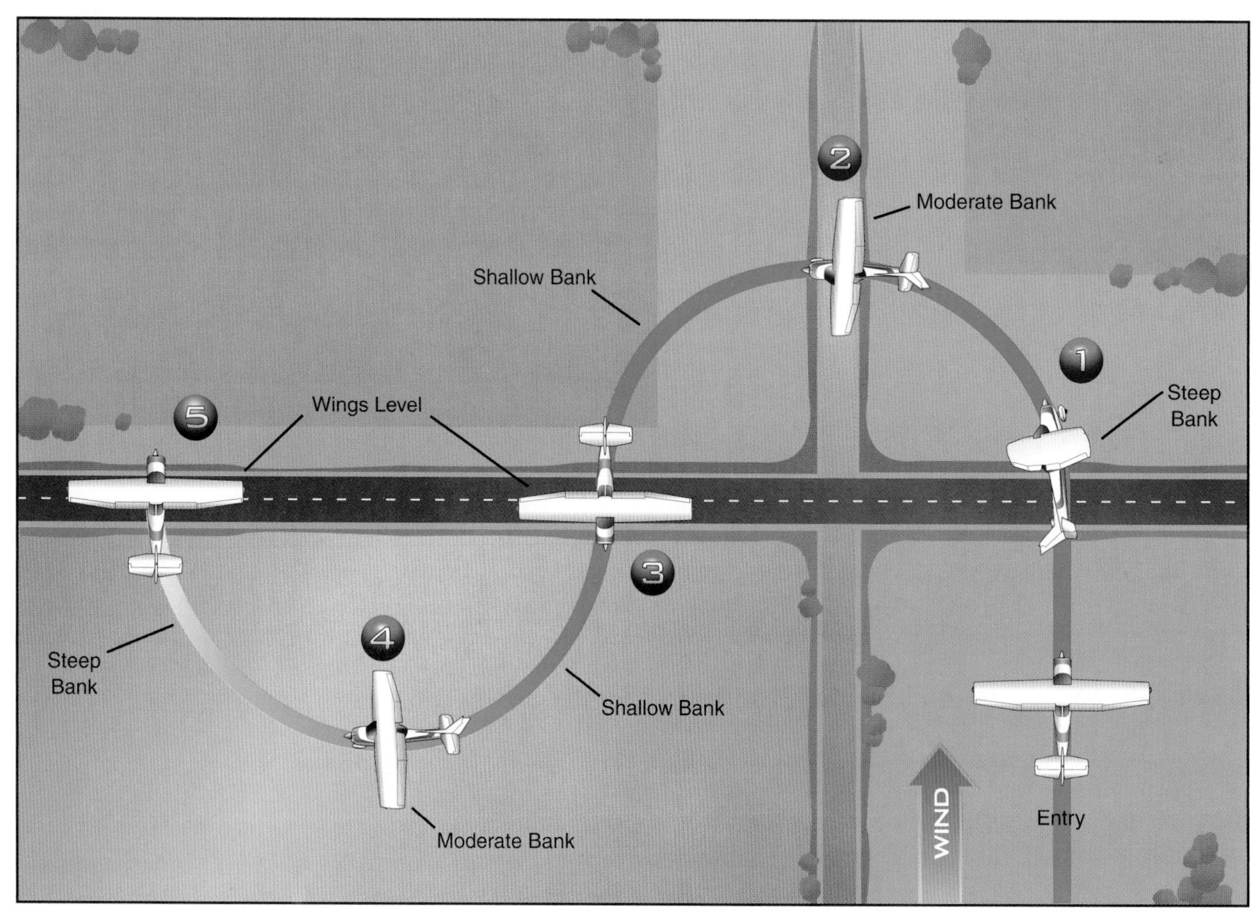

Figure 6-5. S-Turns.

maneuver consists of crossing the road at a 90° angle and immediately beginning a series of 180° turns of uniform radius in opposite directions, re-crossing the road at a 90° angle just as each 180° turn is completed.

To accomplish a constant radius ground track requires a changing roll rate and angle of bank to establish the wind correction angle. Both will increase or decrease as groundspeed increases or decreases.

The bank must be steepest when beginning the turn on the downwind side of the road and must be shallowed gradually as the turn progresses from a downwind heading to an upwind heading. On the upwind side, the turn should be started with a relatively shallow bank and then gradually steepened as the airplane turns from an upwind heading to a downwind heading.

In this maneuver, the airplane should be rolled from one bank directly into the opposite just as the reference line on the ground is crossed.

Before starting the maneuver, a straight ground reference line or road that lies 90° to the direction of the wind should be selected, then the area checked to ensure that no obstructions or other aircraft are in the immediate vicinity. The road should be approached from the upwind side, at the selected altitude on a downwind heading. When directly over the road, the first turn should be started immediately. With the airplane headed downwind, the groundspeed is greatest and the rate of departure from the road will be rapid; so the roll into the steep bank must be fairly rapid to attain the proper wind correction angle. This prevents the airplane from flying too far from the road and from establishing a ground track of excessive radius.

During the latter portion of the first 90° of turn when the airplane's heading is changing from a downwind heading to a crosswind heading, the groundspeed becomes less and the rate of departure from the road decreases. The wind correction angle will be at the maximum when the airplane is headed directly crosswind.

After turning 90°, the airplane's heading becomes more and more an upwind heading, the groundspeed will decrease, and the rate of closure with the road will become slower. If a constant steep bank were maintained, the airplane would turn too quickly for the slower rate of closure, and would be headed perpendicular to the road prematurely. Because of the decreasing groundspeed and rate of closure while approaching the upwind heading, it will be necessary to gradually shallow the bank during the remaining 90° of the semicircle, so that the wind correction angle is removed completely and the wings become level as the 180° turn is completed at the moment the road is reached.

At the instant the road is being crossed again, a turn in the opposite direction should be started. Since the airplane is still flying into the headwind, the groundspeed is relatively slow. Therefore, the turn will have to be started with a shallow bank so as to avoid an excessive rate of turn that would establish the maximum wind correction angle too soon. The degree of bank should be that which is necessary to attain the proper wind correction angle so the ground track describes an arc the same size as the one established on the downwind side.

Since the airplane is turning from an upwind to a downwind heading, the groundspeed will increase and after turning 90°, the rate of closure with the road will increase rapidly. Consequently, the angle of bank and rate of turn must be progressively increased so that the airplane will have turned 180° at the time it reaches the road. Again, the rollout must be timed so the airplane is in straight-and-level flight directly over and perpendicular to the road.

Throughout the maneuver a constant altitude should be maintained, and the bank should be changing constantly to effect a true semicircular ground track.

Often there is a tendency to increase the bank too rapidly during the initial part of the turn on the upwind side, which will prevent the completion of the 180° turn before re-crossing the road. This is apparent when the turn is not completed in time for the airplane to cross the road at a perpendicular angle. To avoid this error, the pilot must visualize the desired half circle ground track, and increase the bank during the early part of this turn. During the latter part of the turn, when approaching the road, the pilot must judge the closure rate properly and increase the bank accordingly, so as to cross the road perpendicular to it just as the rollout is completed.

Common errors in the performance of S-turns across a road are:

- Failure to adequately clear the area.
- Poor coordination.
- Gaining or losing altitude.
- Inability to visualize the half circle ground track.
- Poor timing in beginning and recovering from turns.
- Faulty correction for drift.
- Inadequate visual lookout for other aircraft.

TURNS AROUND A POINT

Turns around a point, as a training maneuver, is a logical extension of the principles involved in the performance of S-turns across a road. Its purposes as a training maneuver are:

- To further perfect turning technique.
- To perfect the ability to subconsciously control the airplane while dividing attention between the flightpath and ground references.
- To teach the student that the radius of a turn is a distance which is affected by the degree of bank used when turning with relation to a definite object.
- To develop a keen perception of altitude.
- To perfect the ability to correct for wind drift while in turns.

In turns around a point, the airplane is flown in two or more complete circles of uniform radii or distance from a prominent ground reference point using a maximum bank of approximately 45° while maintaining a constant altitude.

The factors and principles of drift correction that are involved in S-turns are also applicable in this maneuver. As in other ground track maneuvers, a constant radius around a point will, if any wind exists, require a constantly changing angle of bank and angles of wind correction. The closer the airplane is to a direct downwind heading where the groundspeed is greatest, the steeper the bank and the faster the rate of turn required to establish the proper wind correction angle. The more nearly it is to a direct upwind heading where the groundspeed is least, the shallower the bank and the slower the rate of turn required to establish the proper wind correction angle. It follows, then, that throughout the maneuver the bank and rate of turn must be gradually varied in proportion to the groundspeed.

The point selected for turns around a point should be prominent, easily distinguished by the pilot, and yet small enough to present precise reference. [Figure 6-6] Isolated trees, crossroads, or other similar small landmarks are usually suitable.

To enter turns around a point, the airplane should be flown on a downwind heading to one side of the selected point at a distance equal to the desired radius of turn. In a high-wing airplane, the distance from the point must permit the pilot to see the point throughout the maneuver even with the wing lowered in a bank. If the radius is too large, the lowered wing will block the pilot's view of the point.

When any significant wind exists, it will be necessary to roll into the initial bank at a rapid rate so that the steep-

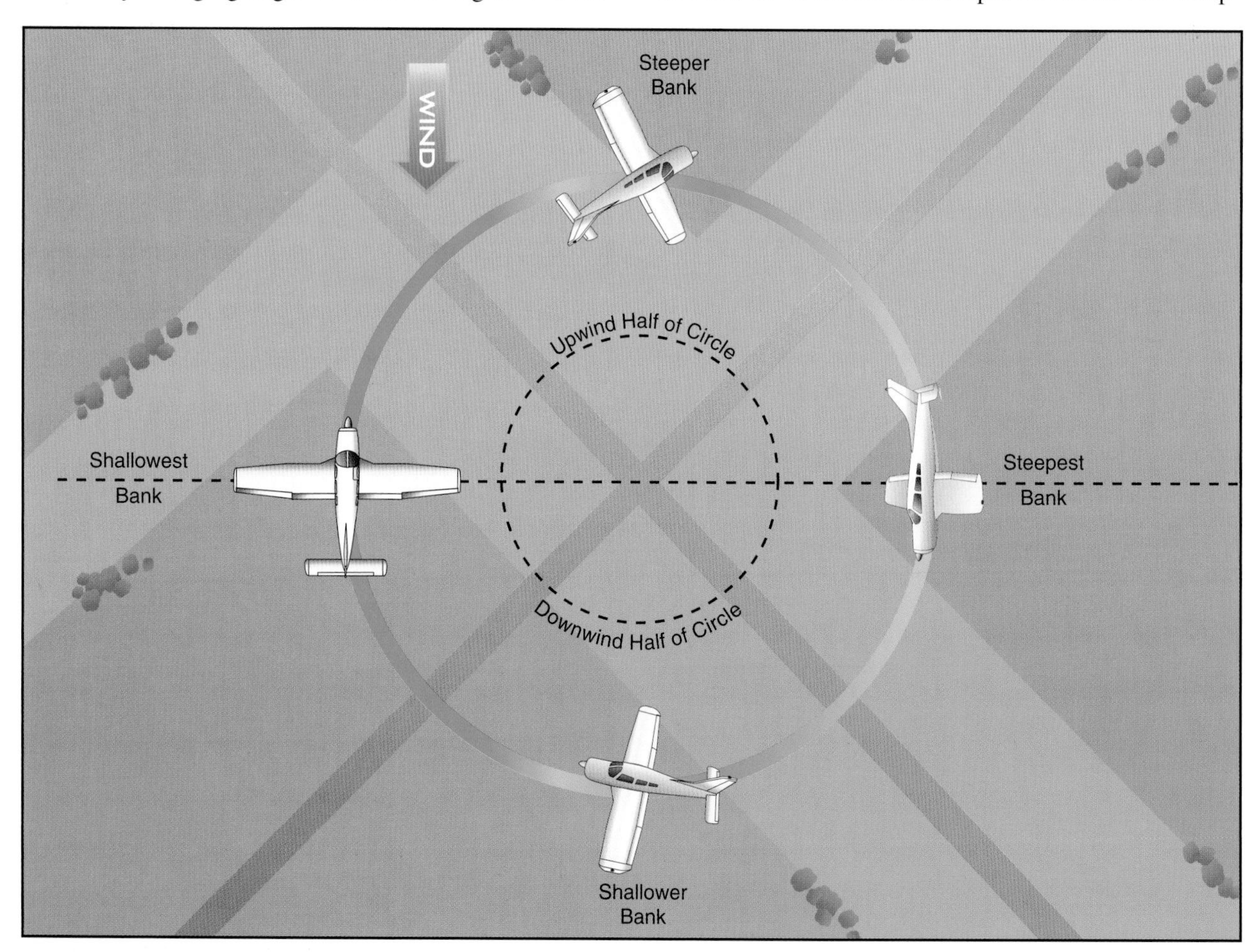

Figure 6-6. Turns around a point.

est bank is attained abeam of the point when the airplane is headed directly downwind. By entering the maneuver while heading directly downwind, the steepest bank can be attained immediately. Thus, if a maximum bank of 45° is desired, the initial bank will be 45° if the airplane is at the correct distance from the point. Thereafter, the bank is shallowed gradually until the point is reached where the airplane is headed directly upwind. At this point, the bank should be gradually steepened until the steepest bank is again attained when heading downwind at the initial point of entry.

Just as S-turns require that the airplane be turned into the wind in addition to varying the bank, so do turns around a point. During the downwind half of the circle, the airplane's nose is progressively turned toward the inside of the circle; during the upwind half, the nose is progressively turned toward the outside. The downwind half of the turn around the point may be compared to the downwind side of the S-turn across a road; the upwind half of the turn around a point may be compared to the upwind side of the S-turn across a road.

As the pilot becomes experienced in performing turns around a point and has a good understanding of the effects of wind drift and varying of the bank angle and wind correction angle as required, entry into the maneuver may be from any point. When entering the maneuver at a point other than downwind, however, the radius of the turn should be carefully selected, taking into account the wind velocity and groundspeed so that an excessive bank is not required later on to maintain the proper ground track. The flight instructor should place particular emphasis on the effect of an incorrect initial bank. This emphasis should continue in the performance of elementary eights.

Common errors in the performance of turns around a point are:

- Failure to adequately clear the area.
- Failure to establish appropriate bank on entry.
- Failure to recognize wind drift.
- Excessive bank and/or inadequate wind correction angle on the downwind side of the circle resulting in drift towards the reference point.
- Inadequate bank angle and/or excessive wind correction angle on the upwind side of the circle resulting in drift away from the reference point.
- Skidding turns when turning from downwind to crosswind.
- Slipping turns when turning from upwind to crosswind.
- Gaining or losing altitude.
- Inadequate visual lookout for other aircraft.
- Inability to direct attention outside the airplane while maintaining precise airplane control.

ELEMENTARY EIGHTS

An "eight" is a maneuver in which the airplane describes a path over the ground more or less in the shape of a figure "8". In all eights except "lazy eights" the path is horizontal as though following a marked path over the ground. There are various types of eights, progressing from the elementary types to very difficult types in the advanced maneuvers. Each has its special use in teaching the student to solve a particular problem of turning with relation to the Earth, or an object on the Earth's surface. Each type, as they advance in difficulty of accomplishment, further perfects the student's coordination technique and requires a higher degree of subconscious flying ability. Of all the training maneuvers available to the instructor, only eights require the progressively higher degree of conscious attention to outside objects. However, the real importance of eights is in the requirement for the perfection and display of subconscious flying.

Elementary eights, specifically eights along a road, eights across a road, and eights around pylons, are variations of turns around a point, which use two points about which the airplane circles in either direction. Elementary eights are designed for the following purposes.

- To perfect turning technique.
- To develop the ability to divide attention between the actual handling of controls and an outside objective.
- To perfect the knowledge of the effect of angle of bank on radius of turn.
- To demonstrate how wind affects the path of the airplane over the ground.
- To gain experience in the visualization of the results of planning before the execution of the maneuver.
- To train the student to think and plan ahead of the airplane.

EIGHTS ALONG A ROAD

An eight along a road is a maneuver in which the ground track consists of two complete adjacent circles of equal radii on each side of a straight road or other reference line on the ground. The ground track resembles a figure 8. [Figure 6-7 on next page]

Like the other ground reference maneuvers, its objective is to develop division of attention while

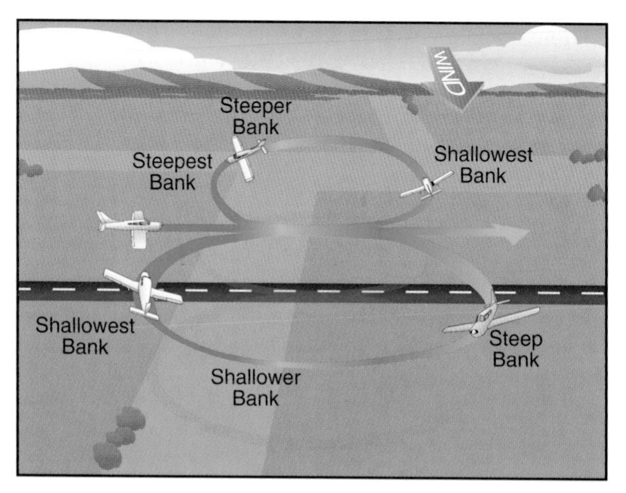

Figure 6-7. Eights along a road.

compensating for drift, maintaining orientation with ground references, and maintaining a constant altitude.

Although eights along a road may be performed with the wind blowing parallel to the road or directly across the road, for simplification purposes, only the latter situation is explained since the principles involved in either case are common.

A reference line or road which is perpendicular to the wind should be selected and the airplane flown parallel to and directly above the road. Since the wind is blowing across the flightpath, the airplane will require some wind correction angle to stay directly above the road during the initial straight and level portion. Before starting the maneuver, the area should be checked to ensure clearance of obstructions and avoidance of other aircraft.

Usually, the first turn should be made toward a downwind heading starting with a medium bank. Since the airplane will be turning more and more directly downwind, the groundspeed will be gradually increasing and the rate of departing the road will tend to become faster. Thus, the bank and rate of turn is increased to establish a wind correction angle to keep the airplane from exceeding the desired distance from the road when 180° of change in direction is completed. The steepest bank is attained when the airplane is headed directly downwind.

As the airplane completes 180° of change in direction, it will be flying parallel to and using a wind correction angle toward the road with the wind acting directly perpendicular to the ground track. At this point, the pilot should visualize the remaining 180° of ground track required to return to the same place over the road from which the maneuver started.

While the turn is continued toward an upwind heading, the wind will tend to keep the airplane from reaching the road, with a decrease in groundspeed and rate of closure. The rate of turn and wind correction angle are decreased proportionately so that the road will be reached just as the 360° turn is completed. To accomplish this, the bank is decreased so that when headed directly upwind, it will be at the shallowest angle. In the last 90° of the turn, the bank may be varied to correct any previous errors in judging the returning rate and closure rate. The rollout should be timed so that the airplane will be straight and level over the starting point, with enough drift correction to hold it over the road.

After momentarily flying straight and level along the road, the airplane is then rolled into a medium bank turn in the opposite direction to begin the circle on the upwind side of the road. The wind will still be decreasing the groundspeed and trying to drift the airplane back toward the road; therefore, the bank must be decreased slowly during the first 90° change in direction in order to reach the desired distance from the road and attain the proper wind correction angle when 180° change in direction has been completed.

As the remaining 180° of turn continues, the wind becomes more of a tailwind and increases the airplane's groundspeed. This causes the rate of closure to become faster; consequently, the angle of bank and rate of turn must be increased further to attain sufficient wind correction angle to keep the airplane from approaching the road too rapidly. The bank will be at its steepest angle when the airplane is headed directly downwind.

In the last 90° of the turn, the rate of turn should be reduced to bring the airplane over the starting point on the road. The rollout must be timed so the airplane will be straight and level, turned into the wind, and flying parallel to and over the road.

The measure of a student's progress in the performance of eights along a road is the smoothness and accuracy of the change in bank used to counteract drift. The sooner the drift is detected and correction applied, the smaller will be the required changes. The more quickly the student can anticipate the corrections needed, the less obvious the changes will be and the more attention can be diverted to the maintenance of altitude and operation of the airplane.

Errors in coordination must be eliminated and a constant altitude maintained. Flying technique must not be allowed to suffer from the fact that the student's attention is diverted. This technique should improve as the student becomes able to divide attention between the operation of the airplane controls and following a designated flightpath.

EIGHTS ACROSS A ROAD

This maneuver is a variation of eights along a road and involves the same principles and techniques. The primary difference is that at the completion of each loop of the figure eight, the airplane should cross an intersection of roads or a specific point on a straight road. [Figure 6-8]

The loops should be across the road and the wind should be perpendicular to the road. Each time the road is crossed, the crossing angle should be the same and the wings of the airplane should be level. The eights also may be performed by rolling from one bank immediately to the other, directly over the road.

EIGHTS AROUND PYLONS

This training maneuver is an application of the same principles and techniques of correcting for wind drift as used in turns around a point and the same objectives as other ground track maneuvers. In this case, two points or pylons on the ground are used as references, and turns around each pylon are made in opposite directions to follow a ground track in the form of a figure 8. [Figure 6-9]

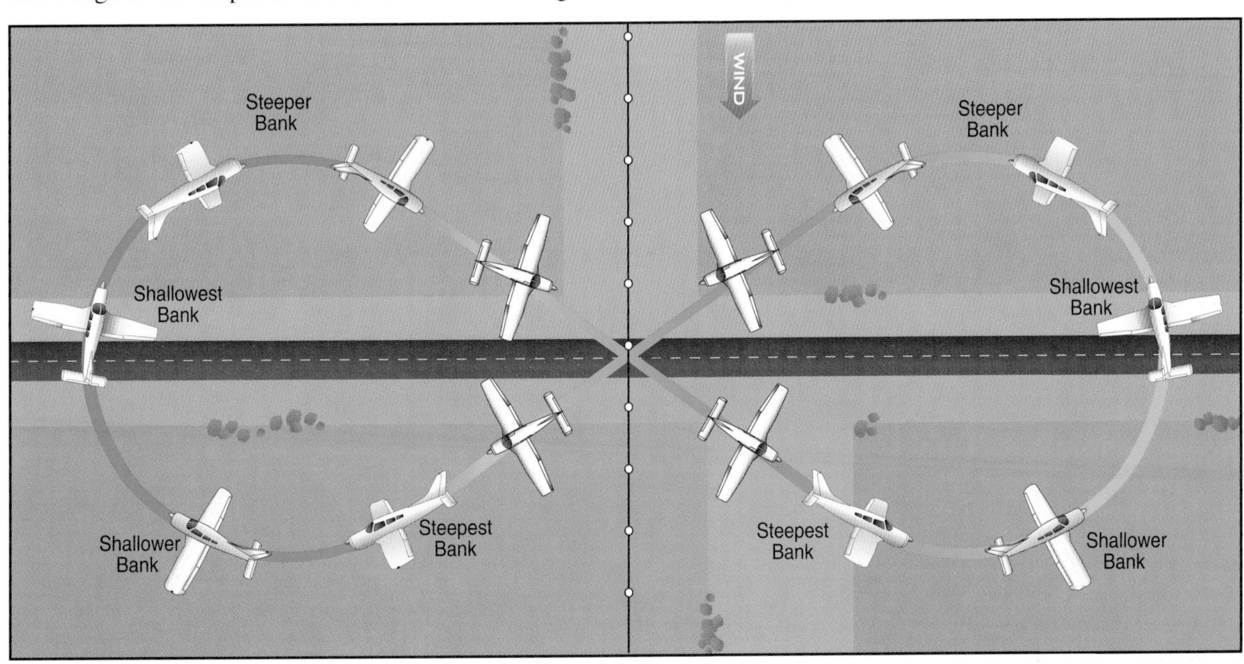

Figure 6-8. Eights across a road.

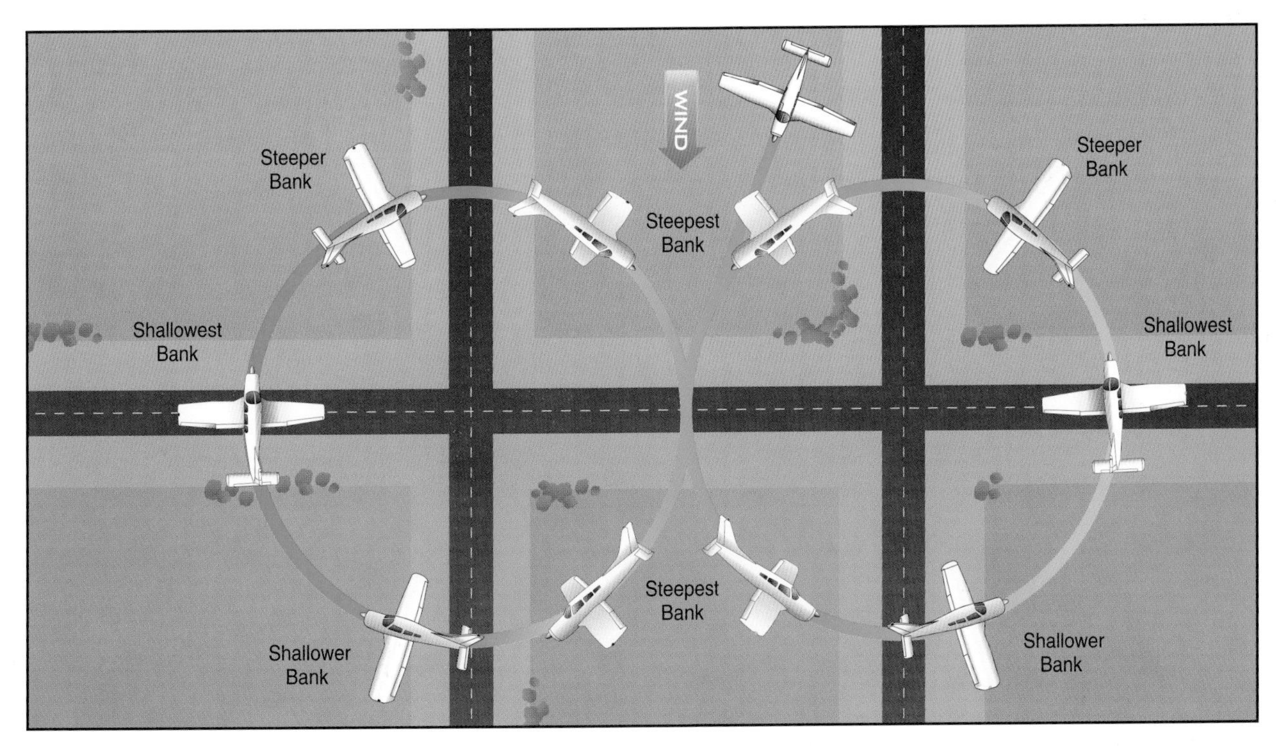

Figure 6-9. Eights around pylons.

The pattern involves flying downwind between the pylons and upwind outside of the pylons. It may include a short period of straight-and-level flight while proceeding diagonally from one pylon to the other.

The pylons selected should be on a line 90° to the direction of the wind and should be in an area away from communities, livestock, or groups of people, to avoid possible annoyance or hazards to others. The area selected should be clear of hazardous obstructions and other air traffic. Throughout the maneuver a constant altitude of at least 500 feet above the ground should be maintained.

The eight should be started with the airplane on a downwind heading when passing between the pylons. The distance between the pylons and the wind velocity will determine the initial angle of bank required to maintain a constant radius from the pylons during each turn. The steepest banks will be necessary just after each turn entry and just before the rollout from each turn where the airplane is headed downwind and the groundspeed is greatest; the shallowest banks will be when the airplane is headed directly upwind and the groundspeed is least.

The rate of bank change will depend on the wind velocity, the same as it does in S-turns and turns around a point, and the bank will be changing continuously during the turns. The adjustment of the bank angle should be gradual from the steepest bank to the shallowest bank as the airplane progressively heads into the wind, followed by a gradual increase until the steepest bank is again reached just prior to rollout. If the airplane is to proceed diagonally from one turn to the other, the rollout from each turn must be completed on the proper heading with sufficient wind correction angle to ensure that after brief straight-and-level flight, the airplane will arrive at the point where a turn of the same radius can be made around the other pylon. The straight-and-level flight segments must be tangent to both circular patterns.

Common errors in the performance of elementary eights are:

- Failure to adequately clear the area.
- Poor choice of ground reference points.
- Improper maneuver entry considering wind direction and ground reference points.
- Incorrect initial bank.
- Poor coordination during turns.
- Gaining or losing altitude.
- Loss of orientation.
- Abrupt rather than smooth changes in bank angle to counteract wind drift in turns.
- Failure to anticipate needed drift correction.
- Failure to apply needed drift correction in a timely manner.
- Failure to roll out of turns on proper heading.
- Inability to divide attention between reference points on the ground, airplane control, and scanning for other aircraft.

EIGHTS-ON-PYLONS (PYLON EIGHTS)

The pylon eight is the most advanced and most difficult of the low altitude flight training maneuvers. Because of the various techniques involved, the pylon eight is unsurpassed for teaching, developing, and testing subconscious control of the airplane.

As the pylon eight is essentially an advanced maneuver in which the pilot's attention is directed at maintaining a pivotal position on a selected pylon, with a minimum of attention within the cockpit, it should not be introduced until the instructor is assured that the student has a complete grasp of the fundamentals. Thus, the prerequisites are the ability to make a coordinated turn without gain or loss of altitude, excellent feel of the airplane, stall recognition, relaxation with low altitude maneuvering, and an absence of the error of over concentration.

Like eights around pylons, this training maneuver also involves flying the airplane in circular paths, alternately left and right, in the form of a figure 8 around two selected points or pylons on the ground. Unlike eights around pylons, however, no attempt is made to maintain a uniform distance from the pylon. In eights-on-pylons, the distance from the pylons varies if there is any wind. Instead, the airplane is flown at such a precise altitude and airspeed that a line parallel to the airplane's lateral axis, and extending from the pilot's eye, appears to pivot on each of the pylons. [Figure 6-10] Also, unlike eights around pylons, in the performance of eights-on-pylons the degree of bank increases as the distance from the pylon decreases.

The altitude that is appropriate for the airplane being flown is called the pivotal altitude and is governed by the groundspeed. While not truly a ground track maneuver as were the preceding maneuvers, the objective is similar—to develop the ability to maneuver the airplane accurately while dividing one's attention between the flightpath and the selected points on the ground.

In explaining the performance of eights-on-pylons, the term "wingtip" is frequently considered as being synonymous with the proper reference line, or pivot point on the airplane. This interpretation is not

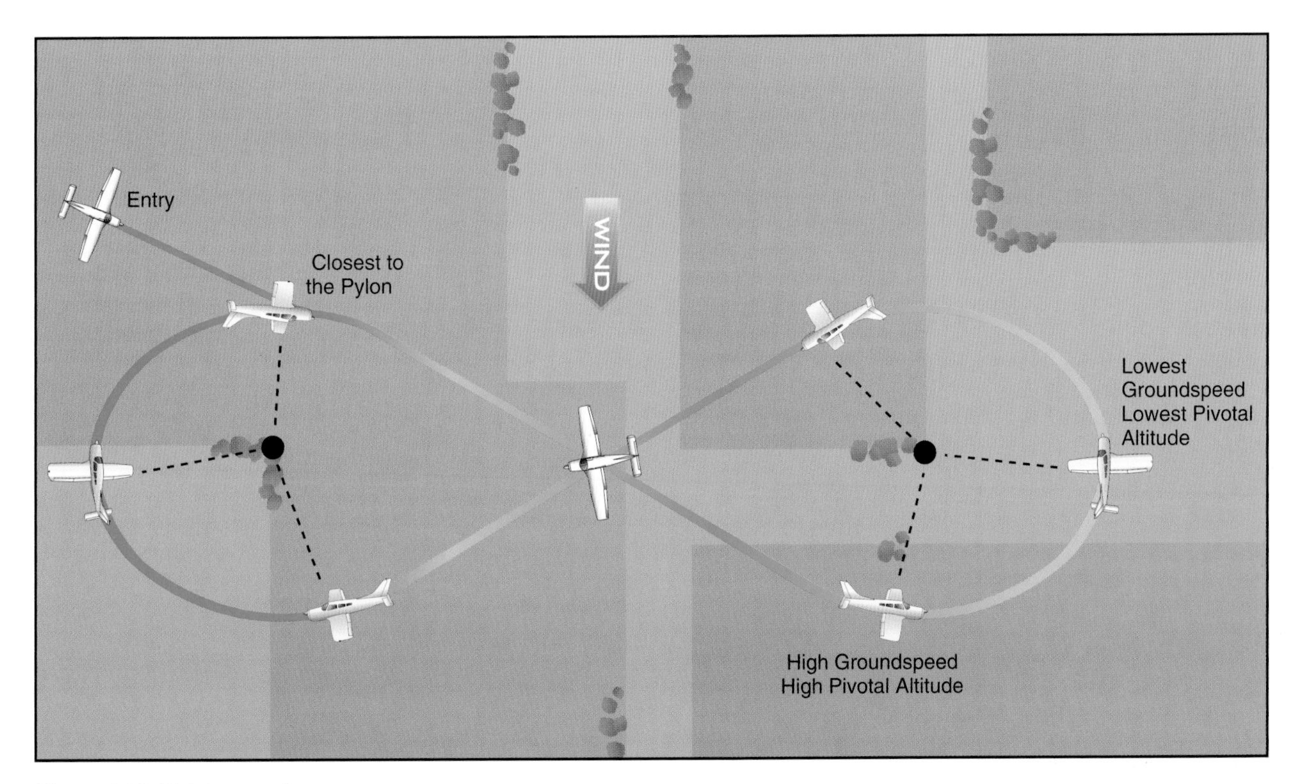

Figure 6-10. Eights-on-pylons.

always correct. High-wing, low-wing, sweptwing, and tapered wing airplanes, as well as those with tandem or side-by-side seating, will all present different angles from the pilot's eye to the wingtip. [Figure 6-11] Therefore, in the correct performance of eights-on-pylons, as in other maneuvers requiring a lateral reference, the pilot should use a sighting reference line that, from eye level, parallels the lateral axis of the airplane.

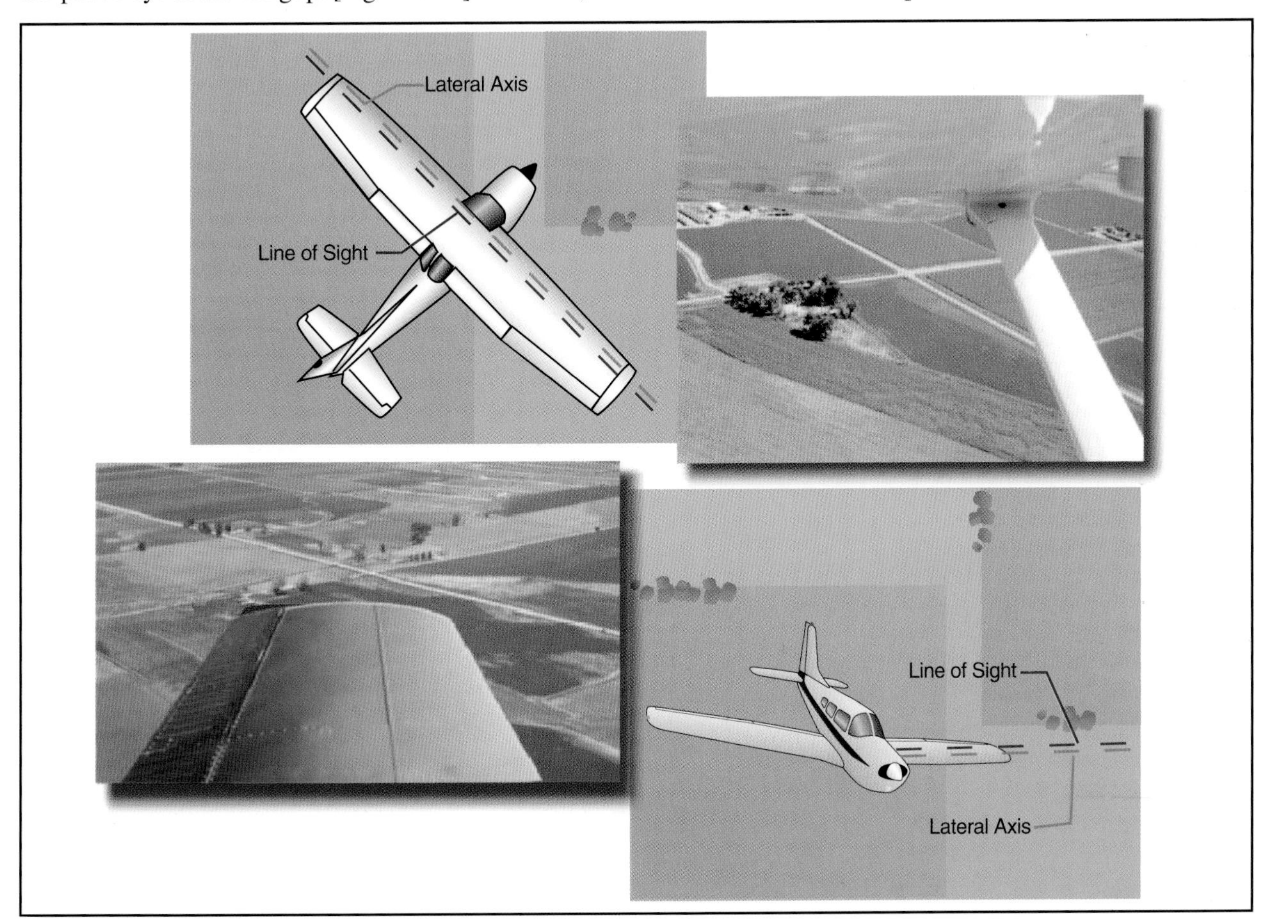

Figure 6-11. Line of sight.

The sighting point or line, while not necessarily on the wingtip itself, may be positioned in relation to the wingtip (ahead, behind, above, or below), but even then it will differ for each pilot, and from each seat in the airplane. This is especially true in tandem (fore and aft) seat airplanes. In side-by-side type airplanes, there will be very little variation in the sighting lines for different persons if those persons are seated so that the eyes of each are at approximately the same level.

An explanation of the pivotal altitude is also essential. There is a specific altitude at which, when the airplane turns at a given groundspeed, a projection of the sighting reference line to the selected point on the ground will appear to pivot on that point. Since different airplanes fly at different airspeeds, the groundspeed will be different. Therefore, each airplane will have its own pivotal altitude. [Figure 6-12] The pivotal altitude does not vary with the angle of bank being used unless the bank is steep enough to affect the groundspeed. A rule of thumb for estimating pivotal altitude in calm wind is to square the true airspeed and divide by 15 for miles per hour (m.p.h.) or 11.3 for knots.

AIRSPEED		APPROXIMATE PIVOTAL ALTITUDE
KNOTS	MPH	
87	100	670
91	105	735
96	110	810
100	115	885
104	120	960
109	125	1050
113	130	1130

Figure 6-12. Speed vs. pivotal altitude.

Distance from the pylon affects the angle of bank. At any altitude above that pivotal altitude, the projected reference line will appear to move rearward in a circular path in relation to the pylon. Conversely, when the airplane is below the pivotal altitude, the projected reference line will appear to move forward in a circular path. [Figure 6-13]

To demonstrate this, the airplane is flown at normal cruising speed, and at an altitude estimated to be below the proper pivotal altitude, and then placed in a medium-banked turn. It will be seen that the projected reference line of sight appears to move forward along the ground (pylon moves back) as the airplane turns.

A climb is then made to an altitude well above the pivotal altitude, and when the airplane is again at normal cruising speed, it is placed in a medium-banked turn. At this higher altitude, the projected reference line of sight now appears to move backward across the ground (pylon moves forward) in a direction opposite that of flight.

After the high altitude extreme has been demonstrated, the power is reduced, and a descent at cruising speed begun in a continuing medium bank around the pylon. The apparent backward travel of the projected reference line with respect to the pylon will slow down as altitude is lost, stop for an instant, then start to reverse itself, and would move forward if the descent were allowed to continue below the pivotal altitude.

The altitude at which the line of sight apparently ceased to move across the ground was the pivotal altitude. If the airplane descended below the pivotal altitude, power should be added to maintain airspeed while altitude is regained to the point at which the projected reference line moves neither backward nor forward but actually pivots on the pylon. In this way the pilot can determine the pivotal altitude of the airplane.

The pivotal altitude is critical and will change with variations in groundspeed. Since the headings throughout the turns continually vary from directly downwind to directly upwind, the groundspeed will constantly change. This will result in the proper pivotal altitude varying slightly throughout the eight. Therefore, adjustment is made for this by climbing or descending, as necessary, to hold the reference line or point on the pylons. This change in altitude will be dependent on how much the wind affects the groundspeed.

The instructor should emphasize that the elevators are the primary control for holding the pylons. Even a very slight variation in altitude effects a double correction, since in losing altitude, speed is gained, and even a slight climb reduces the airspeed. This variation in altitude, although important in holding the pylon, in most cases will be so slight as to be barely perceptible on a sensitive altimeter.

Before beginning the maneuver, the pilot should select two points on the ground along a line which lies 90° to the direction of the wind. The area in which the maneuver is to be performed should be checked for obstructions and any other air traffic, and it should be located where a disturbance to groups of people, livestock, or communities will not result.

The selection of proper pylons is of importance to good eights-on-pylons. They should be sufficiently prominent to be readily seen by the pilot when completing the turn around one pylon and heading for the next, and should be adequately spaced to provide time

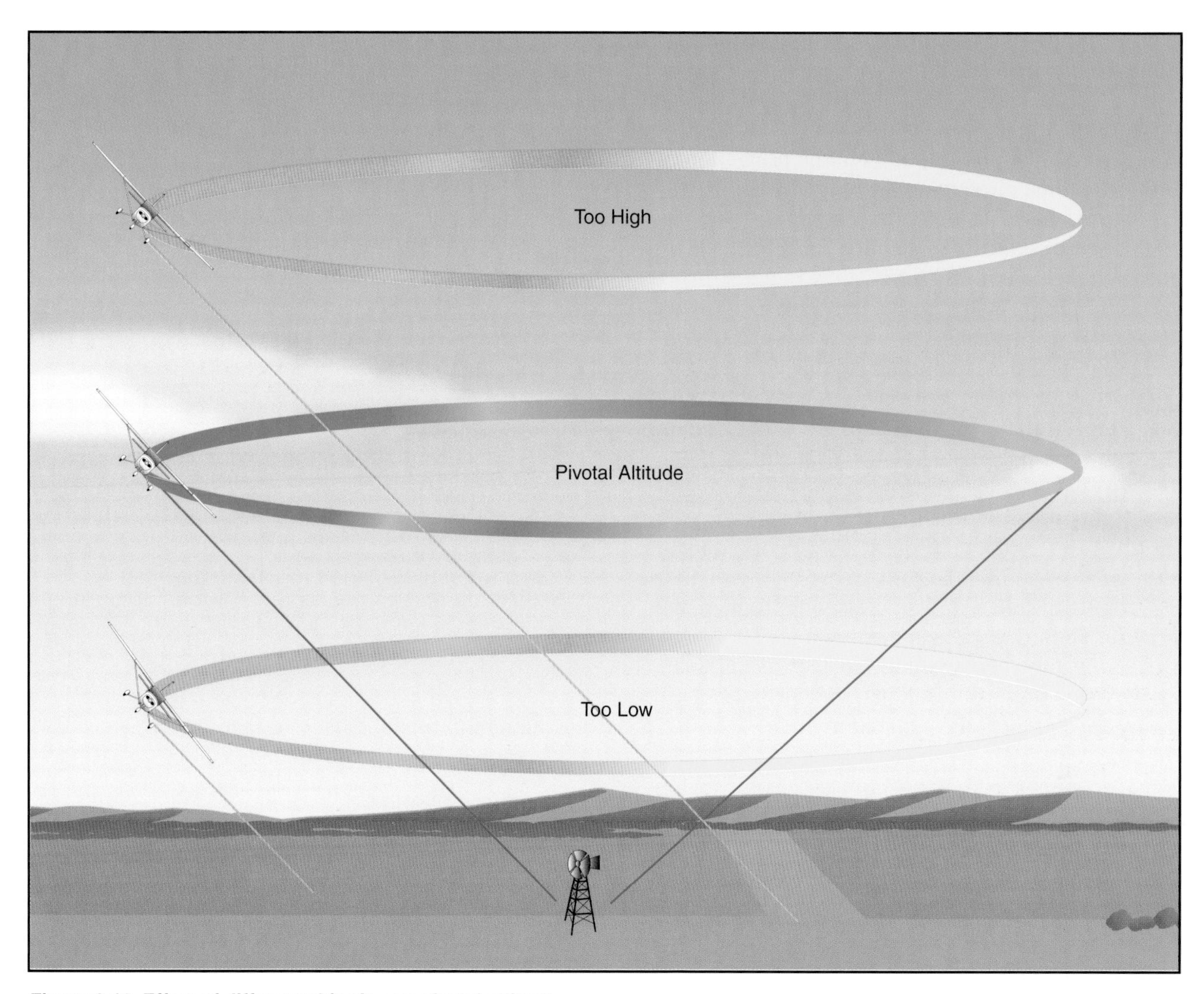

Figure 6-13. Effect of different altitudes on pivotal altitude.

for planning the turns and yet not cause unnecessary straight-and-level flight between the pylons. The selected pylons should also be at the same elevation, since differences of over a very few feet will necessitate climbing or descending between each turn.

For uniformity, the eight is usually begun by flying diagonally crosswind between the pylons to a point downwind from the first pylon so that the first turn can be made into the wind. As the airplane approaches a position where the pylon appears to be just ahead of the wingtip, the turn should be started by lowering the upwind wing to place the pilot's line of sight reference on the pylon. As the turn is continued, the line of sight reference can be held on the pylon by gradually increasing the bank. The reference line should appear to pivot on the pylon. As the airplane heads into the wind, the groundspeed decreases; consequently, the pivotal altitude is lower and the airplane must descend to hold the reference line on the pylon. As the turn progresses on the upwind side of the pylon, the wind becomes more of a crosswind. Since a constant distance from the pylon is not required on this maneuver, no correction to counteract drifting should be applied during the turns.

If the reference line appears to move ahead of the pylon, the pilot should increase altitude. If the reference line appears to move behind the pylon, the pilot should decrease altitude. Varying rudder pressure to yaw the airplane and force the wing and reference line forward or backward to the pylon is a dangerous technique and must not be attempted.

As the airplane turns toward a downwind heading, the rollout from the turn should be started to allow the airplane to proceed diagonally to a point on the downwind side of the second pylon. The rollout must be completed in the proper wind correction angle to correct for wind drift, so that the airplane will arrive at a point downwind from the second pylon the same distance it was from the first pylon at the beginning of the maneuver.

Upon reaching that point, a turn is started in the opposite direction by lowering the upwind wing to again place the pilot's line of sight reference on the pylon. The turn is then continued just as in the turn around the first pylon but in the opposite direction.

With prompt correction, and a very fine control touch, it should be possible to hold the projection of the reference line directly on the pylon even in a stiff wind. Corrections for temporary variations, such as those caused by gusts or inattention, may be made by shallowing the bank to fly relatively straight to bring forward a lagging wing, or by steepening the bank temporarily to turn back a wing which has crept ahead. With practice, these corrections will become so slight as to be barely noticeable. These variations are apparent from the movement of the wingtips long before they are discernable on the altimeter.

Pylon eights are performed at bank angles ranging from shallow to steep. [Figure 6-14] The student should understand that the bank chosen will not alter the pivotal altitude. As proficiency is gained, the instructor should increase the complexity of the maneuver by directing the student to enter at a distance from the pylon that will result in a specific bank angle at the steepest point in the pylon turn.

The most common error in attempting to hold a pylon is incorrect use of the rudder. When the projection of the reference line moves forward with respect to the pylon, many pilots will tend to press the inside rudder to yaw the wing backward. When the reference line moves behind the pylon, they will press the outside rudder to yaw the wing forward. The rudder is to be used only as a coordination control.

Other common errors in the performance of eights-on-pylons (pylon eights) are:

- Failure to adequately clear the area.
- Skidding or slipping in turns (whether trying to hold the pylon with rudder or not).
- Excessive gain or loss of altitude.
- Over concentration on the pylon and failure to observe traffic.
- Poor choice of pylons.
- Not entering the pylon turns into the wind.
- Failure to assume a heading when flying between pylons that will compensate sufficiently for drift.
- Failure to time the bank so that the turn entry is completed with the pylon in position.
- Abrupt control usage.
- Inability to select pivotal altitude.

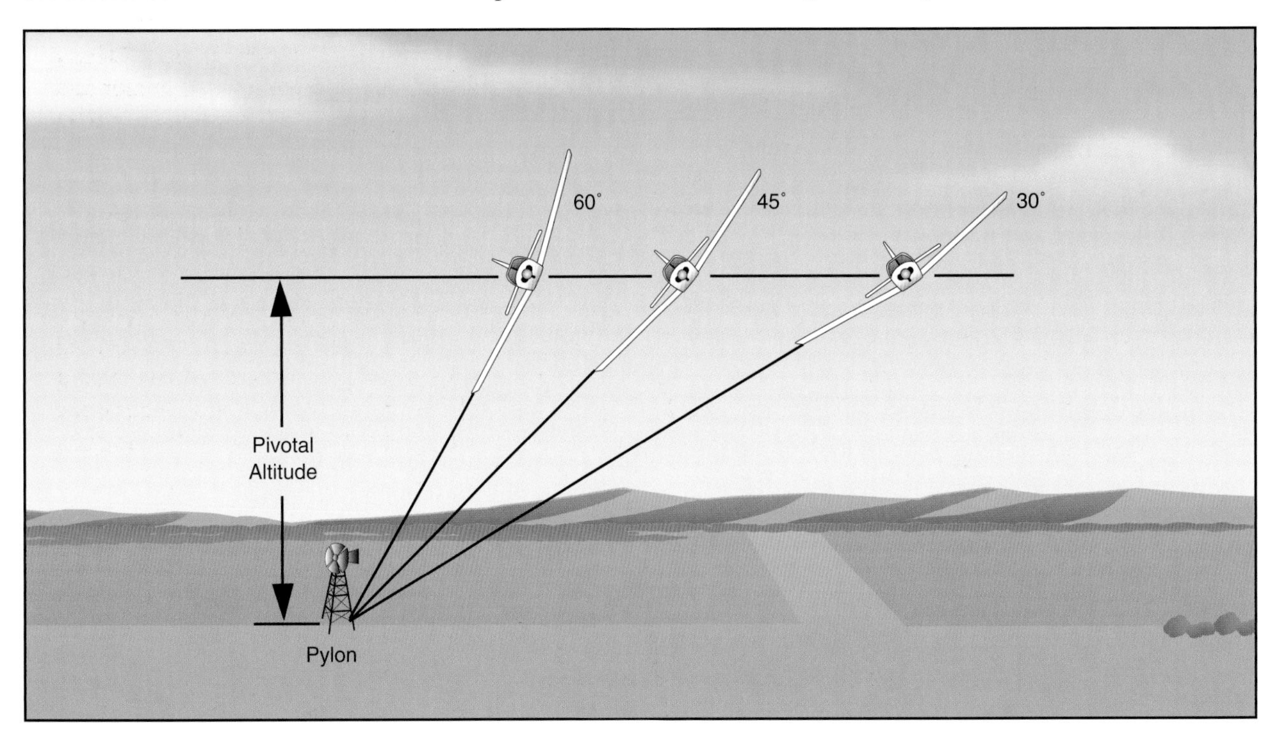

Figure 6-14. Bank angle vs. pivotal altitude.

AIRPORT TRAFFIC PATTERNS AND OPERATIONS

Just as roads and streets are needed in order to utilize automobiles, airports or airstrips are needed to utilize airplanes. Every flight begins and ends at an airport or other suitable landing field. For that reason, it is essential that the pilot learn the traffic rules, traffic procedures, and traffic pattern layouts that may be in use at various airports.

When an automobile is driven on congested city streets, it can be brought to a stop to give way to conflicting traffic; however, an airplane can only be slowed down. Consequently, specific traffic patterns and traffic control procedures have been established at designated airports. The traffic patterns provide specific routes for takeoffs, departures, arrivals, and landings. The exact nature of each airport traffic pattern is dependent on the runway in use, wind conditions, obstructions, and other factors.

Control towers and radar facilities provide a means of adjusting the flow of arriving and departing aircraft, and render assistance to pilots in busy terminal areas. Airport lighting and runway marking systems are used frequently to alert pilots to abnormal conditions and hazards, so arrivals and departures can be made safely.

Airports vary in complexity from small grass or sod strips to major terminals having many paved runways and taxiways. Regardless of the type of airport, the pilot must know and abide by the rules and general operating procedures applicable to the airport being used. These rules and procedures are based not only on logic or common sense, but also on courtesy, and their objective is to keep air traffic moving with maximum safety and efficiency. The use of any traffic pattern, service, or procedure does not alter the responsibility of pilots to see and avoid other aircraft.

STANDARD AIRPORT TRAFFIC PATTERNS

To assure that air traffic flows into and out of an airport in an orderly manner, an airport traffic pattern is established appropriate to the local conditions, including the direction and placement of the pattern, the altitude to be flown, and the procedures for entering and leaving the pattern. Unless the airport displays approved visual markings indicating that turns should be made to the right, the pilot should make all turns in the pattern to the left.

When operating at an airport with an operating control tower, the pilot receives, by radio, a clearance to approach or depart, as well as pertinent information about the traffic pattern. If there is not a control tower, it is the pilot's responsibility to determine the direction of the traffic pattern, to comply with the appropriate traffic rules, and to display common courtesy toward other pilots operating in the area.

The pilot is not expected to have extensive knowledge of all traffic patterns at all airports, but if the pilot is familiar with the basic rectangular pattern, it will be easy to make proper approaches and departures from most airports, regardless of whether they have control towers. At airports with operating control towers, the tower operator may instruct pilots to enter the traffic pattern at any point or to make a straight-in approach without flying the usual rectangular pattern. Many other deviations are possible if the tower operator and the pilot work together in an effort to keep traffic moving smoothly. Jets or heavy airplanes will frequently be flying wider and/or higher patterns than lighter airplanes, and in many cases will make a straight-in approach for landing.

Compliance with the basic rectangular traffic pattern reduces the possibility of conflicts at airports without an operating control tower. It is imperative that the pilot form the habit of exercising constant vigilance in the vicinity of airports even though the air traffic appears to be light.

The standard rectangular traffic pattern is illustrated in figure 7-1 (on next page). The traffic pattern altitude is usually 1,000 feet above the elevation of the airport surface. The use of a common altitude at a given airport is the key factor in minimizing the risk of collisions at airports without operating control towers.

It is recommended that while operating in the traffic pattern at an airport without an operating control tower the pilot maintain an airspeed that conforms with the limits established by Title 14 of the Code of Federal Regulations (14 CFR) part 91 for such an airport: no more than 200 knots (230 miles per hour (m.p.h.)). In any case, the speed should be adjusted,

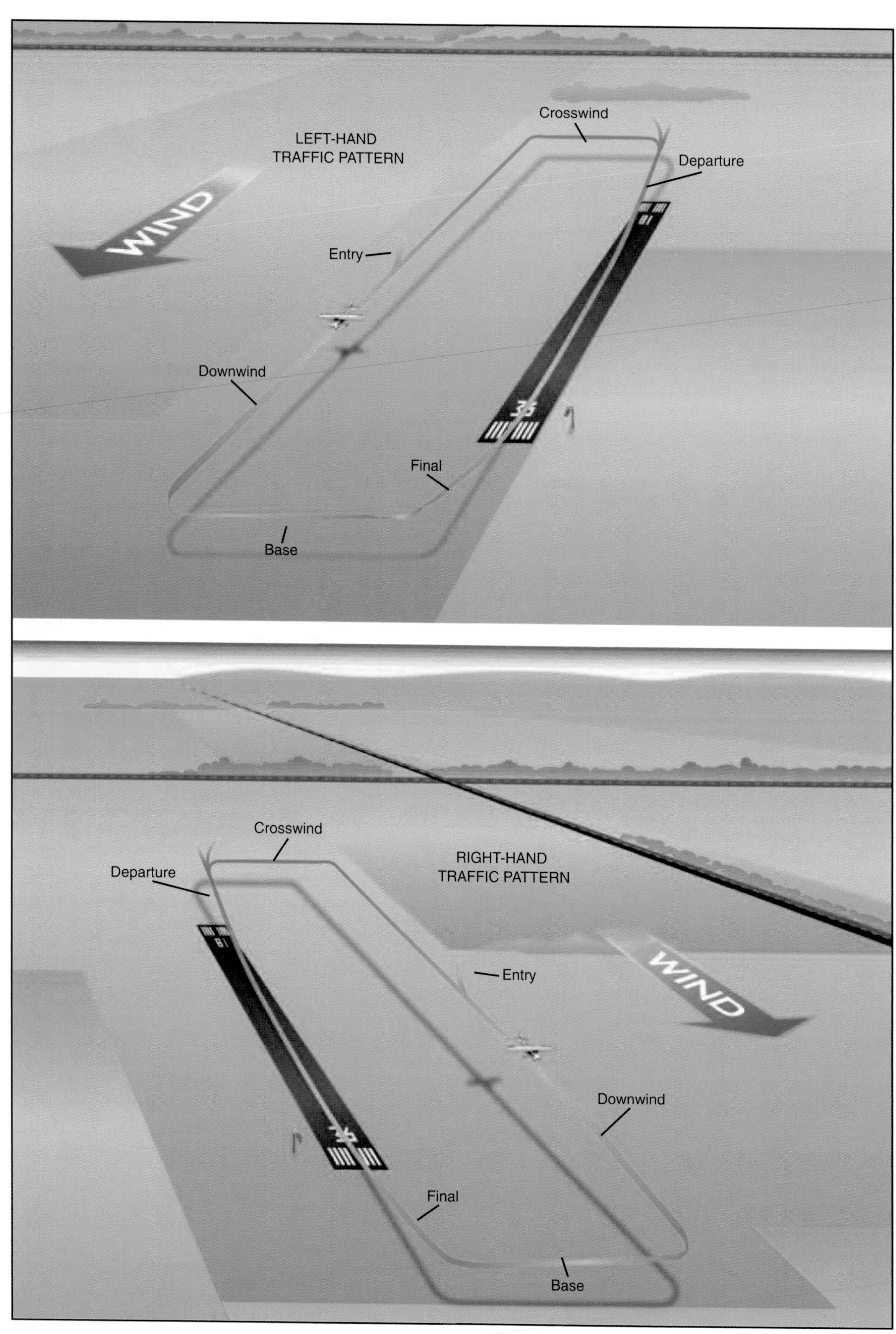

Figure 7-1. Traffic patterns.

when practicable, so that it is compatible with the speed of other airplanes in the pattern.

When entering the traffic pattern at an airport without an operating control tower, inbound pilots are expected to observe other aircraft already in the pattern and to conform to the traffic pattern in use. If other aircraft are not in the pattern, then traffic indicators on the ground and wind indicators must be checked to determine which runway and traffic pattern direction should be used. [Figure 7-2] Many airports have L-shaped traffic pattern indicators displayed with a segmented circle adjacent to the runway. The short member of the L shows the direction in which the traffic pattern turns should be made when using the runway parallel to the long member. These indicators should be checked while at a distance well away from any pattern that might be in use, or while at a safe height well above generally used pattern altitudes. When the proper traffic pattern direction has been determined, the pilot should then proceed to a point well clear of the pattern before descending to the pattern altitude.

When approaching an airport for landing, the traffic pattern should be entered at a 45° angle to the downwind leg, headed toward a point abeam of the midpoint of the runway to be used for landing. Arriving airplanes should be at the proper traffic pattern altitude before entering the pattern, and should stay clear of the traffic flow until established on the entry leg. Entries into traffic patterns while descending create specific collision hazards and should always be avoided.

The entry leg should be of sufficient length to provide a clear view of the entire traffic pattern, and to allow the pilot adequate time for planning the intended path in the pattern and the landing approach.

The downwind leg is a course flown parallel to the landing runway, but in a direction opposite to the intended landing direction. This leg should be approximately 1/2 to 1 mile out from the landing runway, and at the specified traffic pattern altitude. During this leg, the before landing check should be completed and the landing gear extended if retractable. Pattern altitude should be maintained until abeam the approach end of the landing runway. At this point, power should be reduced and a descent begun. The downwind leg continues past a point abeam the approach end of the runway to a point approximately 45° from the approach end of the runway, and a medium bank turn is made onto the base leg.

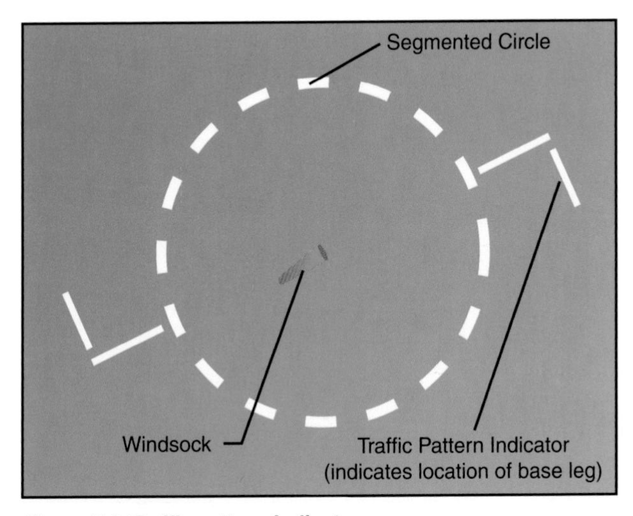

Figure 7-2. Traffic pattern indicators.

The base leg is the transitional part of the traffic pattern between the downwind leg and the final approach leg. Depending on the wind condition, it is established at a sufficient distance from the approach end of the landing runway to permit a gradual descent to the intended touchdown point. The ground track of the airplane while on the base leg should be perpendicular to the extended centerline of the landing runway, although the longitudinal axis of the airplane may not be aligned with the ground track when it is necessary to turn into the wind to counteract drift. While on the base leg, the pilot must ensure, before turning onto the final approach, that there is no danger of colliding with another aircraft that may be already on the final approach.

The final approach leg is a descending flightpath starting from the completion of the base-to-final turn and extending to the point of touchdown. This is probably the most important leg of the entire pattern, because here the pilot's judgment and procedures must be the sharpest to accurately control the airspeed and descent angle while approaching the intended touchdown point.

As stipulated in 14 CFR part 91, aircraft while on final approach to land or while landing, have the right-of-way over other aircraft in flight or operating on the surface. When two or more aircraft are approaching an airport for the purpose of landing, the aircraft at the lower altitude has the right-of-way. Pilots should not take advantage of this rule to cut in front of another aircraft that is on final approach to land, or to overtake that aircraft.

The upwind leg is a course flown parallel to the landing runway, but in the same direction to the intended landing direction. The upwind leg continues past a point abeam of the departure end of the runway to where a medium bank 90° turn is made onto the crosswind leg.

The upwind leg is also the transitional part of the traffic pattern when on the final approach and a go-around is initiated and climb attitude is established. When a safe altitude is attained, the pilot should commence a shallow bank turn to the upwind side of the airport. This will allow better visibility of the runway for departing aircraft.

The departure leg of the rectangular pattern is a straight course aligned with, and leading from, the takeoff runway. This leg begins at the point the airplane leaves the ground and continues until the 90° turn onto the crosswind leg is started.

On the departure leg after takeoff, the pilot should continue climbing straight ahead, and, if remaining in the traffic pattern, commence a turn to the crosswind leg beyond the departure end of the runway within 300 feet of pattern altitude. If departing the traffic pattern, continue straight out or exit with a 45° turn (to the left when in a left-hand traffic pattern; to the right when in a right-hand traffic pattern) beyond the departure end of the runway after reaching pattern altitude.

The crosswind leg is the part of the rectangular pattern that is horizontally perpendicular to the extended centerline of the takeoff runway and is entered by making approximately a 90° turn from the upwind leg. On the crosswind leg, the airplane proceeds to the downwind leg position.

Since in most cases the takeoff is made into the wind, the wind will now be approximately perpendicular to the airplane's flightpath. As a result, the airplane will have to be turned or headed slightly into the wind while on the crosswind leg to maintain a ground track that is perpendicular to the runway centerline extension.

Additional information on airport operations can be found in the *Aeronautical Information Manual* (AIM).

NORMAL APPROACH AND LANDING

A normal approach and landing involves the use of procedures for what is considered a normal situation; that is, when engine power is available, the wind is light or the final approach is made directly into the wind, the final approach path has no obstacles, and the landing surface is firm and of ample length to gradually bring the airplane to a stop. The selected landing point should be beyond the runway's approach threshold but within the first one-third portion of the runway.

The factors involved and the procedures described for the normal approach and landing also have applications to the other-than-normal approaches and landings which are discussed later in this chapter. This being the case, the principles of normal operations are explained first and must be understood before proceeding to the more complex operations. So that the pilot may better understand the factors that will influence judgment and procedures, that last part of the approach pattern and the actual landing will be divided into five phases: **the base leg, the final approach, the roundout, the touchdown,** and **the after-landing roll**.

It must be remembered that the manufacturer's recommended procedures, including airplane configuration and airspeeds, and other information relevant to approaches and landings in a specific make and model airplane are contained in the FAA-approved Airplane Flight Manual and/or Pilot's Operating Handbook (AFM/POH) for that airplane. If any of the information in this chapter differs from the airplane manufacturer's recommendations as contained in the AFM/POH, the airplane manufacturer's recommendations take precedence.

BASE LEG

The placement of the base leg is one of the more important judgments made by the pilot in any landing approach. [Figure 8-1] The pilot must accurately judge the altitude and distance from which a gradual descent will result in landing at the desired spot. The distance will depend on the altitude of the base leg, the effect of wind, and the amount of wing flaps used. When there is a strong wind on final approach or the flaps will be used to produce a steep angle of descent, the base leg must be positioned closer to the approach end of the runway than would be required with a light wind or no

Figure 8-1. Base leg and final approach.

flaps. Normally, the landing gear should be extended and the before landing check completed prior to reaching the base leg.

After turning onto the base leg, the pilot should start the descent with reduced power and airspeed of approximately 1.4 V_{SO}. (V_{SO}—the stalling speed with power off, landing gears and flaps down.) For example, if V_{SO} is 60 knots, the speed should be 1.4 times 60, or 84 knots. Landing flaps may be partially lowered, if desired, at this time. Full flaps are not recommended until the final approach is established. Drift correction should be established and maintained to follow a ground track perpendicular to the extension of the centerline of the runway on which the landing is to be made. Since the final approach and landing will normally be made into the wind, there will be somewhat of a crosswind during the base leg. This requires that the airplane be angled sufficiently into the wind to prevent drifting farther away from the intended landing spot.

The base leg should be continued to the point where a medium to shallow-banked turn will align the airplane's path directly with the centerline of the landing runway. This descending turn should be completed at a safe altitude that will be dependent upon the height of the terrain and any obstructions along the ground track. The turn to the final approach should also be sufficiently above the airport elevation to permit a final approach long enough for the pilot to accurately estimate the resultant point of touchdown, while maintaining the proper approach airspeed. This will require careful planning as to the starting point and the radius of the turn. Normally, it is recommended that the angle of bank not exceed a medium bank because the steeper the angle of bank, the higher the airspeed at which the airplane stalls. Since the base-to-final turn is made at a relatively low altitude, it is important that a stall not occur at this point. If an extremely steep bank is needed to prevent overshooting the proper final approach path, it is advisable to discontinue the approach, go around, and plan to start the turn earlier on the next approach rather than risk a hazardous situation.

FINAL APPROACH

After the base-to-final approach turn is completed, the longitudinal axis of the airplane should be aligned with the centerline of the runway or landing surface, so that drift (if any) will be recognized immediately. On a normal approach, with no wind drift, the longitudinal axis should be kept aligned with the runway centerline throughout the approach and landing. (The proper way to correct for a crosswind will be explained under the section, Crosswind Approach and Landing. For now, only an approach and landing where the wind is straight down the runway will be discussed.)

After aligning the airplane with the runway centerline, the final flap setting should be completed and the pitch attitude adjusted as required for the desired rate of descent. Slight adjustments in pitch and power may be necessary to maintain the descent attitude and the desired approach airspeed. In the absence of the manufacturer's recommended airspeed, a speed equal to 1.3 V_{SO} should be used. If V_{SO} is 60 knots, the speed should be 78 knots. When the pitch attitude and airspeed have been stabilized, the airplane should be retrimmed to relieve the pressures being held on the controls.

The descent angle should be controlled throughout the approach so that the airplane will land in the center of the first third of the runway. The descent angle is affected by all four fundamental forces that act on an airplane (lift, drag, thrust, and weight). If all the forces are constant, the descent angle will be constant in a no-wind condition. The pilot can control these forces by adjusting the airspeed, attitude, power, and drag (flaps or forward slip). The wind also plays a prominent part in the gliding distance over the ground [Figure 8-2]; naturally, the pilot does not have control over the wind but may correct for its effect on the airplane's descent by appropriate pitch and power adjustments.

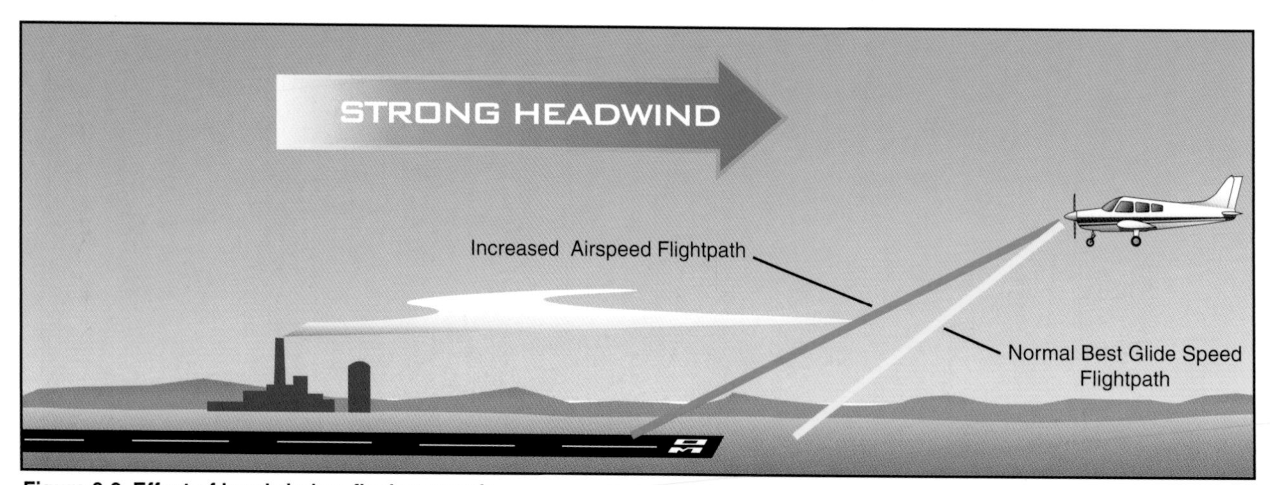

Figure 8-2. Effect of headwind on final approach.

Considering the factors that affect the descent angle on the final approach, for all practical purposes at a given pitch attitude there is only one power setting for one airspeed, one flap setting, and one wind condition. A change in any one of these variables will require an appropriate coordinated change in the other controllable variables. For example, if the pitch attitude is raised too high without an increase of power, the airplane will settle very rapidly and touch down short of the desired spot. For this reason, the pilot should never try to stretch a glide by applying back-elevator pressure alone to reach the desired landing spot. This will shorten the gliding distance if power is not added simultaneously. The proper angle of descent and airspeed should be maintained by coordinating pitch attitude changes and power changes.

The objective of a good final approach is to descend at an angle and airspeed that will permit the airplane to reach the desired touchdown point at an airspeed which will result in minimum floating just before touchdown; in essence, a semi-stalled condition. To accomplish this, it is essential that both the descent angle and the airspeed be accurately controlled. Since on a normal approach the power setting is not fixed as in a power-off approach, the power and pitch attitude should be adjusted simultaneously as necessary, to control the airspeed, and the descent angle, or to attain the desired altitudes along the approach path. By lowering the nose and reducing power to keep approach airspeed constant, a descent at a higher rate can be made to correct for being too high in the approach. This is one reason for performing approaches with partial power; if the approach is too high, merely lower the nose and reduce the power. When the approach is too low, add power and raise the nose.

USE OF FLAPS

The lift/drag factors may also be varied by the pilot to adjust the descent through the use of landing flaps. [Figures 8-3 and 8-4] Flap extension during landings provides several advantages by:

- Producing greater lift and permitting lower landing speed.
- Producing greater drag, permitting a steep descent angle without airspeed increase.
- Reducing the length of the landing roll.

Flap extension has a definite effect on the airplane's pitch behavior. The increased camber from flap deflection produces lift primarily on the rear portion of the wing. This produces a nosedown pitching moment; however, the change in tail loads from the **downwash** deflected by the flaps over the horizontal tail has a significant influence on the pitching moment. Consequently, pitch behavior depends on the design features of the particular airplane.

Flap deflection of up to 15° primarily produces lift with minimal drag. The airplane has a tendency to **balloon**

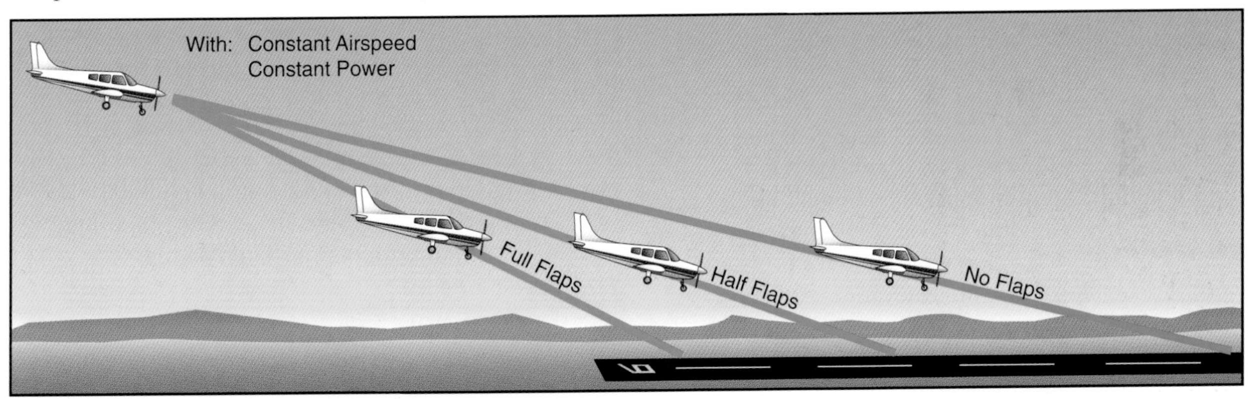

Figure 8-3. Effect of flaps on the landing point.

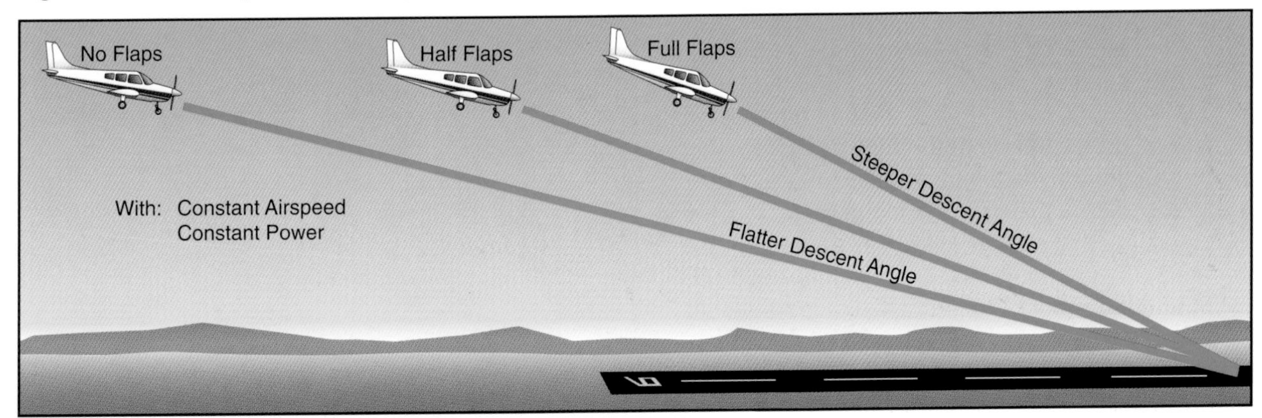

Figure 8-4. Effect of flaps on the approach angle.

up with initial flap deflection because of the lift increase. The nosedown pitching moment, however, tends to offset the balloon. Flap deflection beyond 15° produces a large increase in drag. Also, deflection beyond 15° produces a significant noseup pitching moment in high-wing airplanes because the resulting downwash increases the airflow over the horizontal tail.

The time of flap extension and the degree of deflection are related. Large flap deflections at one single point in the landing pattern produce large lift changes that require significant pitch and power changes in order to maintain airspeed and descent angle. Consequently, the deflection of flaps at certain positions in the landing pattern has definite advantages. Incremental deflection of flaps on downwind, base leg, and final approach allow smaller adjustment of pitch and power compared to extension of full flaps all at one time.

When the flaps are lowered, the airspeed will decrease unless the power is increased or the pitch attitude lowered. On final approach, therefore, the pilot must estimate where the airplane will land through discerning judgment of the descent angle. If it appears that the airplane is going to overshoot the desired landing spot, more flaps may be used if not fully extended or the power reduced further, and the pitch attitude lowered. This will result in a steeper approach. If the desired landing spot is being undershot and a shallower approach is needed, both power and pitch attitude should be increased to readjust the descent angle. Never retract the flaps to correct for undershooting since that will suddenly decrease the lift and cause the airplane to sink even more rapidly.

The airplane must be retrimmed on the final approach to compensate for the change in aerodynamic forces. With the reduced power and with a slower airspeed, the airflow produces less lift on the wings and less downward force on the horizontal stabilizer, resulting in a significant nosedown tendency. The elevator must then be trimmed more noseup.

It will be found that the roundout, touchdown, and landing roll are much easier to accomplish when they are preceded by a proper final approach with precise control of airspeed, attitude, power, and drag resulting in a stabilized descent angle.

ESTIMATING HEIGHT AND MOVEMENT

During the approach, roundout, and touchdown, vision is of prime importance. To provide a wide scope of vision and to foster good judgment of height and movement, the pilot's head should assume a natural, straight-ahead position. The pilot's visual focus should not be fixed on any one side or any one spot ahead of the airplane, but should be changing slowly from a point just over the airplane's nose to the desired touchdown zone and back again, while maintaining a deliberate awareness of distance from either side of the runway within the pilot's peripheral field of vision.

Accurate estimation of distance is, besides being a matter of practice, dependent upon how clearly objects are seen; it requires that the vision be focused properly in order that the important objects stand out as clearly as possible.

Speed blurs objects at close range. For example, most everyone has noted this in an automobile moving at high speed. Nearby objects seem to merge together in a blur, while objects farther away stand out clearly. The driver subconsciously focuses the eyes sufficiently far ahead of the automobile to see objects distinctly.

The distance at which the pilot's vision is focused should be proportionate to the speed at which the airplane is traveling over the ground. Thus, as speed is reduced during the roundout, the distance ahead of the airplane at which it is possible to focus should be brought closer accordingly.

If the pilot attempts to focus on a reference that is too close or looks directly down, the reference will become blurred, [Figure 8-5] and the reaction will be either too abrupt or too late. In this case, the pilot's tendency will be to overcontrol, round out high, and make full-stall, drop-in landings. When the pilot focuses too far ahead, accuracy in judging the closeness of the ground is lost and the consequent reaction will be too slow since there will not appear to be a necessity for action. This will result in the airplane flying into the ground nose first. The change of visual focus from a long distance to a short distance requires a definite time interval and even though the time is brief, the airplane's speed during this interval is such that the airplane travels an appreciable distance, both forward and downward toward the ground.

Figure 8-5. Focusing too close blurs vision.

If the focus is changed gradually, being brought progressively closer as speed is reduced, the time interval

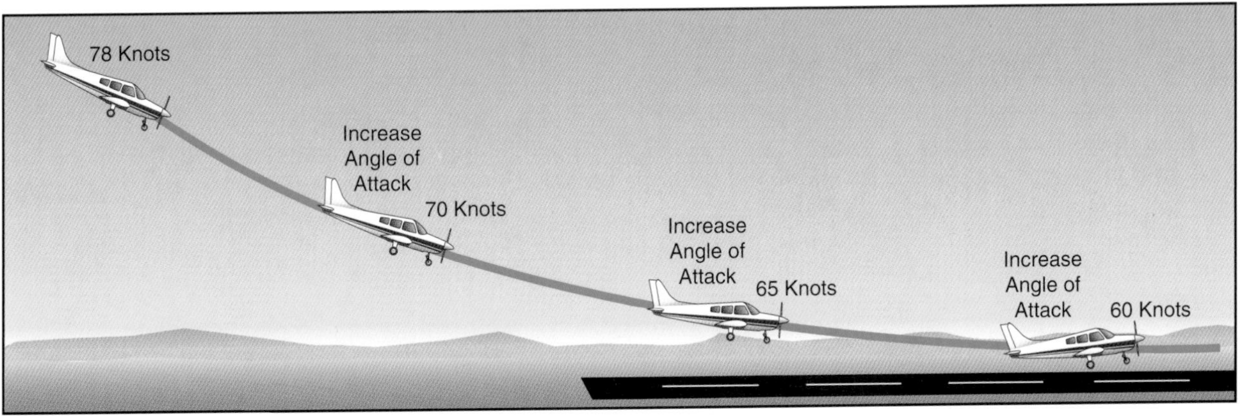

Figure 8-6. Changing angle of attack during roundout.

and the pilot's reaction will be reduced, and the whole landing process smoothed out.

ROUNDOUT (FLARE)

The roundout is a slow, smooth transition from a normal approach attitude to a landing attitude, gradually rounding out the flightpath to one that is parallel with, and within a very few inches above, the runway. When the airplane, in a normal descent, approaches within what appears to be 10 to 20 feet above the ground, the roundout or flare should be started, and once started should be a continuous process until the airplane touches down on the ground.

As the airplane reaches a height above the ground where a timely change can be made into the proper landing attitude, back-elevator pressure should be gradually applied to slowly increase the pitch attitude and angle of attack. [Figure 8-6] This will cause the airplane's nose to gradually rise toward the desired landing attitude. The angle of attack should be increased at a rate that will allow the airplane to continue settling slowly as forward speed decreases.

When the angle of attack is increased, the lift is momentarily increased, which decreases the rate of descent. Since power normally is reduced to idle during the roundout, the airspeed will also gradually decrease. This will cause lift to decrease again, and it must be controlled by raising the nose and further increasing the angle of attack. During the roundout, the airspeed is being decreased to touchdown speed while the lift is being controlled so the airplane will settle gently onto the landing surface. The roundout should be executed at a rate that the proper landing attitude and the proper touchdown airspeed are attained simultaneously just as the wheels contact the landing surface.

The rate at which the roundout is executed depends on the airplane's height above the ground, the rate of descent, and the pitch attitude. A roundout started excessively high must be executed more slowly than one from a lower height to allow the airplane to descend to the ground while the proper landing attitude is being established. The rate of rounding out must also be proportionate to the rate of closure with the ground. When the airplane appears to be descending very slowly, the increase in pitch attitude must be made at a correspondingly slow rate.

Visual cues are important in flaring at the proper altitude and maintaining the wheels a few inches above the runway until eventual touchdown. Flare cues are primarily dependent on the angle at which the pilot's central vision intersects the ground (or runway) ahead and slightly to the side. Proper depth perception is a factor in a successful flare, but the visual cues used most are those related to changes in runway or terrain perspective and to changes in the size of familiar objects near the landing area such as fences, bushes, trees, hangars, and even sod or runway texture. The pilot should direct central vision at a shallow downward angle of from 10° to 15° toward the runway as the roundout/flare is initiated. [Figure 8-7] Maintaining the same viewing angle causes the point

Figure 8-7. To obtain necessary visual cues, the pilot should look toward the runway at a shallow angle.

of visual interception with the runway to move progressively rearward toward the pilot as the airplane loses altitude. This is an important visual cue in assessing the *rate* of altitude loss. Conversely, forward movement of the visual interception point will indicate an increase in altitude, and would mean that the pitch angle was increased too rapidly, resulting in an over flare. Location of the visual interception point in conjunction with assessment of flow velocity of nearby off-runway terrain, as well as the similarity of appearance of height above the runway ahead of the airplane (in comparison to the way it looked when the airplane was taxied prior to takeoff) is also used to judge when the wheels are just a few inches above the runway.

The pitch attitude of the airplane in a full-flap approach is considerably lower than in a no-flap approach. To attain the proper landing attitude before touching down, the nose must travel through a greater pitch change when flaps are fully extended. Since the roundout is usually started at approximately the same height above the ground regardless of the degree of flaps used, the pitch attitude must be increased at a faster rate when full flaps are used; however, the roundout should still be executed at a rate proportionate to the airplane's downward motion.

Once the actual process of rounding out is started, the elevator control should not be pushed forward. If too much back-elevator pressure has been exerted, this pressure should be either slightly relaxed or held constant, depending on the degree of the error. In some cases, it may be necessary to advance the throttle slightly to prevent an excessive rate of sink, or a stall, all of which would result in a hard, drop-in type landing.

It is recommended that the student pilot form the habit of keeping one hand on the throttle throughout the approach and landing, should a sudden and unexpected hazardous situation require an immediate application of power.

TOUCHDOWN

The touchdown is the gentle settling of the airplane onto the landing surface. The roundout and touchdown should be made with the engine idling, and the airplane at minimum controllable airspeed, so that the airplane will touch down on the main gear at approximately stalling speed. As the airplane settles, the proper landing attitude is attained by application of whatever back-elevator pressure is necessary.

Some pilots may try to force or fly the airplane onto the ground without establishing the proper landing attitude. The airplane should never be flown on the runway with excessive speed. It is paradoxical that the way to make an ideal landing is to try to hold the airplane's wheels a few inches off the ground as long as possible with the elevators. In most cases, when the wheels are within 2 or 3 feet off the ground, the airplane will still be settling too fast for a gentle touchdown; therefore, this descent must be retarded by further back-elevator pressure. Since the airplane is already close to its stalling speed and is settling, this added back-elevator pressure will only slow up the settling instead of stopping it. At the same time, it will result in the airplane touching the ground in the proper landing attitude, and the main wheels touching down first so that little or no weight is on the nosewheel. [Figure 8-8]

After the main wheels make initial contact with the ground, back-elevator pressure should be held to maintain a positive angle of attack for aerodynamic braking, and to hold the nosewheel off the ground until the airplane decelerates. As the airplane's momentum decreases, back-elevator pressure may be gradually relaxed to allow the nosewheel to gently settle onto the runway. This will permit steering with the nosewheel. At the same time, it will cause a low angle of attack and negative lift on the wings to prevent floating or skipping, and will allow the full weight of the airplane to rest on the wheels for better braking action.

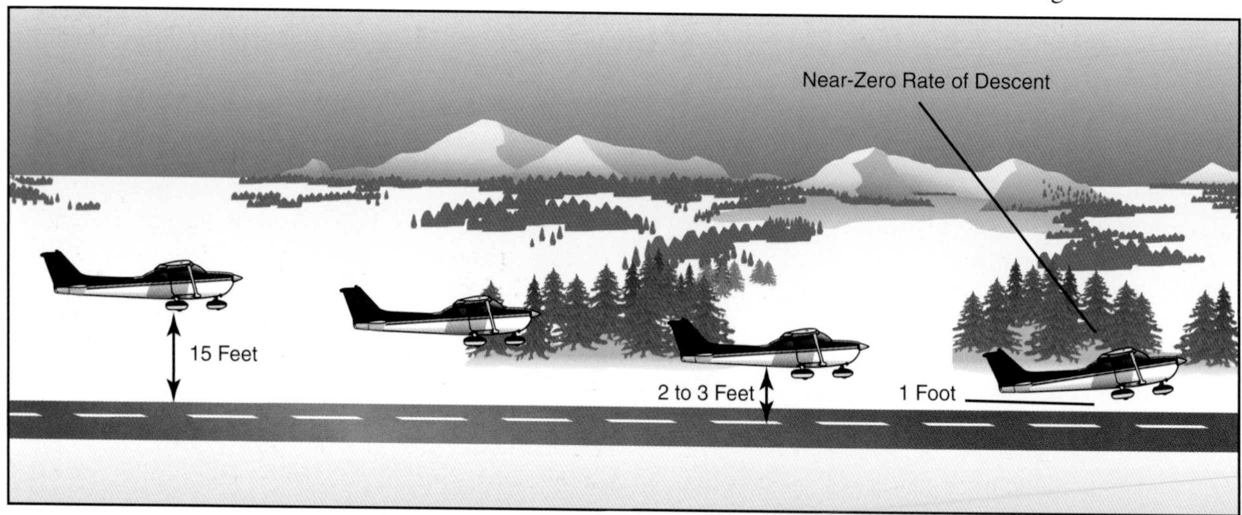

Figure 8-8. A well executed roundout results in attaining the proper landing attitude.

It is extremely important that the touchdown occur with the airplane's longitudinal axis exactly parallel to the direction in which the airplane is moving along the runway. Failure to accomplish this imposes severe side loads on the landing gear. To avoid these side stresses, the pilot should not allow the airplane to touch down while turned into the wind or drifting.

AFTER-LANDING ROLL

The landing process must never be considered complete until the airplane decelerates to the normal taxi speed during the landing roll or has been brought to a complete stop when clear of the landing area. Many accidents have occurred as a result of pilots abandoning their vigilance and positive control after getting the airplane on the ground.

The pilot must be alert for directional control difficulties immediately upon and after touchdown due to the ground friction on the wheels. The friction creates a pivot point on which a moment arm can act. Loss of directional control may lead to an aggravated, uncontrolled, tight turn on the ground, or a **ground loop**. The combination of centrifugal force acting on the center of gravity (CG) and ground friction of the main wheels resisting it during the ground loop may cause the airplane to tip or lean enough for the outside wingtip to contact the ground. This may even impose a sideward force, which could collapse the landing gear.

The rudder serves the same purpose on the ground as it does in the air—it controls the yawing of the airplane. The effectiveness of the rudder is dependent on the airflow, which depends on the speed of the airplane. As the speed decreases and the nosewheel has been lowered to the ground, the steerable nose provides more positive directional control.

The brakes of an airplane serve the same primary purpose as the brakes of an automobile—to reduce speed on the ground. In airplanes, they may also be used as an aid in directional control when more positive control is required than could be obtained with rudder or nosewheel steering alone.

To use brakes, on an airplane equipped with toe brakes, the pilot should slide the toes or feet up from the rudder pedals to the brake pedals. If rudder pressure is being held at the time braking action is needed, that pressure should not be released as the feet or toes are being slid up to the brake pedals, because control may be lost before brakes can be applied.

Putting maximum weight on the wheels after touchdown is an important factor in obtaining optimum braking performance. During the early part of rollout, some lift may continue to be generated by the wing. After touchdown, the nosewheel should be lowered to the runway to maintain directional control. During deceleration, the nose may be pitched down by braking and the weight transferred to the nosewheel from the main wheels. This does not aid in braking action, so back pressure should be applied to the controls without lifting the nosewheel off the runway. This will enable the pilot to maintain directional control while keeping weight on the main wheels.

Careful application of the brakes can be initiated after the nosewheel is on the ground and directional control is established. Maximum brake effectiveness is just short of the point where skidding occurs. If the brakes are applied so hard that skidding takes place, braking becomes ineffective. Skidding can be stopped by releasing the brake pressure. Also, braking effectiveness is not enhanced by alternately applying and reapplying brake pressure. The brakes should be applied firmly and smoothly as necessary.

During the ground roll, the airplane's direction of movement can be changed by carefully applying pressure on one brake or uneven pressures on each brake in the desired direction. Caution must be exercised when applying brakes to avoid overcontrolling.

The ailerons serve the same purpose on the ground as they do in the air—they change the lift and drag components of the wings. During the after-landing roll, they should be used to keep the wings level in much the same way they were used in flight. If a wing starts to rise, aileron control should be applied toward that wing to lower it. The amount required will depend on speed because as the forward speed of the airplane decreases, the ailerons will become less effective. Procedures for using ailerons in crosswind conditions are explained further in this chapter, in the Crosswind Approach and Landing section.

After the airplane is on the ground, back-elevator pressure may be gradually relaxed to place normal weight on the nosewheel to aid in better steering. If available runway permits, the speed of the airplane should be allowed to dissipate in a normal manner. Once the airplane has slowed sufficiently and has turned on to the taxiway and stopped, the pilot should retract the flaps and clean up the airplane. Many accidents have occurred as a result of the pilot unintentionally operating the landing gear control and retracting the gear instead of the flap control when the airplane was still rolling. The habit of positively identifying both of these controls, before actuating them, should be formed from the very beginning of flight training and continued in all future flying activities.

STABILIZED APPROACH CONCEPT

A *stabilized approach* is one in which the pilot establishes and maintains a constant angle glidepath towards a predetermined point on the landing runway. It is based on the pilot's judgment of certain visual clues, and depends on the maintenance of a constant final descent airspeed and configuration.

An airplane descending on final approach at a constant rate and airspeed will be traveling in a straight line toward a spot on the ground ahead. This spot will not be the spot on which the airplane will touch down, because some float will inevitably occur during the roundout (flare). [Figure 8-9] Neither will it be the spot toward which the airplane's nose is pointed, because the airplane is flying at a fairly high angle of attack, and the component of lift exerted parallel to the Earth's surface by the wings tends to carry the airplane forward horizontally.

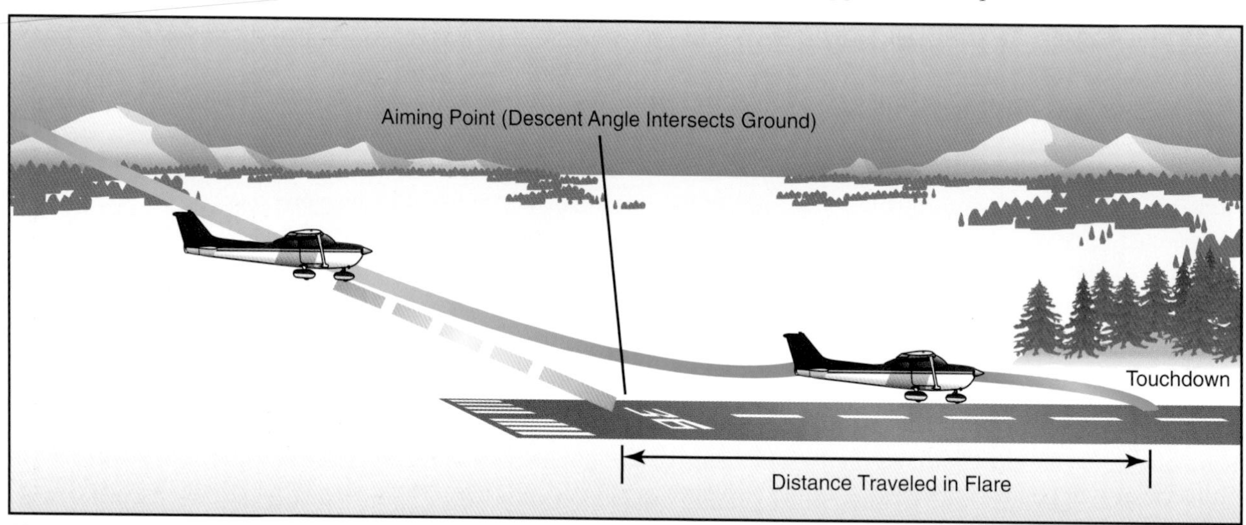

Figure 8-9. Stabilized approach.

The point toward which the airplane is progressing is termed the "aiming point." [Figure 8-9] It is the point on the ground at which, if the airplane maintains a constant glidepath, and was *not* flared for landing, it would strike the ground. To a pilot moving straight ahead toward an object, it appears to be stationary. It does not "move." This is how the aiming point can be distinguished—*it does not move*. However, objects in front of and beyond the aiming point do appear to move as the distance is closed, and they appear to move in opposite directions. During instruction in landings, one of the most important skills a student pilot must acquire is how to use visual cues to accurately determine the true aiming point from any distance out on final approach. From this, the pilot will not only be able to determine if the glidepath will result in an undershoot or overshoot, but, taking into account float during roundout, the pilot will be able to predict the touchdown point to within a very few feet.

For a constant angle glidepath, the distance between the horizon and the aiming point will remain constant. If a final approach descent has been established but the distance between the perceived aiming point and the horizon appears to increase (aiming point moving down away from the horizon), then the true aiming point, and subsequent touchdown point, is farther down the runway. If the distance between the perceived aiming point and the horizon decreases (aiming point moving up toward the horizon), the true aiming point is closer than perceived.

When the airplane is established on final approach, the shape of the runway image also presents clues as to what must be done to maintain a stabilized approach to a safe landing.

A runway, obviously, is normally shaped in the form of an elongated rectangle. When viewed from the air during the approach, the phenomenon known as perspective causes the runway to assume the shape of a trapezoid with the far end looking narrower than the approach end, and the edge lines converging ahead. If the airplane continues down the glidepath *at a constant angle* (stabilized), the image the pilot sees will still be trapezoidal but of proportionately larger dimensions. In other words, *during a stabilized approach the runway shape does not change.* [Figure 8-10]

If the approach becomes shallower, however, the runway will appear to shorten and become wider. Conversely, if the approach is steepened, the runway will appear to become longer and narrower. [Figure 8-11]

The objective of a stabilized approach is to select an appropriate touchdown point on the runway, and adjust the glidepath so that the true aiming point and the desired touchdown point basically coincide. Immediately after rolling out on final approach, the pilot should adjust the pitch attitude and power so that the airplane is descending directly toward the aiming point at the appropriate airspeed. The airplane should

Figure 8-10. Runway shape during stabilized approach.

be in the landing configuration, and trimmed for "hands off" flight. With the approach set up in this manner, the pilot will be free to devote full attention toward outside references. The pilot should not stare at any one place, but rather scan from one point to another, such as from the aiming point to the horizon, to the trees and bushes along the runway, to an area well short of the runway, and back to the aiming point. In this way, the pilot will be more apt to perceive a deviation from the desired glidepath, and whether or not the airplane is proceeding directly toward the aiming point.

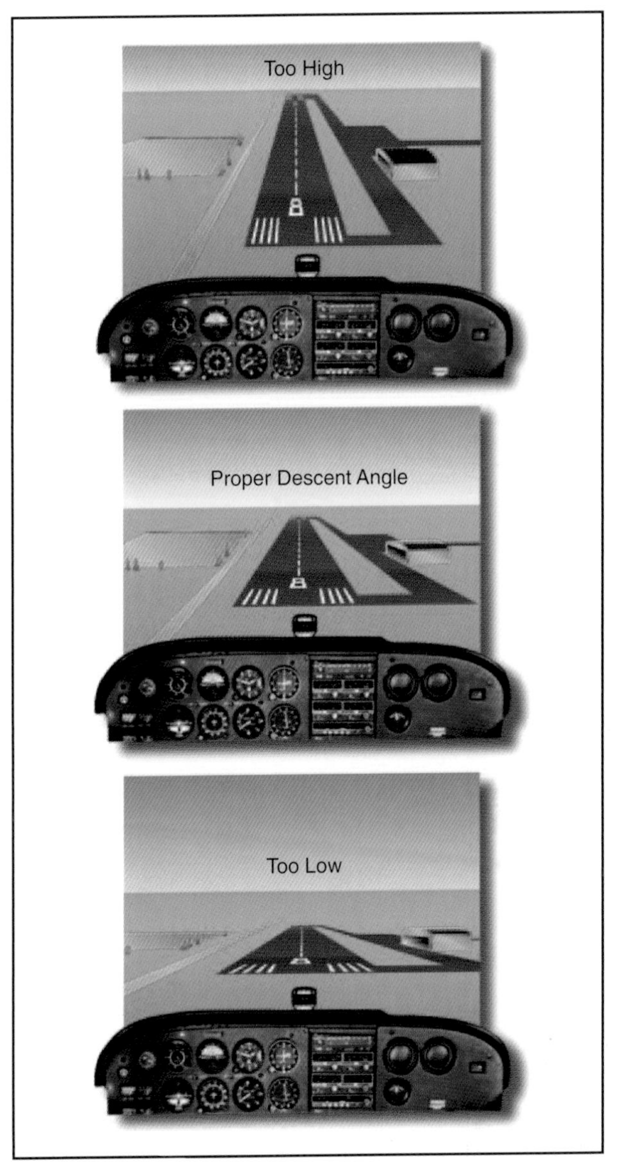

Figure 8-11. Change in runway shape if approach becomes narrow or steep.

If the pilot perceives any indication that the aiming point on the runway is not where desired, an adjustment must be made to the glidepath. This in turn will move the aiming point. For instance, if the pilot perceives that the aiming point is short of the desired touchdown point and will result in an undershoot, an increase in pitch attitude and engine power is warranted. A constant airspeed must be maintained. The pitch and power change, therefore, must be made smoothly and simultaneously. This will result in a shallowing of the glidepath with the resultant aiming point moving towards the desired touchdown point. Conversely, if the pilot perceives that the aiming point is farther down the runway than the desired touchdown point and will result in an overshoot, the glidepath should be steepened by a simultaneous decrease in pitch attitude and power. Once again, the airspeed must be held constant. **It is essential that deviations from the desired glidepath be detected early, so that only slight and infrequent adjustments to glidepath are required.**

The closer the airplane gets to the runway, the larger (and possibly more frequent) the required corrections become, resulting in an *unstabilized* approach.

Common errors in the performance of normal approaches and landings are:

- Inadequate wind drift correction on the base leg.
- Overshooting or undershooting the turn onto final approach resulting in too steep or too shallow a turn onto final approach.
- Flat or skidding turns from base leg to final approach as a result of overshooting/inadequate wind drift correction.
- Poor coordination during turn from base to final approach.
- Failure to complete the landing checklist in a timely manner.
- Unstabilized approach.
- Failure to adequately compensate for flap extension.
- Poor trim technique on final approach.
- Attempting to maintain altitude or reach the runway using elevator alone.
- Focusing too close to the airplane resulting in a too high roundout.
- Focusing too far from the airplane resulting in a too low roundout.
- Touching down prior to attaining proper landing attitude.
- Failure to hold sufficient back-elevator pressure after touchdown.
- Excessive braking after touchdown.

INTENTIONAL SLIPS

A slip occurs when the bank angle of an airplane is too steep for the existing rate of turn. Unintentional slips are most often the result of uncoordinated rudder/aileron application. Intentional slips, however, are used to dissipate altitude without increasing airspeed, and/or to adjust airplane ground track during a crosswind. Intentional slips are especially useful in forced landings, and in situations where obstacles must be cleared during approaches to confined areas. A slip can also be used as an emergency means of rapidly reducing airspeed in situations where wing flaps are inoperative or not installed.

A slip is a combination of forward movement and sideward (with respect to the longitudinal axis of the airplane) movement, the lateral axis being inclined and the sideward movement being toward the low end of this axis (low wing). An airplane in a slip is in fact flying sideways. This results in a change in the direction the relative wind strikes the airplane. Slips are characterized by a marked increase in drag and corresponding decrease in airplane climb, cruise, and glide performance. It is the increase in drag, however, that makes it possible for an airplane in a slip to descend rapidly without an increase in airspeed.

Most airplanes exhibit the characteristic of positive static directional stability and, therefore, have a natural tendency to compensate for slipping. An intentional slip, therefore, requires deliberate cross-controlling ailerons and rudder throughout the maneuver.

A "**sideslip**" is entered by lowering a wing and applying just enough opposite rudder to prevent a turn. In a sideslip, the airplane's longitudinal axis remains parallel to the original flightpath, but the airplane no longer flies straight ahead. Instead the horizontal component of wing lift forces the airplane also to move somewhat sideways toward the low wing. [Figure 8-12] The amount of slip, and therefore the rate of sideward movement, is determined by the bank angle. The steeper the bank—the greater the degree of slip. As bank angle is increased, however, additional opposite rudder is required to prevent turning.

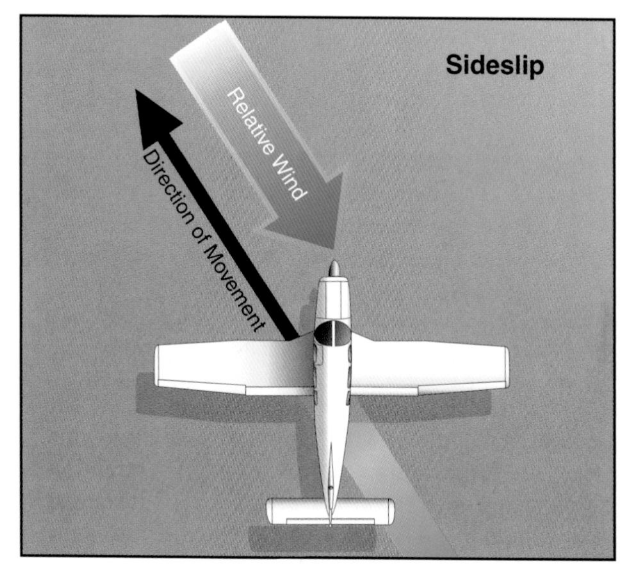

Figure 8-12. Sideslip.

A "**forward slip**" is one in which the airplane's direction of motion continues the same as before the slip was begun. Assuming the airplane is originally in straight flight, the wing on the side toward which the slip is to be made should be lowered by use of the ailerons. Simultaneously, the airplane's nose must be yawed in the opposite direction by applying opposite rudder so that the airplane's longitudinal axis is at an angle to its original flightpath. [Figure 8-13] The degree to which the nose is yawed in the opposite direction from the bank should be such that the original ground track is maintained. In a forward slip, the amount of slip, and therefore the sink rate, is determined by the bank angle. The steeper the bank—the steeper the descent.

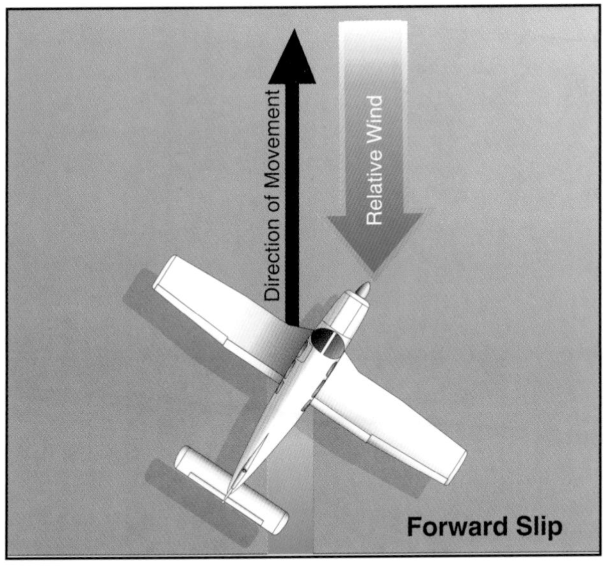

Figure 8-13. Forward slip.

In most light airplanes, the steepness of a slip is limited by the amount of rudder travel available. In both sideslips and forward slips, the point may be reached where full rudder is required to maintain heading even though the ailerons are capable of further steepening the bank angle. This is the **practical slip limit**, because any additional bank would cause the airplane to turn even though full opposite rudder is being applied. If there is a need to descend more rapidly even though the practical slip limit has been reached, lowering the nose will not only increase the sink rate but will also increase airspeed. The increase in airspeed increases rudder effectiveness permitting a steeper slip. Conversely, when the nose is raised, rudder effectiveness decreases and the bank angle must be reduced.

Discontinuing a slip is accomplished by leveling the wings and simultaneously releasing the rudder pressure while readjusting the pitch attitude to the normal glide attitude. If the pressure on the rudder is released abruptly, the nose will swing too quickly into line and the airplane will tend to acquire excess speed.

Because of the location of the pitot tube and static vents, airspeed indicators in some airplanes may have considerable error when the airplane is in a slip. The pilot must be aware of this possibility and recognize a properly performed slip by the attitude of the airplane, the sound of the airflow, and the feel of the flight controls. Unlike skids, however, if an airplane in a slip is made to stall, it displays very little of the yawing tendency that causes a skidding stall to develop into a spin. The airplane in a slip may do little more than tend to roll into a wings level attitude. In fact, in some airplanes stall characteristics may even be improved.

GO-AROUNDS (REJECTED LANDINGS)

Whenever landing conditions are not satisfactory, a go-around is warranted. There are many factors that can contribute to unsatisfactory landing conditions. Situations such as air traffic control requirements, unexpected appearance of hazards on the runway, overtaking another airplane, wind shear, wake turbulence, mechanical failure and/or an unstabilized approach are all examples of reasons to discontinue a landing approach and make another approach under more favorable conditions. The assumption that an aborted landing is invariably the consequence of a poor approach, which in turn is due to insufficient experience or skill, is a fallacy. The go-around is not strictly an emergency procedure. It is a *normal* maneuver that may at times be used in an emergency situation. Like any other normal maneuver, the go-around must be practiced and perfected. The flight instructor should emphasize early on, and the student pilot should be made to understand, that the go-around maneuver is an alternative to any approach and/or landing.

Although the need to discontinue a landing may arise at any point in the landing process, the most critical go-around will be one started when very close to the ground. Therefore, the earlier a condition that warrants a go-around is recognized, the safer the go-around/rejected landing will be. The go-around maneuver is not inherently dangerous in itself. It becomes dangerous only when delayed unduly or executed improperly. Delay in initiating the go-around normally stems from two sources: (1) landing expectancy, or set—the anticipatory belief that conditions are not as threatening as they are and that the approach will surely be terminated with a safe landing, and (2) pride—the mistaken belief that the act of going around is an admission of failure—failure to execute the approach properly. The improper execution of the go-around maneuver stems from a lack of familiarity with the three cardinal principles of the procedure: **power**, **attitude**, and **configuration**.

POWER

Power is the pilot's first concern. The instant the pilot decides to go around, *full* or *maximum allowable takeoff* power must be applied smoothly and without hesitation, and held until flying speed and controllability are restored. Applying only partial power in a go-around is never appropriate. The pilot must be aware of the degree of inertia that must be overcome, before an airplane that is settling towards the ground can regain sufficient airspeed to become fully controllable and capable of turning safely or climbing. The application of power should be smooth as well as positive. Abrupt movements of the throttle in some airplanes will cause the engine to falter. Carburetor heat should be turned off for maximum power.

ATTITUDE

Attitude is always critical when close to the ground, and when power is added, a deliberate effort on the part of the pilot will be required to keep the nose from pitching up prematurely. The airplane executing a go-around must be maintained in an attitude that permits a buildup of airspeed well beyond the stall point before any effort is made to gain altitude, or to execute a turn. Raising the nose too early may produce a stall from which the airplane could not be recovered if the go-around is performed at a low altitude.

A concern for quickly regaining altitude during a go-around produces a natural tendency to pull the nose up. The pilot executing a go-around must accept the fact that an airplane will not climb until it can fly, and it will not fly below stall speed. In some circumstances, it may be desirable to lower the nose briefly to gain airspeed. As soon as the appropriate climb airspeed and pitch attitude are attained, the pilot should "rough trim" the airplane to relieve any adverse control pressures. Later, more precise trim adjustments can be made when flight conditions have stabilized.

CONFIGURATION

In cleaning up the airplane during the go-around, the pilot should be concerned first with flaps and secondly with the landing gear (if retractable). When the decision is made to perform a go-around, takeoff power should be applied immediately and the pitch attitude changed so as to slow or stop the descent. After the descent has been stopped, the landing flaps may be partially retracted or placed in the takeoff position as recommended by the manufacturer. Caution must be used, however, in retracting the flaps. Depending on the airplane's altitude and airspeed, it may be wise to retract the flaps intermittently in small increments to allow time for the airplane to accelerate progressively as they are being raised. A sudden and complete retraction of the flaps could cause a loss of lift resulting in the airplane settling into the ground. [Figure 8-14]

Unless otherwise specified in the AFM/POH, it is generally recommended that the flaps be retracted (at least partially) *before* retracting the landing gear—for two reasons. First, on most airplanes full flaps produce more drag than the landing gear; and second, in case the airplane should inadvertently touch down as the go-around is initiated, it is most desirable to have the landing gear in the down-and-locked position. After a positive rate of climb is established, the landing gear can be retracted.

When takeoff power is applied, it will usually be necessary to hold considerable pressure on the controls to maintain straight flight and a safe climb attitude. Since the airplane has been trimmed for the approach (a low power and low airspeed condition), application of maximum allowable power will require considerable control pressure to maintain a climb pitch attitude. The addition of power will tend to raise the airplane's nose suddenly and veer to the left. Forward elevator pressure must be anticipated and applied to hold the nose in a safe climb attitude. Right rudder pressure must be increased to counteract torque and P-factor, and to keep the nose straight. The airplane must be held in the proper flight attitude regardless of the amount of control pressure that is required. Trim should be used to relieve adverse control pressures and assist the pilot in maintaining a proper pitch attitude. On airplanes that produce high control pressures when using maximum power on go-arounds, pilots should use caution when reaching for the flap handle. Airplane control may become critical during this high workload phase.

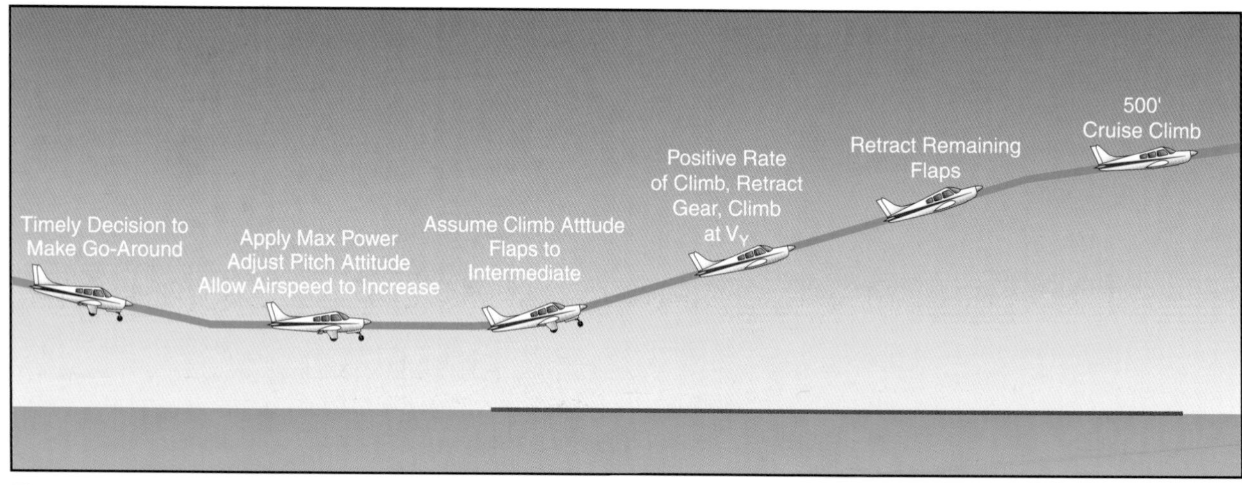

Figure 8-14. Go-around procedure.

The landing gear should be retracted only after the initial or rough trim has been accomplished and when it is certain the airplane will remain airborne. During the initial part of an extremely low go-around, the airplane may settle onto the runway and bounce. This situation is not particularly dangerous if the airplane is kept straight and a constant, safe pitch attitude is maintained. The airplane will be approaching safe flying speed rapidly and the advanced power will cushion any secondary touchdown.

If the pitch attitude is increased excessively in an effort to keep the airplane from contacting the runway, it may cause the airplane to stall. This would be especially likely if no trim correction is made and the flaps remain fully extended. The pilot should not attempt to retract the landing gear until after a rough trim is accomplished and a positive rate of climb is established.

GROUND EFFECT

Ground effect is a factor in every landing and every takeoff in fixed-wing airplanes. Ground effect can also be an important factor in go-arounds. If the go-around is made close to the ground, the airplane may be in the ground effect area. Pilots are often lulled into a sense of false security by the apparent "cushion of air" under the wings that initially assists in the transition from an approach descent to a climb. This "cushion of air," however, is imaginary. The apparent increase in airplane performance is, in fact, due to a reduction in induced drag in the ground effect area. It is "borrowed" performance that must be repaid when the airplane climbs out of the ground effect area. The pilot must factor in ground effect when initiating a go-around close to the ground. An attempt to climb prematurely may result in the airplane not being able to climb, or even maintain altitude at full power.

Common errors in the performance of go-arounds (rejected landings) are:

- Failure to recognize a condition that warrants a rejected landing.
- Indecision.
- Delay in initiating a go-round.
- Failure to apply maximum allowable power in a timely manner.
- Abrupt power application.
- Improper pitch attitude.
- Failure to configure the airplane appropriately.
- Attempting to climb out of ground effect prematurely.
- Failure to adequately compensate for torque/P-factor.

CROSSWIND APPROACH AND LANDING

Many runways or landing areas are such that landings must be made while the wind is blowing across rather than parallel to the landing direction. All pilots should be prepared to cope with these situations when they arise. The same basic principles and factors involved in a normal approach and landing apply to a crosswind approach and landing; therefore, only the additional procedures required for correcting for wind drift are discussed here.

Crosswind landings are a little more difficult to perform than crosswind takeoffs, mainly due to different problems involved in maintaining accurate control of the airplane while its speed is decreasing rather than increasing as on takeoff.

There are two usual methods of accomplishing a crosswind approach and landing—the crab method and the wing-low (sideslip) method. Although the crab method may be easier for the pilot to maintain during final approach, it requires a high degree of judgment and timing in removing the crab immediately prior to touchdown. The wing-low method is recommended in most cases, although a combination of both methods may be used.

CROSSWIND FINAL APPROACH

The crab method is executed by establishing a heading (crab) toward the wind with the wings level so that the airplane's ground track remains aligned with the centerline of the runway. [Figure 8-15] This crab angle is maintained until just prior to touchdown, when the longitudinal axis of the airplane must be aligned with the runway to avoid sideward contact of the wheels with the runway. If a long final approach is being flown, the pilot may use the crab method until just before the roundout is started and then smoothly change to the wing-low method for the remainder of the landing.

Figure 8-15. Crabbed approach.

The wing-low (sideslip) method will compensate for a crosswind from any angle, but more important, it enables the pilot to simultaneously keep the airplane's ground track and longitudinal axis aligned with the runway centerline throughout the final approach, roundout, touchdown, and after-landing roll. This prevents the airplane from touching down in a sideward motion and imposing damaging side loads on the landing gear.

To use the wing-low method, the pilot aligns the airplane's heading with the centerline of the runway, notes the rate and direction of drift, and then promptly applies drift correction by lowering the upwind wing. [Figure 8-16] The amount the wing must be lowered depends on the rate of drift. When the wing is lowered, the airplane will tend to turn in that direction. It is then necessary to simultaneously apply sufficient opposite rudder pressure to prevent the turn and keep the airplane's longitudinal axis aligned with the runway. In other words, the drift is controlled with aileron, and the heading with rudder. The airplane will now be sideslipping into the wind just enough that both the resultant flightpath and the ground track are aligned with the runway. If the crosswind diminishes, this crosswind correction is reduced accordingly, or the airplane will begin slipping away from the desired approach path. [Figure 8-17]

Figure 8-16. Sideslip approach.

To correct for strong crosswind, the slip into the wind is increased by lowering the upwind wing a considerable amount. As a consequence, this will result in a greater tendency of the airplane to turn. Since turning is not desired, considerable opposite rudder must be applied to keep the airplane's longitudinal axis aligned with the runway. In some airplanes, there may not be sufficient rudder travel available to compensate for the strong turning tendency caused by the steep bank. If the required bank is such that full opposite rudder will not prevent a turn, the wind is too strong to safely land the airplane on that particular runway with those wind conditions. Since the airplane's capability will be exceeded, it is imperative that the landing be made on

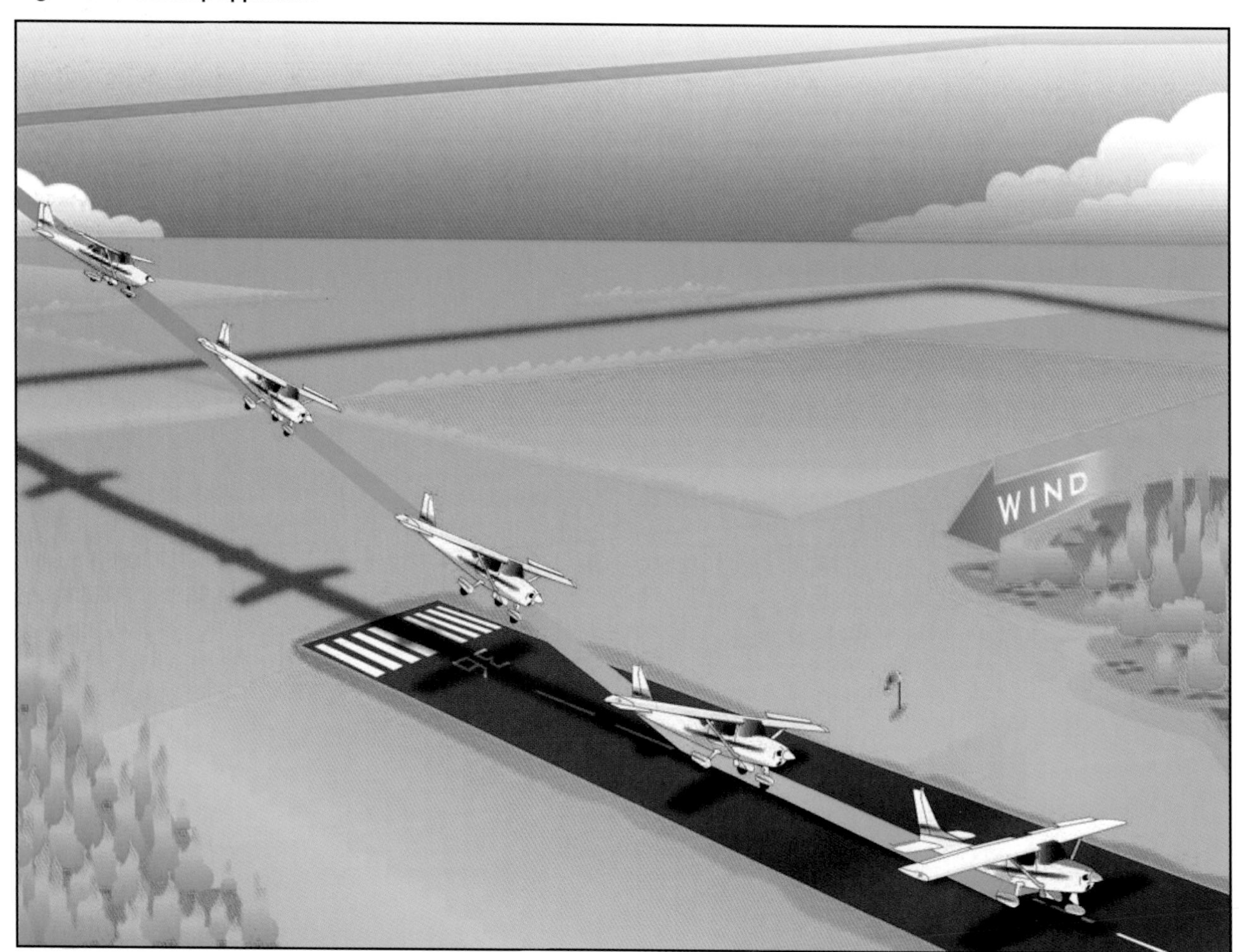

Figure 8-17. Crosswind approach and landing.

a more favorable runway either at that airport or at an alternate airport.

Flaps can and should be used during most approaches since they tend to have a stabilizing effect on the airplane. The degree to which flaps should be extended will vary with the airplane's handling characteristics, as well as the wind velocity.

CROSSWIND ROUNDOUT (FLARE)

Generally, the roundout can be made like a normal landing approach, but the application of a crosswind correction is continued as necessary to prevent drifting.

Since the airspeed decreases as the roundout progresses, the flight controls gradually become less effective. As a result, the crosswind correction being held will become inadequate. When using the wing-low method, it is necessary to gradually increase the deflection of the rudder and ailerons to maintain the proper amount of drift correction.

Do not level the wings; keep the upwind wing down throughout the roundout. If the wings are leveled, the airplane will begin drifting and the touchdown will occur while drifting. Remember, the primary objective is to land the airplane without subjecting it to any side loads that result from touching down while drifting.

CROSSWIND TOUCHDOWN

If the crab method of drift correction has been used throughout the final approach and roundout, the crab must be removed the instant before touchdown by applying rudder to align the airplane's longitudinal axis with its direction of movement. This requires timely and accurate action. Failure to accomplish this will result in severe side loads being imposed on the landing gear.

If the wing-low method is used, the crosswind correction (aileron into the wind and opposite rudder) should be maintained throughout the roundout, and the touchdown made on the upwind main wheel.

During gusty or high wind conditions, prompt adjustments must be made in the crosswind correction to assure that the airplane does not drift as the airplane touches down.

As the forward momentum decreases after initial contact, the weight of the airplane will cause the downwind main wheel to gradually settle onto the runway.

In those airplanes having nosewheel steering interconnected with the rudder, the nosewheel may not be aligned with the runway as the wheels touch down because opposite rudder is being held in the crosswind correction. To prevent swerving in the direction the nosewheel is offset, the corrective rudder pressure must be promptly relaxed just as the nosewheel touches down.

CROSSWIND AFTER-LANDING ROLL

Particularly during the after-landing roll, special attention must be given to maintaining directional control by the use of rudder or nosewheel steering, while keeping the upwind wing from rising by the use of aileron.

When an airplane is airborne, it moves with the air mass in which it is flying regardless of the airplane's heading and speed. When an airplane is on the ground, it is unable to move with the air mass (crosswind) because of the resistance created by ground friction on the wheels.

Characteristically, an airplane has a greater profile or side area, behind the main landing gear than forward of it does. With the main wheels acting as a pivot point and the greater surface area exposed to the crosswind behind that pivot point, the airplane will tend to turn or weathervane into the wind.

Wind acting on an airplane during crosswind landings is the result of two factors. One is the natural wind, which acts in the direction the air mass is traveling, while the other is induced by the movement of the airplane and acts parallel to the direction of movement. Consequently, a crosswind has a headwind component acting along the airplane's ground track and a crosswind component acting 90° to its track. The resultant or relative wind is somewhere between the two components. As the airplane's forward speed decreases during the after-landing roll, the headwind component decreases and the relative wind has more of a crosswind component. The greater the crosswind component, the more difficult it is to prevent weathervaning.

Retaining control on the ground is a critical part of the after-landing roll, because of the weathervaning effect of the wind on the airplane. Additionally, tire side load from runway contact while drifting frequently generates roll-overs in tricycle geared airplanes. The basic factors involved are cornering angle and side load.

Cornering angle is the angular difference between the heading of a tire and its path. Whenever a load bearing tire's path and heading diverge, a side load is created. It is accompanied by tire distortion. Although side load differs in varying tires and air pressures, it is completely independent of speed, and through a considerable range, is directional proportional to the cornering angle and the weight supported by the tire. As little as 10° of cornering angle will create a side load equal to half the supported weight; after 20° the side load does not increase with increasing cornering angle. For each high-wing, tricycle geared airplane, there is a cornering angle at which roll-over is inevitable. The roll-over axis being the line linking the nose and main wheels. At lesser angles, the roll-over may be avoided by use of ailerons, rudder, or steerable nosewheel *but not brakes.*

While the airplane is decelerating during the after-landing roll, more and more aileron is applied to keep the upwind wing from rising. Since the airplane is slowing down, there is less airflow around the ailerons and they become less effective. At the same time, the relative wind is becoming more of a crosswind and exerting a greater lifting force on the upwind wing. When the airplane is coming to a stop, the aileron control must be held fully toward the wind.

MAXIMUM SAFE CROSSWIND VELOCITIES

Takeoffs and landings in certain crosswind conditions are inadvisable or even dangerous. [Figure 8-18] If the crosswind is great enough to warrant an extreme drift correction, a hazardous landing condition may result. Therefore, the takeoff and landing capabilities with respect to the reported surface wind conditions and the available landing directions must be considered.

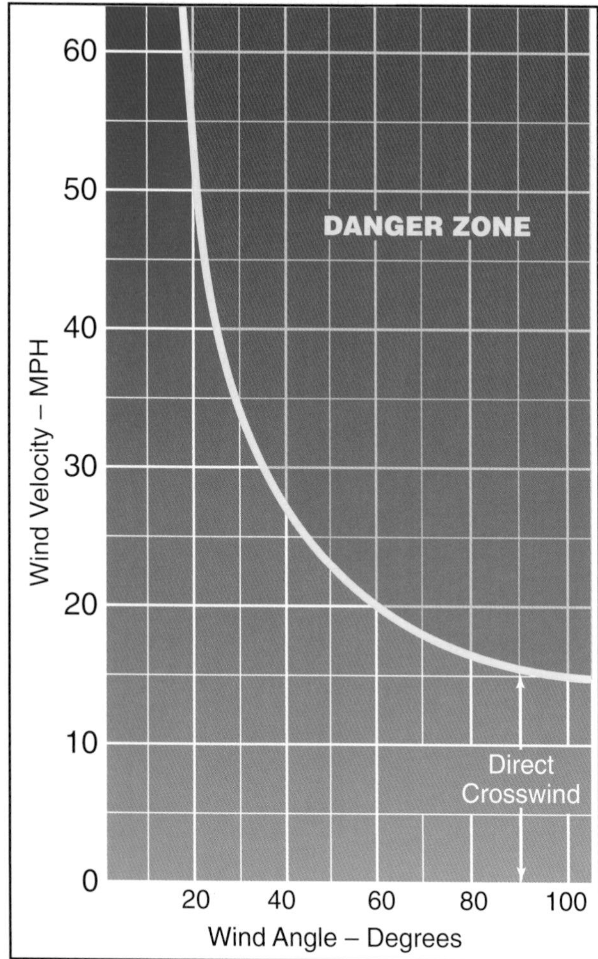

Figure 8-18. Crosswind chart.

Before an airplane is type certificated by the Federal Aviation Administration (FAA), it must be flight tested to meet certain requirements. Among these is the demonstration of being satisfactorily controllable with no exceptional degree of skill or alertness on the part of the pilot in 90° crosswinds up to a velocity equal to 0.2 $V_{SO.}$ This means a windspeed of two-tenths of the airplane's stalling speed with power off and landing gear/flaps down. Regulations require that the demonstrated crosswind velocity be included on a placard in airplanes certificated after May 3, 1962.

The headwind component and the crosswind component for a given situation can be determined by reference to a crosswind component chart. [Figure 8-19] It is imperative that pilots determine the maximum crosswind component of each airplane they fly, and avoid operations in wind conditions that exceed the capability of the airplane.

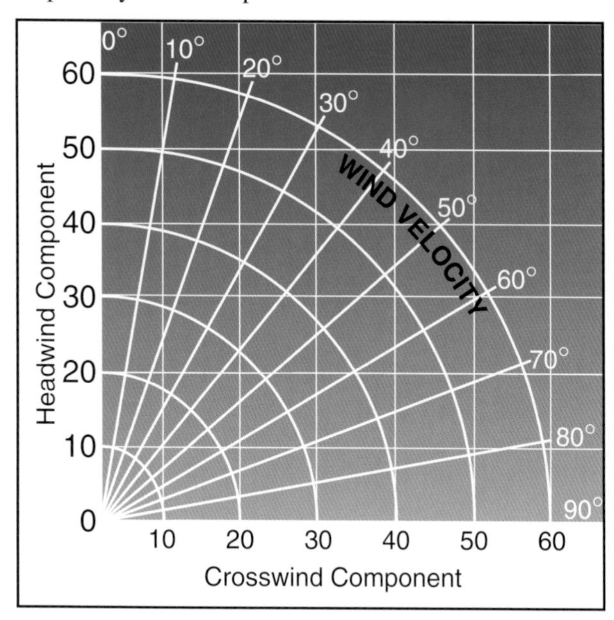

Figure 8-19. Crosswind component chart.

Common errors in the performance of crosswind approaches and landings are:

- Attempting to land in crosswinds that exceed the airplane's maximum demonstrated crosswind component.
- Inadequate compensation for wind drift on the turn from base leg to final approach, resulting in undershooting or overshooting.
- Inadequate compensation for wind drift on final approach.
- Unstabilized approach.
- Failure to compensate for increased drag during sideslip resulting in excessive sink rate and/or too low an airspeed.
- Touchdown while drifting.

- Excessive airspeed on touchdown.
- Failure to apply appropriate flight control inputs during rollout.
- Failure to maintain direction control on rollout.
- Excessive braking.

Turbulent Air Approach and Landing

Power-on approaches at an airspeed slightly above the normal approach speed should be used for landing in turbulent air. This provides for more positive control of the airplane when strong horizontal wind gusts, or up and down drafts, are experienced. Like other power-on approaches (when the pilot can vary the amount of power), a coordinated combination of both pitch and power adjustments is usually required. As in most other landing approaches, the proper approach attitude and airspeed require a minimum roundout and should result in little or no floating during the landing.

To maintain good control, the approach in turbulent air with gusty crosswind may require the use of partial wing flaps. With less than full flaps, the airplane will be in a higher pitch attitude. Thus, it will require less of a pitch change to establish the landing attitude, and the touchdown will be at a higher airspeed to ensure more positive control. The speed should not be so excessive that the airplane will float past the desired landing area.

One procedure is to use the normal approach speed plus one-half of the wind gust factors. If the normal speed is 70 knots, and the wind gusts increase 15 knots, airspeed of 77 knots is appropriate. In any case, the airspeed and the amount of flaps should be as the airplane manufacturer recommends.

An adequate amount of power should be used to maintain the proper airspeed and descent path throughout the approach, and the throttle retarded to idling position only after the main wheels contact the landing surface. Care must be exercised in closing the throttle before the pilot is ready for touchdown. In this situation, the sudden or premature closing of the throttle may cause a sudden increase in the descent rate that could result in a hard landing.

Landings from power approaches in turbulence should be such that the touchdown is made with the airplane in approximately level flight attitude. The pitch attitude at touchdown should be only enough to prevent the nosewheel from contacting the surface before the main wheels have touched the surface. After touchdown, the pilot should avoid the tendency to apply forward pressure on the yoke as this may result in **wheelbarrowing** and possible loss of control. The airplane should be allowed to decelerate normally, assisted by careful use of wheel brakes. Heavy braking should be avoided until the wings are devoid of lift and the airplane's full weight is resting on the landing gear.

Short-Field Approach and Landing

Short-field approaches and landings require the use of procedures for approaches and landings at fields with a relatively short landing area or where an approach is made over obstacles that limit the available landing area. [Figures 8-20 and 8-21] As in short-field takeoffs, it is one of the most critical of the maximum performance operations. It requires that the pilot fly the airplane at one of its crucial performance capabilities while close to the ground in order to safely land within confined areas. This low-speed type of power-on approach is closely related to the performance of flight at minimum controllable airspeeds.

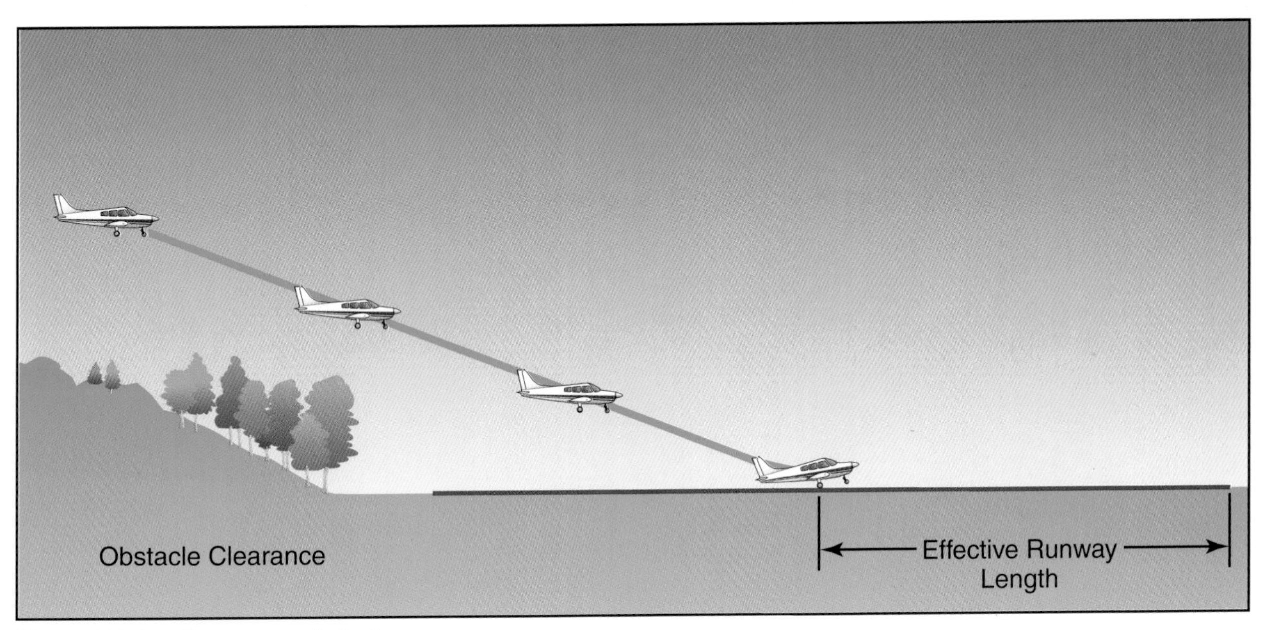

Figure 8-20. Landing over an obstacle.

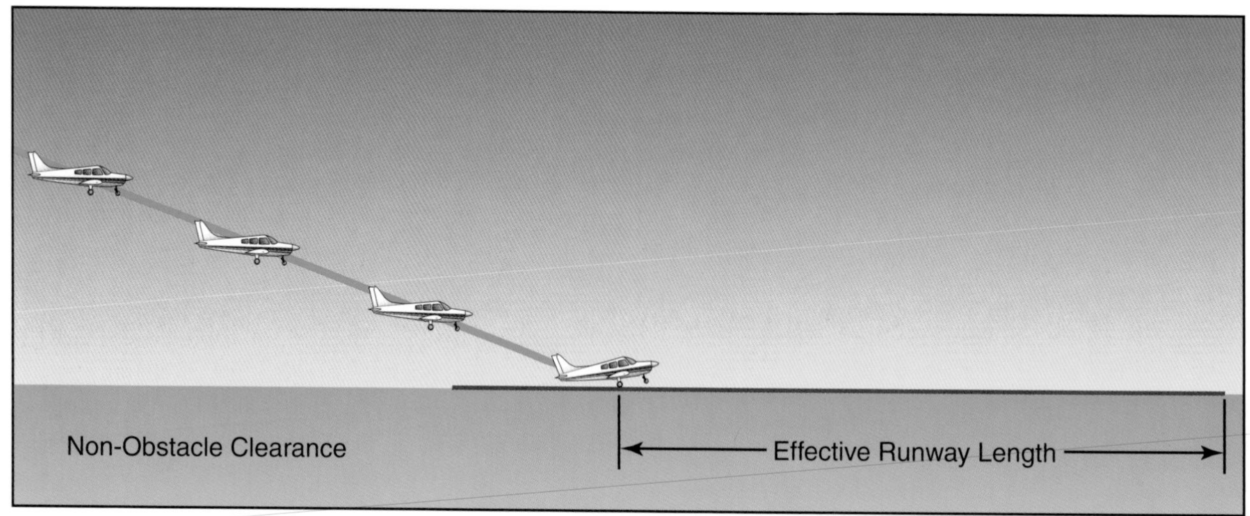

Figure 8-21. Landing on a short-field.

To land within a short-field or a confined area, the pilot must have precise, positive control of the rate of descent and airspeed to produce an approach that will clear any obstacles, result in little or no floating during the roundout, and permit the airplane to be stopped in the shortest possible distance.

The procedures for landing in a short-field or for landing approaches over obstacles, as recommended in the AFM/POH, should be used. A stabilized approach is essential. [Figures 8-22 and 8-23] These procedures generally involve the use of full flaps, and the final approach started from an altitude of at least 500 feet higher than the touchdown area. A wider than normal pattern should be used so that the airplane can be properly configured and trimmed. In the absence of the manufacturer's recommended approach speed, a speed of not more than 1.3 V_{SO} should be used. For example, in an airplane that stalls at 60 knots with power off, and flaps and landing gear extended, the approach speed should not be higher than 78 knots. In gusty air, no more than one-half the gust factor should be added. An excessive amount of airspeed could result in a touchdown too far from the runway threshold or an after-landing roll that exceeds the available landing area.

After the landing gear and full flaps have been extended, the pilot should simultaneously adjust the power and the pitch attitude to establish and maintain the proper descent angle and airspeed. A coordinated combination of both pitch and power adjustments is required. When this is done properly, very little change in the airplane's pitch attitude and power setting is necessary to make corrections in the angle of descent and airspeed.

The short-field approach and landing is in reality an accuracy approach to a spot landing. The procedures previously outlined in the section on the stabilized approach concept should be used. If it appears that the obstacle clearance is excessive and touchdown will occur well beyond the desired spot, leaving insufficient room to stop, power may be reduced while lowering the pitch attitude to steepen the descent path and increase the rate of descent. If it appears that the descent angle will not ensure safe clearance of obstacles, power should be increased while simultaneously raising the pitch attitude to shallow the descent path and decrease the rate of descent. Care must be taken to avoid an excessively low airspeed. If the speed is allowed to become too slow, an increase in pitch and application of full power

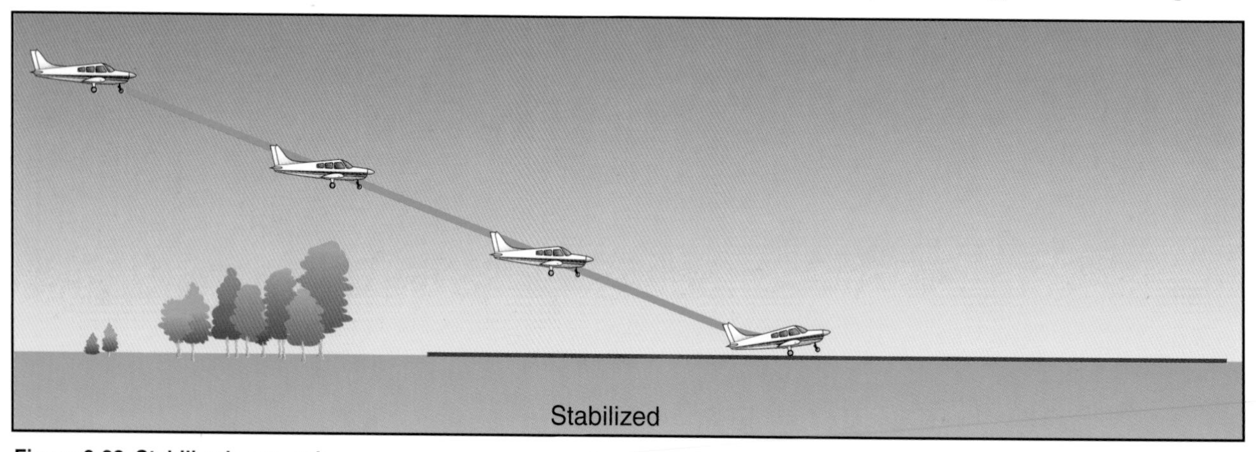

Figure 8-22. Stabilized approach.

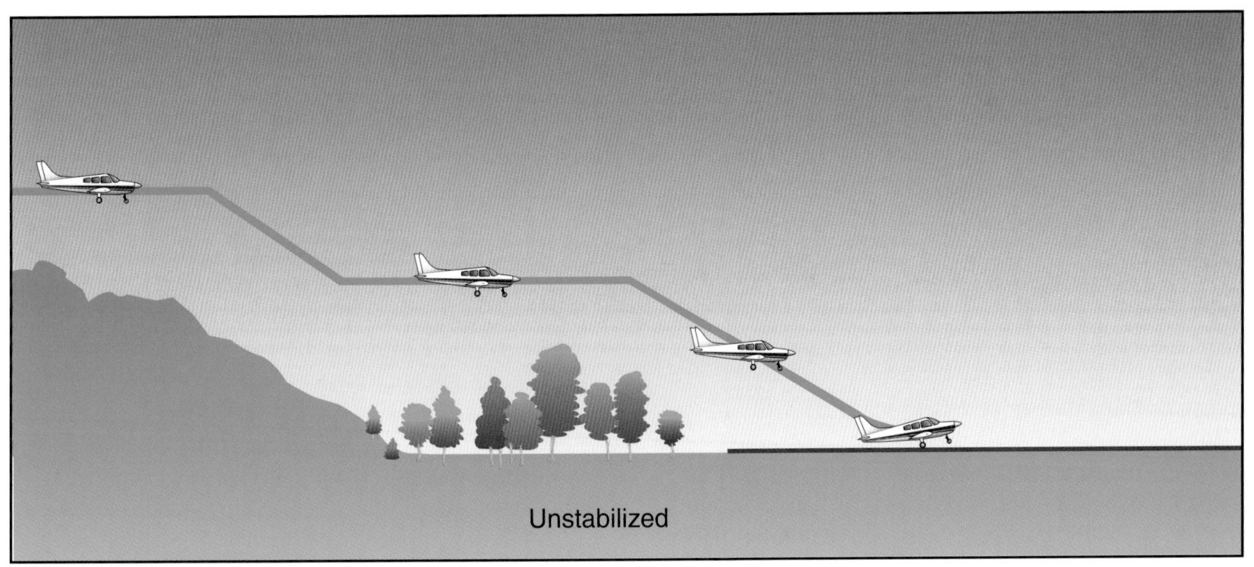

Figure 8-23. Unstabilized approach.

may only result in a further rate of descent. This occurs when the angle of attack is so great and creating so much drag that the maximum available power is insufficient to overcome it. This is generally referred to as operating in the **region of reversed command** or operating on the **back side of the power curve**.

Because the final approach over obstacles is made at a relatively steep approach angle and close to the airplane's stalling speed, the initiation of the roundout or flare must be judged accurately to avoid flying into the ground, or stalling prematurely and sinking rapidly. A lack of floating during the flare, with sufficient control to touch down properly, is one verification that the approach speed was correct.

Touchdown should occur at the minimum controllable airspeed with the airplane in approximately the pitch attitude that will result in a power-off stall when the throttle is closed. Care must be exercised to avoid closing the throttle too rapidly before the pilot is ready for touchdown, as closing the throttle may result in an immediate increase in the rate of descent and a hard landing.

Upon touchdown, the airplane should be held in this positive pitch attitude as long as the elevators remain effective. This will provide aerodynamic braking to assist in deceleration.

Immediately upon touchdown, and closing the throttle, appropriate braking should be applied to minimize the after-landing roll. The airplane should be stopped within the shortest possible distance consistent with safety and controllability. If the proper approach speed has been maintained, resulting in minimum float during the roundout, and the touchdown made at minimum control speed, minimum braking will be required.

Common errors in the performance of short-field approaches and landings are:

- Failure to allow enough room on final to set up the approach, necessitating an overly steep approach and high sink rate.
- Unstabilized approach.
- Undue delay in initiating glidepath corrections.
- Too low an airspeed on final resulting in inability to flare properly and landing hard.
- Too high an airspeed resulting in floating on roundout.
- Prematurely reducing power to idle on roundout resulting in hard landing.
- Touchdown with excessive airspeed.
- Excessive and/or unnecessary braking after touchdown.
- Failure to maintain directional control.

SOFT-FIELD APPROACH AND LANDING

Landing on fields that are rough or have soft surfaces, such as snow, sand, mud, or tall grass requires unique procedures. When landing on such surfaces, the objective is to touch down as smoothly as possible, and at the slowest possible landing speed. The pilot must control the airplane in a manner that the wings support the weight of the airplane as long as practical, to minimize drag and stresses imposed on the landing gear by the rough or soft surface.

The approach for the soft-field landing is similar to the normal approach used for operating into long, firm landing areas. The major difference between the two is that, during the soft-field landing, the airplane is held 1 to 2 feet off the surface in ground effect as long as possible. This permits a more gradual dissipation of forward speed to allow the wheels to touch down gently at minimum speed. This technique minimizes the nose-over forces that suddenly affect the airplane at the moment of touchdown. Power can be used throughout the level-off and touchdown to ensure touchdown at the slowest possible airspeed, and the airplane should be *flown* onto the ground with the weight fully supported by the wings. [Figure 8-24]

The use of flaps during soft-field landings will aid in touching down at minimum speed and is recommended whenever practical. In low-wing airplanes, the flaps may suffer damage from mud, stones, or slush thrown up by the wheels. If flaps are used, it is generally inadvisable to retract them during the after-landing roll because the need for flap retraction is usually less important than the need for total concentration on maintaining full control of the airplane.

The final approach airspeed used for short-field landings is equally appropriate to soft-field landings. The use of higher approach speeds may result in excessive float in ground effect, and floating makes a smooth, controlled touchdown even more difficult. There is, however, no reason for a steep angle of descent unless obstacles are present in the approach path.

Touchdown on a soft or rough field should be made at the lowest possible airspeed with the airplane in a nose-high pitch attitude. In nosewheel-type airplanes, after the main wheels touch the surface, the pilot should hold sufficient back-elevator pressure to keep the nosewheel off the surface. Using back-elevator pressure and engine power, the pilot can control the rate at which the weight of the airplane is transferred from the wings to the wheels.

Field conditions may warrant that the pilot maintain a flight condition in which the main wheels are just touching the surface but the weight of the airplane is still being supported by the wings, until a suitable taxi surface is reached. At any time during this transition phase, before the weight of the airplane is being supported by the wheels, and before the nosewheel is on the surface, the pilot should be able to apply full power and perform a safe takeoff (obstacle clearance and field length permitting) should the pilot elect to abandon the landing. Once committed to a landing, the pilot should gently lower the nosewheel to the surface. A slight addition of power usually will aid in easing the nosewheel down.

The use of brakes on a soft field is not needed and should be avoided as this may tend to impose a heavy load on the nose gear due to premature or hard contact with the landing surface, causing the nosewheel to dig in. The soft or rough surface itself will provide sufficient reduction in the airplane's forward speed. Often it will be found that upon landing on a very soft field, the pilot will need to increase power to keep the airplane moving and from becoming stuck in the soft surface.

Common errors in the performance of soft-field approaches and landings are:

- Excessive descent rate on final approach.
- Excessive airspeed on final approach.
- Unstabilized approach.
- Roundout too high above the runway surface.
- Poor power management during roundout and touchdown.
- Hard touchdown.
- Inadequate control of the airplane weight transfer from wings to wheels after touchdown.
- Allowing the nosewheel to "fall" to the runway after touchdown rather than controlling its descent.

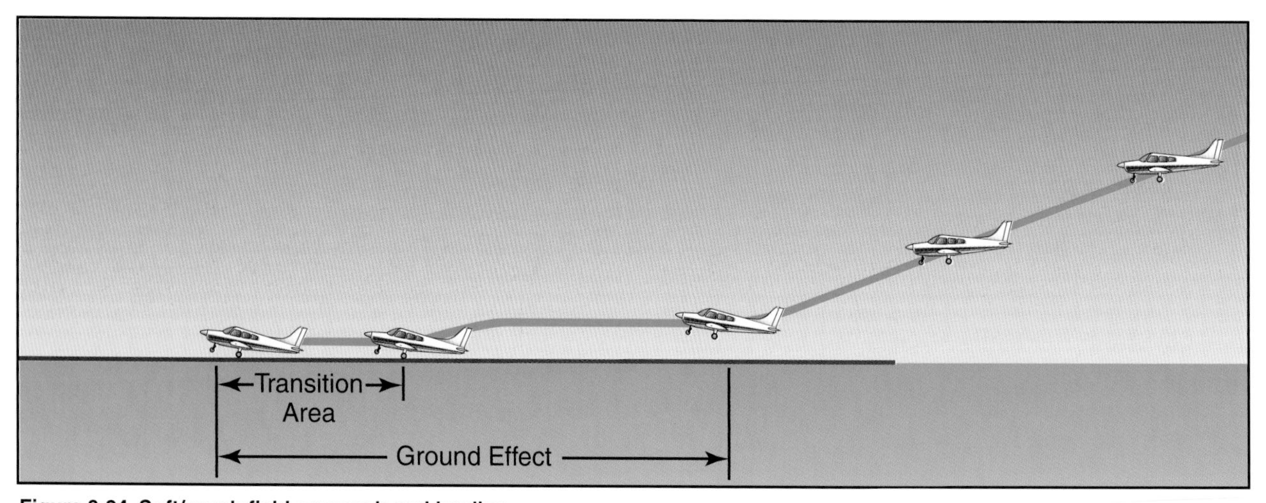

Figure 8-24. Soft/rough field approach and landing.

POWER-OFF ACCURACY APPROACHES

Power-off accuracy approaches are approaches and landings made by gliding with the engine idling, through a specific pattern to a touchdown beyond and within 200 feet of a designated line or mark on the runway. The objective is to instill in the pilot the judgment and procedures necessary for accurately flying the airplane, without power, to a safe landing.

The ability to estimate the distance an airplane will glide to a landing is the real basis of all power-off accuracy approaches and landings. This will largely determine the amount of maneuvering that may be done from a given altitude. In addition to the ability to estimate distance, it requires the ability to maintain the proper glide while maneuvering the airplane.

With experience and practice, altitudes up to approximately 1,000 feet can be estimated with fair accuracy, while above this level the accuracy in judgment of height above the ground decreases, since all features tend to merge. The best aid in perfecting the ability to judge height above this altitude is through the indications of the altimeter and associating them with the general appearance of the Earth.

The judgment of altitude in feet, hundreds of feet, or thousands of feet is not as important as the ability to estimate gliding angle and its resultant distance. The pilot who knows the normal glide angle of the airplane can estimate with reasonable accuracy, the approximate spot along a given ground path at which the airplane will land, regardless of altitude. The pilot, who also has the ability to accurately estimate altitude, can judge how much maneuvering is possible during the glide, which is important to the choice of landing areas in an actual emergency.

The objective of a good final approach is to descend at an angle that will permit the airplane to reach the desired landing area, and at an airspeed that will result in minimum floating just before touchdown. To accomplish this, it is essential that both the descent angle and the airspeed be accurately controlled.

Unlike a normal approach when the power setting is variable, on a power-off approach the power is fixed at the idle setting. Pitch attitude is adjusted to control the airspeed. This will also change the glide or descent angle. By lowering the nose to keep the approach airspeed constant, the descent angle will steepen. If the airspeed is too high, raise the nose, and when the airspeed is too low, lower the nose. If the pitch attitude is raised too high, the airplane will settle rapidly due to a slow airspeed and insufficient lift. For this reason, never try to stretch a glide to reach the desired landing spot.

Uniform approach patterns such as the 90°, 180°, or 360° power-off approaches are described further in this chapter. Practice in these approaches provides the pilot with a basis on which to develop judgment in gliding distance and in planning an approach.

The basic procedure in these approaches involves closing the throttle at a given altitude, and gliding to a key position. This position, like the pattern itself, must not be allowed to become the primary objective; it is merely a convenient point in the air from which the pilot can judge whether the glide will safely terminate at the desired spot. The selected key position should be one that is appropriate for the available altitude and the wind condition. From the key position, the pilot must constantly evaluate the situation.

It must be emphasized that, although accurate spot touchdowns are important, safe and properly executed approaches and landings are vital. The pilot must never sacrifice a good approach or landing just to land on the desired spot.

90° POWER-OFF APPROACH

The 90° power-off approach is made from a base leg and requires only a 90° turn onto the final approach. The approach path may be varied by positioning the base leg closer to or farther out from the approach end of the runway according to wind conditions. [Figure 8-25]

The glide from the key position on the base leg through the 90° turn to the final approach is the final part of all accuracy landing maneuvers.

The 90° power-off approach usually begins from a rectangular pattern at approximately 1,000 feet above the ground or at normal traffic pattern altitude. The airplane should be flown onto a downwind leg at the same distance from the landing surface as in a normal traffic pattern. The before landing checklist should be completed on the downwind leg, including extension of the landing gear if the airplane is equipped with retractable gear.

After a medium-banked turn onto the base leg is completed, the throttle should be retarded slightly and the airspeed allowed to decrease to the normal base-leg speed. [Figure 8-26] On the base leg, the airspeed, wind drift correction, and altitude should be maintained while proceeding to the 45° key position. At this position, the intended landing spot will appear to be on a 45° angle from the airplane's nose.

The pilot can determine the strength and direction of the wind from the amount of crab necessary to hold the desired ground track on the base leg. This will help in planning the turn onto the final approach and in lowering the correct amount of flaps.

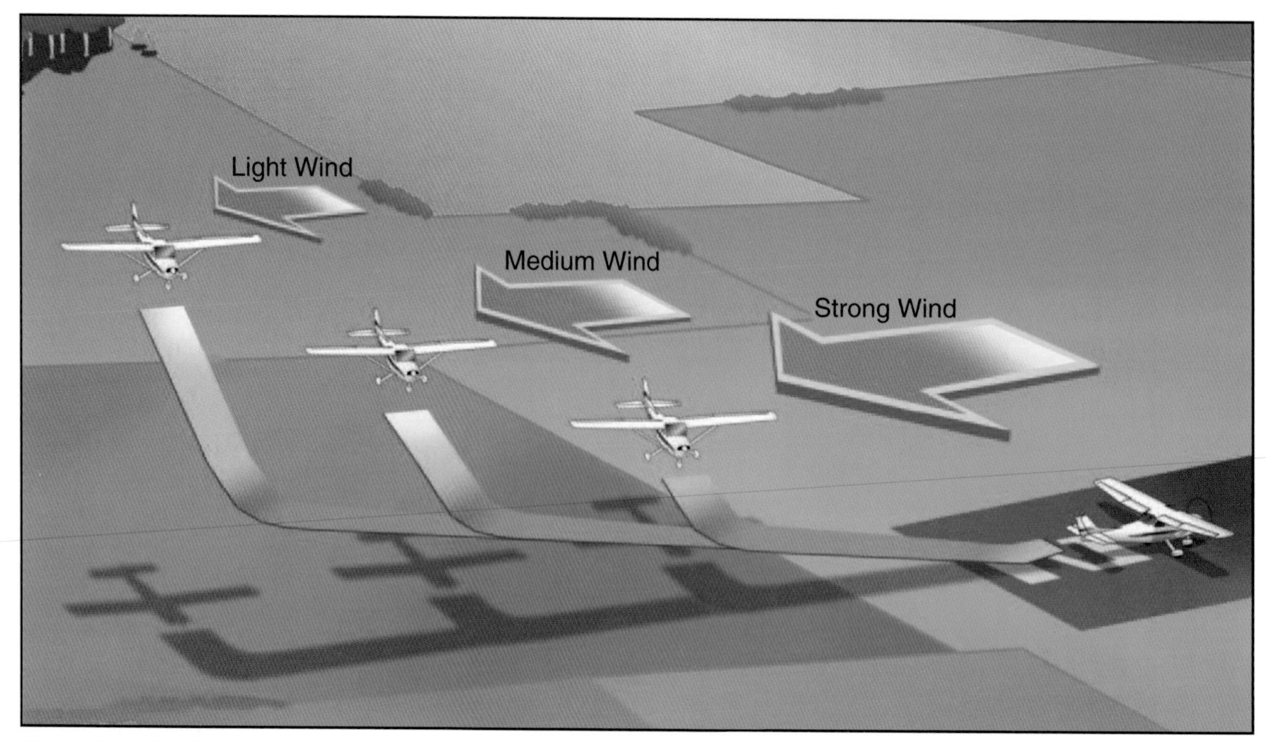

Figure 8-25. Plan the base leg for wind conditions.

At the 45° key position, the throttle should be closed completely, the propeller control (if equipped) advanced to the full increase r.p.m. position, and altitude maintained until the airspeed decreases to the manufacturer's recommended glide speed. In the absence of a recommended speed, use 1.4 V_{SO}. When this airspeed is attained, the nose should be lowered to maintain the gliding speed and the controls retrimmed.

The base-to-final turn should be planned and accomplished so that upon rolling out of the turn the airplane will be aligned with the runway centerline. When on final approach, the wing flaps are lowered and the pitch attitude adjusted, as necessary, to establish the proper descent angle and airspeed (1.3 V_{SO}), then the controls retrimmed. Slight adjustments in pitch attitude or flaps setting may be necessary to control the glide

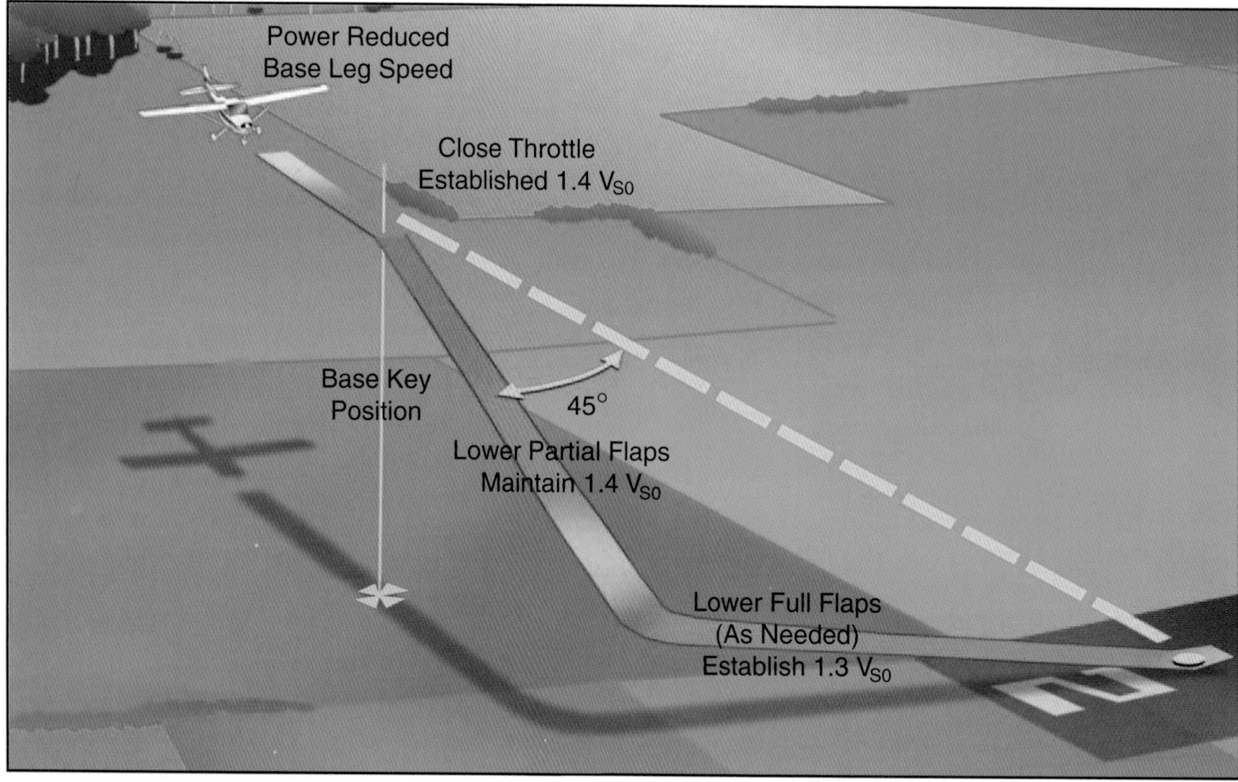

Figure 8-26. 90° power-off approach.

angle and airspeed. However, NEVER TRY TO STRETCH THE GLIDE OR RETRACT THE FLAPS to reach the desired landing spot. The final approach may be made with or without the use of slips.

After the final approach glide has been established, full attention is then given to making a good, safe landing rather than concentrating on the selected landing spot. The base-leg position and the flap setting already determined the probability of landing on the spot. In any event, it is better to execute a good landing 200 feet from the spot than to make a poor landing precisely on the spot.

180° POWER-OFF APPROACH

The 180° power-off approach is executed by gliding with the power off from a given point on a downwind leg to a preselected landing spot. [Figure 8-27] It is an extension of the principles involved in the 90° power-off approach just described. Its objective is to further develop judgment in estimating distances and glide ratios, in that the airplane is flown without power from a higher altitude and through a 90° turn to reach the base-leg position at a proper altitude for executing the 90° approach.

The 180° power-off approach requires more planning and judgment than the 90° power-off approach. In the execution of 180° power-off approaches, the airplane is flown on a downwind heading parallel to the landing runway. The altitude from which this type of approach should be started will vary with the type of airplane, but it should usually not exceed 1,000 feet above the ground, except with large airplanes. Greater accuracy in judgment and maneuvering is required at higher altitudes.

When abreast of or opposite the desired landing spot, the throttle should be closed and altitude maintained while decelerating to the manufacturer's recommended glide speed, or 1.4 V_{SO}. The point at which the throttle is closed is the downwind key position.

The turn from the downwind leg to the base leg should be a uniform turn with a medium or slightly steeper bank. The degree of bank and amount of this initial turn will depend upon the glide angle of the airplane and the velocity of the wind. Again, the base leg should be positioned as needed for the altitude, or wind condition. Position the base leg to conserve or dissipate altitude so as to reach the desired landing spot.

The turn onto the base leg should be made at an altitude high enough and close enough to permit the airplane to glide to what would normally be the base key position in a 90° power-off approach.

Although the key position is important, it must not be overemphasized nor considered as a fixed point on the ground. Many inexperienced pilots may gain a conception of it as a particular landmark, such as a tree, crossroad, or other visual reference, to be reached at a certain altitude. This will result in a mechanical conception and leave the pilot at a total

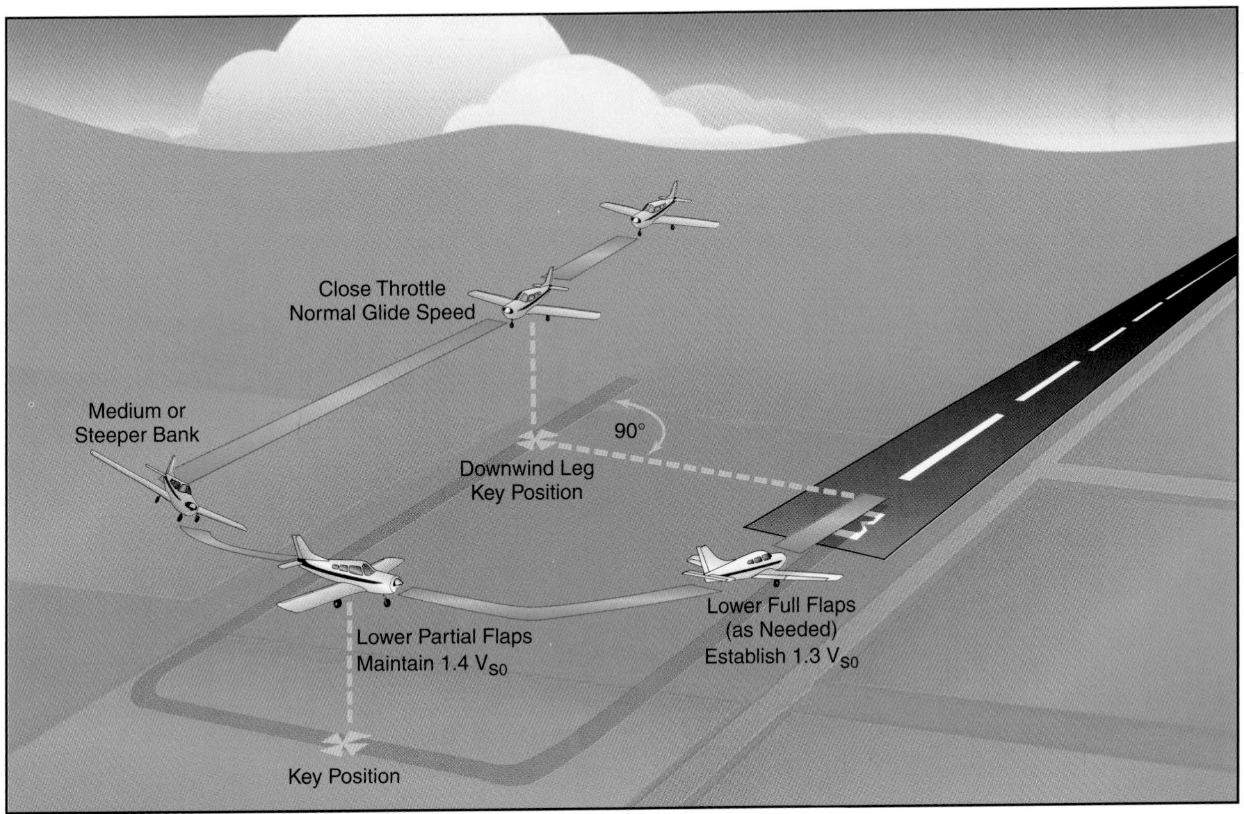

Figure 8-27. 180° power-off approach.

loss any time such objects are not present. Both altitude and geographical location should be varied as much as is practical to eliminate any such conception. After reaching the base key position, the approach and landing are the same as in the 90° power-off approach.

360° POWER-OFF APPROACH

The 360° power-off approach is one in which the airplane glides through a 360° change of direction to the preselected landing spot. The entire pattern is designed to be circular, but the turn may be shallowed, steepened, or discontinued at any point to adjust the accuracy of the flightpath.

The 360° approach is started from a position over the approach end of the landing runway or slightly to the side of it, with the airplane headed in the proposed landing direction and the landing gear and flaps retracted. [Figure 8-28]

It is usually initiated from approximately 2,000 feet or more above the ground—where the wind may vary significantly from that at lower altitudes. This must be taken into account when maneuvering the airplane to a point from which a 90° or 180° power-off approach can be completed.

After the throttle is closed over the intended point of landing, the proper glide speed should immediately be established, and a medium-banked turn made in the desired direction so as to arrive at the downwind key position opposite the intended landing spot. At or just beyond the downwind key position, the landing gear may be extended if the airplane is equipped with retractable gear. The altitude at the downwind key position should be approximately 1,000 to 1,200 feet above the ground.

After reaching that point, the turn should be continued to arrive at a base-leg key position, at an altitude of about 800 feet above the terrain. Flaps may be used at this position, as necessary, but full flaps should not be used until established on the final approach.

The angle of bank can be varied as needed throughout the pattern to correct for wind conditions and to align the airplane with the final approach. The turn-to-final should be completed at a minimum altitude of 300 feet above the terrain.

Common errors in the performance of power-off accuracy approaches are:

- Downwind leg too far from the runway/landing area.
- Overextension of downwind leg resulting from tailwind.
- Inadequate compensation for wind drift on base leg.
- Skidding turns in an effort to increase gliding distance.

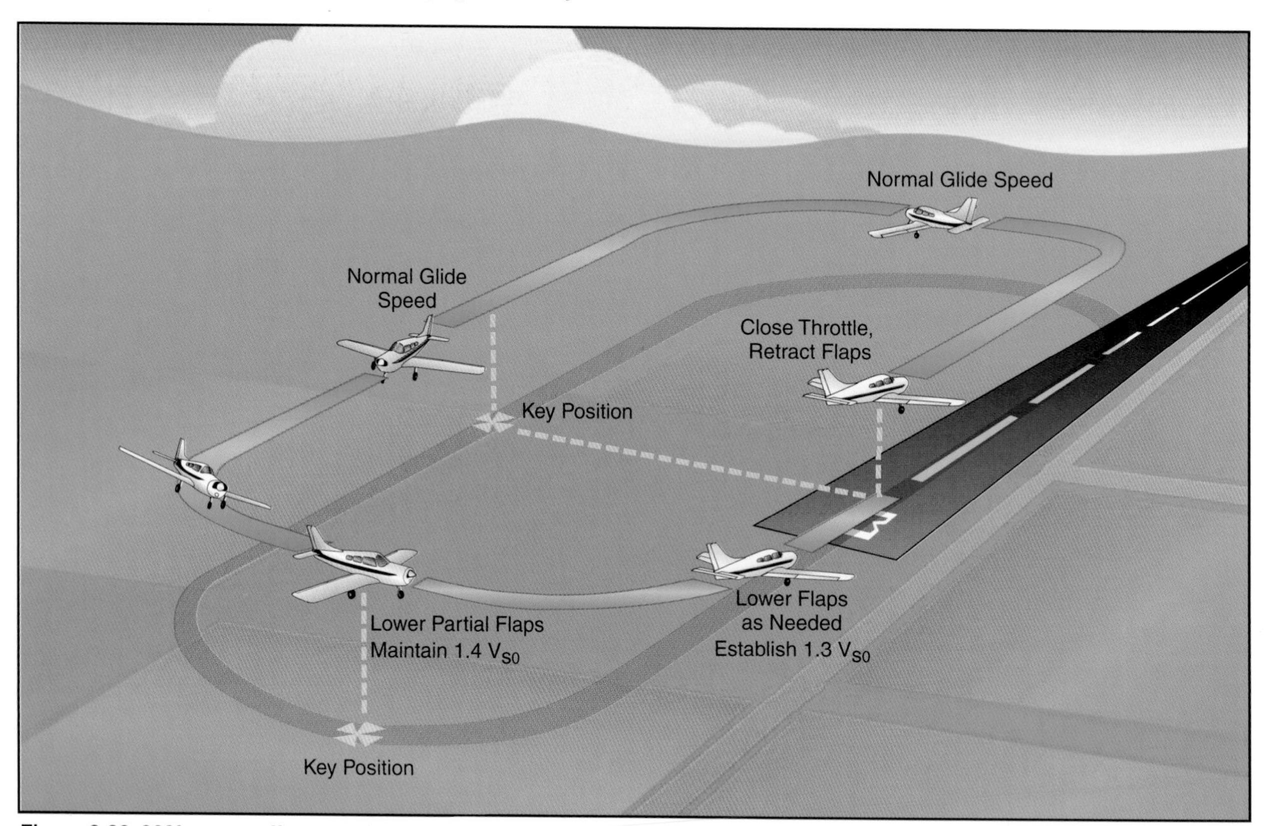

Figure 8-28. 360° power-off approach.

- Failure to lower landing gear in retractable gear airplanes.
- Attempting to "stretch" the glide during undershoot.
- Premature flap extension/landing gear extension.
- Use of throttle to increase the glide instead of merely clearing the engine.
- Forcing the airplane onto the runway in order to avoid overshooting the designated landing spot.

EMERGENCY APPROACHES AND LANDINGS (SIMULATED)

From time to time on dual flights, the instructor should give simulated emergency landings by retarding the throttle and calling "simulated emergency landing." The objective of these simulated emergency landings is to develop the pilot's accuracy, judgment, planning, procedures, and confidence when little or no power is available.

A simulated emergency landing may be given with the airplane in any configuration. When the instructor calls "simulated emergency landing," the pilot should immediately establish a glide attitude and ensure that the flaps and landing gear are in the proper configuration for the existing situation. When the proper glide speed is attained, the nose should then be lowered and the airplane trimmed to maintain that speed.

A constant gliding speed should be maintained because variations of gliding speed nullify all attempts at accuracy in judgment of gliding distance and the landing spot. The many variables, such as altitude, obstruction, wind direction, landing direction, landing surface and gradient, and landing distance requirements of the airplane will determine the pattern and approach procedures to use.

Utilizing any combination of normal gliding maneuvers, from wings level to spirals, the pilot should eventually arrive at the normal key position at a normal traffic pattern altitude for the selected landing area. From this point on, the approach will be as nearly as possible a normal power-off approach. [Figure 8-29]

With the greater choice of fields afforded by higher altitudes, the inexperienced pilot may be inclined to delay making a decision, and with considerable altitude in which to maneuver, errors in maneuvering and estimation of glide distance may develop.

All pilots should learn to determine the wind direction and estimate its speed from the windsock at the airport, smoke from factories or houses, dust, brush fires, and windmills.

Once a field has been selected, the student pilot should always be required to indicate it to the instructor. Normally, the student should be required to plan and fly a pattern for landing on the field first elected until the instructor terminates the simulated emergency

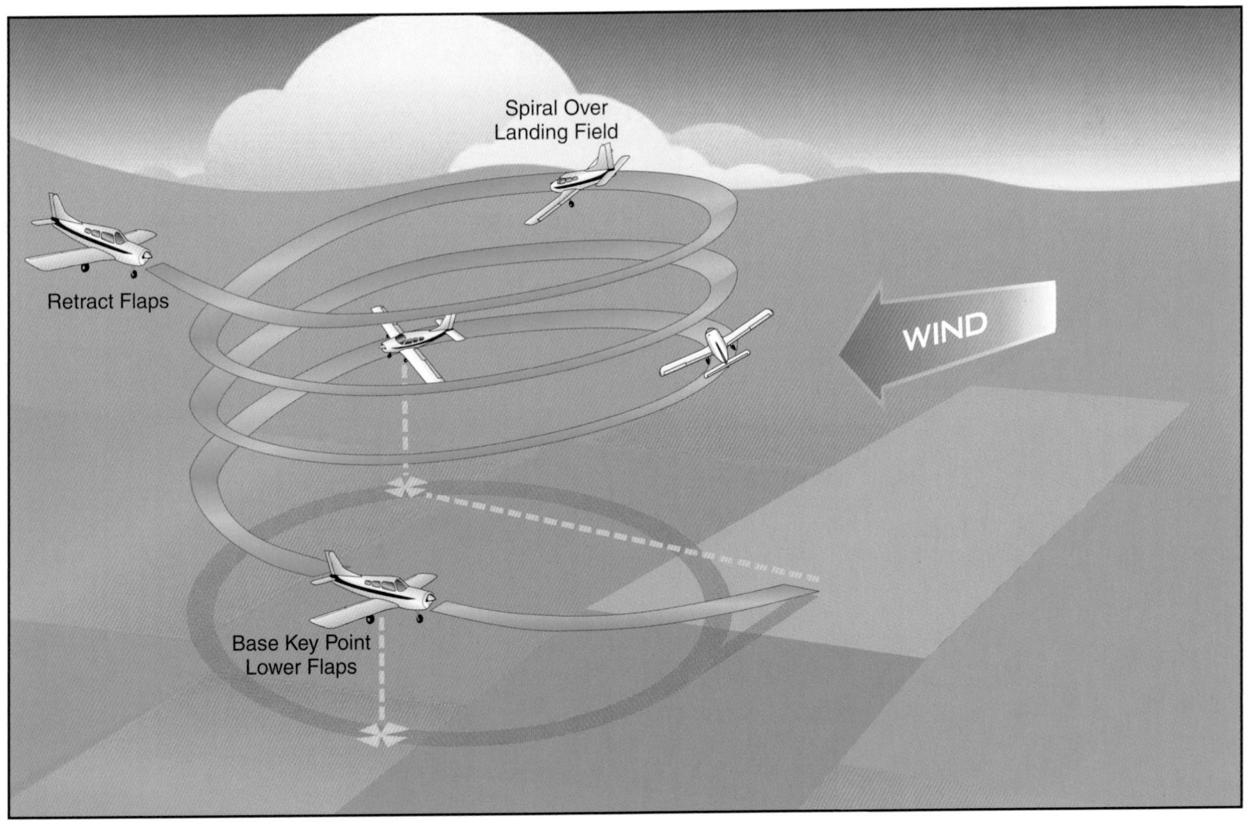

Figure 8-29. Remain over intended landing area.

landing. This will give the instructor an opportunity to explain and correct any errors; it will also give the student an opportunity to see the results of the errors. However, if the student realizes during the approach that a poor field has been selected—one that would obviously result in disaster if a landing were to be made—and there is a more advantageous field within gliding distance, a change to the better field should be permitted. The hazards involved in these last-minute decisions, such as excessive maneuvering at very low altitudes, should be thoroughly explained by the instructor.

Slipping the airplane, using flaps, varying the position of the base leg, and varying the turn onto final approach should be stressed as ways of correcting for misjudgment of altitude and glide angle.

Eagerness to get down is one of the most common faults of inexperienced pilots during simulated emergency landings. In giving way to this, they forget about speed and arrive at the edge of the field with too much speed to permit a safe landing. Too much speed may be just as dangerous as too little; it results in excessive floating and overshooting the desired landing spot. It should be impressed on the students that they cannot dive at a field and expect to land on it.

During all simulated emergency landings, the engine should be kept warm and cleared. During a simulated emergency landing, either the instructor or the student should have complete control of the throttle. There should be no doubt as to who has control since many near accidents have occurred from such misunderstandings.

Every simulated emergency landing approach should be terminated as soon as it can be determined whether a safe landing could have been made. In no case should it be continued to a point where it creates an undue hazard or an annoyance to persons or property on the ground.

In addition to flying the airplane from the point of simulated engine failure to where a reasonable safe landing could be made, the student should also be taught certain emergency cockpit procedures. The habit of performing these cockpit procedures should be developed to such an extent that, when an engine failure actually occurs, the student will check the critical items that would be necessary to get the engine operating again while selecting a field and planning an approach. Combining the two operations—accomplishing emergency procedures and planning

65 KIAS (flaps DOWN).
2. Mixture -- IDLE CUT-OFF.
3. Fuel Selector Valve -- OFF.
4. Ignition Switch -- OFF.
5. Wing Flaps -- AS REQUIRED.
6. Master Switch -- OFF.

ENGINE FAILURE DURING FLIGHT (RESTART PROCEDURES)

1. **Airspeed -- 70 KIAS.**
2. **Carburetor Heat -- ON.**
3. **Fuel Selector Valve -- BOTH.**
4. Mixture -- RICH.
5. Ignition Switch -- BOTH (or START if propeller is stopped).
6. Primer -- IN and LOCKED.

FORCED LANDINGS

EMERGENCY LANDING WITHOUT ENGINE POWER

1. Airspeed -- 70 KIAS (flaps UP).
   65 KIAS (flaps DOWN).
2. Mixture -- IDLE CUT-OFF.
3. Fuel Selector Valve -- OFF.
4. Ignition Switch -- OFF.
5. Wing Flaps -- AS REQUIRED (30° recommended).
6. Master Switch -- OFF.
7. Doors -- UNLATCH PRIOR TO TOUCHDOWN.
8. Touchdown -- SLIGHTLY TAIL LOW.
9. Brakes -- APPLY HEAVILY.

Figure 8-30. Sample emergency checklist.

and flying the approach—will be difficult for the student during the early training in emergency landings.

There are definite steps and procedures to be followed in a simulated emergency landing. Although they may differ somewhat from the procedures used in an actual emergency, they should be learned thoroughly by the student, and each step called out to the instructor. The use of a checklist is strongly recommended. Most airplane manufacturers provide a checklist of the appropriate items. [Figure 8-30]

Critical items to be checked should include the position of the fuel tank selector, the quantity of fuel in the tank selected, the fuel pressure gauge to see if the electric fuel pump is needed, the position of the mixture control, the position of the magneto switch, and the use of carburetor heat. Many actual emergency landings have been made and later found to be the result of the fuel selector valve being positioned to an empty tank while the other tank had plenty of fuel. It may be wise to change the position of the fuel selector valve even though the fuel gauge indicates fuel in all tanks because fuel gauges can be inaccurate. Many actual emergency landings could have been prevented if the pilots had developed the habit of checking these critical items during flight training to the extent that it carried over into later flying.

Instruction in emergency procedures should not be limited to simulated emergency landings caused by power failures. Other emergencies associated with the operation of the airplane should be explained, demonstrated, and practiced if practicable. Among these emergencies are such occurrences as fire in flight, electrical or hydraulic system malfunctions, unexpected severe weather conditions, engine overheating, imminent fuel exhaustion, and the emergency operation of airplane systems and equipment.

FAULTY APPROACHES AND LANDINGS

LOW FINAL APPROACH

When the base leg is too low, insufficient power is used, landing flaps are extended prematurely, or the velocity of the wind is misjudged, sufficient altitude may be lost, which will cause the airplane to be well below the proper final approach path. In such a situation, the pilot would have to apply considerable power to fly the airplane (at an excessively low altitude) up to the runway threshold. When it is realized the runway will not be reached unless appropriate action is taken, power must be applied immediately to maintain the airspeed while the pitch attitude is raised to increase lift and stop the descent. When the proper approach path has been intercepted, the correct approach attitude should be reestablished and the power reduced and a stabilized approach maintained. [Figure 8-31] DO NOT increase the pitch attitude without increasing the power, since the airplane will decelerate rapidly and may approach the critical angle of attack and stall. DO NOT retract the flaps; this will suddenly decrease lift and cause the airplane to sink more rapidly. If there is any doubt about the approach being safely completed, it is advisable to EXECUTE AN IMMEDIATE GO-AROUND.

HIGH FINAL APPROACH

When the final approach is too high, lower the flaps as required. Further reduction in power may be necessary, while lowering the nose simultaneously to maintain approach airspeed and steepen the approach path. [Figure 8-32] When the proper approach path has been intercepted, adjust the power as required to maintain a

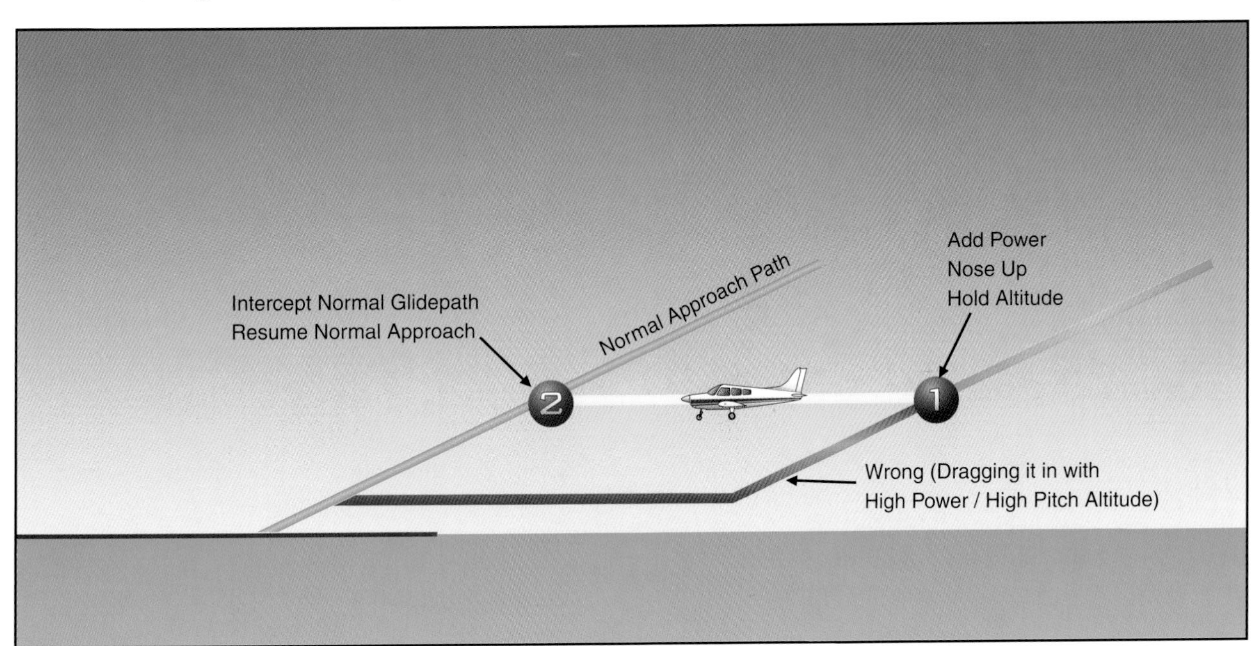

Figure 8-31. Right and wrong methods of correction for low final approach.

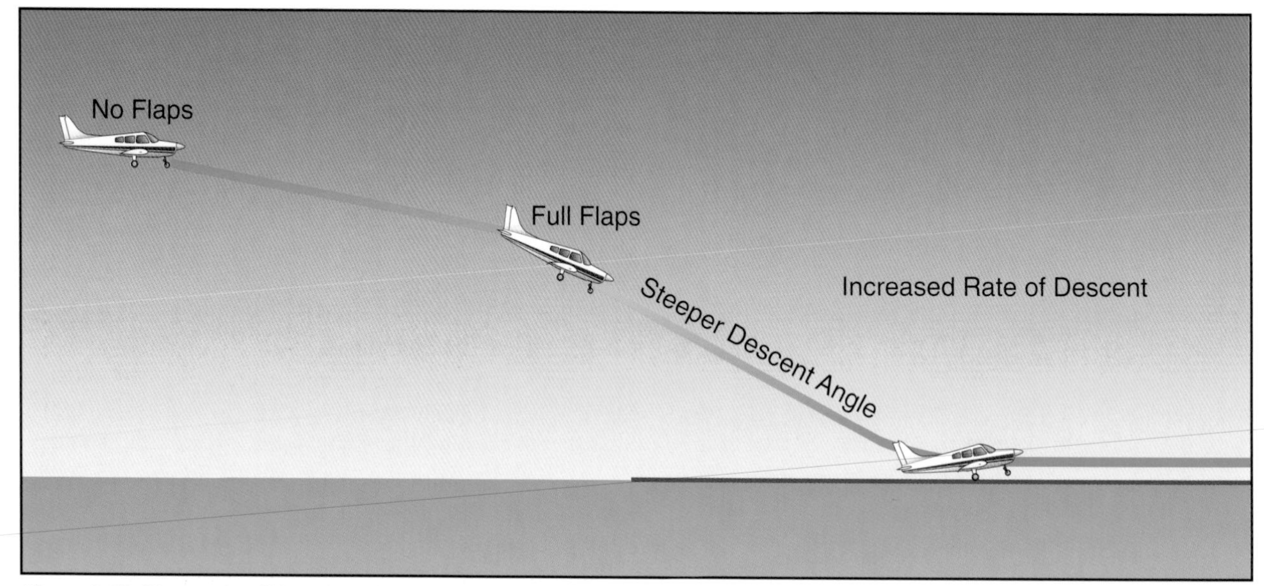

Figure 8-32. Change in glidepath and increase in descent rate for high final approach.

stabilized approach. When steepening the approach path, however, care must be taken that the descent does not result in an excessively high sink rate. If a high sink rate is continued close to the surface, it may be difficult to slow to a proper rate prior to ground contact. Any sink rate in excess of 800 - 1,000 feet per minute is considered excessive. A go-around should be initiated if the sink rate becomes excessive.

SLOW FINAL APPROACH

When the airplane is flown at a slower-than-normal airspeed on the final approach, the pilot's judgment of the rate of sink (descent) and the height of roundout will be difficult. During an excessively slow approach, the wing is operating near the critical angle of attack and, depending on the pitch attitude changes and control usage, the airplane may stall or sink rapidly, contacting the ground with a hard impact.

Whenever a slow-speed approach is noted, the pilot should apply power to accelerate the airplane and increase the lift to reduce the sink rate and to prevent a stall. This should be done while still at a high enough altitude to reestablish the correct approach airspeed and attitude. If too slow and too low, it is best to EXECUTE A GO-AROUND.

USE OF POWER

Power can be used effectively during the approach and roundout to compensate for errors in judgment. Power can be added to accelerate the airplane to increase lift without increasing the angle of attack; thus, the descent can be slowed to an acceptable rate. If the proper landing attitude has been attained and the airplane is only slightly high, the landing attitude should be held constant and sufficient power applied to help ease the airplane onto the ground. After the airplane has touched down, it will be necessary to close the throttle so the additional thrust and lift will be removed and the airplane will stay on the ground.

HIGH ROUNDOUT

Sometimes when the airplane appears to temporarily stop moving downward, the roundout has been made too rapidly and the airplane is flying level, too high above the runway. Continuing the roundout would further reduce the airspeed, resulting in an increase in angle of attack to the critical angle. This would result in the airplane stalling and dropping hard onto the runway. To prevent this, the pitch attitude should be held constant until the airplane decelerates enough to again start descending. Then the roundout can be continued to establish the proper landing attitude. This procedure should only be used when there is adequate airspeed. It may be necessary to add a slight amount of power to keep the airspeed from decreasing excessively and to avoid losing lift too rapidly.

Although back-elevator pressure may be relaxed slightly, the nose should not be lowered any perceptible amount to make the airplane descend when fairly close to the runway unless some power is added momentarily. The momentary decrease in lift that would result from lowering the nose and decreasing the angle of attack may be so great that the airplane might contact the ground with the nosewheel first, which could collapse.

When the proper landing attitude is attained, the airplane is approaching a stall because the airspeed is decreasing and the critical angle of attack is being approached, even though the pitch attitude is no longer being increased. [Figure 8-33]

It is recommended that a GO-AROUND be executed any time it appears the nose must be lowered significantly or that the landing is in any other way uncertain.

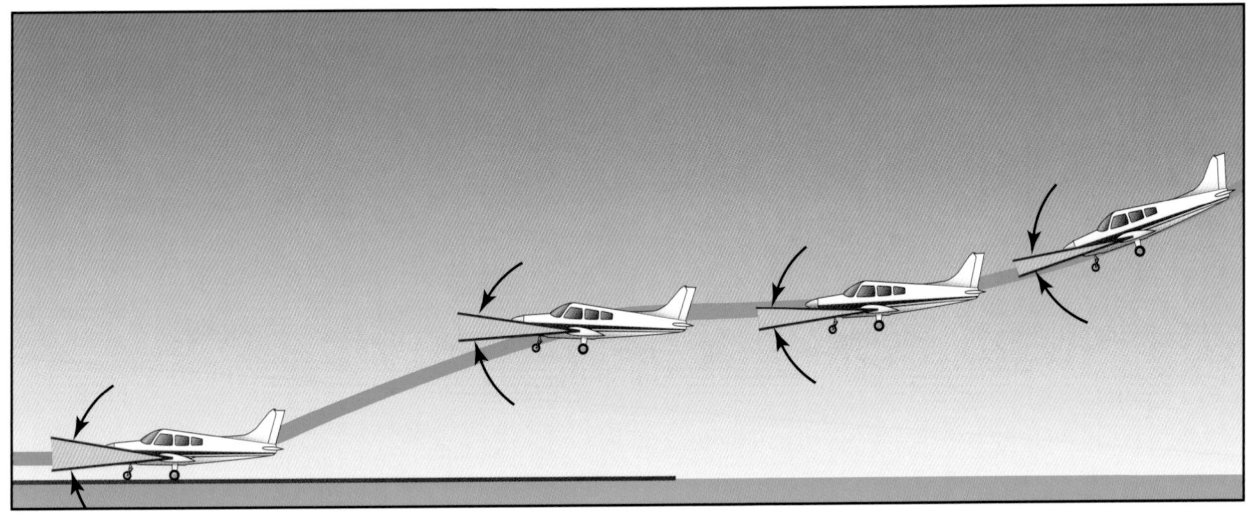

Figure 8-33. Rounding out too high.

LATE OR RAPID ROUNDOUT

Starting the roundout too late or pulling the elevator control back too rapidly to prevent the airplane from touching down prematurely can impose a heavy load factor on the wing and cause an accelerated stall.

Suddenly increasing the angle of attack and stalling the airplane during a roundout is a dangerous situation since it may cause the airplane to land extremely hard on the main landing gear, and then bounce back into the air. As the airplane contacts the ground, the tail will be forced down very rapidly by the back-elevator pressure and by inertia acting downward on the tail.

Recovery from this situation requires prompt and positive application of power prior to occurrence of the stall. This may be followed by a normal landing if sufficient runway is available—otherwise the pilot should EXECUTE A GO-AROUND immediately.

If the roundout is late, the nosewheel may strike the runway first, causing the nose to bounce upward. No attempt should be made to force the airplane back onto the ground; a GO-AROUND should be executed immediately.

FLOATING DURING ROUNDOUT

If the airspeed on final approach is excessive, it will usually result in the airplane floating. [Figure 8-34] Before touchdown can be made, the airplane may be well past the desired landing point and the available runway may be insufficient. When diving an airplane on final approach to land at the proper point, there will be an appreciable increase in airspeed. The proper touchdown attitude cannot be established without producing an excessive angle of attack and lift. This will cause the airplane to gain altitude or balloon.

Any time the airplane floats, judgment of speed, height, and rate of sink must be especially acute. The pilot must smoothly and gradually adjust the pitch attitude as the airplane decelerates to touchdown speed and starts to settle, so the proper landing attitude is attained at the moment of touchdown. The slightest

Figure 8-34. Floating during roundout.

error in judgment and timing will result in either ballooning or bouncing.

The recovery from floating will depend on the amount of floating and the effect of any crosswind, as well as the amount of runway remaining. Since prolonged floating utilizes considerable runway length, it should be avoided especially on short runways or in strong crosswinds. If a landing cannot be made on the first third of the runway, or the airplane drifts sideways, the pilot should EXECUTE A GO-AROUND.

BALLOONING DURING ROUNDOUT

If the pilot misjudges the rate of sink during a landing and thinks the airplane is descending faster than it should, there is a tendency to increase the pitch attitude and angle of attack too rapidly. This not only stops the descent, but actually starts the airplane climbing. This climbing during the roundout is known as ballooning. [Figure 8-35] Ballooning can be dangerous because the height above the ground is increasing and the airplane may be rapidly approaching a stalled condition. The altitude gained in each instance will depend on the airspeed or the speed with which the pitch attitude is increased.

When ballooning is slight, a constant landing attitude should be held and the airplane allowed to gradually decelerate and settle onto the runway. Depending on the severity of ballooning, the use of throttle may be helpful in cushioning the landing. By adding power, thrust can be increased to keep the airspeed from decelerating too rapidly and the wings from suddenly losing lift, but throttle must be closed immediately after touchdown. Remember that torque will be created as power is applied; therefore, it will be necessary to use rudder pressure to keep the airplane straight as it settles onto the runway.

When ballooning is excessive, it is best to EXECUTE A GO-AROUND IMMEDIATELY; DO NOT ATTEMPT TO SALVAGE THE LANDING. Power must be applied before the airplane enters a stalled condition.

The pilot must be extremely cautious of ballooning when there is a crosswind present because the crosswind correction may be inadvertently released or it may become inadequate. Because of the lower airspeed after ballooning, the crosswind affects the airplane more. Consequently, the wing will have to be lowered even further to compensate for the increased drift. It is imperative that the pilot makes certain that the appropriate wing is down and that directional control is maintained with opposite rudder. If there is any doubt, or the airplane starts to drift, EXECUTE A GO-AROUND.

BOUNCING DURING TOUCHDOWN

When the airplane contacts the ground with a sharp impact as the result of an improper attitude or an excessive rate of sink, it tends to bounce back into the air. Though the airplane's tires and shock struts provide some springing action, the airplane does not bounce like a rubber ball. Instead, it rebounds into the air because the wing's angle of attack was abruptly increased, producing a sudden addition of lift. [Figure 8-36]

The abrupt change in angle of attack is the result of inertia instantly forcing the airplane's tail downward when the main wheels contact the ground sharply. The severity of the bounce depends on the airspeed at the moment of contact and the degree to which the angle of attack or pitch attitude was increased.

Since a bounce occurs when the airplane makes contact with the ground before the proper touchdown

Figure 8-35. Ballooning during roundout.

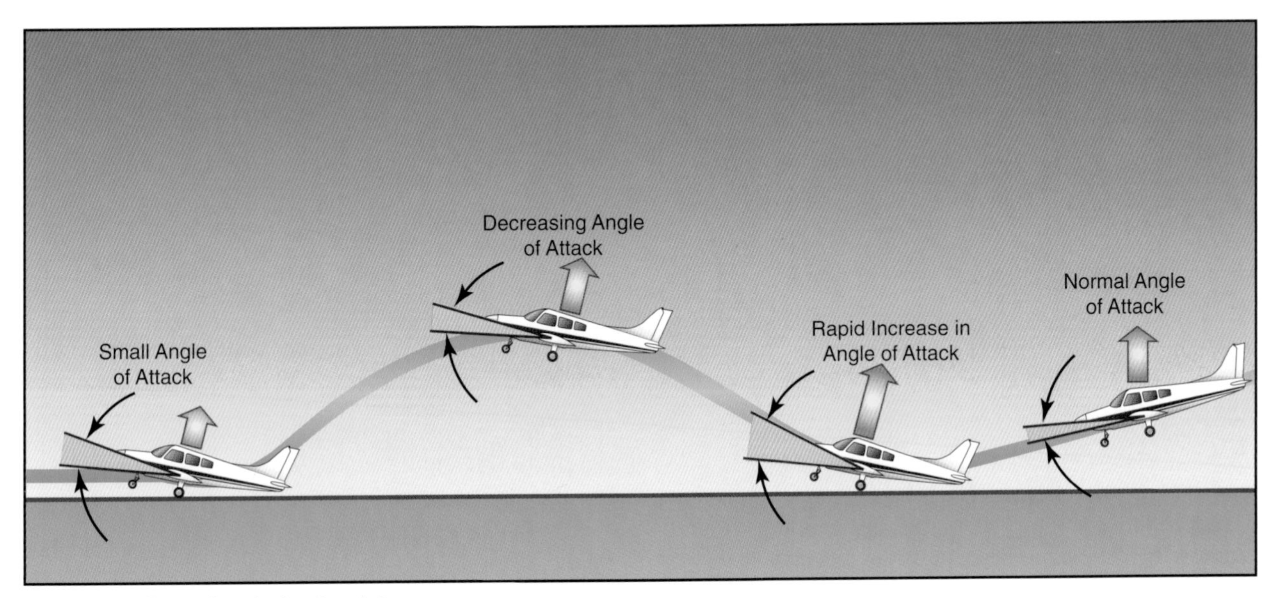

Figure 8-36. Bouncing during touchdown.

attitude is attained, it is almost invariably accompanied by the application of excessive back-elevator pressure. This is usually the result of the pilot realizing too late that the airplane is not in the proper attitude and attempting to establish it just as the second touchdown occurs.

The corrective action for a bounce is the same as for ballooning and similarly depends on its severity. When it is very slight and there is no extreme change in the airplane's pitch attitude, a follow-up landing may be executed by applying sufficient power to cushion the subsequent touchdown, and smoothly adjusting the pitch to the proper touchdown attitude.

In the event a very slight bounce is encountered while landing with a crosswind, crosswind correction must be maintained while the next touchdown is made. Remember that since the subsequent touchdown will be made at a slower airspeed, the upwind wing will have to be lowered even further to compensate for drift.

Extreme caution and alertness must be exercised any time a bounce occurs, but particularly when there is a crosswind. Inexperienced pilots will almost invariably release the crosswind correction. When one main wheel of the airplane strikes the runway, the other wheel will touch down immediately afterwards, and the wings will become level. Then, with no crosswind correction as the airplane bounces, the wind will cause the airplane to roll with the wind, thus exposing even more surface to the crosswind and drifting the airplane more rapidly.

When a bounce is severe, the safest procedure is to EXECUTE A GO-AROUND IMMEDIATELY. No attempt to salvage the landing should be made. Full power should be applied while simultaneously maintaining directional control, and lowering the nose to a safe climb attitude. The go-around procedure should be continued even though the airplane may descend and another bounce may be encountered. It would be extremely foolish to attempt a landing from a bad bounce since airspeed diminishes very rapidly in the nose-high attitude, and a stall may occur before a subsequent touchdown could be made.

PORPOISING

In a bounced landing that is improperly recovered, the airplane comes in nose first setting off a series of motions that imitate the jumps and dives of a porpoise—hence the name. [Figure 8-37] The problem is improper airplane attitude at touchdown, sometimes caused by inattention, not knowing where the ground is, mistrimming or forcing the airplane onto the runway.

Ground effect decreases elevator control effectiveness and increases the effort required to raise the nose. Not enough elevator or stabilator trim can result in a nose-low contact with the runway and a porpoise develops.

Porpoising can also be caused by improper airspeed control. Usually, if an approach is too fast, the airplane floats and the pilot tries to force it on the runway when the airplane still wants to fly. A gust of wind, a bump in the runway, or even a slight tug on the control wheel will send the airplane aloft again.

The corrective action for a porpoise is the same as for a bounce and similarly depends on its severity. When it is very slight and there is no extreme change in the airplane's pitch attitude, a follow-up landing may be executed by applying sufficient power to cushion the subsequent touchdown, and smoothly adjusting the pitch to the proper touchdown attitude.

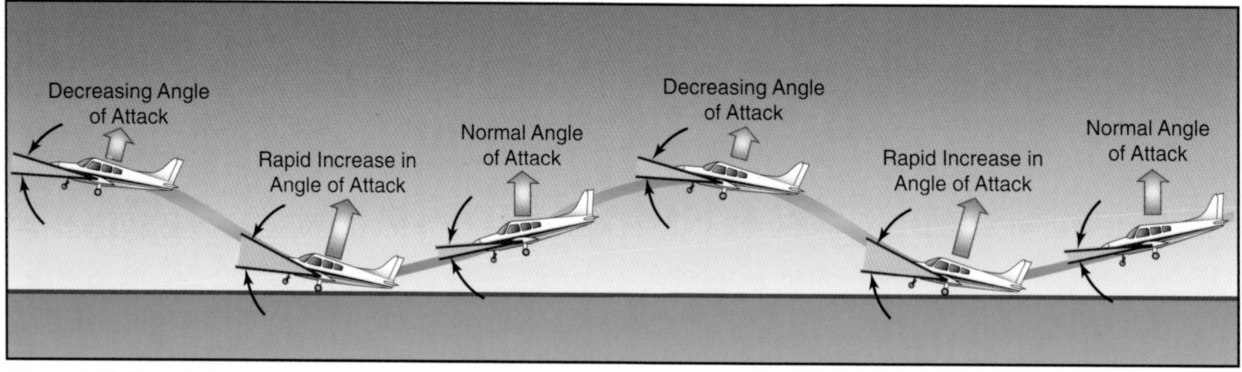

Figure 8-37. Porpoising.

When a porpoise is severe, the safest procedure is to EXECUTE A GO-AROUND IMMEDIATELY. In a severe porpoise, the airplane's pitch oscillations can become progressively worse, until the airplane strikes the runway nose first with sufficient force to collapse the nose gear. Pilot attempts to correct a severe porpoise with flight control and power inputs will most likely be untimely and out of sequence with the oscillations, and only make the situation worse. No attempt to salvage the landing should be made. Full power should be applied while simultaneously maintaining directional control, and lowering the nose to a safe climb attitude.

WHEELBARROWING

When a pilot permits the airplane weight to become concentrated about the nosewheel during the takeoff or landing roll, a condition known as wheelbarrowing will occur. Wheelbarrowing may cause loss of directional control during the landing roll because braking action is ineffective, and the airplane tends to swerve or pivot on the nosewheel, particularly in crosswind conditions. One of the most common causes of wheelbarrowing during the landing roll is a simultaneous touchdown of the main and nosewheel, with excessive speed, followed by application of forward pressure on the elevator control. Usually, the situation can be corrected by smoothly applying back-elevator pressure. However, if wheelbarrowing is encountered and runway and other conditions permit, it may be advisable to promptly initiate a go-around. Wheelbarrowing will not occur if the pilot achieves and maintains the correct landing attitude, touches down at the proper speed, and gently lowers the nosewheel while losing speed on rollout. If the pilot decides to stay on the ground rather than attempt a go-around or if directional control is lost, the throttle should be closed and the pitch attitude smoothly but firmly rotated to the proper landing attitude. Raise the flaps to reduce lift and to increase the load on the main wheels for better braking action.

HARD LANDING

When the airplane contacts the ground during landings, its vertical speed is instantly reduced to zero. Unless provisions are made to slow this vertical speed and cushion the impact of touchdown, the force of contact with the ground may be so great it could cause structural damage to the airplane.

The purpose of pneumatic tires, shock absorbing landing gears, and other devices is to cushion the impact and to increase the time in which the airplane's vertical descent is stopped. The importance of this cushion may be understood from the computation that a 6-inch free fall on landing is roughly equal, to a 340-foot-per-minute descent. Within a fraction of a second, the airplane must be slowed from this rate of vertical descent to zero, without damage.

During this time, the landing gear together with some aid from the lift of the wings must supply whatever force is needed to counteract the force of the airplane's inertia and weight. The lift decreases rapidly as the airplane's forward speed is decreased, and the force on the landing gear increases by the impact of touchdown. When the descent stops, the lift will be practically zero, leaving the landing gear alone to carry both the airplane's weight and inertia force. The load imposed at the instant of touchdown may easily be three or four times the actual weight of the airplane depending on the severity of contact.

TOUCHDOWN IN A DRIFT OR CRAB

At times the pilot may correct for wind drift by crabbing on the final approach. If the roundout and touchdown are made while the airplane is drifting or in a crab, it will contact the ground while moving sideways. This will impose extreme side loads on the landing gear, and if severe enough, may cause structural failure.

The most effective method to prevent drift in primary training airplanes is the wing-low method. This technique keeps the longitudinal axis of the airplane aligned with both the runway and the direction of motion throughout the approach and touchdown.

There are three factors that will cause the longitudinal axis and the direction of motion to be misaligned during touchdown: drifting, crabbing, or a combination of both.

If the pilot has not taken adequate corrective action to avoid drift during a crosswind landing, the main wheels' tire tread offers resistance to the airplane's sideward movement in respect to the ground. Consequently, any sidewise velocity of the airplane is abruptly decelerated, with the result that the inertia force is as shown in figure 8-38. This creates a moment around the main wheel when it contacts the ground, tending to overturn or tip the airplane. If the windward wingtip is raised by the action of this moment, all the weight and shock of landing will be borne by one main wheel. This could cause structural damage.

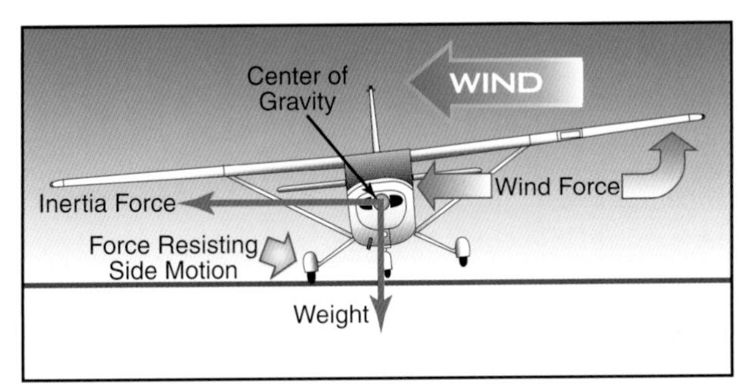

Figure 8-38. Drifting during touchdown.

Not only are the same factors present that are attempting to raise a wing, but the crosswind is also acting on the fuselage surface behind the main wheels, tending to yaw (weathervane) the airplane into the wind. This often results in a ground loop.

GROUND LOOP

A ground loop is an uncontrolled turn during ground operation that may occur while taxiing or taking off, but especially during the after-landing roll. Drift or weathervaning does not always cause a ground loop, although these things may cause the initial swerve. Careless use of the rudder, an uneven ground surface, or a soft spot that retards one main wheel of the airplane may also cause a swerve. In any case, the initial swerve tends to make the airplane ground loop, whether it is a tailwheel-type or nosewheel-type. [Figure 8-39]

Nosewheel-type airplanes are somewhat less prone to ground loop than tailwheel-type airplanes. Since the center of gravity (CG) is located forward of the main landing gear on these airplanes, any time a swerve develops, centrifugal force acting on the CG will tend to stop the swerving action.

If the airplane touches down while drifting or in a crab, the pilot should apply aileron toward the high wing and stop the swerve with the rudder. Brakes should be used to correct for turns or swerves only when the rudder is inadequate. The pilot must exercise caution when applying corrective brake action because it is very easy to overcontrol and aggravate the situation.

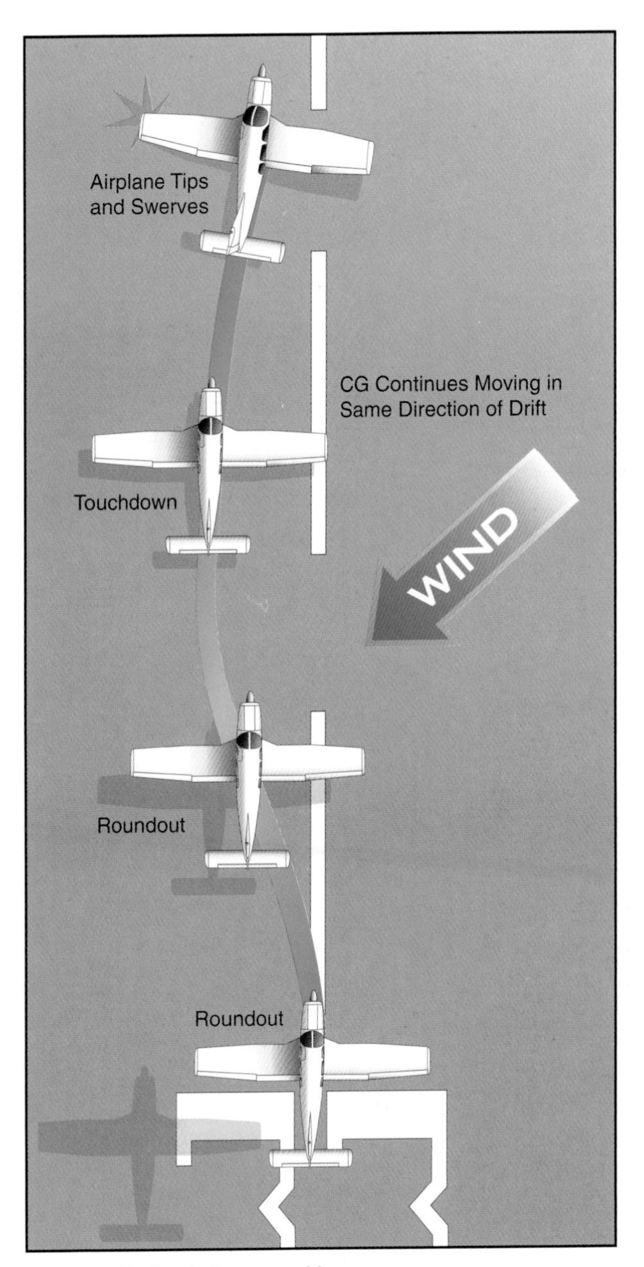

Figure 8-39. Start of a ground loop.

If brakes are used, sufficient brake should be applied on the low-wing wheel (outside of the turn) to stop the swerve. When the wings are approximately level, the new direction must be maintained until the airplane has slowed to taxi speed or has stopped.

In nosewheel airplanes, a ground loop is almost always a result of wheelbarrowing. The pilot must be aware that even though the nosewheel-type airplane is less prone than the tailwheel-type airplane, virtually every type of airplane, including large multiengine airplanes, can be made to ground loop when sufficiently mishandled.

WING RISING AFTER TOUCHDOWN

When landing in a crosswind, there may be instances when a wing will rise during the after-landing roll. This may occur whether or not there is a loss of directional control, depending on the amount of crosswind and the degree of corrective action.

Any time an airplane is rolling on the ground in a crosswind condition, the upwind wing is receiving a greater force from the wind than the downwind wing. This causes a lift differential. Also, as the upwind wing rises, there is an increase in the angle of attack, which increases lift on the upwind wing, rolling the airplane downwind.

When the effects of these two factors are great enough, the upwind wing may rise even though directional control is maintained. If no correction is applied, it is possible that the upwind wing will rise sufficiently to cause the downwind wing to strike the ground.

In the event a wing starts to rise during the landing roll, the pilot should immediately apply more aileron pressure toward the high wing and continue to maintain direction. The sooner the aileron control is applied, the more effective it will be. The further a wing is allowed to rise before taking corrective action, the more airplane surface is exposed to the force of the crosswind. This diminishes the effectiveness of the aileron.

HYDROPLANING

Hydroplaning is a condition that can exist when an airplane is landed on a runway surface contaminated with standing water, slush, and/or wet snow. Hydroplaning can have serious adverse effects on ground controllability and braking efficiency. The three basic types of hydroplaning are dynamic hydroplaning, reverted rubber hydroplaning, and viscous hydroplaning. Any one of the three can render an airplane partially or totally uncontrollable anytime during the landing roll.

DYNAMIC HYDROPLANING

Dynamic hydroplaning is a relatively high-speed phenomenon that occurs when there is a film of water on the runway that is at least one-tenth inch deep. As the speed of the airplane and the depth of the water increase, the water layer builds up an increasing resistance to displacement, resulting in the formation of a wedge of water beneath the tire. At some speed, termed the hydroplaning speed (V_P), the water pressure equals the weight of the airplane and the tire is lifted off the runway surface. In this condition, the tires no longer contribute to directional control and braking action is nil.

Dynamic hydroplaning is related to tire inflation pressure. Data obtained during hydroplaning tests have shown the minimum dynamic hydroplaning speed (V_P) of a tire to be 8.6 times the square root of the tire pressure in pounds per square inch (PSI). For an airplane with a main tire pressure of 24 pounds, the calculated hydroplaning speed would be approximately 42 knots. It is important to note that the calculated speed referred to above is for the start of dynamic hydroplaning. Once hydroplaning has started, it may persist to a significantly slower speed depending on the type being experienced.

REVERTED RUBBER HYDROPLANING

Reverted rubber (steam) hydroplaning occurs during heavy braking that results in a prolonged locked-wheel skid. Only a thin film of water on the runway is required to facilitate this type of hydroplaning.

The tire skidding generates enough heat to cause the rubber in contact with the runway to revert to its original uncured state. The reverted rubber acts as a seal between the tire and the runway, and delays water exit from the tire footprint area. The water heats and is converted to steam which supports the tire off the runway.

Reverted rubber hydroplaning frequently follows an encounter with dynamic hydroplaning, during which time the pilot may have the brakes locked in an attempt to slow the airplane. Eventually the airplane slows enough to where the tires make contact with the runway surface and the airplane begins to skid. The remedy for this type of hydroplane is for the pilot to release the brakes and allow the wheels to spin up and apply moderate braking. Reverted rubber hydroplaning is insidious in that the pilot may not know when it begins, and it can persist to very slow groundspeeds (20 knots or less).

VISCOUS HYDROPLANING

Viscous hydroplaning is due to the viscous properties of water. A thin film of fluid no more than one thousandth of an inch in depth is all that is needed. The tire cannot penetrate the fluid and the tire rolls on top of the film. This can occur at a much lower speed than dynamic hydroplane, but requires a smooth or smooth acting surface such as asphalt or a touchdown area coated with the accumulated rubber of past landings. Such a surface can have the same friction coefficient as wet ice.

When confronted with the possibility of hydroplaning, it is best to land on a grooved runway (if available). Touchdown speed should be as slow as possible consistent with safety. After the nosewheel is lowered to the runway, moderate braking should be applied. If deceleration is not detected and hydroplaning is suspected, the nose should be raised and aerodynamic drag utilized to decelerate to a point where the brakes do become effective.

Proper braking technique is essential. The brakes should be applied firmly until reaching a point just short of a skid. At the first sign of a skid, the pilot should release brake pressure and allow the wheels to spin up. Directional control should be maintained as far as possible with the rudder. Remember that in a crosswind, if hydroplaning should occur, the crosswind will cause the airplane to simultaneously weathervane into the wind as well as slide downwind.

Chapter 9 Performance Maneuvers

PERFORMANCE MANEUVERS

Performance maneuvers are used to develop a high degree of pilot skill. They aid the pilot in analyzing the forces acting on the airplane and in developing a fine control touch, coordination, timing, and division of attention for precise maneuvering of the airplane. Performance maneuvers are termed "advanced" maneuvers because the degree of skill required for proper execution is normally not acquired until a pilot has obtained a sense of orientation and control feel in "normal" maneuvers. An important benefit of performance maneuvers is the sharpening of fundamental skills to the degree that the pilot can cope with unusual or unforeseen circumstances occasionally encountered in normal flight.

Advanced maneuvers are variations and/or combinations of the basic maneuvers previously learned. They embody the same principles and techniques as the basic maneuvers, but require a higher degree of skill for proper execution. The student, therefore, who demonstrates a lack of progress in the performance of advanced maneuvers, is more than likely deficient in one or more of the basic maneuvers. The flight instructor should consider breaking the advanced maneuver down into its component basic maneuvers in an attempt to identify and correct the deficiency before continuing with the advanced maneuver.

STEEP TURNS

The objective of the maneuver is to develop the smoothness, coordination, orientation, division of attention, and control techniques necessary for the execution of maximum performance turns when the airplane is near its performance limits. Smoothness of control use, coordination, and accuracy of execution are the important features of this maneuver.

The steep turn maneuver consists of a turn in either direction, using a bank angle between 45 to 60°. This will cause an overbanking tendency during which maximum turning performance is attained and relatively high load factors are imposed. Because of the high load factors imposed, these turns should be performed at an airspeed that does not exceed the airplane's design maneuvering speed (V_A). The principles of an ordinary steep turn apply, but as a practice maneuver the steep turns should be continued until 360° or 720° of turn have been completed. [Figure 9-1]

Figure 9-1. Steep turns.

An airplane's maximum turning performance is its fastest rate of turn and its shortest radius of turn, which change with both airspeed and angle of bank. Each airplane's turning performance is limited by the amount of power its engine is developing, its limit load factor (structural strength), and its aerodynamic characteristics.

The limiting load factor determines the maximum bank, which can be maintained without stalling or exceeding the airplane's structural limitations. In most small planes, the maximum bank has been found to be approximately 50° to 60°.

The pilot should realize the tremendous additional load that is imposed on an airplane as the bank is increased beyond 45°. During a coordinated turn with a 70° bank, a load factor of approximately 3 Gs is placed on the airplane's structure. Most general aviation type airplanes are stressed for approximately 3.8 Gs.

Regardless of the airspeed or the type of airplanes involved, a given angle of bank in a turn, during which altitude is maintained, will always produce the same load factor. Pilots must be aware that an additional load factor increases the stalling speed at a significant rate—stalling speed increases with the square root of the load factor. For example, a light plane that stalls at 60 knots in level flight will stall at nearly 85 knots in a 60° bank. The pilot's understanding and observance of this fact is an indispensable safety precaution for the performance of all maneuvers requiring turns.

Before starting the steep turn, the pilot should ensure that the area is clear of other air traffic since the rate of turn will be quite rapid. After establishing the manufacturer's recommended entry speed or the design maneuvering speed, the airplane should be smoothly rolled into a selected bank angle between 45 to 60°. As the turn is being established, back-elevator pressure should be smoothly increased to increase the angle of attack. This provides the additional wing lift required to compensate for the increasing load factor.

After the selected bank angle has been reached, the pilot will find that considerable force is required on the elevator control to hold the airplane in level flight—to maintain altitude. Because of this increase in the force applied to the elevators, the load factor increases rapidly as the bank is increased. Additional back-elevator pressure increases the angle of attack, which results in an increase in drag. Consequently, power must be added to maintain the entry altitude and airspeed.

Eventually, as the bank approaches the airplane's maximum angle, the maximum performance or structural limit is being reached. If this limit is exceeded, the airplane will be subjected to excessive structural loads, and will lose altitude, or stall. The limit load factor must not be exceeded, to prevent structural damage.

During the turn, the pilot should not stare at any one object. To maintain altitude, as well as orientation, requires an awareness of the relative position of the nose, the horizon, the wings, and the amount of bank. The pilot who references the aircraft's turn by watching only the nose will have difficulty holding altitude constant; on the other hand, the pilot who watches the nose, the horizon, and the wings can usually hold altitude within a few feet. If the altitude begins to increase, or decrease, relaxing or increasing the back-elevator pressure will be required as appropriate. This may also require a power adjustment to maintain the selected airspeed. A small increase or decrease of 1 to 3° of bank angle may be used to control small altitude deviations. All bank angle changes should be done with coordinated use of aileron and rudder.

The rollout from the turn should be timed so that the wings reach level flight when the airplane is exactly on the heading from which the maneuver was started. While the recovery is being made, back-elevator pressure is gradually released and power reduced, as necessary, to maintain the altitude and airspeed.

Common errors in the performance of steep turns are:

- Failure to adequately clear the area.
- Excessive pitch change during entry or recovery.
- Attempts to start recovery prematurely.
- Failure to stop the turn on a precise heading.
- Excessive rudder during recovery, resulting in skidding.
- Inadequate power management.
- Inadequate airspeed control.
- Poor coordination.
- Gaining altitude in right turns and/or losing altitude in left turns.
- Failure to maintain constant bank angle.
- Disorientation.
- Attempting to perform the maneuver by instrument reference rather than visual reference.
- Failure to scan for other traffic during the maneuver.

STEEP SPIRAL

The objective of this maneuver is to improve pilot techniques for airspeed control, wind drift control, planning, orientation, and division of attention. The steep spiral is not only a valuable flight training maneuver, but it has practical application in providing a procedure for dissipating altitude while remaining over a selected spot in preparation for landing, especially for emergency forced landings.

A steep spiral is a constant gliding turn, during which a constant radius around a point on the ground is maintained similar to the maneuver, turns around a point. The radius should be such that the steepest bank will not exceed 60°. Sufficient altitude must be obtained before starting this maneuver so that the spiral may be continued through a series of at least three 360° turns. [Figure 9-2] The maneuver should not be continued below 1,000 feet above the surface unless performing an emergency landing in conjunction with the spiral.

Operating the engine at idle speed for a prolonged period during the glide may result in excessive engine cooling or spark plug fouling. The engine should be cleared periodically by briefly advancing the throttle to normal cruise power, while adjusting the pitch attitude to maintain a constant airspeed. Preferably, this should be done while headed into the wind to minimize any variation in groundspeed and radius of turn.

After the throttle is closed and gliding speed is established, a gliding spiral should be started and a turn of constant radius maintained around the selected spot on the ground. This will require correction for wind drift by steepening the bank on downwind headings and shallowing the bank on upwind headings, just as in the maneuver, turns around a point. During the descending spiral, the pilot must judge the direction and speed of the wind at different altitudes and make appropriate changes in the angle of bank to maintain a uniform radius.

A constant airspeed should also be maintained throughout the maneuver. Failure to hold the airspeed constant will cause the radius of turn and necessary angle of bank to vary excessively. On the downwind side of the maneuver, the steeper the bank angle, the lower the pitch attitude must be to maintain a given airspeed. Conversely, on the upwind side, as the bank angle becomes shallower, the pitch attitude must be raised to maintain the proper airspeed. This is necessary because the airspeed tends to change as the bank is changed from shallow to steep to shallow.

During practice of the maneuver, the pilot should execute three turns and roll out toward a definite object or on a specific heading. During the rollout, smoothness is essential, and the use of controls must be so coordinated that no increase or decrease of speed results when the straight glide is resumed.

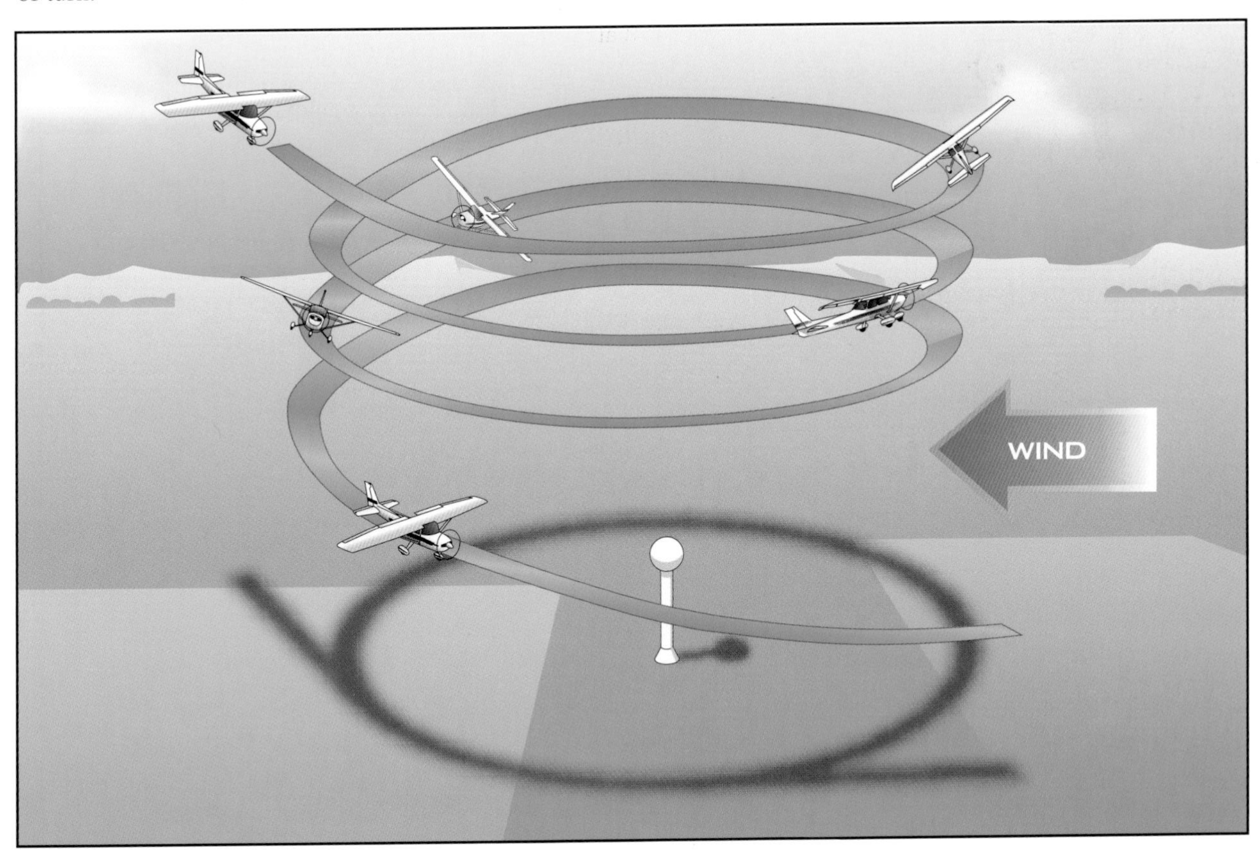

Figure 9-2. Steep spiral.

Common errors in the performance of steep spirals are:

- Failure to adequately clear the area.
- Failure to maintain constant airspeed.
- Poor coordination, resulting in skidding and/or slipping.
- Inadequate wind drift correction.
- Failure to coordinate the controls so that no increase/decrease in speed results when straight glide is resumed.
- Failure to scan for other traffic.
- Failure to maintain orientation.

CHANDELLE

The objective of this maneuver is to develop the pilot's coordination, orientation, planning, and accuracy of control during maximum performance flight.

A chandelle is a maximum performance climbing turn beginning from approximately straight-and-level flight, and ending at the completion of a precise 180° of turn in a wings-level, nose-high attitude at the minimum controllable airspeed. [Figure 9-3] The maneuver demands that the maximum flight performance of the airplane be obtained; the airplane should gain the most altitude possible for a given degree of bank and power setting without stalling. Since numerous atmospheric variables beyond control of the pilot will affect the specific amount of altitude gained, the quality of the performance of the maneuver is not judged solely on the altitude gain, but by the pilot's overall proficiency as it pertains to climb performance for the power/bank combination used, and to the elements of piloting skill demonstrated.

Prior to starting a chandelle, the flaps and gear (if retractable) should be in the UP position, power set to cruise condition, and the airspace behind and above clear of other air traffic. The maneuver should be entered from straight-and-level flight (or a shallow dive) and at a speed no greater than the maximum entry speed recommended by the manufacturer—in most cases not above the airplane's design maneuvering speed (V_A).

After the appropriate airspeed and power setting have been established, the chandelle is started by smoothly entering a coordinated turn with an angle of bank appropriate for the airplane being flown. Normally, this angle of bank should not exceed approximately 30°. After the appropriate bank is established, a climbing turn should be started by smoothly applying back-elevator pressure to increase the pitch attitude at a constant rate and to attain the highest pitch attitude as 90° of turn is completed. As the climb is initiated in airplanes with fixed-pitch propellers, full throttle may be applied, but is applied gradually so that the maximum allowable r.p.m. is not exceeded. In airplanes with constant-speed propellers, power may be left at the normal cruise setting.

Figure 9-3. Chandelle.

Once the bank has been established, the angle of bank should remain constant until 90° of turn is completed. Although the degree of bank is fixed during this climbing turn, it may appear to increase and, in fact, actually will tend to increase if allowed to do so as the maneuver continues.

When the turn has progressed 90° from the original heading, the pilot should begin rolling out of the bank at a constant rate while maintaining a constant-pitch attitude. Since the angle of bank will be decreasing during the rollout, the vertical component of lift will increase slightly. For this reason, it may be necessary to release a slight amount of back-elevator pressure in order to keep the nose of the airplane from rising higher.

As the wings are being leveled at the completion of 180° of turn, the pitch attitude should be noted by checking the outside references and the attitude indicator. This pitch attitude should be held momentarily while the airplane is at the minimum controllable airspeed. Then the pitch attitude may be gently reduced to return to straight-and-level cruise flight.

Since the airspeed is constantly decreasing throughout the maneuver, the effects of engine torque become more and more prominent. Therefore, right-rudder pressure is gradually increased to control yaw and maintain a constant rate of turn and to keep the airplane in coordinated flight. The pilot should maintain coordinated flight by the feel of pressures being applied on the controls and by the ball instrument of the turn-and-slip indicator. If coordinated flight is being maintained, the ball will remain in the center of the race.

To roll out of a left chandelle, the left aileron must be lowered to raise the left wing. This creates more drag than the aileron on the right wing, resulting in a tendency for the airplane to yaw to the left. With the low airspeed at this point, torque effect tries to make the airplane yaw to the left even more. Thus, there are two forces pulling the airplane's nose to the left—aileron drag and torque. To maintain coordinated flight, considerable right-rudder pressure is required during the rollout to overcome the effects of aileron drag and torque.

In a chandelle to the right, when control pressure is applied to begin the rollout, the aileron on the right wing is lowered. This creates more drag on that wing and tends to make the airplane yaw to the right. At the same time, the effect of torque at the lower airspeed is causing the airplane's nose to yaw to the left. Thus, aileron drag pulling the nose to the right and torque pulling to the left, tend to neutralize each other. If excessive left-rudder pressure is applied, the rollout will be uncoordinated.

The rollout to the left can usually be accomplished with very little left rudder, since the effects of aileron drag and torque tend to neutralize each other. Releasing some right rudder, which has been applied to correct for torque, will normally give the same effect as applying left-rudder pressure. When the wings become level and the ailerons are neutralized, the aileron drag disappears. Because of the low airspeed and high power, the effects of torque become the more prominent force and must continue to be controlled with rudder pressure.

A rollout to the left is accomplished mainly by applying aileron pressure. During the rollout, right-rudder pressure should be gradually released, and left rudder applied only as necessary to maintain coordination. Even when the wings are level and aileron pressure is released, right-rudder pressure must be held to counteract torque and hold the nose straight.

Common errors in the performance of chandelles are:

- Failure to adequately clear the area.
- Too shallow an initial bank, resulting in a stall.
- Too steep an initial bank, resulting in failure to gain maximum performance.
- Allowing the actual bank to increase after establishing initial bank angle.
- Failure to start the recovery at the 90° point in the turn.
- Allowing the pitch attitude to increase as the bank is rolled out during the second 90° of turn.
- Removing all of the bank before the 180° point is reached.
- Nose low on recovery, resulting in too much airspeed.
- Control roughness.
- Poor coordination (slipping or skidding).
- Stalling at any point during the maneuver.
- Execution of a steep turn instead of a climbing maneuver.
- Failure to scan for other aircraft.
- Attempting to perform the maneuver by instrument reference rather than visual reference.

LAZY EIGHT

The lazy eight is a maneuver designed to develop perfect coordination of controls through a wide range of airspeeds and altitudes so that certain accuracy points are reached with planned attitude and airspeed. In its execution, the dive, climb, and turn are all combined, and the combinations are varied and applied throughout the performance range of the airplane. It is the only standard flight training maneuver during which at no time do the forces on the controls remain constant.

The lazy eight as a training maneuver has great value since constantly varying forces and attitudes are required. These forces must be constantly coordinated, due not only to the changing combinations of banks, dives, and climbs, but also to the constantly varying airspeed. The maneuver helps develop subconscious feel, planning, orientation, coordination, and speed sense. It is not possible to do a lazy eight mechanically, because the control pressures required for perfect coordination are never exactly the same.

This maneuver derives its name from the manner in which the extended longitudinal axis of the airplane is made to trace a flight pattern in the form of a figure 8 lying on its side (a lazy 8). [Figure 9-4]

A lazy eight consists of two 180° turns, in opposite directions, while making a climb and a descent in a symmetrical pattern during each of the turns. At no time throughout the lazy eight is the airplane flown straight and level; instead, it is rolled directly from one bank to the other with the wings level only at the moment the turn is reversed at the completion of each 180° change in heading.

As an aid to making symmetrical loops of the 8 during each turn, prominent reference points should be selected on the horizon. The reference points selected should be 45°, 90°, and 135° from the direction in which the maneuver is begun.

Prior to performing a lazy eight, the airspace behind and above should be clear of other air traffic. The maneuver should be entered from straight-and-level flight at normal cruise power and at the airspeed recommended by the manufacturer or at the airplane's design maneuvering speed.

The maneuver is started from level flight with a gradual climbing turn in the direction of the 45° reference point. The climbing turn should be planned and controlled so that the maximum pitch-up attitude is reached at the 45° point. The rate of rolling into the bank must be such as to prevent the rate of turn from becoming too rapid. As the pitch attitude is raised, the airspeed decreases, causing the rate of turn to increase. Since the bank also is being increased, it too causes the rate of turn to increase. Unless the maneuver is begun with a slow rate of roll, the combination of increasing pitch and increasing bank will cause the rate of turn to be so rapid that the 45° reference point will be reached before the highest pitch attitude is attained.

At the 45° point, the pitch attitude should be at maximum and the angle of bank continuing to

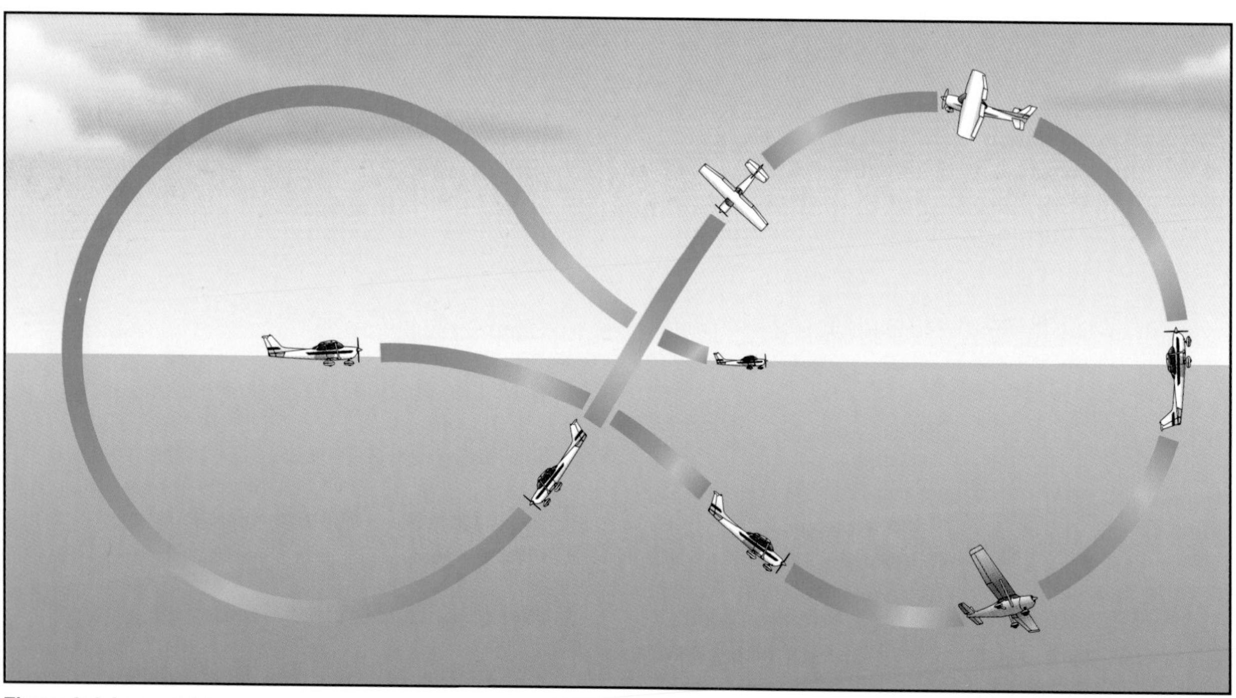

Figure 9-4. Lazy eight.

increase. Also, at the 45° point, the pitch attitude should start to decrease slowly toward the horizon and the 90° reference point. Since the airspeed is still decreasing, right-rudder pressure will have to be applied to counteract torque.

As the airplane's nose is being lowered toward the 90° reference point, the bank should continue to increase. Due to the decreasing airspeed, a slight amount of opposite aileron pressure may be required to prevent the bank from becoming too steep. When the airplane completes 90° of the turn, the bank should be at the maximum angle (approximately 30°), the airspeed should be at its minimum (5 to 10 knots above stall speed), and the airplane pitch attitude should be passing through level flight. It is at this time that an imaginary line, extending from the pilot's eye and parallel to the longitudinal axis of the airplane, passes through the 90° reference point.

Lazy eights normally should be performed with no more than approximately a 30° bank. Steeper banks may be used, but control touch and technique must be developed to a much higher degree than when the maneuver is performed with a shallower bank.

The pilot should not hesitate at this point but should continue to fly the airplane into a descending turn so that the airplane's nose describes the same size loop below the horizon as it did above. As the pilot's reference line passes through the 90° point, the bank should be decreased gradually, and the airplane's nose allowed to continue lowering. When the airplane has turned 135°, the nose should be in its lowest pitch attitude. The airspeed will be increasing during this descending turn, so it will be necessary to gradually relax rudder and aileron pressure and to simultaneously raise the nose and roll the wings level. As this is being accomplished, the pilot should note the amount of turn remaining and adjust the rate of rollout and pitch change so that the wings become level and the original airspeed is attained in level flight just as the 180° point is reached. Upon returning to the starting altitude and the 180° point, a climbing turn should be started immediately in the opposite direction toward the selected reference points to complete the second half of the eight in the same manner as the first half. [Figure 9-5]

Due to the decreasing airspeed, considerable right-rudder pressure is gradually applied to counteract torque at the top of the eight in both the right and left turns. The pressure will be greatest at the point of lowest airspeed.

More right-rudder pressure will be needed during the climbing turn to the right than in the turn to the left because more torque correction is needed to prevent yaw from decreasing the rate of turn. In the left climbing turn, the torque will tend to contribute to the

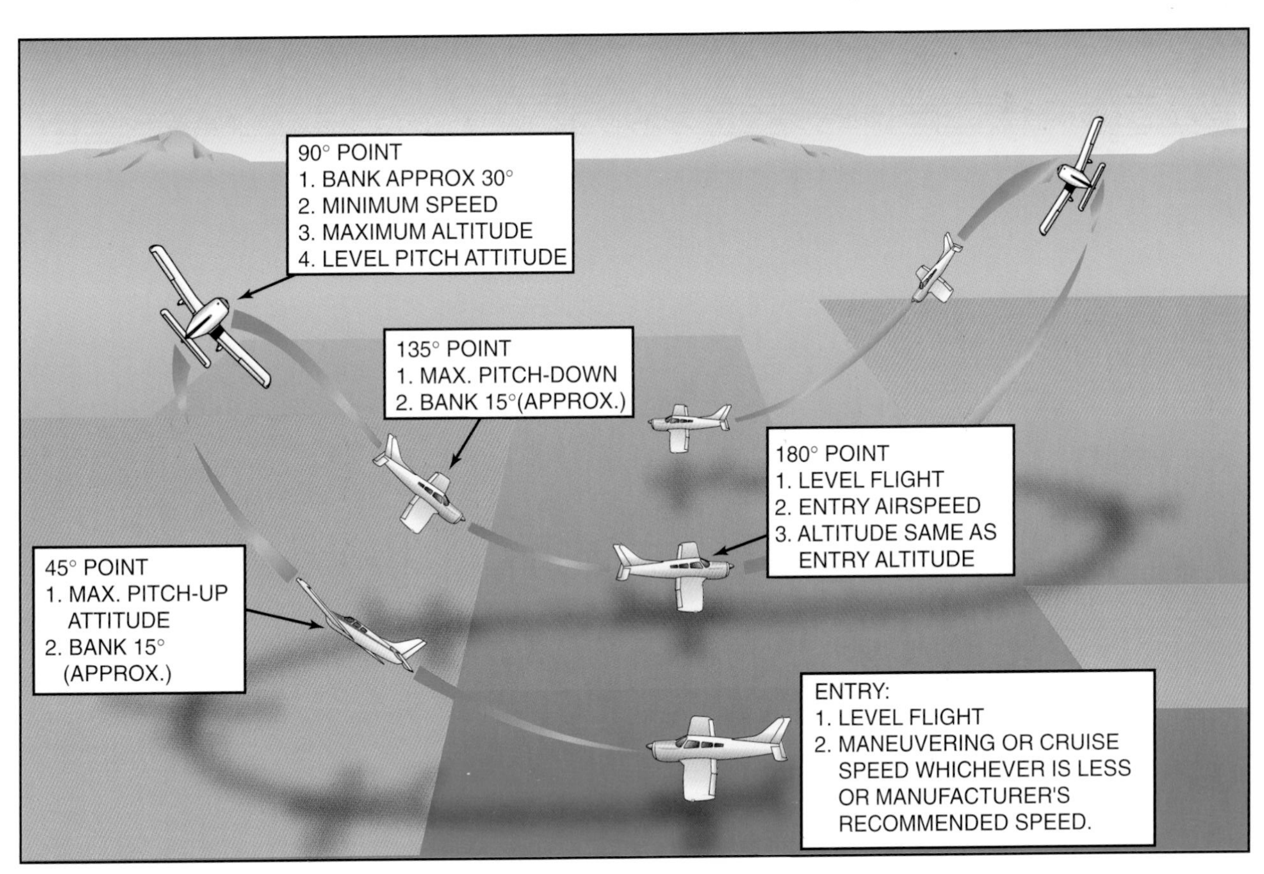

Figure 9-5. Lazy eight.

turn; consequently, less rudder pressure is needed. It will be noted that the controls are slightly crossed in the right climbing turn because of the need for left aileron pressure to prevent overbanking and right rudder to overcome torque.

The correct power setting for the lazy eight is that which will maintain the altitude for the maximum and minimum airspeeds used during the climbs and descents of the eight. Obviously, if excess power were used, the airplane would have gained altitude when the maneuver is completed; if insufficient power were used, altitude would have been lost.

Common errors in the performance of lazy eights are:

- Failure to adequately clear the area.
- Using the nose, or top of engine cowl, instead of the true longitudinal axis, resulting in unsymmetrical loops.
- Watching the airplane instead of the reference points.
- Inadequate planning, resulting in the peaks of the loops both above and below the horizon not coming in the proper place.
- Control roughness, usually caused by attempts to counteract poor planning.
- Persistent gain or loss of altitude with the completion of each eight.
- Attempting to perform the maneuver rhythmically, resulting in poor pattern symmetry.
- Allowing the airplane to "fall" out of the tops of the loops rather than flying the airplane through the maneuver.
- Slipping and/or skidding.
- Failure to scan for other traffic.

NIGHT VISION

Generally, most pilots are poorly informed about night vision. Human eyes never function as effectively at night as the eyes of animals with nocturnal habits, but if humans learn how to use their eyes correctly and know their limitations, night vision can be improved significantly. There are several reasons for training to use the eyes correctly.

One reason is the mind and eyes act as a team for a person to see well; both team members must be used effectively. The construction of the eyes is such that to see at night they are used differently than during the day. Therefore, it is important to understand the eye's construction and how the eye is affected by darkness.

Innumerable light-sensitive nerves, called "cones" and "rods," are located at the back of the eye or retina, a layer upon which all images are focused. These nerves connect to the cells of the optic nerve, which transmits messages directly to the brain. The cones are located in the center of the retina, and the rods are concentrated in a ring around the cones. [Figure 10-1]

The function of the cones is to detect color, details, and faraway objects. The rods function when something is seen out of the corner of the eye or peripheral vision. They detect objects, particularly those that are moving, but do not give detail or color—only shades of gray. Both the cones and the rods are used for vision during daylight.

Although there is not a clear-cut division of function, the rods make night vision possible. The rods and cones function in daylight and in moonlight, but in the absence of normal light, the process of night vision is placed almost entirely on the rods.

The fact that the rods are distributed in a band around the cones and do not lie directly behind the pupils makes off-center viewing (looking to one side of an object) important during night flight. During daylight, an object can be seen best by looking directly at it, but at night a scanning procedure to permit off-center viewing of the object is more effective. Therefore, the pilot should consciously practice this scanning procedure to improve night vision.

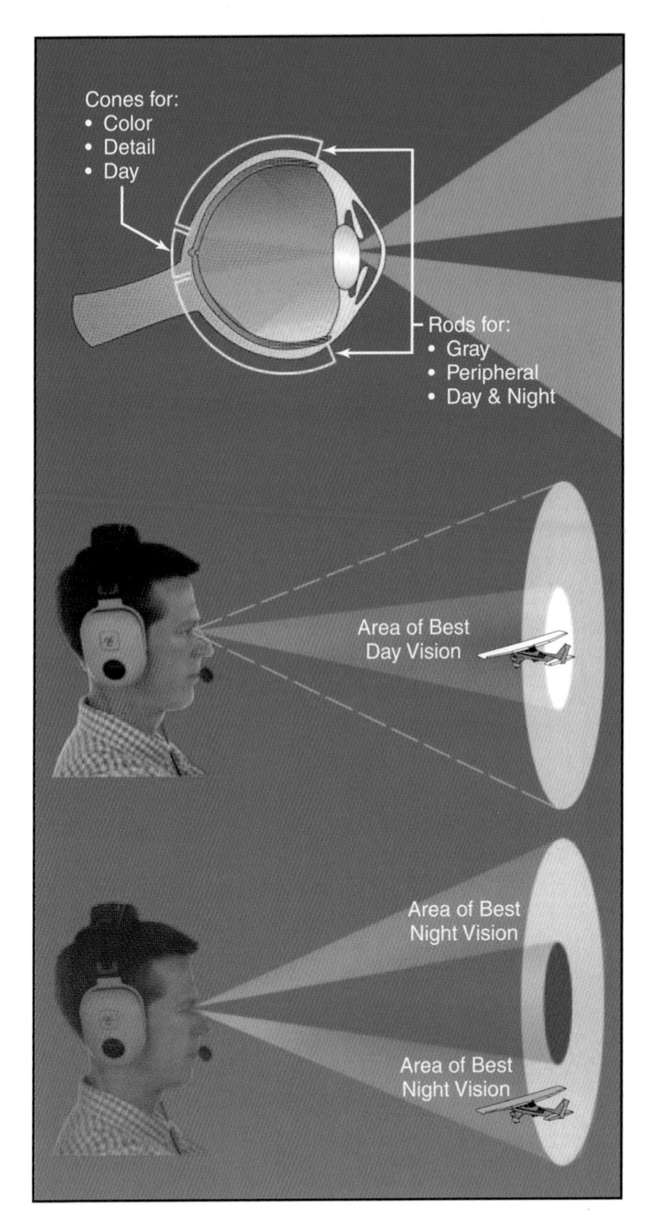

Figure 10-1. Rods and cones.

The eye's adaptation to darkness is another important aspect of night vision. When a dark room is entered, it is difficult to see anything until the eyes become adjusted to the darkness. Most everyone has experienced this after entering a darkened movie theater. In this process, the pupils of the eyes first enlarge to receive as much of the available light as possible. After approximately 5 to 10 minutes, the cones become adjusted to the dim light and the eyes become 100 times more sensitive to the light than they were before the dark room was entered. Much more time, about 30 minutes, is needed for the rods to become adjusted to darkness, but when they do adjust, they are about 100,000 times more sensitive to light than they were in the lighted area. After the adaptation process is complete, much more can be seen, especially if the eyes are used correctly.

After the eyes have adapted to the dark, the entire process is reversed when entering a lighted room. The eyes are first dazzled by the brightness, but become completely adjusted in a very few seconds, thereby losing their adaptation to the dark. Now, if the dark room is reentered, the eyes again go through the long process of adapting to the darkness.

The pilot before and during night flight must consider the adaptation process of the eyes. First, the eyes should be allowed to adapt to the low level of light and then they should be kept adapted. After the eyes have become adapted to the darkness, the pilot should avoid exposing them to any bright white light that will cause temporary blindness and could result in serious consequences.

Temporary blindness, caused by an unusually bright light, may result in illusions or after images until the eyes recover from the brightness. The brain creates these illusions reported by the eyes. This results in misjudging or incorrectly identifying objects, such as mistaking slanted clouds for the horizon or populated areas for a landing field. Vertigo is experienced as a feeling of dizziness and imbalance that can create or increase illusions. The illusions seem very real and pilots at every level of experience and skill can be affected. Recognizing that the brain and eyes can play tricks in this manner is the best protection for flying at night.

Good eyesight depends upon physical condition. Fatigue, colds, vitamin deficiency, alcohol, stimulants, smoking, or medication can seriously impair vision. Keeping these facts in mind and taking adequate precautions should safeguard night vision.

In addition to the principles previously discussed, the following items will aid in increasing night vision effectiveness.

- Adapt the eyes to darkness prior to flight and keep them adapted. About 30 minutes is needed to adjust the eyes to maximum efficiency after exposure to a bright light.
- If oxygen is available, use it during night flying. Keep in mind that a significant deterioration in night vision can occur at cabin altitudes as low as 5,000 feet.
- Close one eye when exposed to bright light to help avoid the blinding effect.
- Do not wear sunglasses after sunset.
- Move the eyes more slowly than in daylight.
- Blink the eyes if they become blurred.
- Concentrate on seeing objects.
- Force the eyes to view off center.
- Maintain good physical condition.
- Avoid smoking, drinking, and using drugs that may be harmful.

Night Illusions

In addition to night vision limitations, pilots should be aware that night illusions could cause confusion and concerns during night flying. The following discussion covers some of the common situations that cause illusions associated with night flying.

On a clear night, distant stationary lights can be mistaken for stars or other aircraft. Even the northern lights can confuse a pilot and indicate a false horizon. Certain geometrical patterns of ground lights, such as a freeway, runway, approach, or even lights on a moving train can cause confusion. Dark nights tend to eliminate reference to a visual horizon. As a result, pilots need to rely less on outside references at night and more on flight and navigation instruments.

Visual autokinesis can occur when a pilot stares at a single light source for several seconds on a dark night. The result is that the light will appear to be moving. The autokinesis effect will not occur if the pilot expands the visual field. It is a good procedure not to become fixed on one source of light.

Distractions and problems can result from a flickering light in the cockpit, anticollision light, strobe lights, or other aircraft lights and can cause flicker vertigo. If continuous, the possible physical reactions can be nausea, dizziness, grogginess, unconsciousness, headaches, or confusion. The pilot should try to eliminate any light source causing blinking or flickering problems in the cockpit.

A black-hole approach occurs when the landing is made from over water or non-lighted terrain where the runway lights are the only source of light. Without peripheral visual cues to help, pilots will have trouble orientating themselves relative to Earth. The runway can seem out of position (downsloping or upsloping) and in the worse case, results in landing short of the runway. If an electronic glide slope or visual approach slope indicator (VASI) is available, it should be used. If navigation aids (NAVAIDs) are unavailable, careful attention should be given to using the flight instruments to assist in maintaining orientation and a normal approach. If at any time the pilot is unsure of his or her position or attitude, a go-around should be executed.

Bright runway and approach lighting systems, especially where few lights illuminate the surrounding terrain, may create the illusion of less distance to the runway. In this situation, the tendency is to fly a higher approach. Also, when flying over terrain with only a few lights, it will make the runway recede or appear farther away. With this situation, the tendency is common to fly a lower-than-normal approach. If the runway has a city in the distance on higher terrain, the tendency will be to fly a lower-than-normal approach. A good review of the airfield layout and boundaries before initiating any approach will help the pilot maintain a safe approach angle.

Illusions created by runway lights result in a variety of problems. Bright lights or bold colors advance the runway, making it appear closer.

Night landings are further complicated by the difficulty of judging distance and the possibility of confusing approach and runway lights. For example, when a double row of approach lights joins the boundary lights of the runway, there can be confusion where the approach lights terminate and runway lights begin. Under certain conditions, approach lights can make the aircraft seem higher in a turn to final, than when its wings are level.

Pilot Equipment

Before beginning a night flight, carefully consider personal equipment that should be readily available during the flight. At least one reliable flashlight is recommended as standard equipment on all night flights. Remember to place a spare set of batteries in the flight kit. A D-cell size flashlight with a bulb switching mechanism that can be used to select white or red light is preferable. The white light is used while performing the preflight visual inspection of the airplane, and the red light is used when performing cockpit operations. Since the red light is nonglaring, it will not impair night vision. Some pilots prefer two flashlights, one with a white light for preflight, and the other a penlight type with a red light. The latter can be suspended by a string from around the neck to ensure the light is always readily available. One word of caution; if a red light is used for reading an aeronautical chart, the red features of the chart will not show up.

Aeronautical charts are essential for night cross-country flight and, if the intended course is near the edge of the chart, the adjacent chart should also be available. The lights of cities and towns can be seen at surprising distances at night, and if this adjacent chart is not available to identify those landmarks, confusion could result. Regardless of the equipment used, organization of the cockpit eases the burden on the pilot and enhances safety.

Airplane Equipment and Lighting

Title 14 of the Code of Federal Regulations (14 CFR) part 91 specifies the basic minimum airplane equipment required for night flight. This equipment includes only basic instruments, lights, electrical energy source, and spare fuses.

The standard instruments required for instrument flight under 14 CFR part 91 are a valuable asset for aircraft control at night. An anticollision light system, including a flashing or rotating beacon and position lights, is required airplane equipment. Airplane position lights are arranged similar to those of boats and ships. A red light is positioned on the left wingtip, a green light on the right wingtip, and a white light on the tail. [Figure 10-2]

Figure 10-2. Position lights.

This arrangement provides a means by which pilots can determine the general direction of movement of other airplanes in flight. If both a red and green light of another aircraft were observed, the airplane would be flying toward the pilot, and could be on a collision course.

Landing lights are not only useful for taxi, takeoffs, and landings, but also provide a means by which airplanes can be seen at night by other pilots. The Federal Aviation Administration (FAA) has initiated a voluntary pilot safety program called "Operation Lights ON." The "lights on" idea is to enhance the "see and be seen" concept of averting collisions both in the air and on the ground, and to reduce the potential for bird strikes. Pilots are encouraged to turn on their landing lights when operating within 10 miles of an airport. This is for both day and night, or in conditions of reduced visibility. This should also be done in areas where flocks of birds may be expected.

Although turning on aircraft lights supports the see and be seen concept, pilots should not become complacent about keeping a sharp lookout for other aircraft. Most aircraft lights blend in with the stars or the lights of the cities at night and go unnoticed unless a conscious effort is made to distinguish them from other lights.

AIRPORT AND NAVIGATION LIGHTING AIDS

The lighting systems used for airports, runways, obstructions, and other visual aids at night are other important aspects of night flying.

Lighted airports located away from congested areas can be identified readily at night by the lights outlining the runways. Airports located near or within large cities are often difficult to identify in the maze of lights. It is important not to only know the exact location of an airport relative to the city, but also to be able to identify these airports by the characteristics of their lighting pattern.

Aeronautical lights are designed and installed in a variety of colors and configurations, each having its own purpose. Although some lights are used only during low ceiling and visibility conditions, this discussion includes only the lights that are fundamental to visual flight rules (VFR) night operation.

It is recommended that prior to a night flight, and particularly a cross-country night flight, the pilot check the availability and status of lighting systems at the destination airport. This information can be found on aeronautical charts and in the *Airport/Facility Directory*. The status of each facility can be determined by reviewing pertinent *Notices to Airmen* (NOTAMs).

A rotating beacon is used to indicate the location of most airports. The beacon rotates at a constant speed, thus producing what appears to be a series of light flashes at regular intervals. These flashes may be one or two different colors that are used to identify various types of landing areas. For example:

- Lighted civilian land airports—alternating white and green.
- Lighted civilian water airports—alternating white and yellow.
- Lighted military airports—alternating white and green, but are differentiated from civil airports by dual peaked (two quick) white flashes, then green.

Beacons producing red flashes indicate obstructions or areas considered hazardous to aerial navigation. Steady burning red lights are used to mark obstructions on or near airports and sometimes to supplement flashing lights on en route obstructions. High intensity flashing white lights are used to mark some supporting structures of overhead transmission lines that stretch across rivers, chasms, and gorges. These high intensity lights are also used to identify tall structures, such as chimneys and towers.

As a result of the technological advancements in aviation, runway lighting systems have become quite sophisticated to accommodate takeoffs and landings in various weather conditions. However, the pilot whose flying is limited to VFR only needs to be concerned with the following basic lighting of runways and taxiways.

The basic runway lighting system consists of two straight parallel lines of runway-edge lights defining the lateral limits of the runway. These lights are aviation white, although aviation yellow may be substituted for a distance of 2,000 feet from the far end of the runway to indicate a caution zone. At some airports, the intensity of the runway-edge lights can be adjusted to satisfy the individual needs of the pilot. The length limits of the runway are defined by straight lines of lights across the runway ends. At some airports, the runway threshold lights are aviation green, and the runway end lights are aviation red.

At many airports, the taxiways are also lighted. A taxiway-edge lighting system consists of blue lights that outline the usable limits of taxi paths.

PREPARATION AND PREFLIGHT

Night flying requires that pilots be aware of, and operate within, their abilities and limitations. Although careful planning of any flight is essential, night flying demands more attention to the details of preflight preparation and planning.

Preparation for a night flight should include a thorough review of the available weather reports and forecasts with particular attention given to temperature/dewpoint spread. A narrow temperature/dewpoint spread may indicate the possibility of ground fog. Emphasis should also be placed on wind direction and speed, since its effect on the airplane cannot be as easily detected at night as during the day.

On night cross-country flights, appropriate aeronautical charts should be selected, including the appropriate adjacent charts. Course lines should be drawn in black to be more distinguishable.

Prominently lighted checkpoints along the prepared course should be noted. Rotating beacons at airports, lighted obstructions, lights of cities or towns, and lights from major highway traffic all provide excellent visual checkpoints. The use of radio navigation aids and communication facilities add significantly to the safety and efficiency of night flying.

All personal equipment should be checked prior to flight to ensure proper functioning. It is very disconcerting to find, at the time of need, that a flashlight, for example, does not work.

All airplane lights should be turned ON momentarily and checked for operation. Position lights can be checked for loose connections by tapping the light fixture. If the lights blink while being tapped, further investigation to determine the cause should be made prior to flight.

The parking ramp should be examined prior to entering the airplane. During the day, it is quite easy to see stepladders, chuckholes, wheel chocks, and other obstructions, but at night it is more difficult. A check of the area can prevent taxiing mishaps.

STARTING, TAXIING, AND RUNUP

After the pilot is seated in the cockpit and prior to starting the engine, all items and materials to be used on the flight should be arranged in such a manner that they will be readily available and convenient to use.

Extra caution should be taken at night to assure the propeller area is clear. Turning the rotating beacon ON, or flashing the airplane position lights will serve to alert persons nearby to remain clear of the propeller. To avoid excessive drain of electrical current from the battery, it is recommended that unnecessary electrical equipment be turned OFF until after the engine has been started.

After starting and before taxiing, the taxi or landing light should be turned ON. Continuous use of the landing light with r.p.m. power settings normally used for taxiing may place an excessive drain on the airplane's electrical system. Also, overheating of the landing light could become a problem because of inadequate airflow to carry the heat away. Landing lights should be used as necessary while taxiing. When using landing lights, consideration should be given to not blinding other pilots. Taxi slowly, particularly in congested areas. If taxi lines are painted on the ramp or taxiway, these lines should be followed to ensure a proper path along the route.

The before takeoff and runup should be performed using the checklist. During the day, forward movement of the airplane can be detected easily. At night, the airplane could creep forward without being noticed unless the pilot is alert for this possibility. Hold or lock the brakes during the runup and be alert for any forward movement.

TAKEOFF AND CLIMB

Night flying is very different from day flying and demands more attention of the pilot. The most noticeable difference is the limited availability of outside visual references. Therefore, flight instruments should be used to a greater degree in controlling the airplane. This is particularly true on night takeoffs and climbs. The cockpit lights should be adjusted to a minimum brightness that will allow the pilot to read the instruments and switches and yet not hinder the pilot's outside vision. This will also eliminate light reflections on the windshield and windows.

After ensuring that the final approach and runway are clear of other air traffic, or when cleared for takeoff by the tower, the landing lights and taxi lights should be turned ON and the airplane lined up with the centerline of the runway. If the runway does not have centerline lighting, use the painted centerline and the runway-edge lights. After the airplane is aligned, the heading indicator should be noted or set to correspond to the known runway direction. To begin the takeoff, the brakes should be released and the throttle smoothly advanced to maximum allowable power. As the airplane accelerates, it should be kept moving straight ahead between and parallel to the runway-edge lights.

The procedure for night takeoffs is the same as for normal daytime takeoffs except that many of the runway visual cues are not available. Therefore, the flight instruments should be checked frequently during the takeoff to ensure the proper pitch attitude, heading, and airspeed are being attained. As the airspeed reaches the normal lift-off speed, the pitch attitude should be adjusted to that which will establish a normal climb. This should be accomplished by referring to both outside visual references, such as lights, and to the flight instruments. [Figure 10-3]

Figure 10-3. Establish a positive climb.

After becoming airborne, the darkness of night often makes it difficult to note whether the airplane is getting closer to or farther from the surface. To ensure the airplane continues in a positive climb, be sure a climb is indicated on the attitude indicator, vertical speed indicator (VSI), and altimeter. It is also important to ensure the airspeed is at best climb speed.

Necessary pitch and bank adjustments should be made by referencing the attitude and heading indicators. It is recommended that turns not be made until reaching a safe maneuvering altitude.

Although the use of the landing [illegible] during the takeoff, they become [illegible] airplane has climbed to an altit[illegible] beam no longer extends to the s[illegible] cause distortion when it is reflect[illegible] fog that might exist in the climb. [illegible] landing light is used for the take[illegible] off after the climb is well establ[illegible] traffic in the area does not requi[illegible] avoidance.

ORIENTATION AND NA[illegible]

Generally, at night it is diffic[illegible] restrictions to visibility, particul[illegible] under overcast. The pilot flying [illegible] cise caution to avoid flying int[illegible] fog. Usually, the first indication [illegible] visibility conditions is the gradual disappearance of lights on the ground. If the lights begin to take on an appearance of being surrounded by a halo or glow, the pilot should use caution in attempting further flight in that same direction. Such a halo or glow around lights on the ground is indicative of ground fog. Remember that if a descent must be made through fog, smoke, or haze in order to land, the horizontal visibility is considerably less when looking through the restriction than it is when looking straight down through it from above. Under no circumstances should a VFR night-flight be made during poor or marginal weather conditions unless both the pilot and aircraft are certificated and equipped for flight under instrument flight rules (IFR).

The pilot should practice and acquire competency in straight-and-level flight, climbs and descents, level turns, climbing and descending turns, and steep turns. Recovery from unusual attitudes should also be practiced, but only on dual flights with a flight instructor. The pilot should also practice these maneuvers with all the cockpit lights turned OFF. This blackout training is necessary if the pilot experiences an electrical or instrument light failure. Training should also include using the navigation equipment and local NAVAIDs.

In spite of fewer references or checkpoints, night cross-country flights do not present particular problems if preplanning is adequate, and the pilot continues to monitor position, time estimates, and fuel consumed. NAVAIDs, if available, should be used to assist in monitoring en route progress.

Crossing large bodies of water at night in single-engine airplanes could be potentially hazardous, not only from the standpoint of landing (ditching) in the water, but also because with little or no lighting the horizon blends with the water, in which case, depth perception and orientation become difficult. During poor visibility conditions over water, the horizon will become obscure, and may result in a loss of orienta-[illegible]s, the stars may be reflected [illegible]ch could appear as a contin-[illegible] making the horizon difficult [illegible]

[illegible]lings, or other objects may [illegible]lot when seen from different [illegible]f 2,000 feet, a group of lights [illegible]n individually, while at 5,000 [illegible] lights could appear to be one [illegible] illusions may become quite [illegible]ges and if not overcome could [illegible]pect to approaches to lighted [illegible]

[illegible]ND LANDINGS

[illegible] airport to enter the traffic pat-[illegible]portant that the runway lights and other airport lighting be identified as early as possible. If the airport layout is unfamiliar to the pilot, sighting of the runway may be difficult until very close-in due to the maze of lights observed in the area. [Figure 10-4] The pilot should fly toward the rotating beacon until the lights outlining the runway are distinguishable. To fly a traffic pattern of proper size and direction, the runway threshold and runway-edge lights must be positively identified. Once the airport lights are seen, these lights should be kept in sight throughout the approach.

Figure 10-4. Use light patterns for orientation.

Distance may be deceptive at night due to limited lighting conditions. A lack of intervening references on the ground and the inability of the pilot to compare the size and location of different ground objects cause this. This also applies to the estimation of altitude and speed. Consequently, more dependence must be placed on flight instruments, particularly the altimeter and the airspeed indicator.

Figure 10-5. VASI.

When entering the traffic pattern, allow for plenty of time to complete the before landing checklist. If the heading indicator contains a heading bug, setting it to the runway heading will be an excellent reference for the pattern legs.

Every effort should be made to maintain the recommended airspeeds and execute the approach and landing in the same manner as during the day. A low, shallow approach is definitely inappropriate during a night operation. The altimeter and VSI should be constantly cross-checked against the airplane's position along the base leg and final approach. A visual approach slope indicator (VASI) is an indispensable aid in establishing and maintaining a proper glidepath. [Figure 10-5]

After turning onto the final approach and aligning the airplane midway between the two rows of runway-edge lights, the pilot should note and correct for any wind drift. Throughout the final approach, pitch and power should be used to maintain a stabilized approach. Flaps should be used the same as in a normal approach. Usually, halfway through the final approach, the landing light should be turned on. Earlier use of the landing light may be necessary because of "Operation Lights ON" or for local traffic considerations. The landing light is sometimes ineffective since the light beam will usually not reach the ground from higher altitudes. The light may even be reflected back into the pilot's eyes by any existing haze, smoke, or fog. This disadvantage is overshadowed by the safety considerations provided by using the "Operation Lights ON" procedure around other traffic.

The roundout and touchdown should be made in the same manner as in day landings. At night, the judgment of height, speed, and sink rate is impaired by the scarcity of observable objects in the landing area. The inexperienced pilot may have a tendency to round out too high until attaining familiarity with the proper height for the correct roundout. To aid in determining the proper roundout point, continue a constant approach descent until the landing lights reflect on the runway and tire marks on the runway can be seen clearly. At this point the roundout should be started smoothly and the throttle gradually reduced to idle as the airplane is touching down. [Figure 10-6] During landings without the use of landing lights, the roundout may be started when the runway lights at the

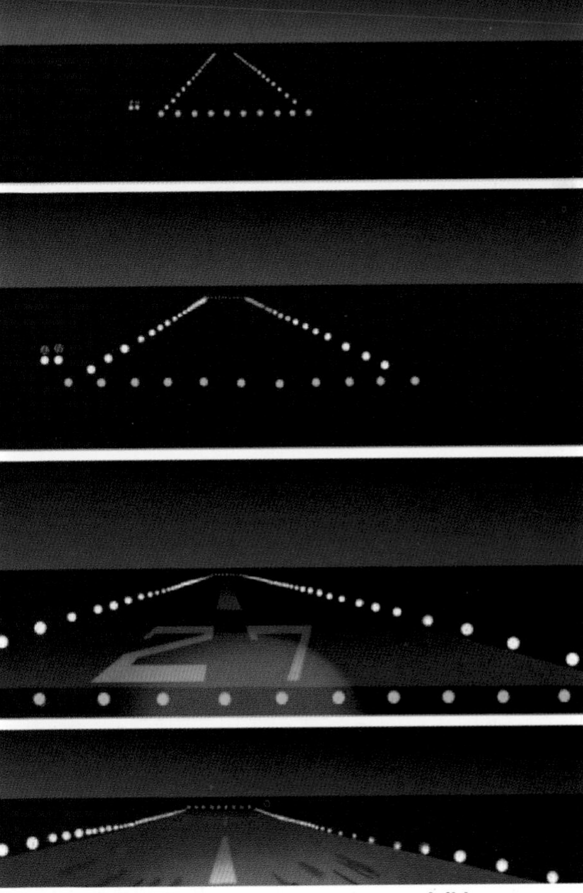

Figure 10-6. Roundout when tire marks are visible.

far end of the runway first appear to be rising higher than the nose of the airplane. This demands a smooth and very timely roundout, and requires that the pilot feel for the runway surface using power and pitch changes, as necessary, for the airplane to settle slowly to the runway. Blackout landings should always be included in night pilot training as an emergency procedure.

Night Emergencies

Perhaps the pilot's greatest concern about flying a single-engine airplane at night is the possibility of a complete engine failure and the subsequent emergency landing. This is a legitimate concern, even though continuing flight into adverse weather and poor pilot judgment account for most serious accidents.

If the engine fails at night, several important procedures and considerations to keep in mind are:

- Maintain positive control of the airplane and establish the best glide configuration and airspeed. Turn the airplane towards an airport or away from congested areas.

- Check to determine the cause of the engine malfunction, such as the position of fuel selectors, magneto switch, or primer. If possible, the cause of the malfunction should be corrected immediately and the engine restarted.

- Announce the emergency situation to Air Traffic Control (ATC) or UNICOM. If already in radio contact with a facility, do not change frequencies, unless instructed to change.

- If the condition of the nearby terrain is known, turn towards an unlighted portion of the area. Plan an emergency approach to an unlighted portion.

- Consider an emergency landing area close to public access if possible. This may facilitate rescue or help, if needed.

- Maintain orientation with the wind to avoid a downwind landing.

- Complete the before landing checklist, and check the landing lights for operation at altitude and turn ON in sufficient time to illuminate the terrain or obstacles along the flightpath. The landing should be completed in the normal landing attitude at the slowest possible airspeed. If the landing lights are unusable and outside visual references are not available, the airplane should be held in level-landing attitude until the ground is contacted.

- After landing, turn off all switches and evacuate the airplane as quickly as possible.

Chapter 11
Transition to Complex Airplanes

HIGH PERFORMANCE AND COMPLEX AIRPLANES

Transition to a complex airplane, or a high performance airplane, can be demanding for most pilots without previous experience. Increased performance and increased complexity both require additional planning, judgment, and piloting skills. Transition to these types of airplanes, therefore, should be accomplished in a systematic manner through a structured course of training administered by a qualified flight instructor.

A complex airplane is defined as an airplane equipped with a retractable landing gear, wing flaps, and a controllable-pitch propeller. For a seaplane to be considered complex, it is required to have wing flaps and a controllable-pitch propeller. A high performance airplane is defined as an airplane with an engine of more than 200 horsepower.

WING FLAPS

Airplanes can be designed to fly fast or slow. High speed requires thin, moderately **cambered** airfoils with a small wing area, whereas the high lift needed for low speeds is obtained with thicker highly cambered airfoils with a larger wing area. [Figure 11-1] Many attempts have been made to compromise this conflicting requirement of high cruise and slow landing speeds.

Since an airfoil cannot have two different cambers at the same time, one of two things must be done. Either the airfoil can be a compromise, or a cruise airfoil can be combined with a device for increasing the camber of the airfoil for low-speed flight. One method for varying an airfoil's camber is the addition of trailing edge flaps. Engineers call these devices a high-lift system.

FUNCTION OF FLAPS

Flaps work primarily by changing the camber of the airfoil since deflection adds aft camber. Flap deflection does not increase the critical (stall) angle of attack, and in some cases flap deflection actually decreases the critical angle of attack.

Deflection of trailing edge control surfaces, such as the aileron, alters both lift and drag. With aileron deflection, there is asymmetrical lift (rolling moment) and drag (adverse yaw). Wing flaps differ in that deflection acts symmetrically on the airplane. There is no roll or yaw effect, and pitch changes depend on the airplane design.

Figure 11-1. Airfoil types.

Pitch behavior depends on flap type, wing position, and horizontal tail location. The increased camber from flap deflection produces lift primarily on the rear portion of the wing. This produces a nosedown pitching moment; however, the change in tail load from the downwash deflected by the flaps over the horizontal tail has a significant influence on the pitching moment. Consequently, pitch behavior depends on the design features of the particular airplane.

Flap deflection of up to 15° primarily produces lift with minimal drag. The tendency to balloon up with initial flap deflection is because of lift increase, but the nosedown pitching moment tends to offset the balloon. Deflection beyond 15° produces a large increase in drag. Drag from flap deflection is **parasite drag**, and as such is proportional to the square of the speed. Also, deflection beyond 15° produces a significant noseup pitching moment in most high-wing airplanes because the resulting downwash increases the airflow over the horizontal tail.

FLAP EFFECTIVENESS

Flap effectiveness depends on a number of factors, but the most noticeable are size and type. For the purpose of this chapter, trailing edge flaps are classified as four basic types: plain (hinge), split, slotted, and Fowler. [Figure 11-2]

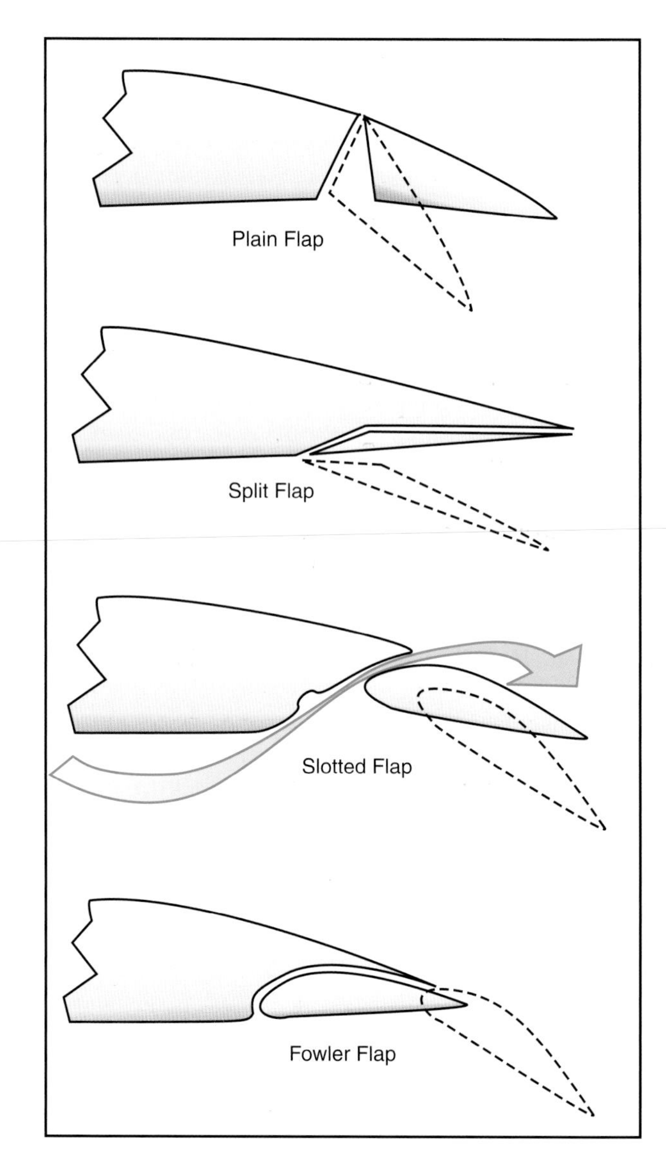

Figure 11-2. Four basic types of flaps.

The plain or hinge flap is a hinged section of the wing. The structure and function are comparable to the other control surfaces—ailerons, rudder, and elevator. The split flap is more complex. It is the lower or underside portion of the wing; deflection of the flap leaves the trailing edge of the wing undisturbed. It is, however, more effective than the hinge flap because of greater lift and less pitching moment, but there is more drag. Split flaps are more useful for landing, but the partially deflected hinge flaps have the advantage in takeoff. The split flap has significant drag at small deflections, whereas the hinge flap does not because airflow remains "attached" to the flap.

The slotted flap has a gap between the wing and the leading edge of the flap. The slot allows high pressure airflow on the wing undersurface to energize the lower pressure over the top, thereby delaying flow separation. The slotted flap has greater lift than the hinge flap but less than the split flap; but, because of a higher lift-drag ratio, it gives better takeoff and climb performance. Small deflections of the slotted flap give a higher drag than the hinge flap but less than the split. This allows the slotted flap to be used for takeoff.

The Fowler flap deflects down and aft to increase the wing area. This flap can be multi-slotted making it the most complex of the trailing edge systems. This system does, however, give the maximum **lift coefficient**. Drag characteristics at small deflections are much like the slotted flap. Because of structural complexity and difficulty in sealing the slots, Fowler flaps are most commonly used on larger airplanes.

OPERATIONAL PROCEDURES

It would be impossible to discuss all the many airplane design and flap combinations. This emphasizes the importance of the FAA-approved Airplane Flight Manual and/or Pilot's Operating Handbook (AFM/POH) for a given airplane. However, while some AFM/POHs are specific as to operational use of flaps, many are lacking. Hence, flap operation makes pilot judgment of critical importance. In addition, flap operation is used for landings and takeoffs, during which the airplane is in close proximity to the ground where the margin for error is small.

Since the recommendations given in the AFM/POH are based on the airplane and the flap design combination, the pilot must relate the manufacturer's recommendation to aerodynamic effects of flaps. This requires that the pilot have a basic background knowledge of flap aerodynamics and geometry. With this information, the pilot must make a decision as to the degree of flap deflection and time of deflection based on runway and approach conditions relative to the wind conditions.

The time of flap extension and degree of deflection are related. Large flap deflections at one single point in the landing pattern produce large lift changes that require significant pitch and power changes in order to maintain airspeed and glide slope. Incremental deflection of flaps on downwind, base, and final approach allow smaller adjustment of pitch and power compared to extension of full flaps all at one time. This procedure facilitates a more stabilized approach.

A soft- or short-field landing requires minimal speed at touchdown. The flap deflection that results in minimal groundspeed, therefore, should be used. If obstacle clearance is a factor, the flap deflection that results in the steepest angle of approach should be used. It should be noted, however, that the flap setting that gives the minimal speed at touchdown does not necessarily give the steepest angle of approach; however, maximum flap extension gives the steepest angle of approach and minimum speed at touchdown. Maximum flap extension, particularly beyond 30 to 35°, results in a large amount of drag. This requires higher power settings than used with partial flaps. Because of the steep approach angle combined with power to offset drag, the flare with full flaps becomes critical. The drag produces a high sink rate that must be controlled with power, yet failure to reduce power at a rate so that the power is idle at touchdown allows the airplane to float down the runway. A reduction in power too early results in a hard landing.

Crosswind component is another factor to be considered in the degree of flap extension. The deflected flap presents a surface area for the wind to act on. In a crosswind, the "flapped" wing on the upwind side is more affected than the downwind wing. This is, however, eliminated to a slight extent in the crabbed approach since the airplane is more nearly aligned with the wind. When using a wing low approach, however, the lowered wing partially blankets the upwind flap, but the dihedral of the wing combined with the flap and wind make lateral control more difficult. Lateral control becomes more difficult as flap extension reaches maximum and the crosswind becomes perpendicular to the runway.

Crosswind effects on the "flapped" wing become more pronounced as the airplane comes closer to the ground. The wing, flap, and ground form a "container" that is filled with air by the crosswind. With the wind striking the deflected flap and fuselage side and with the flap located behind the main gear, the upwind wing will tend to rise and the airplane will tend to turn into the wind. Proper control position, therefore, is essential for maintaining runway alignment. Also, it may be necessary to retract the flaps upon positive ground contact.

The go-around is another factor to consider when making a decision about degree of flap deflection and about where in the landing pattern to extend flaps. Because of the nosedown pitching moment produced with flap extension, trim is used to offset this pitching moment. Application of full power in the go-around increases the airflow over the "flapped" wing. This produces additional lift causing the nose to pitch up. The pitch-up tendency does not diminish completely with flap retraction because of the trim setting. Expedient retraction of flaps is desirable to eliminate drag, thereby allowing rapid increase in airspeed; however, flap retraction also decreases lift so that the airplane sinks rapidly.

The degree of flap deflection combined with design configuration of the horizontal tail relative to the wing requires that the pilot carefully monitor pitch and airspeed, carefully control flap retraction to minimize altitude loss, and properly use the rudder for coordination. Considering these factors, the pilot should extend the same degree of deflection at the same point in the landing pattern. This requires that a consistent traffic pattern be used. Therefore, the pilot can have a preplanned go-around sequence based on the airplane's position in the landing pattern.

There is no single formula to determine the degree of flap deflection to be used on landing, because a landing involves variables that are dependent on each other. The AFM/POH for the particular airplane will contain the manufacturer's recommendations for some landing situations. On the other hand, AFM/POH information on flap usage for takeoff is more precise. The manufacturer's requirements are based on the climb performance produced by a given flap design. Under no circumstances should a flap setting given in the AFM/POH be exceeded for takeoff.

CONTROLLABLE-PITCH PROPELLER

Fixed-pitch propellers are designed for best efficiency at one speed of rotation and forward speed. This type of propeller will provide suitable performance in a narrow range of airspeeds; however, efficiency would suffer considerably outside this range. To provide high propeller efficiency through a wide range of operation, the **propeller blade angle** must be controllable. The most convenient way of controlling the propeller blade angle is by means of a constant-speed governing system.

CONSTANT-SPEED PROPELLER

The constant-speed propeller keeps the blade angle adjusted for maximum efficiency for most conditions of flight. When an engine is running at constant speed, the torque (power) exerted by the engine at the propeller shaft must equal the opposing load provided by the resistance of the air. The r.p.m. is controlled by regulating the torque absorbed by the propeller—in other words by increasing or decreasing the resistance offered by the air to the propeller. In the case of a fixed-pitch propeller, the torque absorbed by the propeller is a function of speed, or r.p.m. If the power output of the engine is changed, the engine will accelerate or decelerate until an r.p.m. is reached at which the power delivered is equal to the power absorbed. In the case of a constant-speed propeller, the power absorbed is independent of the r.p.m., for by varying the **pitch** of the blades, the air resistance and hence the torque or load, can be changed without reference to propeller speed. This is accomplished with a constant-speed propeller by means of a governor. The governor, in most cases, is geared to the engine crankshaft and thus is sensitive to changes in engine r.p.m.

The pilot controls the engine r.p.m. indirectly by means of a propeller control in the cockpit, which is connected to the governor. For maximum takeoff power, the propeller control is moved all the way forward to the low pitch/high r.p.m. position, and the throttle is moved forward to the maximum allowable manifold pressure position. To reduce power for climb or cruise, manifold pressure is reduced to the desired value with the throttle, and the engine r.p.m. is reduced by moving the propeller control back toward the high pitch/low r.p.m. position until the desired r.p.m. is observed on the tachometer. Pulling back on the propeller control causes the propeller blades to move to a higher angle. Increasing the propeller blade angle (of attack) results in an increase in the resistance of the air. This puts a load on the engine so it slows down. In other words, the resistance of the air at the higher blade angle is greater than the torque, or power, delivered to the propeller by the engine, so it slows down to a point where the two forces are in balance.

When an airplane is nosed up into a climb from level flight, the engine will tend to slow down. Since the governor is sensitive to small changes in engine r.p.m., it will decrease the blade angle just enough to keep the engine speed from falling off. If the airplane is nosed down into a dive, the governor will increase the blade angle enough to prevent the engine from overspeeding. This allows the engine to maintain a constant r.p.m., and thus maintain the power output. Changes in airspeed and power can be obtained by changing r.p.m. at a constant manifold pressure; by changing the manifold pressure at a constant r.p.m.; or by changing both r.p.m. and manifold pressure. Thus the constant-speed propeller makes it possible to obtain an infinite number of power settings.

TAKEOFF, CLIMB, AND CRUISE

During takeoff, when the forward motion of the airplane is at low speeds and when maximum power and thrust are required, the constant-speed propeller sets up a low propeller blade angle (pitch). The low blade angle keeps the angle of attack, with respect to the relative wind, small and efficient at the low speed. [Figure 11-3]

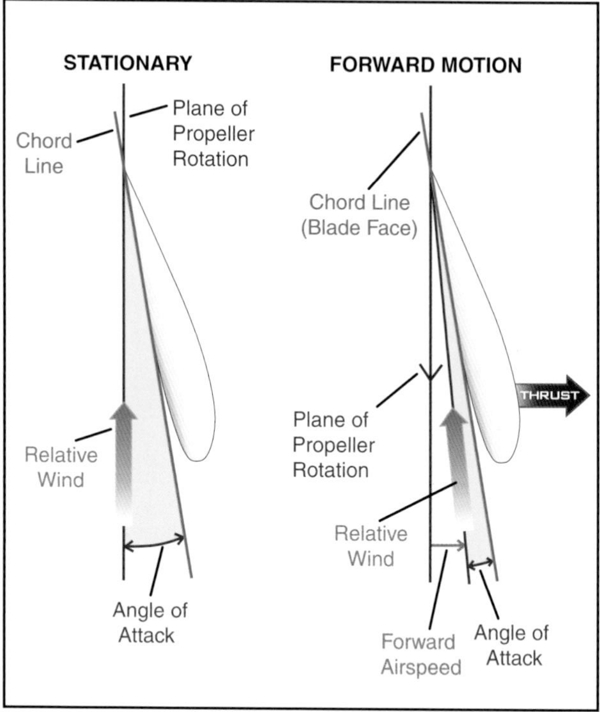

Figure 11-3. Propeller blade angle.

At the same time, it allows the propeller to "slice it thin" and handle a smaller mass of air per revolution. This light load allows the engine to turn at maximum r.p.m. and develop maximum power. Although the mass of air per revolution is small, the number of revolutions per minute is high. Thrust is maximum at the beginning of the takeoff and then decreases as the airplane gains speed and the airplane drag increases. Due to the high slipstream velocity during takeoff, the effective lift of the wing behind the propeller(s) is increased.

As the airspeed increases after lift-off, the load on the engine is lightened because of the small blade angle. The governor senses this and increases the blade angle slightly. Again, the higher blade angle, with the higher speeds, keeps the angle of attack with respect to the relative wind small and efficient.

For climb after takeoff, the power output of the engine is reduced to climb power by decreasing the manifold pressure and lowering r.p.m. by increasing the blade angle. At the higher (climb) airspeed and the higher blade angle, the propeller is handling a greater mass of air per second at a lower slipstream velocity. This reduction in power is offset by the increase in propeller efficiency. The angle of attack is again kept small by the increase in the blade angle with an increase in airspeed.

At cruising altitude, when the airplane is in level flight, less power is required to produce a higher airspeed than is used in climb. Consequently, engine power is again reduced by lowering the manifold pressure and increasing the blade angle (to decrease r.p.m.). The higher airspeed and higher blade angle enable the propeller to handle a still greater mass of air per second at still smaller slipstream velocity. At normal cruising speeds, propeller efficiency is at, or near maximum efficiency. Due to the increase in blade angle and airspeed, the angle of attack is still small and efficient.

BLADE ANGLE CONTROL

Once the pilot selects the r.p.m. settings for the propeller, the propeller governor automatically adjusts the blade angle to maintain the selected r.p.m. It does this by using oil pressure. Generally, the oil pressure used for pitch change comes directly from the engine lubricating system. When a governor is employed, engine oil is used and the oil pressure is usually boosted by a pump, which is integrated with the governor. The higher pressure provides a quicker blade angle change. The r.p.m. at which the propeller is to operate is adjusted in the governor head. The pilot changes this setting by changing the position of the governor rack through the cockpit propeller control.

On some constant-speed propellers, changes in pitch are obtained by the use of an inherent centrifugal twisting moment of the blades that tends to flatten the blades toward low pitch, and oil pressure applied to a hydraulic piston connected to the propeller blades which moves them toward high pitch. Another type of constant-speed propeller uses counterweights attached to the blade shanks in the hub. Governor oil pressure and the blade twisting moment move the blades toward the low pitch position, and centrifugal force acting on the counterweights moves them (and the blades) toward the high pitch position. In the first case above, governor oil pressure moves the blades towards high pitch, and in the second case, governor oil pressure and the blade twisting moment move the blades toward low pitch. A loss of governor oil pressure, therefore, will affect each differently.

GOVERNING RANGE

The blade angle range for constant-speed propellers varies from about 11 1/2 to 40°. The higher the speed of the airplane, the greater the blade angle range. [Figure 11-4]

The range of possible blade angles is termed the propeller's governing range. The governing range is defined by the limits of the propeller blade's travel between high and low blade angle pitch stops. As long as the propeller blade angle is within the governing range and not against either pitch stop, a constant engine r.p.m. will be maintained. However, once the propeller blade reaches its pitch-stop limit, the engine r.p.m. will increase or decrease with changes in airspeed and propeller load similar to a fixed-pitch propeller. For example, once a specific r.p.m. is selected, if the airspeed decreases enough, the propeller blades will reduce pitch, in an attempt to maintain the selected r.p.m., until they contact their low pitch stops. From that point, any further reduction in airspeed will cause the engine r.p.m. to decrease. Conversely, if the airspeed increases, the propeller blade angle will increase until the high pitch stop is reached. The engine r.p.m. will then begin to increase.

CONSTANT-SPEED PROPELLER OPERATION

The engine is started with the propeller control in the low pitch/high r.p.m. position. This position reduces the load or drag of the propeller and the result is easier starting and warm-up of the engine. During warm-up, the propeller blade changing mechanism should be operated slowly and smoothly through a full cycle. This is done by moving the propeller control (with the

Aircraft Type	Design Speed (m.p.h.)	Blade Angle Range	Pitch Low	Pitch High
Fixed Gear	160	11 1/2°	10 1/2°	22°
Retractable	180	15°	11°	26°
Turbo Retractable	225/240	20°	14°	34°
Turbine Retractable	250/300	30°	10°	40°
Transport Retractable	325	40°	10/15°	50/55°

Figure 11-4. Blade angle range (values are approximate).

manifold pressure set to produce about 1,600 r.p.m.) to the high pitch/low r.p.m. position, allowing the r.p.m. to stabilize, and then moving the propeller control back to the low pitch takeoff position. This should be done for two reasons: to determine whether the system is operating correctly, and to circulate fresh warm oil through the propeller governor system. It should be remembered that the oil has been trapped in the propeller cylinder since the last time the engine was shut down. There is a certain amount of leakage from the propeller cylinder, and the oil tends to congeal, especially if the outside air temperature is low. Consequently, if the propeller isn't exercised before takeoff, there is a possibility that the engine may **overspeed** on takeoff.

An airplane equipped with a constant-speed propeller has better takeoff performance than a similarly powered airplane equipped with a fixed-pitch propeller. This is because with a constant-speed propeller, an airplane can develop its maximum rated horsepower (red line on the tachometer) while motionless. An airplane with a fixed-pitch propeller, on the other hand, must accelerate down the runway to increase airspeed and aerodynamically unload the propeller so that r.p.m. and horsepower can steadily build up to their maximum. With a constant-speed propeller, the tachometer reading should come up to within 40 r.p.m. of the red line as soon as full power is applied, and should remain there for the entire takeoff.

Excessive manifold pressure raises the cylinder compression pressure, resulting in high stresses within the engine. Excessive pressure also produces high engine temperatures. A combination of high manifold pressure and low r.p.m. can induce damaging **detonation**. In order to avoid these situations, the following sequence should be followed when making power changes.

- When increasing power, increase the r.p.m. first, and then the manifold pressure.
- When decreasing power, decrease the manifold pressure first, and then decrease the r.p.m.

It is a fallacy that (in non-turbocharged engines) the manifold pressure in inches of mercury (inches Hg) should never exceed r.p.m. in hundreds for cruise power settings. The cruise power charts in the AFM/POH should be consulted when selecting cruise power settings. Whatever the combinations of r.p.m. and manifold pressure listed in these charts—they have been flight tested and approved by the airframe and powerplant engineers for the respective airframe and engine manufacturer. Therefore, if there are power settings such as 2,100 r.p.m. and 24 inches manifold pressure in the power chart, they are approved for use.

With a constant-speed propeller, a power descent can be made without overspeeding the engine. The system compensates for the increased airspeed of the descent by increasing the propeller blade angles. If the descent is too rapid, or is being made from a high altitude, the maximum blade angle limit of the blades is not sufficient to hold the r.p.m. constant. When this occurs, the r.p.m. is responsive to any change in throttle setting.

Some pilots consider it advisable to set the propeller control for maximum r.p.m. during the approach to have full horsepower available in case of emergency. If the governor is set for this higher r.p.m. early in the approach when the blades have not yet reached their minimum angle stops, the r.p.m. may increase to unsafe limits. However, if the propeller control is not readjusted for the takeoff r.p.m. until the approach is almost completed, the blades will be against, or very near their minimum angle stops and there will be little if any change in r.p.m. In case of emergency, both throttle and propeller controls should be moved to takeoff positions.

Many pilots prefer to feel the airplane respond immediately when they give short bursts of the throttle during approach. By making the approach under a little power and having the propeller control set at or near cruising r.p.m., this result can be obtained.

Although the governor responds quickly to any change in throttle setting, a sudden and large increase in the throttle setting will cause a momentary overspeeding of the engine until the blades become adjusted to absorb the increased power. If an emergency demanding full power should arise during approach, the sudden advancing of the throttle will cause momentary overspeeding of the engine beyond the r.p.m. for which the governor is adjusted. This temporary increase in engine speed acts as an emergency power reserve.

Some important points to remember concerning constant-speed propeller operation are:

- The red line on the tachometer not only indicates maximum allowable r.p.m.; it also indicates the r.p.m. required to obtain the engine's rated horsepower.
- A momentary propeller overspeed may occur when the throttle is advanced rapidly for takeoff. This is usually not serious if the rated r.p.m. is not exceeded by 10 percent for more than 3 seconds.
- The green arc on the tachometer indicates the normal operating range. When developing

power in this range, the engine drives the propeller. Below the green arc, however, it is usually the **windmilling** propeller that powers the engine. Prolonged operation below the green arc can be detrimental to the engine.

- On takeoffs from low elevation airports, the manifold pressure in inches of mercury may exceed the r.p.m. This is normal in most cases. The pilot should consult the AFM/POH for limitations.
- All power changes should be made smoothly and slowly to avoid **overboosting** and/or overspeeding.

TURBOCHARGING

The turbocharged engine allows the pilot to maintain sufficient cruise power at high altitudes where there is less drag, which means faster true airspeeds and increased range with fuel economy. At the same time, the powerplant has flexibility and can be flown at a low altitude without the increased fuel consumption of a turbine engine. When attached to the standard powerplant, the turbocharger does not take any horsepower from the powerplant to operate; it is relatively simple mechanically, and some models can pressurize the cabin as well.

The turbocharger is an exhaust-driven device, which raises the pressure and density of the induction air delivered to the engine. It consists of two separate components: a compressor and a turbine connected by a common shaft. The compressor supplies pressurized air to the engine for high altitude operation. The compressor and its housing are between the ambient air intake and the induction air manifold. The turbine and its housing are part of the exhaust system and utilize the flow of exhaust gases to drive the compressor. [Figure 11-5]

The turbine has the capability of producing **manifold pressure** in excess of the maximum allowable for the particular engine. In order not to exceed the maximum allowable manifold pressure, a bypass or waste gate is used so that some of the exhaust will be diverted overboard before it passes through the turbine.

The position of the waste gate regulates the output of the turbine and therefore, the compressed air available to the engine. When the waste gate is closed, all of the exhaust gases pass through and drive the turbine. As the waste gate opens, some of the exhaust gases are routed around the turbine, through the exhaust bypass and overboard through the exhaust pipe.

The waste gate actuator is a spring-loaded piston, operated by engine oil pressure. The actuator, which adjusts the waste gate position, is connected to the waste gate by a mechanical linkage.

The control center of the turbocharger system is the pressure controller. This device simplifies turbocharging to one control: the throttle. Once the pilot has set the desired manifold pressure, virtually no throttle adjustment is required with changes in altitude. The controller senses compressor discharge requirements for various altitudes and controls the oil pressure to the waste gate actuator which adjusts the waste gate accordingly. Thus the turbocharger maintains only the manifold pressure called for by the throttle setting.

GROUND BOOSTING VS. ALTITUDE TURBOCHARGING

Altitude turbocharging (sometimes called "normalizing") is accomplished by using a turbocharger that will maintain maximum allowable sea level manifold pressure (normally 29 – 30 inches Hg) up to a certain altitude. This altitude is specified by the airplane manufacturer and is referred to as the airplane's **critical altitude**. Above the critical altitude,

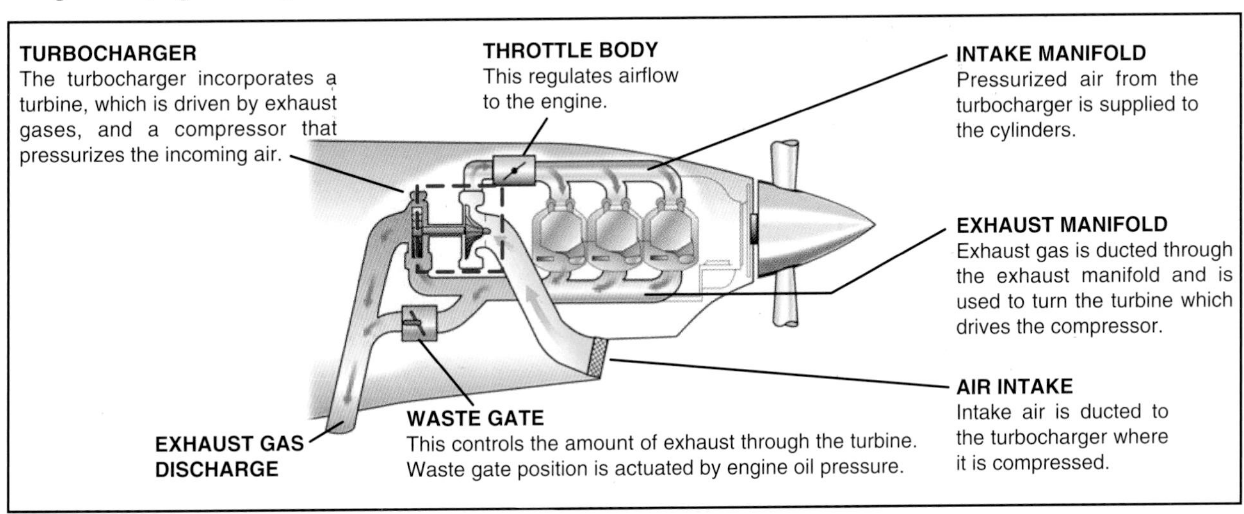

Figure 11-5. Turbocharging system.

the manifold pressure decreases as additional altitude is gained. Ground boosting, on the other hand, is an application of turbocharging where more than the standard 29 inches of manifold pressure is used in flight. In various airplanes using ground boosting, takeoff manifold pressures may go as high as 45 inches of mercury.

Although a sea level power setting and maximum r.p.m. can be maintained up to the critical altitude, this does not mean that the engine is developing sea level power. Engine power is not determined just by manifold pressure and r.p.m. Induction air temperature is also a factor. Turbocharged induction air is heated by compression. This temperature rise decreases induction air density which causes a power loss. Maintaining the equivalent horsepower output will require a somewhat higher manifold pressure at a given altitude than if the induction air were not compressed by turbocharging. If, on the other hand, the system incorporates an automatic density controller which, instead of maintaining a constant manifold pressure, automatically positions the waste gate so as to maintain constant air density to the engine, a near constant horsepower output will result.

OPERATING CHARACTERISTICS

First and foremost, all movements of the power controls on turbocharged engines should be slow and gentle. Aggressive and/or abrupt throttle movements increase the possibility of **overboosting**. The pilot should carefully monitor engine indications when making power changes.

When the waste gate is open, the turbocharged engine will react the same as a normally aspirated engine when the r.p.m. is varied. That is, when the r.p.m. is increased, the manifold pressure will decrease slightly. When the engine r.p.m. is decreased, the manifold pressure will increase slightly. However, when the waste gate is closed, manifold pressure variation with engine r.p.m. is just the opposite of the normally aspirated engine. An increase in engine r.p.m. will result in an increase in manifold pressure, and a decrease in engine r.p.m. will result in a decrease in manifold pressure.

Above the critical altitude, where the waste gate is closed, any change in airspeed will result in a corresponding change in manifold pressure. This is true because the increase in ram air pressure with an increase in airspeed is magnified by the compressor resulting in an increase in manifold pressure. The increase in manifold pressure creates a higher mass flow through the engine, causing higher turbine speeds and thus further increasing manifold pressure.

When running at high altitudes, aviation gasoline may tend to vaporize prior to reaching the cylinder. If this occurs in the portion of the fuel system between the fuel tank and the engine-driven fuel pump, an auxiliary positive pressure pump may be needed in the tank. Since engine-driven pumps pull fuel, they are easily vapor locked. A boost pump provides positive pressure—pushes the fuel—reducing the tendency to vaporize.

HEAT MANAGEMENT

Turbocharged engines must be thoughtfully and carefully operated, with continuous monitoring of pressures and temperatures. There are two temperatures that are especially important—turbine inlet temperature (TIT) or in some installations exhaust gas temperature (EGT), and cylinder head temperature. TIT or EGT limits are set to protect the elements in the hot section of the turbocharger, while cylinder head temperature limits protect the engine's internal parts.

Due to the heat of compression of the induction air, a turbocharged engine runs at higher operating temperatures than a non-turbocharged engine. Because turbocharged engines operate at high altitudes, their environment is less efficient for cooling. At altitude the air is less dense and therefore, cools less efficiently. Also, the less dense air causes the compressor to work harder. Compressor turbine speeds can reach 80,000 – 100,000 r.p.m., adding to the overall engine operating temperatures. Turbocharged engines are also operated at higher power settings a greater portion of the time.

High heat is detrimental to piston engine operation. Its cumulative effects can lead to piston, ring, and cylinder head failure, and place thermal stress on other operating components. Excessive cylinder head temperature can lead to detonation, which in turn can cause catastrophic engine failure. Turbocharged engines are especially heat sensitive. The key to turbocharger operation, therefore, is effective heat management.

The pilot monitors the condition of a turbocharged engine with manifold pressure gauge, tachometer, exhaust gas temperature/turbine inlet temperature gauge, and cylinder head temperature. The pilot manages the "heat system" with the throttle, propeller r.p.m., mixture, and cowl flaps. At any given cruise power, the mixture is the most influential control over the exhaust gas/turbine inlet temperature. The throttle regulates total fuel flow, but the mixture governs the fuel to air ratio. The mixture, therefore, controls temperature.

Exceeding temperature limits in an after takeoff climb is usually not a problem since a full rich mixture cools with excess fuel. At cruise, however, the pilot normally reduces power to 75 percent or less and simultaneously adjusts the mixture. Under cruise conditions, temperature limits should be monitored most closely because it's there that the temperatures are most likely to reach the maximum, even though the engine is producing less power. Overheating in an enroute climb, however, may require fully open cowl flaps and a higher airspeed.

Since turbocharged engines operate hotter at altitude than do normally aspirated engines, they are more prone to damage from cooling stress. Gradual reductions in power, and careful monitoring of temperatures are essential in the descent phase. The pilot may find it helpful to lower the landing gear to give the engine something to work against while power is reduced and provide time for a slow cool down. It may also be necessary to lean the mixture slightly to eliminate roughness at the lower power settings.

TURBOCHARGER FAILURE

Because of the high temperatures and pressures produced in the turbine exhaust systems, any malfunction of the turbocharger must be treated with extreme caution. In all cases of turbocharger operation, the manufacturer's recommended procedures should be followed. This is especially so in the case of turbocharger malfunction. However, in those instances where the manufacturer's procedures do not adequately describe the actions to be taken in the event of a turbocharger failure, the following procedures should be used.

OVERBOOST CONDITION

If an excessive rise in manifold pressure occurs during normal advancement of the throttle (possibly owing to faulty operation of the waste gate):

- Immediately retard the throttle smoothly to limit the manifold pressure below the maximum for the r.p.m. and mixture setting.
- Operate the engine in such a manner as to avoid a further overboost condition.

LOW MANIFOLD PRESSURE

Although this condition may be caused by a minor fault, it is quite possible that a serious exhaust leak has occurred creating a potentially hazardous situation:

- Shut down the engine in accordance with the recommended engine failure procedures, unless a greater emergency exists that warrants continued engine operation.
- If continuing to operate the engine, use the lowest power setting demanded by the situation and land as soon as practicable.

It is very important to ensure that corrective maintenance is undertaken following any turbocharger malfunction.

RETRACTABLE LANDING GEAR

The primary benefits of being able to retract the landing gear are increased climb performance and higher cruise airspeeds due to the resulting decrease in drag. Retractable landing gear systems may be operated either hydraulically or electrically, or may employ a combination of the two systems. Warning indicators are provided in the cockpit to show the pilot when the wheels are down and locked and when they are up and locked or if they are in intermediate positions. Systems for emergency operation are also provided. The complexity of the retractable landing gear system requires that specific operating procedures be adhered to and that certain operating limitations not be exceeded.

LANDING GEAR SYSTEMS

An **electrical** landing gear retraction system utilizes an electrically driven motor for gear operation. The system is basically an electrically driven jack for raising and lowering the gear. When a switch in the cockpit is moved to the UP position, the electric motor operates. Through a system of shafts, gears, adapters, an actuator screw, and a torque tube, a force is transmitted to the drag strut linkages. Thus, the gear retracts and locks. Struts are also activated that open and close the gear doors. If the switch is moved to the DOWN position, the motor reverses and the gear moves down and locks. Once activated the gear motor will continue to operate until an up or down limit switch on the motor's gearbox is tripped.

A **hydraulic** landing gear retraction system utilizes pressurized hydraulic fluid to actuate linkages to raise and lower the gear. When a switch in the cockpit is moved to the UP position, hydraulic fluid is directed into the gear up line. The fluid flows through sequenced valves and downlocks to the gear actuating cylinders. A similar process occurs during gear extension. The pump which pressurizes the fluid in the system can be either engine driven or electrically powered. If an electrically powered pump is used to pressurize the fluid, the system is referred to as an **electrohydraulic** system. The system also incorporates a hydraulic reservoir to contain excess fluid, and to provide a means of determining system fluid level.

Regardless of its power source, the hydraulic pump is designed to operate within a specific range. When a sensor detects excessive pressure, a relief valve within the pump opens, and hydraulic pressure is routed back to the reservoir. Another type of relief valve prevents excessive pressure that may result from thermal expansion. Hydraulic pressure is also regulated by limit switches. Each gear has two limit switches—one dedicated to extension and one dedicated to retraction. These switches de-energize the hydraulic pump after the landing gear has completed its gear cycle. In the event of limit switch failure, a backup pressure relief valve activates to relieve excess system pressure.

CONTROLS AND POSITION INDICATORS

Landing gear position is controlled by a switch in the cockpit. In most airplanes, the gear switch is shaped like a wheel in order to facilitate positive identification and to differentiate it from other cockpit controls. [Figure 11-6]

Landing gear position indicators vary with different make and model airplanes. The most common types of landing gear position indicators utilize a group of lights. One type consists of a group of three green lights, which illuminate when the landing gear is down and locked. [Figure 11-6] Another type consists of one green light to indicate when the landing gear is down and an amber light to indicate when the gear is up. Still other systems incorporate a red or amber light to indicate when the gear is in transit or unsafe for landing. [Figure 11-7] The lights are usually of the "press to test" type, and the bulbs are interchangeable. [Figure 11-6]

Other types of landing gear position indicators consist of tab-type indicators with markings "UP" to indicate the gear is up and locked, a display of red and white diagonal stripes to show when the gear is unlocked, or a silhouette of each gear to indicate when it locks in the DOWN position.

LANDING GEAR SAFETY DEVICES

Most airplanes with a retractable landing gear have a gear warning horn that will sound when the airplane is configured for landing and the landing gear is not down and locked. Normally, the horn is linked to the throttle or flap position, and/or the airspeed indicator so that when the airplane is below a certain airspeed, configuration, or power setting with the gear retracted, the warning horn will sound.

Accidental retraction of a landing gear may be prevented by such devices as mechanical downlocks, safety switches, and ground locks. Mechanical downlocks are built-in components of a gear retraction system and are operated automatically by the gear retraction system. To prevent accidental operation of the downlocks, and inadvertent landing gear retraction while the airplane is on the ground, electrically operated safety switches are installed.

A landing gear safety switch, sometimes referred to as a squat switch, is usually mounted in a bracket on one of the main gear shock struts. [Figure 11-8] When the strut is compressed by the weight of the airplane, the switch opens the electrical circuit to the motor or mechanism that powers retraction. In this way, if the landing gear switch in the cockpit is placed in the RETRACT position when weight is on the gear, the gear will remain extended, and the warning horn may sound as an alert to the unsafe condition. Once the weight is off the gear, however, such as on takeoff, the safety switch will release and the gear will retract.

Many airplanes are equipped with additional safety devices to prevent collapse of the gear when the airplane is on the ground. These devices are called ground locks. One common type is a pin installed in aligned holes drilled in two or more units of the landing gear support structure. Another type is a spring-loaded clip designed to fit around and hold two or more units of the support structure together. All types of ground locks usually have red streamers permanently attached to them to readily indicate whether or not they are installed.

EMERGENCY GEAR EXTENSION SYSTEMS

The emergency extension system lowers the landing gear if the main power system fails. Some airplanes

Figure 11-6. Typical landing gear switches and position indicators.

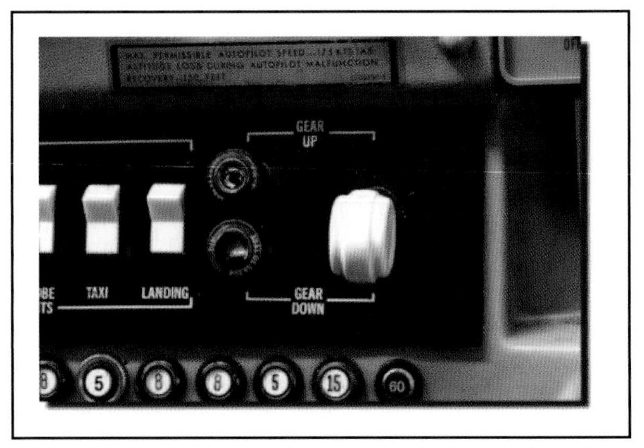

Figure 11-7. Typical landing gear switches and position indicators.

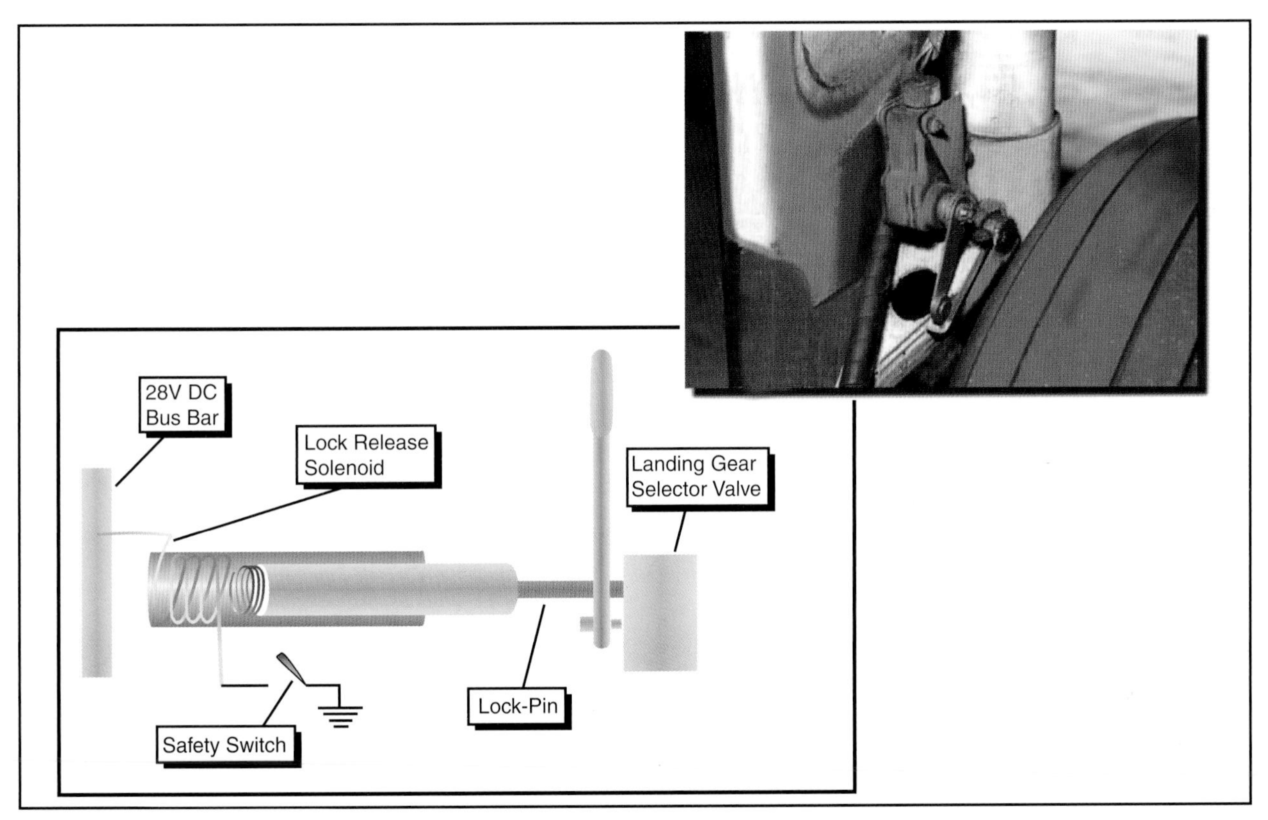

Figure 11-8. Landing gear safety switch.

have an emergency release handle in the cockpit, which is connected through a mechanical linkage to the gear uplocks. When the handle is operated, it releases the uplocks and allows the gears to free fall, or extend under their own weight. [Figure 11-9]

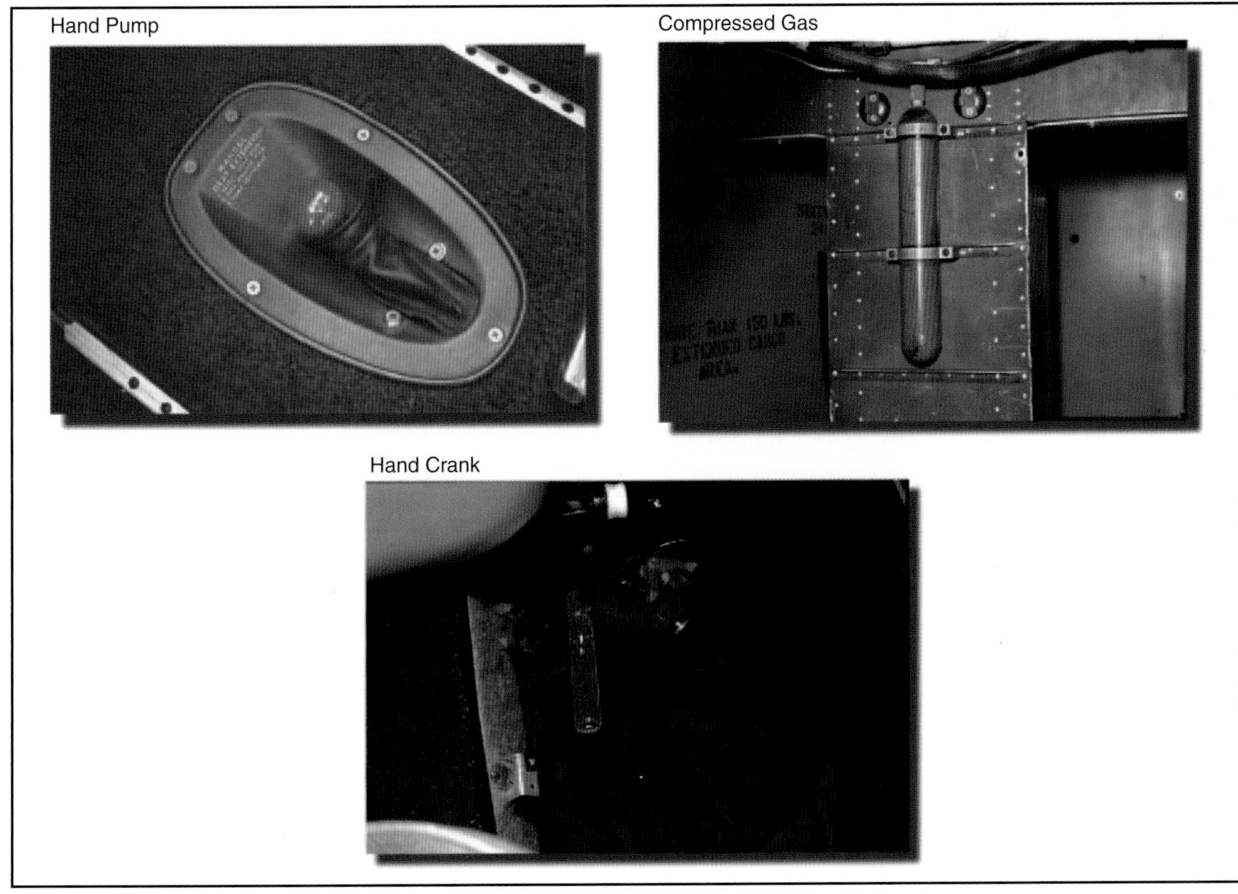

Figure 11-9. Typical emergency gear extension systems.

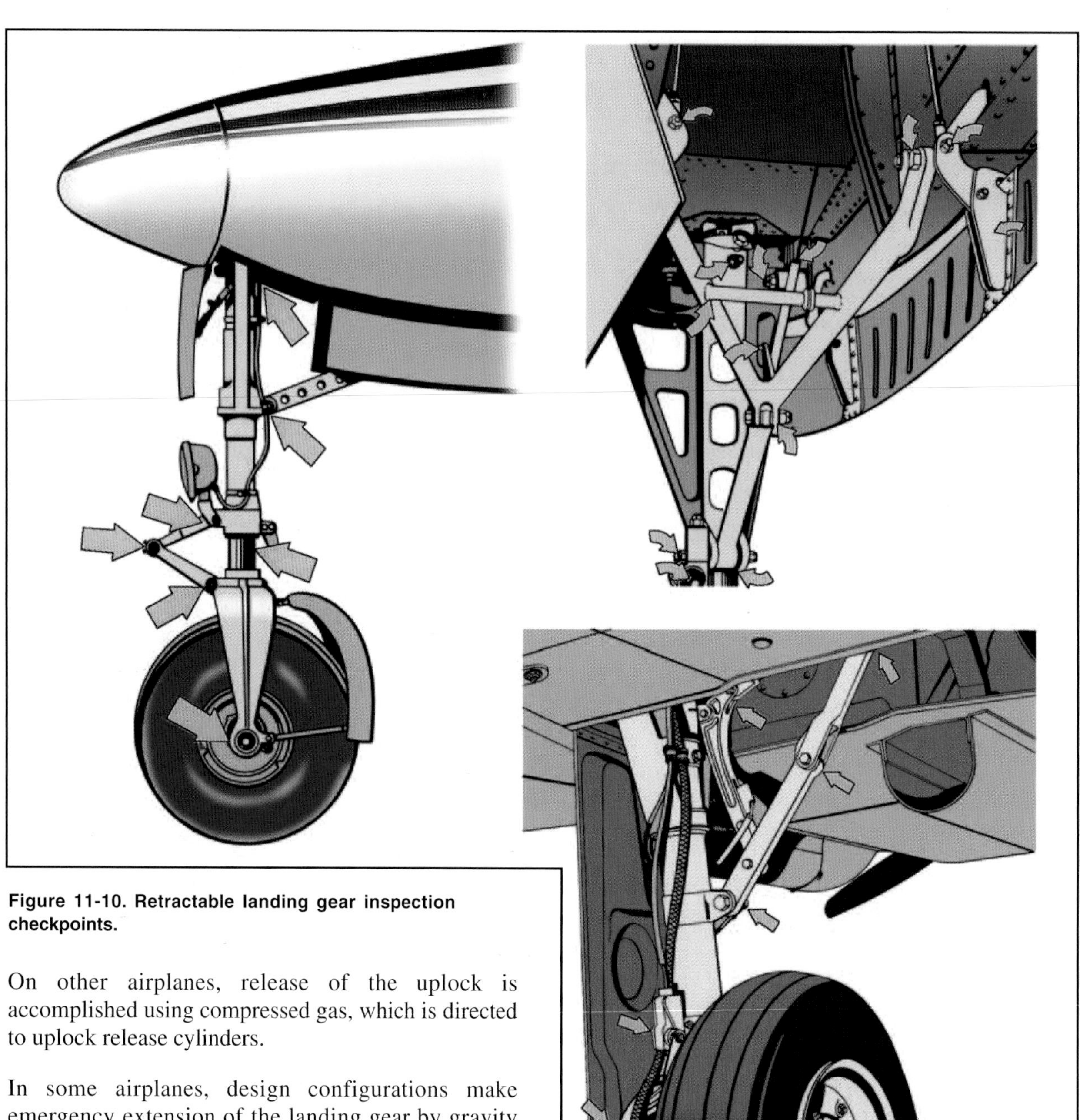

Figure 11-10. Retractable landing gear inspection checkpoints.

On other airplanes, release of the uplock is accomplished using compressed gas, which is directed to uplock release cylinders.

In some airplanes, design configurations make emergency extension of the landing gear by gravity and air loads alone impossible or impractical. In these airplanes, provisions are included for forceful gear extension in an emergency. Some installations are designed so that either hydraulic fluid or compressed gas provides the necessary pressure, while others use a manual system such as a hand crank for emergency gear extension. [Figure 11-9] Hydraulic pressure for emergency operation of the landing gear may be provided by an auxiliary hand pump, an accumulator, or an electrically powered hydraulic pump depending on the design of the airplane.

OPERATIONAL PROCEDURES

PREFLIGHT

Because of their complexity, retractable landing gears demand a close inspection prior to every flight. The inspection should begin inside the cockpit. The pilot should first make certain that the landing gear selector switch is in the GEAR DOWN position. The pilot should then turn on the battery master switch and ensure that the landing gear position indicators show that the gear is down and locked.

External inspection of the landing gear should consist of checking individual system components. [Figure 11-10] The landing gear, wheel well, and adjacent areas should be clean and free of mud and debris. Dirty switches and valves may cause false safe light indications or interrupt the extension cycle before the landing gear is completely down and locked. The wheel wells should be clear of any obstructions, as foreign objects may damage the gear or interfere with its operation. Bent gear doors may be an indication of possible problems with normal gear operation.

Shock struts should be properly inflated and the pistons clean. Main gear and nose gear uplock and downlock mechanisms should be checked for general condition. Power sources and retracting mechanisms should be checked for general condition, obvious defects, and security of attachment. Hydraulic lines should be checked for signs of chafing, and leakage at attach points. Warning system micro switches (squat switches) should be checked for cleanliness and security of attachment. Actuating cylinders, sprockets, universals, drive gears, linkages and any other accessible components should be checked for condition and obvious defects. The airplane structure to which the landing gear is attached should be checked for distortion, cracks, and general condition. All bolts and rivets should be intact and secure.

TAKEOFF AND CLIMB

Normally, the landing gear should be retracted after lift-off when the airplane has reached an altitude where, in the event of an engine failure or other emergency requiring an aborted takeoff, the airplane could no longer be landed on the runway. This procedure, however, may not apply to *all* situations. Landing gear retraction should be preplanned, taking into account the length of the runway, climb gradient, obstacle clearance requirements, the characteristics of the terrain beyond the departure end of the runway, and the climb characteristics of the particular airplane. For example, in some situations it may be preferable, in the event of an engine failure, to make an off airport forced landing with the gear extended in order to take advantage of the energy absorbing qualities of terrain (see Chapter 16). In which case, a delay in retracting the landing gear after takeoff from a short runway may be warranted. In other situations, obstacles in the climb path may warrant a timely gear retraction after takeoff. Also, in some airplanes the initial climb pitch attitude is such that any view of the runway remaining is blocked, making an assessment of the feasibility of touching down on the remaining runway difficult.

Premature landing gear retraction should be avoided. The landing gear should not be retracted until a positive rate of climb is indicated on the flight instruments. If the airplane has not attained a positive rate of climb, there is always the chance it may settle back onto the runway with the gear retracted. This is especially so in cases of premature lift-off. The pilot should also remember that leaning forward to reach the landing gear selector may result in inadvertent forward pressure on the yoke, which will cause the airplane to descend.

As the landing gear retracts, airspeed will increase and the airplane's pitch attitude may change. The gear may take several seconds to retract. Gear retraction and locking (and gear extension and locking) is accompanied by sound and feel that are unique to the specific make and model airplane. The pilot should become familiar with the sounds and feel of normal gear retraction so that any abnormal gear operation can be readily discernable. Abnormal landing gear retraction is most often a clear sign that the gear extension cycle will also be abnormal.

APPROACH AND LANDING

The operating loads placed on the landing gear at higher airspeeds may cause structural damage due to the forces of the airstream. Limiting speeds, therefore, are established for gear operation to protect the gear components from becoming overstressed during flight. These speeds are not found on the airspeed indicator. They are published in the AFM/POH for the particular airplane and are usually listed on placards in the cockpit. [Figure 11-11] The maximum landing extended speed (V_{LE}) is the maximum speed at which the airplane can be flown with the landing gear extended. The maximum landing gear operating speed (V_{LO}) is the maximum speed at which the landing gear may be operated through its cycle.

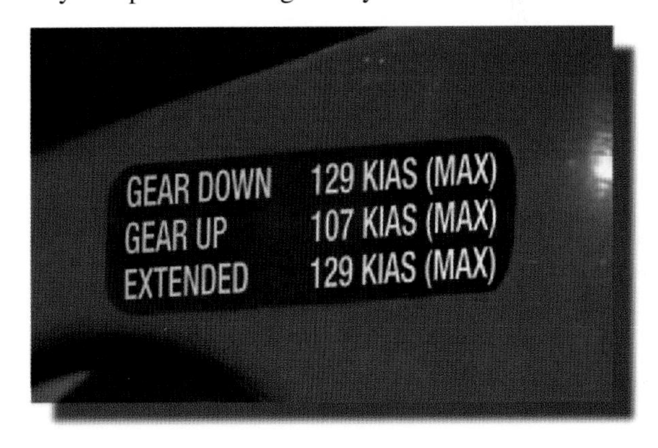

Figure 11-11. Placarded gear speeds in the cockpit.

The landing gear is extended by placing the gear selector switch in the GEAR DOWN position. As the landing gear extends, the airspeed will decrease and the pitch attitude may increase. During the several seconds it takes for the gear to extend, the pilot should be attentive to any abnormal sounds or feel. The pilot should confirm that the landing gear has extended and locked by the normal sound and feel of the system operation as well as by the gear position indicators in the cockpit. Unless the landing gear has been previously extended to aid in a descent to traffic pattern altitude, the landing gear should be extended by the time the airplane reaches a point on the downwind leg that is opposite the point of intended landing. The pilot should establish a standard procedure consisting of a specific position on the downwind leg at which to lower the landing gear. Strict adherence to this procedure will aid the pilot in avoiding unintentional gear up landings.

Operation of an airplane equipped with a retractable landing gear requires the deliberate, careful, and continued use of an appropriate checklist. When on the downwind leg, the pilot should make it a habit to **complete** the landing gear checklist for that airplane. This accomplishes two purposes. It ensures that action has been taken to lower the gear, and it increases the pilot's awareness so that the gear down indicators can be **rechecked** prior to landing.

Unless good operating practices dictate otherwise, the landing roll should be completed and the airplane clear of the runway before any levers or switches are operated. This will accomplish the following: The landing gear strut safety switches will be actuated, deactivating the landing gear retract system. After rollout and clearing the runway, the pilot will be able to focus attention on the after landing checklist and to identify the proper controls.

Pilots transitioning to retractable gear airplanes should be aware that the most common pilot operational factors involved in retractable gear airplane accidents are:

- Neglected to extend landing gear.
- Inadvertently retracted landing gear.
- Activated gear, but failed to check gear position.
- Misused emergency gear system.
- Retracted gear prematurely on takeoff.
- Extended gear too late.

In order to minimize the chances of a landing gear related mishap, the pilot should:

- Use an appropriate checklist. (A condensed checklist mounted in view of the pilot as a reminder for its use and easy reference can be especially helpful.)
- Be familiar with, and periodically review, the landing gear emergency extension procedures for the particular airplane.
- Be familiar with the landing gear warning horn and warning light systems for the particular airplane. Use the horn system to cross-check the warning light system when an unsafe condition is noted.
- Review the procedure for replacing light bulbs in the landing gear warning light displays for the particular airplane, so that you can properly replace a bulb to determine if the bulb(s) in the display is good. Check to see if spare bulbs are available in the airplane spare bulb supply as part of the preflight inspection.
- Be familiar with and aware of the sounds and feel of a properly operating landing gear system.

TRANSITION TRAINING

Transition to a complex airplane or a high performance airplane should be accomplished through a structured course of training administered by a competent and qualified flight instructor. The training should be accomplished in accordance with a ground and flight training syllabus. [Figure 11-12]

This sample syllabus for transition training is to be considered flexible. The arrangement of the subject matter may be changed and the emphasis may be shifted to fit the qualifications of the transitioning pilot, the airplane involved, and the circumstances of the training situation, provided the prescribed proficiency standards are achieved. These standards are contained in the practical test standards appropriate for the certificate that the transitioning pilot holds or is working towards.

The training times indicated in the syllabus are based on the capabilities of a pilot who is currently active and fully meets the present requirements for the issuance of at least a private pilot certificate. The time periods may be reduced for pilots with higher qualifications or increased for pilots who do not meet the current certification requirements or who have had little recent flight experience.

Ground Instruction	Flight Instruction	Directed Practice*
1 Hour	**1 Hour**	
1. Operations sections of flight manual 2. Line inspection 3. Cockpit familiarization	1. Flight training maneuvers 2. Takeoffs, landings and go-arounds	
1 Hour	**1 Hour**	**1 Hour**
1. Aircraft loading, limitations and servicing 2. Instruments, radio and special equipment 3. Aircraft systems	1. Emergency operations 2. Control by reference to instruments 3. Use of radio and autopilot	As assigned by flight instructor
1 Hour	**1 Hour**	**1 Hour**
1. Performance section of flight manual 2. Cruise control 3. Review	1. Short and soft-field takeoffs and landings 2. Maximum performance operations	As assigned by flight instructor
1 Hour—CHECKOUT		

* The directed practice indicated may be conducted solo or with a safety pilot at the discretion of the instructor.

Figure 11-12. Transition training syllabus.

Chapter 12

Transition to Multiengine Airplanes

MULTIENGINE FLIGHT

This chapter is devoted to the factors associated with the operation of small multiengine airplanes. For the purpose of this handbook, a "small" multiengine airplane is a reciprocating or turbopropeller-powered airplane with a maximum certificated takeoff weight of 12,500 pounds or less. This discussion assumes a conventional design with two engines—one mounted on each wing. Reciprocating engines are assumed unless otherwise noted. The term "light-twin," although not formally defined in the regulations, is used herein as a small multiengine airplane with a maximum certificated takeoff weight of 6,000 pounds or less.

There are several unique characteristics of multiengine airplanes that make them worthy of a separate class rating. Knowledge of these factors and proficient flight skills are a key to safe flight in these airplanes. This chapter deals extensively with the numerous aspects of one engine inoperative (OEI) flight. However, pilots are strongly cautioned not to place undue emphasis on mastery of OEI flight as the sole key to flying multiengine airplanes safely. The inoperative engine information that follows is extensive only because this chapter emphasizes the differences between flying multiengine airplanes as contrasted to single-engine airplanes.

The modern, well-equipped multiengine airplane can be remarkably capable under many circumstances. But, as with single-engine airplanes, it must be flown prudently by a current and competent pilot to achieve the highest possible level of safety.

This chapter contains information and guidance on the performance of certain maneuvers and procedures in small multiengine airplanes for the purposes of flight training and pilot certification testing. The final authority on the operation of a particular make and model airplane, however, is the airplane manufacturer. Both the flight instructor and the student should be aware that if any of the guidance in this handbook conflicts with the airplane manufacturer's recommended procedures and guidance as contained in the FAA-approved Airplane Flight Manual and/or Pilot's Operating Handbook (AFM/POH), it is the airplane manufacturer's guidance and procedures that take precedence.

GENERAL

The basic difference between operating a multiengine airplane and a single-engine airplane is the potential problem involving an engine failure. The penalties for loss of an engine are twofold: performance and control. The most obvious problem is the loss of 50 percent of power, which reduces climb performance 80 to 90 percent, sometimes even more. The other is the control problem caused by the remaining thrust, which is now asymmetrical. Attention to both these factors is crucial to safe OEI flight. The performance and systems redundancy of a multiengine airplane is a safety advantage only to a trained and proficient pilot.

TERMS AND DEFINITIONS

Pilots of single-engine airplanes are already familiar with many performance "V" speeds and their definitions. Twin-engine airplanes have several additional V speeds unique to OEI operation. These speeds are differentiated by the notation "$_{SE}$", for single engine. A review of some key V speeds and several new V speeds unique to twin-engine airplanes follows.

- V_R – Rotation speed. The speed at which back pressure is applied to rotate the airplane to a takeoff attitude.
- V_{LOF} – Lift-off speed. The speed at which the airplane leaves the surface. (Note: some manufacturers reference takeoff performance data to V_R, others to V_{LOF}.)
- V_X – Best angle of climb speed. The speed at which the airplane will gain the greatest altitude for a given distance of forward travel.
- V_{XSE} – Best angle-of-climb speed with one engine inoperative.
- V_Y – Best rate of climb speed. The speed at which the airplane will gain the most altitude for a given unit of time.
- V_{YSE} – Best rate-of-climb speed with one engine inoperative. Marked with a blue radial line on most airspeed indicators. Above the single-engine absolute ceiling, V_{YSE} yields the minimum rate of sink.
- V_{SSE} – Safe, intentional one-engine-inoperative speed. Originally known as safe single-engine

speed. Now formally defined in Title 14 of the Code of Federal Regulations (14 CFR) part 23, Airworthiness Standards, and required to be established and published in the AFM/POH. It is the minimum speed to intentionally render the critical engine inoperative.

- **V_{MC}** – Minimum control speed with the critical engine inoperative. Marked with a red radial line on most airspeed indicators. The minimum speed at which directional control can be maintained under a very specific set of circumstances outlined in 14 CFR part 23, Airworthiness Standards. Under the small airplane certification regulations currently in effect, the flight test pilot must be able to (1) stop the turn that results when the critical engine is suddenly made inoperative within 20° of the original heading, using maximum rudder deflection and a maximum of 5° bank, and (2) thereafter, maintain straight flight with not more than a 5° bank. There is no requirement in this determination that the airplane be capable of climbing at this airspeed. V_{MC} only addresses directional control. Further discussion of V_{MC} as determined during airplane certification and demonstrated in pilot training follows in minimum control airspeed (V_{MC}) demonstration. [Figure 12-1]

Figure 12-1. Airspeed indicator markings for a multiengine airplane.

Unless otherwise noted, when V speeds are given in the AFM/POH, they apply to sea level, standard day conditions at maximum takeoff weight. Performance speeds vary with aircraft weight, configuration, and atmospheric conditions. The speeds may be stated in statute miles per hour (m.p.h.) or knots (kts), and they may be given as calibrated airspeeds (CAS) or indicated airspeeds (IAS). As a general rule, the newer AFM/POHs show V speeds in knots indicated airspeed (KIAS). Some V speeds are also stated in knots calibrated airspeed (KCAS) to meet certain regulatory requirements. Whenever available, pilots should operate the airplane from published indicated airspeeds.

With regard to climb performance, the multiengine airplane, particularly in the takeoff or landing configuration, may be considered to be a single-engine airplane with its powerplant divided into two units. There is nothing in 14 CFR part 23 that requires a multiengine airplane to maintain altitude while in the takeoff or landing configuration with one engine inoperative. In fact, many twins are not required to do this in any configuration, even at sea level.

The current 14 CFR part 23 single-engine climb performance requirements for reciprocating engine-powered multiengine airplanes are as follows.

- More than 6,000 pounds maximum weight and/or V_{SO} more than 61 knots: the single-engine rate of climb in feet per minute (f.p.m.) at 5,000 feet MSL must be equal to at least .027 V_{SO}^2. For airplanes type certificated February 4, 1991, or thereafter, the climb requirement is expressed in terms of a climb gradient, 1.5 percent. The climb gradient is not a direct equivalent of the .027 V_{SO}^2 formula. Do not confuse the date of type certification with the airplane's model year. The type certification basis of many multiengine airplanes dates back to CAR 3 (the Civil Aviation Regulations, forerunner of today's Code of Federal Regulations).

- 6,000 pounds or less maximum weight and V_{SO} 61 knots or less: the single-engine rate of climb at 5,000 feet MSL must simply be determined. The rate of climb could be a negative number. There is no requirement for a single-engine positive rate of climb at 5,000 feet or any other altitude. For light-twins type certificated February 4, 1991, or thereafter, the single-engine climb gradient (positive or negative) is simply determined.

Rate of climb is the altitude gain per unit of time, while climb gradient is the actual measure of altitude gained per 100 feet of horizontal travel, expressed as a percentage. An altitude gain of 1.5 feet per 100 feet of travel (or 15 feet per 1,000, or 150 feet per 10,000) is a climb gradient of 1.5 percent.

There is a dramatic performance loss associated with the loss of an engine, particularly just after takeoff. Any airplane's climb performance is a function of thrust horsepower which is in excess of that required for level flight. In a hypothetical twin with each engine producing 200 thrust horsepower, assume that the total level-flight thrust horsepower required is 175. In this situation, the airplane would ordinarily have a reserve of 225 thrust horsepower available for climb. Loss of one engine would leave only 25 (200 minus 175) thrust horsepower available for climb, a drastic reduction. Sea level rate-of-climb performance losses of at least 80 to 90 percent, even under ideal circumstances, are typical for multiengine airplanes in OEI flight.

OPERATION OF SYSTEMS

This section will deal with systems that are generally found on multiengine airplanes. Multiengine airplanes share many features with complex single-engine airplanes. There are certain systems and features covered here, however, that are generally unique to airplanes with two or more engines.

PROPELLERS

The propellers of the multiengine airplane may outwardly appear to be identical in operation to the constant-speed propellers of many single-engine airplanes, but this is not the case. The propellers of multiengine airplanes are featherable, to minimize drag in the event of an engine failure. Depending upon single-engine performance, this feature often permits continued flight to a suitable airport following an engine failure. To feather a propeller is to stop engine rotation with the propeller blades streamlined with the airplane's relative wind, thus to minimize drag. [Figure 12-2]

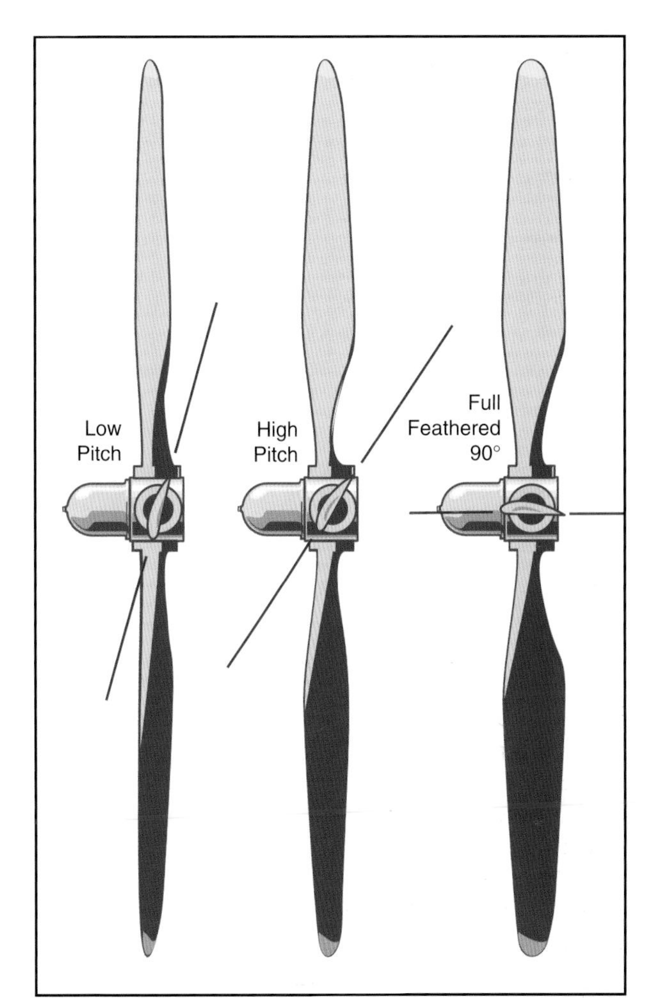

Figure 12-2. Feathered propeller.

Feathering is necessary because of the change in parasite drag with propeller blade angle. [Figure 12-3] When the propeller blade angle is in the feathered position, the change in parasite drag is at a minimum and, in the case of a typical multiengine airplane, the added parasite drag from a single feathered propeller is a relatively small contribution to the airplane total drag.

At the smaller blade angles near the flat pitch position, the drag added by the propeller is very large. At these small blade angles, the propeller windmilling at high r.p.m. can create such a tremendous amount of drag that the airplane may be uncontrollable. The propeller windmilling at high speed in the low range of blade angles can produce an increase in parasite drag which may be as great as the parasite drag of the basic airplane.

As a review, the constant-speed propellers on almost all single-engine airplanes are of the non-feathering, oil-pressure-to-increase-pitch design. In this design, increased oil pressure from the propeller governor drives the blade angle towards high pitch, low r.p.m.

In contrast, the constant-speed propellers installed on most multiengine airplanes are full feathering, counterweighted, oil-pressure-to-decrease-pitch designs. In this design, increased oil pressure from the propeller governor drives the blade angle towards low pitch, high r.p.m.—away from the feather blade angle. In effect, the only thing that keeps these propellers from feathering is a constant supply of high pressure engine oil. This is a necessity to enable propeller feathering in the event of a loss of oil pressure or a propeller governor failure.

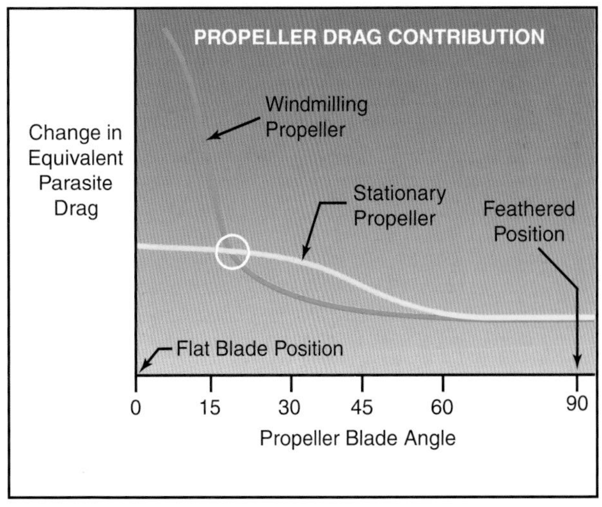

Figure 12-3. Propeller drag contribution.

The aerodynamic forces alone acting upon a windmilling propeller tend to drive the blades to low pitch, high r.p.m. Counterweights attached to the shank of each blade tend to drive the blades to high pitch, low r.p.m. Inertia, or apparent force called centrifugal force acting through the counterweights is generally slightly greater than the aerodynamic forces. Oil pressure from the propeller governor is used to counteract the counterweights and drives the blade angles to low pitch, high r.p.m. A reduction in oil pressure causes the r.p.m. to be reduced from the influence of the counterweights. [Figure 12-4]

To feather the propeller, the propeller control is brought fully aft. All oil pressure is dumped from the governor, and the counterweights drive the propeller blades towards feather. As centrifugal force acting on the counterweights decays from decreasing r.p.m., additional forces are needed to completely feather the blades. This additional force comes from either a spring or high pressure air stored in the propeller dome, which forces the blades into the feathered position. The entire process may take up to 10 seconds.

Feathering a propeller only alters blade angle and stops engine rotation. To completely secure the engine, the pilot must still turn off the fuel (mixture, electric boost pump, and fuel selector), ignition, alternator/generator, and close the cowl flaps. If the airplane is pressurized, there may also be an air bleed to close for the failed engine. Some airplanes are equipped with firewall shutoff valves that secure several of these systems with a single switch.

Completely securing a failed engine may not be necessary or even desirable depending upon the failure mode, altitude, and time available. The position of the fuel controls, ignition, and alternator/generator switches of the failed engine has no effect on aircraft performance. There is always the distinct possibility of manipulating the incorrect switch under conditions of haste or pressure.

To unfeather a propeller, the engine must be rotated so that oil pressure can be generated to move the propeller blades from the feathered position. The ignition is turned on prior to engine rotation with the throttle at low idle and the mixture rich. With the propeller control in a high r.p.m. position, the starter is engaged. The engine will begin to windmill, start, and run as oil pressure moves the blades out of feather. As the engine starts, the propeller r.p.m. should be immediately reduced until the engine has had several minutes to warm up; the pilot should monitor cylinder head and oil temperatures.

Should the r.p.m. obtained with the starter be insufficient to unfeather the propeller, an increase in airspeed

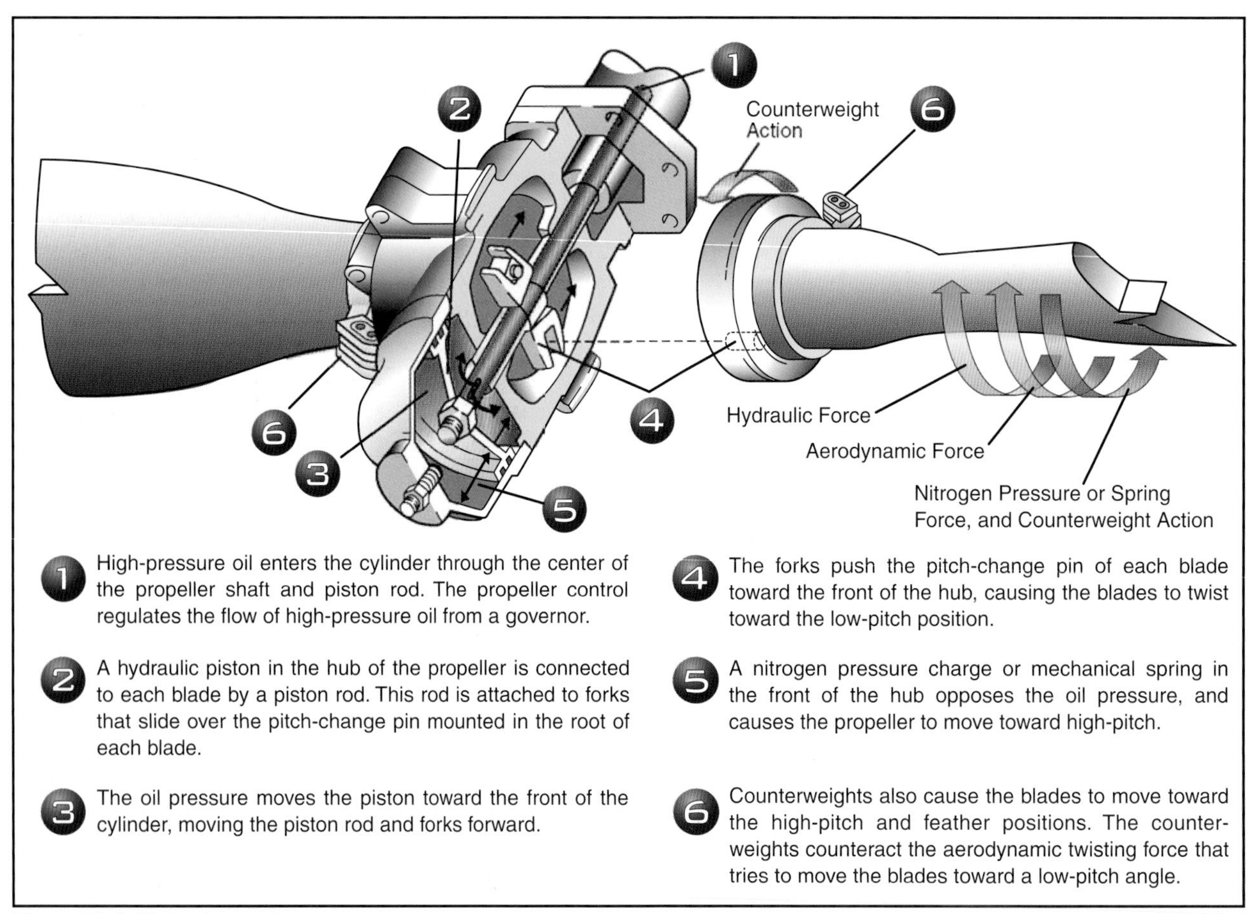

Figure 12-4. Pitch change forces.

from a shallow dive will usually help. In any event, the AFM/POH procedures should be followed for the exact unfeathering procedure. Both feathering and starting a feathered reciprocating engine on the ground are strongly discouraged by manufacturers due to the excessive stress and vibrations generated.

As just described, a loss of oil pressure from the propeller governor allows the counterweights, spring and/or dome charge to drive the blades to feather. Logically then, the propeller blades should feather every time an engine is shut down as oil pressure falls to zero. Yet, this does not occur. Preventing this is a small pin in the pitch changing mechanism of the propeller hub that will not allow the propeller blades to feather once r.p.m. drops below approximately 800. The pin senses a lack of centrifugal force from propeller rotation and falls into place, preventing the blades from feathering. Therefore, if a propeller is to be feathered, it must be done before engine r.p.m. decays below approximately 800. On one popular model of turboprop engine, the propeller blades do, in fact, feather with each shutdown. This propeller is not equipped with such centrifugally-operated pins, due to a unique engine design.

An unfeathering accumulator is an optional device that permits starting a feathered engine in flight without the use of the electric starter. An accumulator is any device that stores a reserve of high pressure. On multiengine airplanes, the unfeathering accumulator stores a small reserve of engine oil under pressure from compressed air or nitrogen. To start a feathered engine in flight, the pilot moves the propeller control out of the feather position to release the accumulator pressure. The oil flows under pressure to the propeller hub and drives the blades toward the high r.p.m., low pitch position, whereupon the propeller will usually begin to windmill. (On some airplanes, an assist from the electric starter may be necessary to initiate rotation and completely unfeather the propeller.) If fuel and ignition are present, the engine will start and run. For airplanes used in training, this saves much electric starter and battery wear. High oil pressure from the propeller governor recharges the accumulator just moments after engine rotation begins.

PROPELLER SYNCHRONIZATION

Many multiengine airplanes have a propeller synchronizer (prop sync) installed to eliminate the annoying "drumming" or "beat" of propellers whose r.p.m. are close, but not precisely the same. To use prop sync, the propeller r.p.m. are coarsely matched by the pilot and the system is engaged. The prop sync adjusts the r.p.m. of the "slave" engine to precisely match the r.p.m. of the "master" engine, and then maintains that relationship. The prop sync should be disengaged when the pilot selects a new propeller r.p.m., then re-engaged after the new r.p.m. is set. The prop sync should always be off for takeoff, landing, and single-engine operation. The AFM/POH should be consulted for system description and limitations.

A variation on the propeller synchronizer is the propeller synchrophaser. Prop sychrophase acts much like a synchronizer to precisely match r.p.m., but the synchrophaser goes one step further. It not only matches r.p.m. but actually compares and adjusts the positions of the individual blades of the propellers in their arcs. There can be significant propeller noise and vibration reductions with a propeller synchrophaser. From the pilot's perspective, operation of a propeller synchronizer and a propeller syncrophaser are very similar. A synchrophaser is also commonly referred to as prop sync, although that is not entirely correct nomenclature from a technical standpoint.

As a pilot aid to manually synchronizing the propellers, some twins have a small gauge mounted in or by the tachometer(s) with a propeller symbol on a disk that spins. The pilot manually fine tunes the engine r.p.m. so as to stop disk rotation, thereby synchronizing the propellers. This is a useful backup to synchronizing engine r.p.m. using the audible propeller beat. This gauge is also found installed with most propeller synchronizer and synchrophase systems. Some synchrophase systems use a knob for the pilot to control the phase angle.

FUEL CROSSFEED

Fuel crossfeed systems are also unique to multiengine airplanes. Using crossfeed, an engine can draw fuel from a fuel tank located in the opposite wing.

On most multiengine airplanes, operation in the crossfeed mode is an emergency procedure used to extend airplane range and endurance in OEI flight. There are a few models that permit crossfeed as a normal, fuel balancing technique in normal operation, but these are not common. The AFM/POH will describe crossfeed limitations and procedures, which vary significantly among multiengine airplanes.

Checking crossfeed operation on the ground with a quick repositioning of the fuel selectors does nothing more than ensure freedom of motion of the handle. To actually check crossfeed operation, a complete, functional crossfeed system check should be accomplished. To do this, each engine should be operated from its crossfeed position during the runup. The engines should be checked individually, and allowed to run at moderate power (1,500 r.p.m. minimum) for at least 1 minute to ensure that fuel flow can be established from the crossfeed source. Upon completion of the check, each engine should be operated for at least 1 minute at moderate power from the main (takeoff) fuel tanks to reconfirm fuel flow prior to takeoff.

This suggested check is not required prior to every flight. Infrequently used, however, crossfeed lines are ideal places for water and debris to accumulate unless they are used from time to time and drained using their external drains during preflight. Crossfeed is ordinarily not used for completing single-engine flights when an alternate airport is readily at hand, and it is never used during takeoff or landings.

COMBUSTION HEATER

Combustion heaters are common on multiengine airplanes. A combustion heater is best described as a small furnace that burns gasoline to produce heated air for occupant comfort and windshield defogging. Most are thermostatically operated, and have a separate hour meter to record time in service for maintenance purposes. Automatic overtemperature protection is provided by a thermal switch mounted on the unit, which cannot be accessed in flight. This requires the pilot or mechanic to actually visually inspect the unit for possible heat damage in order to reset the switch.

When finished with the combustion heater, a cool down period is required. Most heaters require that outside air be permitted to circulate through the unit for at least 15 seconds in flight, or that the ventilation fan be operated for at least 2 minutes on the ground. Failure to provide an adequate cool down will usually trip the thermal switch and render the heater inoperative until the switch is reset.

FLIGHT DIRECTOR/AUTOPILOT

Flight director/autopilot (FD/AP) systems are common on the better-equipped multiengine airplanes. The system integrates pitch, roll, heading, altitude, and radio navigation signals in a computer. The outputs, called computed commands, are displayed on a flight command indicator, or FCI. The FCI replaces the conventional attitude indicator on the instrument panel. The FCI is occasionally referred to as a flight director indicator (FDI), or as an attitude director indicator (ADI). The entire flight director/autopilot system is sometimes called an integrated flight control system (IFCS) by some manufacturers. Others may use the term "automatic flight control system (AFCS)."

The FD/AP system may be employed at three different levels.

- Off (raw data).
- Flight director (computed commands).
- Autopilot.

With the system off, the FCI operates as an ordinary attitude indicator. On most FCIs, the command bars are biased out of view when the flight director is off. The pilot maneuvers the airplane as though the system were not installed.

To maneuver the airplane using the flight director, the pilot enters the desired modes of operation (heading, altitude, nav intercept, and tracking) on the FD/AP mode controller. The computed flight commands are then displayed to the pilot through either a single-cue or dual-cue system in the FCI. On a single-cue system, the commands are indicated by "V" bars. On a dual-cue system, the commands are displayed on two separate command bars, one for pitch and one for roll. To maneuver the airplane using computed commands, the pilot "flies" the symbolic airplane of the FCI to match the steering cues presented.

On most systems, to engage the autopilot the flight director must first be operating. At any time thereafter, the pilot may engage the autopilot through the mode controller. The autopilot then maneuvers the airplane to satisfy the computed commands of the flight director.

Like any computer, the FD/AP system will only do what it is told. The pilot must ensure that it has been properly programmed for the particular phase of flight desired. The armed and/or engaged modes are usually displayed on the mode controller or separate annunciator lights. When the airplane is being hand-flown, if the flight director is not being used at any particular moment, it should be off so that the command bars are pulled from view.

Prior to system engagement, all FD/AP computer and trim checks should be accomplished. Many newer systems cannot be engaged without the completion of a self-test. The pilot must also be very familiar with various methods of disengagement, both normal and emergency. System details, including approvals and limitations, can be found in the supplements section of the AFM/POH. Additionally, many avionics manufacturers can provide informative pilot operating guides upon request.

YAW DAMPER

The yaw damper is a servo that moves the rudder in response to inputs from a gyroscope or accelerometer that detects yaw rate. The yaw damper minimizes motion about the vertical axis caused by turbulence. (Yaw dampers on sweptwing airplanes provide another, more vital function of damping dutch roll characteristics.) Occupants will feel a smoother ride, particularly if seated in the rear of the airplane, when the yaw damper is engaged. The yaw damper should be off for takeoff and landing. There may be additional restrictions against its use during single-engine operation. Most yaw dampers can be engaged independently of the autopilot.

ALTERNATOR/GENERATOR

Alternator or generator paralleling circuitry matches the output of each engine's alternator/generator so that the electrical system load is shared equally between them. In the event of an alternator/generator failure, the inoperative unit can be isolated and the entire electrical system powered from the remaining one. Depending upon the electrical capacity of the alternator/generator, the pilot may need to reduce the electrical load (referred to as load shedding) when operating on a single unit. The AFM/POH will contain system description and limitations.

NOSE BAGGAGE COMPARTMENT

Nose baggage compartments are common on multiengine airplanes (and are even found on a few single-engine airplanes). There is nothing strange or exotic about a nose baggage compartment, and the usual guidance concerning observation of load limits applies. They are mentioned here in that pilots occasionally neglect to secure the latches properly, and therein lies the danger. When improperly secured, the door will open and the contents may be drawn out, usually into the propeller arc, and usually just after takeoff. Even when the nose baggage compartment is empty, airplanes have been lost when the pilot became distracted by the open door. Security of the nose baggage compartment latches and locks is a vital preflight item.

Most airplanes will continue to fly with a nose baggage door open. There may be some buffeting from the disturbed airflow and there will be an increase in noise. Pilots should never become so preoccupied with an open door (of any kind) that they fail to fly the airplane.

Inspection of the compartment interior is also an important preflight item. More than one pilot has been surprised to find a supposedly empty compartment packed to capacity or loaded with ballast. The tow bars, engine inlet covers, windshield sun screens, oil containers, spare chocks, and miscellaneous small hand tools that find their way into baggage compartments should be secured to prevent damage from shifting in flight.

ANTI-ICING/DEICING

Anti-icing/deicing equipment is frequently installed on multiengine airplanes and consists of a combination of different systems. These may be classified as either anti-icing or deicing, depending upon function. The presence of anti-icing and deicing equipment, even though it may appear elaborate and complete, does not necessarily mean that the airplane is approved for flight in icing conditions. The AFM/POH, placards, and even the manufacturer should be consulted for specific determination of approvals and limitations.

Anti-icing equipment is provided to prevent ice from forming on certain protected surfaces. Anti-icing equipment includes heated pitot tubes, heated or non-icing static ports and fuel vents, propeller blades with electrothermal boots or alcohol slingers, windshields with alcohol spray or electrical resistance heating, windshield defoggers, and heated stall warning lift detectors. On many turboprop engines, the "lip" surrounding the air intake is heated either electrically or with bleed air. In the absence of AFM/POH guidance to the contrary, anti-icing equipment is actuated prior to flight into known or suspected icing conditions.

Deicing equipment is generally limited to pneumatic boots on wing and tail leading edges. Deicing equipment is installed to remove ice that has already formed on protected surfaces. Upon pilot actuation, the boots inflate with air from the pneumatic pumps to break off accumulated ice. After a few seconds of inflation, they are deflated back to their normal position with the assistance of a vacuum. The pilot monitors the buildup of ice and cycles the boots as directed in the AFM/POH. An ice light on the left engine nacelle allows the pilot to monitor wing ice accumulation at night.

Other airframe equipment necessary for flight in icing conditions includes an alternate induction air source and an alternate static system source. Ice tolerant antennas will also be installed.

In the event of impact ice accumulating over normal engine air induction sources, carburetor heat (carbureted engines) or alternate air (fuel injected engines) should be selected. Ice buildup on normal induction sources can be detected by a loss of engine r.p.m. with fixed-pitch propellers and a loss of manifold pressure with constant-speed propellers. On some fuel injected engines, an alternate air source is automatically activated with blockage of the normal air source.

An alternate static system provides an alternate source of static air for the pitot-static system in the unlikely event that the primary static source becomes blocked. In non-pressurized airplanes, most alternate static sources are plumbed to the cabin. On pressurized airplanes, they are usually plumbed to a non-pressurized baggage compartment. The pilot must activate the alternate static source by opening a valve or a fitting in the cockpit. Upon activation, the airspeed indicator, altimeter, and the vertical speed indicator (VSI) will be affected and will read somewhat in error. A correction table is frequently provided in the AFM/POH.

Anti-icing/deicing equipment only eliminates ice from the protected surfaces. Significant ice accumulations may form on unprotected areas, even with proper use of anti-ice and deice systems. Flight at high angles of attack or even normal climb speeds will permit significant ice accumulations on lower wing surfaces, which are unprotected. Many AFM/POHs mandate minimum speeds to be maintained in icing conditions. Degradation of all flight characteristics and large performance losses can be expected with ice accumulations. Pilots should not rely upon the stall warning devices for adequate stall warning with ice accumulations.

Ice will accumulate unevenly on the airplane. It will add weight and drag (primarily drag), and decrease thrust and lift. Even wing shape affects ice accumulation; thin airfoil sections are more prone to ice accumulation than thick, highly-cambered sections. For this reason certain surfaces, such as the horizontal stabilizer, are more prone to icing than the wing. With ice accumulations, landing approaches should be made with a minimum wing flap setting (flap extension increases the angle of attack of the horizontal stabilizer) and with an added margin of airspeed. Sudden and large configuration and airspeed changes should be avoided.

Unless otherwise recommended in the AFM/POH, the autopilot should not be used in icing conditions. Continuous use of the autopilot will mask trim and handling changes that will occur with ice accumulation. Without this control feedback, the pilot may not be aware of ice accumulation building to hazardous levels. The autopilot will suddenly disconnect when it reaches design limits and the pilot may find the airplane has assumed unsatisfactory handling characteristics.

The installation of anti-ice/deice equipment on airplanes without AFM/POH approval for flight into icing conditions is to facilitate escape when such conditions are inadvertently encountered. Even with AFM/POH approval, the prudent pilot will avoid icing conditions to the maximum extent practicable, and avoid extended flight in any icing conditions. No multiengine airplane is approved for flight into severe icing conditions, and none are intended for indefinite flight in continuous icing conditions.

PERFORMANCE AND LIMITATIONS

Discussion of performance and limitations requires the definition of several terms.

- **Accelerate-stop distance** is the runway length required to accelerate to a specified speed (either V_R or V_{LOF}, as specified by the manufacturer), experience an engine failure, and bring the airplane to a complete stop.

- **Accelerate-go distance** is the horizontal distance required to continue the takeoff and climb to 50 feet, assuming an engine failure at V_R or V_{LOF}, as specified by the manufacturer.

- **Climb gradient** is a slope most frequently expressed in terms of altitude gain per 100 feet of horizontal distance, whereupon it is stated as a percentage. A 1.5 percent climb gradient is an altitude gain of one and one-half feet per 100 feet of horizontal travel. Climb gradient may also be expressed as a function of altitude gain per nautical mile, or as a ratio of the horizontal distance to the vertical distance (50:1, for example). Unlike rate of climb, climb gradient is affected by wind. Climb gradient is improved with a headwind component, and reduced with a tailwind component. [Figure 12-5]

- The **all-engine service ceiling** of multiengine airplanes is the highest altitude at which the airplane can maintain a steady rate of climb of 100 f.p.m. with both engines operating. The airplane has reached its **absolute ceiling** when climb is no longer possible.

- The **single-engine service ceiling** is reached when the multiengine airplane can no longer maintain a 50 f.p.m. rate of climb with one engine inoperative, and its **single-engine absolute ceiling** when climb is no longer possible.

The takeoff in a multiengine airplane should be planned in sufficient detail so that the appropriate action is taken in the event of an engine failure. The pilot should be thoroughly familiar with the airplane's performance capabilities and limitations in order to make an informed takeoff decision as part of the preflight planning. That decision should be reviewed as the last item of the "before takeoff" checklist.

In the event of an engine failure shortly after takeoff, the decision is basically one of continuing flight or landing, even off-airport. If single-engine climb performance is adequate for continued flight, and the airplane has been promptly and correctly configured, the climb after takeoff may be continued. If single-engine climb performance is such that climb is unlikely or impossible, a landing will have to be made in the most suitable area. To be avoided above all is attempting to continue flight when it is not within the airplane's performance capability to do so. [Figure 12-6]

Takeoff planning factors include weight and balance, airplane performance (both single and multiengine), runway length, slope and contamination, terrain and obstacles in the area, weather conditions, and pilot proficiency. Most multiengine airplanes have AFM/POH performance charts and the pilot should be highly proficient in their use. Prior to takeoff, the multiengine pilot should ensure that the weight and balance limitations have been observed, the runway

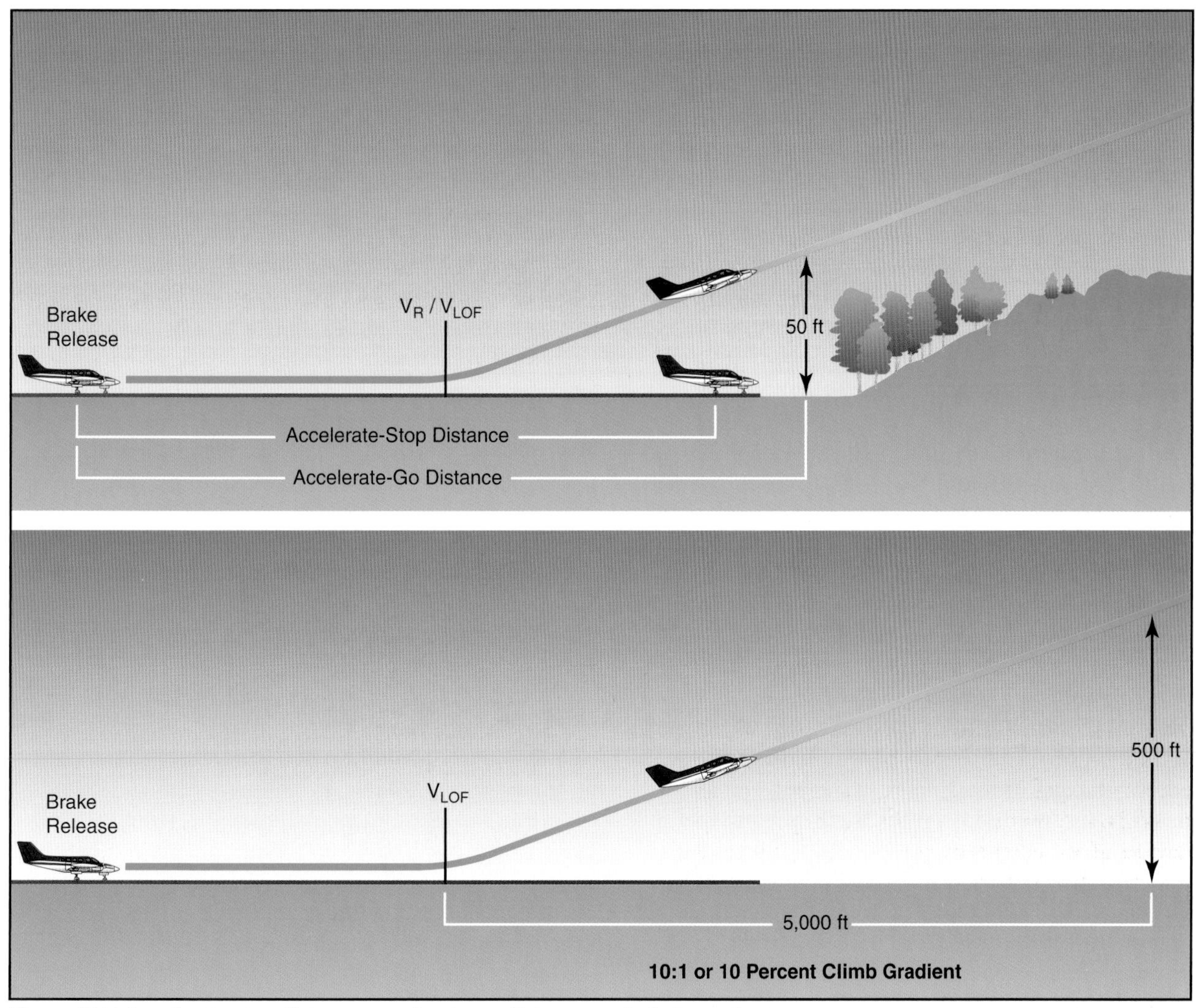

Figure 12-5. Accelerate-stop distance, accelerate-go distance, and climb gradient.

length is adequate, the normal flightpath will clear obstacles and terrain, and that a definitive course of action has been planned in the event of an engine failure.

The regulations do not specifically require that the runway length be equal to or greater than the accelerate-stop distance. Most AFM/POHs publish accelerate-stop distances only as an advisory. It becomes a limitation only when published in the limitations section of the AFM/POH. Experienced multiengine pilots, however, recognize the safety margin of runway lengths in excess of the bare minimum required for normal takeoff. They will insist on runway lengths of at least accelerate-stop distance as a matter of safety and good operating practice.

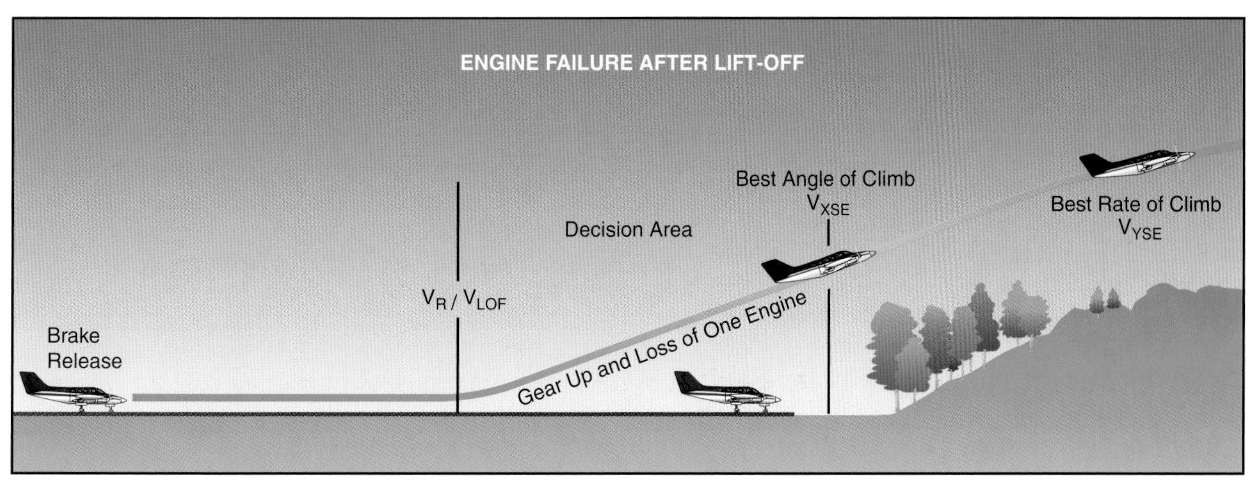

Figure 12-6. Area of decision.

The multiengine pilot must keep in mind that the accelerate-go distance, as long as it is, has only brought the airplane, under ideal circumstances, to a point a mere 50 feet above the takeoff elevation. To achieve even this meager climb, the pilot had to instantaneously recognize and react to an unanticipated engine failure, retract the landing gear, identify and feather the correct engine, all the while maintaining precise airspeed control and bank angle as the airspeed is nursed to V_{YSE}. Assuming flawless airmanship thus far, the airplane has now arrived at a point little more than one wingspan above the terrain, assuming it was absolutely level and without obstructions.

With (for the purpose of illustration) a net 150 f.p.m. rate of climb at a 90-knot V_{YSE}, it will take approximately 3 minutes to climb an additional 450 feet to reach 500 feet AGL. In doing so, the airplane will have traveled an additional 5 nautical miles beyond the original accelerate-go distance, with a climb gradient of about 1.6 percent. A turn of any consequence, such as to return to the airport, will seriously degrade the already marginal climb performance.

Not all multiengine airplanes have published accelerate-go distances in their AFM/POH, and fewer still publish climb gradients. When such information is published, the figures will have been determined under ideal flight testing conditions. It is unlikely that this performance will be duplicated in service conditions.

The point of the foregoing is to illustrate the marginal climb performance of a multiengine airplane that suffers an engine failure shortly after takeoff, even under ideal conditions. The prudent multiengine pilot should pick a point in the takeoff and climb sequence in advance. If an engine fails before this point, the takeoff should be rejected, even if airborne, for a landing on whatever runway or surface lies essentially ahead. If an engine fails after this point, the pilot should promptly execute the appropriate engine failure procedure and continue the climb, assuming the performance capability exists. As a general recommendation, if the landing gear has not been selected up, the takeoff should be rejected, even if airborne.

As a practical matter for planning purposes, the option of continuing the takeoff probably does not exist unless the published single-engine rate-of-climb performance is at least 100 to 200 f.p.m. Thermal turbulence, wind gusts, engine and propeller wear, or poor technique in airspeed, bank angle, and rudder control can easily negate even a 200 f.p.m. rate of climb.

WEIGHT AND BALANCE

The weight and balance concept is no different than that of a single-engine airplane. The actual execution, however, is almost invariably more complex due to a number of new loading areas, including nose and aft baggage compartments, nacelle lockers, main fuel tanks, aux fuel tanks, nacelle fuel tanks, and numerous seating options in a variety of interior configurations. The flexibility in loading offered by the multiengine airplane places a responsibility on the pilot to address weight and balance prior to each flight.

The terms "empty weight, licensed empty weight, standard empty weight, and basic empty weight" as they appear on the manufacturer's original weight and balance documents are sometimes confused by pilots.

In 1975, the General Aviation Manufacturers Association (GAMA) adopted a standardized format for AFM/POHs. It was implemented by most manufacturers in model year 1976. Airplanes whose manufacturers conform to the GAMA standards utilize the following terminology for weight and balance:

  Standard empty weight
\+ Optional equipment
= Basic empty weight

Standard empty weight is the weight of the standard airplane, full hydraulic fluid, unusable fuel, and full oil. Optional equipment includes the weight of all equipment installed beyond standard. Basic empty weight is the standard empty weight plus optional equipment. Note that basic empty weight includes no usable fuel, but full oil.

Airplanes manufactured prior to the GAMA format generally utilize the following terminology for weight and balance, although the exact terms may vary somewhat:

  Empty weight
\+ Unusable fuel
= Standard empty weight

  Standard empty weight
\+ Optional equipment
= Licensed empty weight

Empty weight is the weight of the standard airplane, full hydraulic fluid and undrainable oil. Unusable fuel is the fuel remaining in the airplane not available to the engines. Standard empty weight is the empty weight plus unusable fuel. When optional equipment is added to the standard empty weight, the result is licensed empty weight. Licensed empty weight, therefore, includes the standard airplane, optional equipment, full hydraulic fluid, unusable fuel, and undrainable oil.

The major difference between the two formats (GAMA and the old) is that basic empty weight includes full oil, and licensed empty weight does not.

Oil must always be added to any weight and balance utilizing a licensed empty weight.

When the airplane is placed in service, amended weight and balance documents are prepared by appropriately rated maintenance personnel to reflect changes in installed equipment. The old weight and balance documents are customarily marked "superseded" and retained in the AFM/POH. Maintenance personnel are under no regulatory obligation to utilize the GAMA terminology, so weight and balance documents subsequent to the original may use a variety of terms. Pilots should use care to determine whether or not oil has to be added to the weight and balance calculations or if it is already included in the figures provided.

The multiengine airplane is where most pilots encounter the term "zero fuel weight" for the first time. Not all multiengine airplanes have a zero fuel weight limitation published in their AFM/POH, but many do. Zero fuel weight is simply the maximum allowable weight of the airplane and payload, assuming there is no usable fuel on board. The actual airplane is not devoid of fuel at the time of loading, of course. This is merely a calculation that assumes it was. If a zero fuel weight limitation is published, then all weight in excess of that figure must consist of usable fuel. The purpose of a zero fuel weight is to limit load forces on the wing spars with heavy fuselage loads.

Assume a hypothetical multiengine airplane with the following weights and capacities:

Basic empty weight . . . . .3,200 lb.

Zero fuel weight .4,400 lb.

Maximum takeoff weight . . . . .5,200 lb.

Maximum usable fuel . . . . .180 gal.

1. Calculate the useful load:

Maximum takeoff weight . . . . .5,200 lb.

Basic empty weight . . . .-3,200 lb.

Useful load . . . . .2,000 lb.

The useful load is the maximum combination of usable fuel, passengers, baggage, and cargo that the airplane is capable of carrying.

2. Calculate the payload:

Zero fuel weight . . . . 4,400 lb.

Basic empty weight. . . . . -3,200 lb.

Payload . . . . 1,200 lb.

The payload is the maximum combination of passengers, baggage, and cargo that the airplane is capable of carrying. A zero fuel weight, if published, is the limiting weight.

3. Calculate the fuel capacity at maximum payload (1,200 lb.):

Maximum takeoff weight . . . . .5,200 lb.

Zero fuel weight . . . .-4,400 lb.

Fuel allowed . . . . .800 lb.

Assuming maximum payload, the only weight permitted in excess of the zero fuel weight must consist of usable fuel. In this case, 133.3 gallons.

4. Calculate the payload at maximum fuel capacity (180 gal.):

Basic empty weight . . . . .3,200 lb.

Maximum usable fuel . . . .+1,080 lb.

Weight with max. fuel . . . . .4,280 lb.

Maximum takeoff weight . . . . .5,200 lb.

Weight with max. fuel . . . .-4,280 lb.

Payload allowed . . . . .920 lb.

Assuming maximum fuel, the payload is the difference between the weight of the fueled airplane and the maximum takeoff weight.

Some multiengine airplanes have a ramp weight, which is in excess of the maximum takeoff weight. The ramp weight is an allowance for fuel that would be burned during taxi and runup, permitting a takeoff at full maximum takeoff weight. The airplane must weigh no more than maximum takeoff weight at the beginning of the takeoff roll.

A maximum landing weight is a limitation against landing at a weight in excess of the published value. This requires preflight planning of fuel burn to ensure that the airplane weight upon arrival at destination will be at or below the maximum landing weight. In the event of an emergency requiring an immediate landing, the pilot should recognize that the structural margins designed into the airplane are not fully available when over landing weight. An overweight landing inspection may be advisable—the service manual or manufacturer should be consulted.

Although the foregoing problems only dealt with weight, the balance portion of weight and balance is equally vital. The flight characteristics of the multiengine airplane will vary significantly with shifts of the center of gravity (CG) within the approved envelope.

At forward CGs, the airplane will be more stable, with a slightly higher stalling speed, a slightly slower cruising speed, and favorable stall characteristics. At aft CGs, the airplane will be less stable, with a slightly lower stalling speed, a slightly faster cruising speed, and less desirable stall characteristics. Forward CG limits are usually determined in certification by elevator/stabilator authority in the landing round-out. Aft CG limits are determined by the minimum acceptable longitudinal stability. It is contrary to the airplane's operating limitations and the Code of Federal Regulations (CFR) to exceed any weight and balance parameter.

Some multiengine airplanes may require ballast to remain within CG limits under certain loading conditions. Several models require ballast in the aft baggage compartment with only a student and instructor on board to avoid exceeding the forward CG limit. When passengers are seated in the aft-most seats of some models, ballast or baggage may be required in the nose baggage compartment to avoid exceeding the aft CG limit. The pilot must direct the seating of passengers and placement of baggage and cargo to achieve a center of gravity within the approved envelope. Most multiengine airplanes have general loading recommendations in the weight and balance section of the AFM/POH. When ballast is added, it must be securely tied down and it must not exceed the maximum allowable floor loading.

Some airplanes make use of a special weight and balance plotter. It consists of several movable parts that can be adjusted over a plotting board on which the CG envelope is printed. The reverse side of the typical plotter contains general loading recommendations for the particular airplane. A pencil line plot can be made directly on the CG envelope imprinted on the working side of the plotting board. This plot can easily be erased and recalculated anew for each flight. This plotter is to be used only for the make and model airplane for which it was designed.

GROUND OPERATION

Good habits learned with single-engine airplanes are directly applicable to multiengine airplanes for preflight and engine start. Upon placing the airplane in motion to taxi, the new multiengine pilot will notice several differences, however. The most obvious is the increased wingspan and the need for even greater vigilance while taxiing in close quarters. Ground handling may seem somewhat ponderous and the multiengine airplane will not be as nimble as the typical two- or four-place single-engine airplane. As always, use care not to ride the brakes by keeping engine power to a minimum. One ground handling advantage of the multiengine airplane over single-engine airplanes is the differential power capability. Turning with an assist from differential power minimizes both the need for brakes during turns and the turning radius.

The pilot should be aware, however, that making a sharp turn assisted by brakes and differential power can cause the airplane to pivot about a stationary inboard wheel and landing gear. This is abuse for which the airplane was not designed and should be guarded against.

Unless otherwise directed by the AFM/POH, all ground operations should be conducted with the cowl flaps fully open. The use of strobe lights is normally deferred until taxiing onto the active runway.

NORMAL AND CROSSWIND TAKEOFF AND CLIMB

With the "before takeoff" checklist complete and air traffic control (ATC) clearance received, the airplane should be taxied into position on the runway centerline. If departing from an airport without an operating control tower, a careful check for approaching aircraft should be made along with a radio advisory on the appropriate frequency. Sharp turns onto the runway combined with a rolling takeoff are not a good operating practice and may be prohibited by the AFM/POH due to the possibility of "unporting" a fuel tank pickup. (The takeoff itself may be prohibited by the AFM/POH under any circumstances below certain fuel levels.) The flight controls should be positioned for a crosswind, if present. Exterior lights such as landing and taxi lights, and wingtip strobes should be illuminated immediately prior to initiating the takeoff roll, day or night. If holding in takeoff position for any length of time, particularly at night, the pilot should activate all exterior lights upon taxiing into position.

Takeoff power should be set as recommended in the AFM/POH. With normally aspirated (non-turbocharged) engines, this will be full throttle. Full throttle is also used in most turbocharged engines. There are some turbocharged engines, however, that require the pilot to set a specific power setting, usually just below red line manifold pressure. This yields takeoff power with less than full throttle travel.

Turbocharged engines often require special consideration. Throttle motion with turbocharged engines should be exceptionally smooth and deliberate. It is acceptable, and may even be desirable, to hold the airplane in position with brakes as the throttles are advanced. Brake release customarily occurs after significant boost from the turbocharger is established. This prevents wasting runway with slow, partial throttle acceleration as the engine power is increased. If runway length or obstacle clearance is critical, full power should be set before brake release, as specified in the performance charts.

As takeoff power is established, initial attention should be divided between tracking the runway centerline and monitoring the engine gauges. Many novice multiengine pilots tend to fixate on the airspeed indicator just as soon as the airplane begins its takeoff roll. Instead, the pilot should confirm that both engines are developing full-rated manifold pressure and r.p.m., and that the fuel flows, fuel pressures, exhaust gas temperatures (EGTs), and oil pressures are matched in their normal ranges. A directed and purposeful scan of the engine gauges can be accomplished well before the airplane approaches rotation speed. If a crosswind is present, the aileron displacement in the direction of the crosswind may be reduced as the airplane accelerates. The elevator/stabilator control should be held neutral throughout.

Full rated takeoff power should be used for every takeoff. Partial power takeoffs are not recommended. There is no evidence to suggest that the life of modern reciprocating engines is prolonged by partial power takeoffs. Paradoxically, excessive heat and engine wear can occur with partial power as the fuel metering system will fail to deliver the slightly over-rich mixture vital for engine cooling during takeoff.

There are several key airspeeds to be noted during the takeoff and climb sequence in any twin. The first speed to consider is V_{MC}. If an engine fails below V_{MC} while the airplane is on the ground, the takeoff must be rejected. Directional control can only be maintained by promptly closing both throttles and using rudder and brakes as required. If an engine fails below V_{MC} while airborne, directional control is not possible with the remaining engine producing takeoff power. On takeoffs, therefore, the airplane should never be airborne before the airspeed reaches and exceeds V_{MC}. Pilots should use the manufacturer's recommended rotation speed (V_R) or lift-off speed (V_{LOF}). If no such speeds are published, a minimum of V_{MC} plus 5 knots should be used for V_R.

The rotation to a takeoff pitch attitude is done smoothly. With a crosswind, the pilot should ensure that the landing gear does not momentarily touch the runway after the airplane has lifted off, as a side drift will be present. The rotation may be accomplished more positively and/or at a higher speed under these conditions. However, the pilot should keep in mind that the AFM/POH performance figures for accelerate-stop distance, takeoff ground roll, and distance to clear an obstacle were calculated at the recommended V_R and/or V_{LOF} speed.

After lift-off, the next consideration is to gain altitude as rapidly as possible. After leaving the ground, altitude gain is more important than achieving an excess of airspeed. Experience has shown that excessive speed cannot be effectively converted into altitude in the event of an engine failure. Altitude gives the pilot time to think and react. Therefore, the airplane should be allowed to accelerate in a shallow climb to attain V_Y, the best all-engine rate-of-climb speed. V_Y should then be maintained until a safe single-engine maneuvering altitude, considering terrain and obstructions, is achieved.

To assist the pilot in takeoff and initial climb profile, some AFM/POHs give a "50-foot" or "50-foot barrier" speed to use as a target during rotation, lift-off, and acceleration to V_Y.

Landing gear retraction should normally occur after a positive rate of climb is established. Some AFM/POHs direct the pilot to apply the wheel brakes momentarily after lift-off to stop wheel rotation prior to landing gear retraction. If flaps were extended for takeoff, they should be retracted as recommended in the AFM/POH.

Once a safe single-engine maneuvering altitude has been reached, typically a minimum of 400-500 feet AGL, the transition to an enroute climb speed should be made. This speed is higher than V_Y and is usually maintained to cruising altitude. Enroute climb speed gives better visibility, increased engine cooling, and a higher groundspeed. Takeoff power can be reduced, if desired, as the transition to enroute climb speed is made.

Some airplanes have a climb power setting published in the AFM/POH as a recommendation (or sometimes as a limitation), which should then be set for enroute climb. If there is no climb power setting published, it is customary, but not a requirement, to reduce manifold pressure and r.p.m. somewhat for enroute climb. The propellers are usually synchronized after the first power reduction and the yaw damper, if installed, engaged. The AFM/POH may also recommend leaning

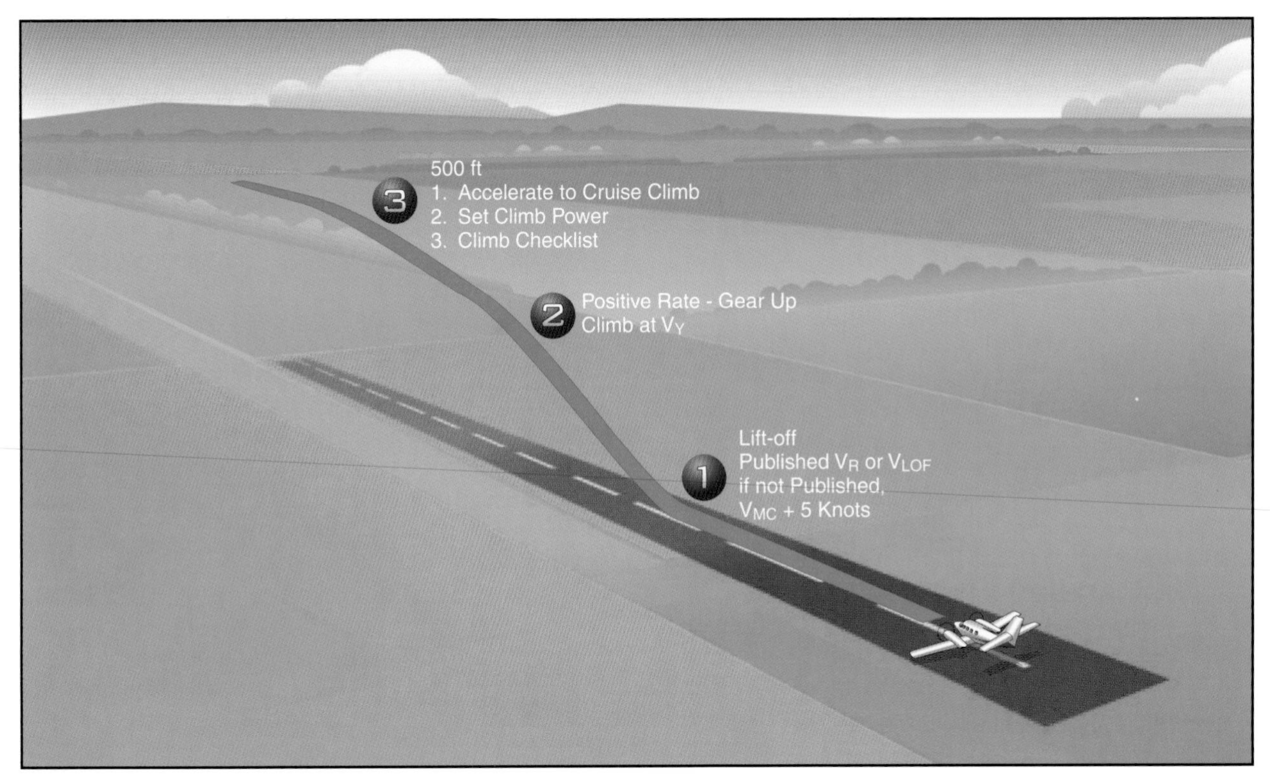

Figure 12-7. Takeoff and climb profile.

the mixtures during climb. The "climb" checklist should be accomplished as traffic and work load allow. [Figure 12-7]

Level Off and Cruise

Upon leveling off at cruising altitude, the pilot should allow the airplane to accelerate at climb power until cruising airspeed is achieved, then cruise power and r.p.m. should be set. To extract the maximum cruise performance from any airplane, the power setting tables provided by the manufacturer should be closely followed. If the cylinder head and oil temperatures are within their normal ranges, the cowl flaps may be closed. When the engine temperatures have stabilized, the mixtures may be leaned per AFM/POH recommendations. The remainder of the "cruise" checklist should be completed by this point.

Fuel management in multiengine airplanes is often more complex than in single-engine airplanes. Depending upon system design, the pilot may need to select between main tanks and auxiliary tanks, or even employ fuel transfer from one tank to another. In complex fuel systems, limitations are often found restricting the use of some tanks to level flight only, or requiring a reserve of fuel in the main tanks for descent and landing. Electric fuel pump operation can vary widely among different models also, particularly during tank switching or fuel transfer. Some fuel pumps are to be on for takeoff and landing; others are to be off. There is simply no substitute for thorough systems and AFM/POH knowledge when operating complex aircraft.

Normal Approach and Landing

Given the higher cruising speed (and frequently, altitude) of multiengine airplanes over most single-engine airplanes, the descent must be planned in advance. A hurried, last minute descent with power at or near idle is inefficient and can cause excessive engine cooling. It may also lead to passenger discomfort, particularly if the airplane is unpressurized. As a rule of thumb, if terrain and passenger conditions permit, a maximum of a 500 f.p.m. rate of descent should be planned. Pressurized airplanes can plan for higher descent rates, if desired.

In a descent, some airplanes require a minimum EGT, or may have a minimum power setting or cylinder head temperature to observe. In any case, combinations of very low manifold pressure and high r.p.m. settings are strongly discouraged by engine manufacturers. If higher descent rates are necessary, the pilot should consider extending partial flaps or lowering the landing gear before retarding the power excessively. The "descent" checklist should be initiated upon leaving cruising altitude and completed before arrival in the terminal area. Upon arrival in the terminal area, pilots are encouraged to turn on their landing and recognition lights when operating below 10,000 feet, day or night, and especially when operating within 10 miles of any airport or in conditions of reduced visibility.

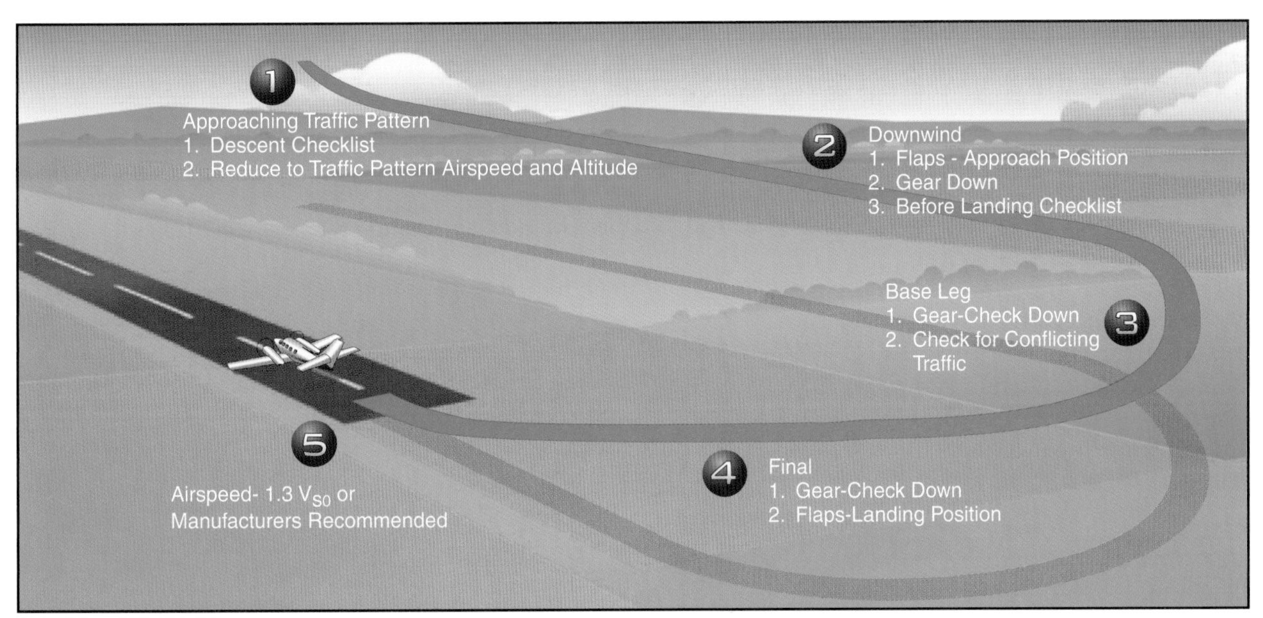

Figure 12-8. Normal two-engine approach and landing.

The traffic pattern and approach are typically flown at somewhat higher indicated airspeeds in a multiengine airplane contrasted to most single-engine airplanes. The pilot may allow for this through an early start on the "before landing" checklist. This provides time for proper planning, spacing, and thinking well ahead of the airplane. Many multiengine airplanes have partial flap extension speeds above V_{FE}, and partial flaps can be deployed prior to traffic pattern entry. Normally, the landing gear should be selected and confirmed down when abeam the intended point of landing as the downwind leg is flown. [Figure 12-8]

The Federal Aviation Administration (FAA) recommends a stabilized approach concept. To the greatest extent practical, on final approach and within 500 feet AGL, the airplane should be on speed, in trim, configured for landing, tracking the extended centerline of the runway, and established in a constant angle of descent towards an aim point in the touchdown zone. Absent unusual flight conditions, only minor corrections will be required to maintain this approach to the roundout and touchdown.

The final approach should be made with power and at a speed recommended by the manufacturer; if a recommended speed is not furnished, the speed should be no slower than the single-engine best rate-of-climb speed (V_{YSE}) until short final with the landing assured, but in no case less than critical engine-out minimum control speed (V_{MC}). Some multiengine pilots prefer to delay full flap extension to short final with the landing assured. This is an acceptable technique with appropriate experience and familiarity with the airplane.

In the roundout for landing, residual power is gradually reduced to idle. With the higher wing loading of multiengine airplanes and with the drag from two windmilling propellers, there will be minimal float. Full stall landings are generally undesirable in twins. The airplane should be held off as with a high performance single-engine model, allowing touchdown of the main wheels prior to a full stall.

Under favorable wind and runway conditions, the nosewheel can be held off for best aerodynamic braking. Even as the nosewheel is gently lowered to the runway centerline, continued elevator back pressure will greatly assist the wheel brakes in stopping the airplane.

If runway length is critical, or with a strong crosswind, or if the surface is contaminated with water, ice or snow, it is undesirable to rely solely on aerodynamic braking after touchdown. The full weight of the airplane should be placed on the wheels as soon as practicable. The wheel brakes will be more effective than aerodynamic braking alone in decelerating the airplane.

Once on the ground, elevator back pressure should be used to place additional weight on the main wheels and to add additional drag. When necessary, wing flap retraction will also add additional weight to the wheels and improve braking effectivity. Flap retraction during the landing rollout is discouraged, however, unless there is a clear, operational need. It should not be accomplished as routine with each landing.

Some multiengine airplanes, particularly those of the cabin class variety, can be flown through the roundout and touchdown with a small amount of power. This is an acceptable technique to prevent high sink rates and to cushion the touchdown. The pilot should keep in mind, however, that the primary purpose in landing is to get the airplane down and stopped. This technique should only be attempted when there is a generous margin of runway length. As propeller blast flows directly over the wings, lift as well as thrust is produced. The pilot should taxi clear of the runway as soon as speed and safety permit, and then accomplish the "after landing" checklist. Ordinarily, no attempt should be made to retract the wing flaps or perform other checklist duties until the airplane has been brought to a halt when clear of the active runway. Exceptions to this would be the rare operational needs discussed above, to relieve the weight from the wings and place it on the wheels. In these cases, AFM/POH guidance should be followed. The pilot should not indiscriminately reach out for any switch or control on landing rollout. An inadvertent landing gear retraction while meaning to retract the wing flaps may result.

CROSSWIND APPROACH AND LANDING

The multiengine airplane is often easier to land in a crosswind than a single-engine airplane due to its higher approach and landing speed. In any event, the principles are no different between singles and twins. Prior to touchdown, the longitudinal axis must be aligned with the runway centerline to avoid landing gear side loads.

The two primary methods, crab and wing-low, are typically used in conjunction with each other. As soon as the airplane rolls out onto final approach, the crab angle to track the extended runway centerline is established. This is coordinated flight with adjustments to heading to compensate for wind drift either left or right. Prior to touchdown, the transition to a sideslip is made with the upwind wing lowered and opposite rudder applied to prevent a turn. The airplane touches down on the landing gear of the upwind wing first, followed by that of the downwind wing, and then the nose gear. Follow-through with the flight controls involves an increasing application of aileron into the wind until full control deflection is reached.

The point at which the transition from the crab to the sideslip is made is dependent upon pilot familiarity with the airplane and experience. With high skill and experience levels, the transition can be made during the roundout just before touchdown. With lesser skill and experience levels, the transition is made at increasing distances from the runway. Some multiengine airplanes (as some single-engine airplanes) have AFM/POH limitations against slips in excess of a certain time period; 30 seconds, for example. This is to prevent engine power loss from fuel starvation as the fuel in the tank of the lowered wing flows towards the wingtip, away from the fuel pickup point. This time limit must be observed if the wing-low method is utilized.

Some multiengine pilots prefer to use differential power to assist in crosswind landings. The asymmetrical thrust produces a yawing moment little different from that produced by the rudder. When the upwind wing is lowered, power on the upwind engine is increased to prevent the airplane from turning. This alternate technique is completely acceptable, but most pilots feel they can react to changing wind conditions quicker with rudder and aileron than throttle movement. This is especially true with turbocharged engines where the throttle response may lag momentarily. The differential power technique should be practiced with an instructor familiar with it before being attempted alone.

SHORT-FIELD TAKEOFF AND CLIMB

The short-field takeoff and climb differs from the normal takeoff and climb in the airspeeds and initial climb profile. Some AFM/POHs give separate short-field takeoff procedures and performance charts that recommend specific flap settings and airspeeds. Other AFM/POHs do not provide separate short-field procedures. In the absence of such specific procedures, the airplane should be operated only as recommended in the AFM/POH. No operations should be conducted contrary to the recommendations in the AFM/POH.

On short-field takeoffs in general, just after rotation and lift-off, the airplane should be allowed to accelerate to V_X, making the initial climb over obstacles at V_X and transitioning to V_Y as obstacles are cleared. [Figure 12-9]

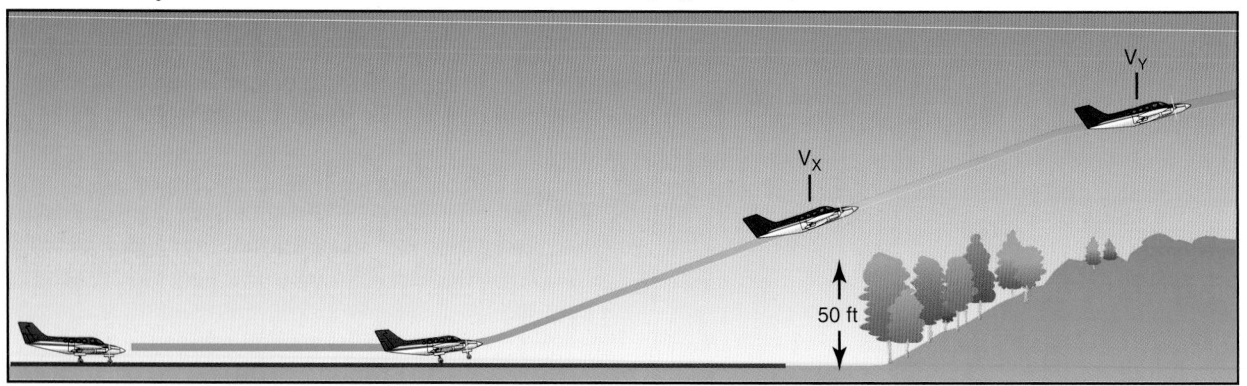

Figure 12-9. Short-field takeoff and climb.

When partial flaps are recommended for short-field takeoffs, many light-twins have a strong tendency to become airborne prior to V_{MC} plus 5 knots. Attempting to prevent premature lift-off with forward elevator pressure results in wheelbarrowing. To prevent this, allow the airplane to become airborne, but only a few inches above the runway. The pilot should be prepared to promptly abort the takeoff and land in the event of engine failure on takeoff with landing gear and flaps extended at airspeeds below V_X.

Engine failure on takeoff, particularly with obstructions, is compounded by the low airspeeds and steep climb attitudes utilized in short-field takeoffs. V_X and V_{XSE} are often perilously close to V_{MC}, leaving scant margin for error in the event of engine failure as V_{XSE} is assumed. If flaps were used for takeoff, the engine failure situation becomes even more critical due to the additional drag incurred. If V_X is less than 5 knots higher than V_{MC}, give strong consideration to reducing useful load or using another runway in order to increase the takeoff margins so that a short-field technique will not be required.

SHORT-FIELD APPROACH AND LANDING

The primary elements of a short-field approach and landing do not differ significantly from a normal approach and landing. Many manufacturers do not publish short-field landing techniques or performance charts in the AFM/POH. In the absence of specific short-field approach and landing procedures, the airplane should be operated as recommended in the AFM/POH. No operations should be conducted contrary to the AFM/POH recommendations.

The emphasis in a short-field approach is on configuration (full flaps), a stabilized approach with a constant angle of descent, and precise airspeed control. As part of a short-field approach and landing procedure, some AFM/POHs recommend a slightly slower than normal approach airspeed. If no such slower speed is published, use the AFM/POH-recommended normal approach speed.

Full flaps are used to provide the steepest approach angle. If obstacles are present, the approach should be planned so that no drastic power reductions are required after they are cleared. The power should be smoothly reduced to idle in the roundout prior to touchdown. Pilots should keep in mind that the propeller blast blows over the wings, providing some lift in addition to thrust. Significantly reducing power just after obstacle clearance usually results in a sudden, high sink rate that may lead to a hard landing.

After the short-field touchdown, maximum stopping effort is achieved by retracting the wing flaps, adding back pressure to the elevator/stabilator, and applying heavy braking. However, if the runway length permits, the wing flaps should be left in the extended position until the airplane has been stopped clear of the runway. There is always a significant risk of retracting the landing gear instead of the wing flaps when flap retraction is attempted on the landing rollout.

Landing conditions that involve either a short-field, high-winds or strong crosswinds are just about the only situations where flap retraction on the landing rollout should be considered. When there is an operational need to retract the flaps just after touchdown, it must be done deliberately, with the flap handle positively identified before it is moved.

GO-AROUND

When the decision to go around is made, the throttles should be advanced to takeoff power. With adequate airspeed, the airplane should be placed in a climb pitch attitude. These actions, which are accomplished simultaneously, will arrest the sink rate and place the airplane in the proper attitude for transition to a climb. The initial target airspeed will be V_Y, or V_X if obstructions are present. With sufficient airspeed, the flaps should be retracted from full to an intermediate position and the landing gear retracted when there is a positive rate of climb and no chance of runway contact. The remaining flaps should then be retracted. [Figure 12-10]

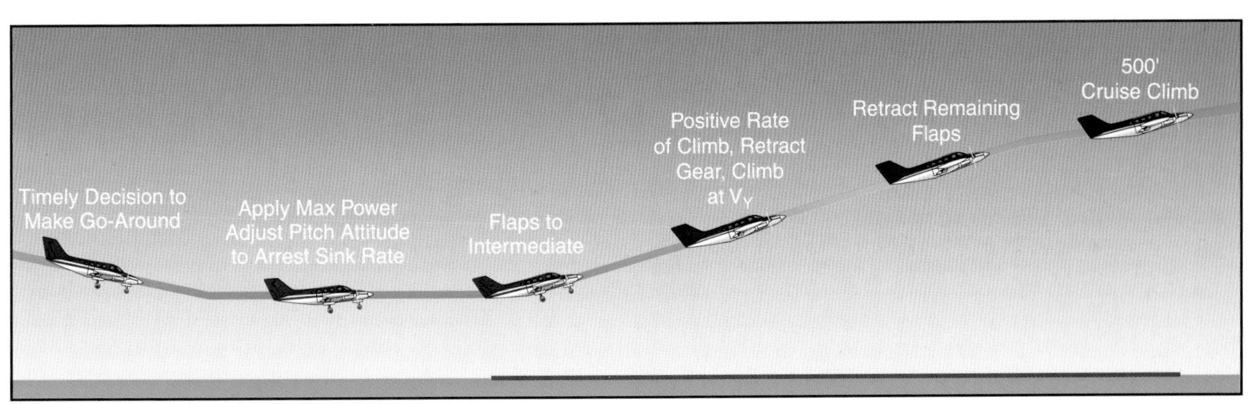

Figure 12-10. Go-around procedure.

If the go-around was initiated due to conflicting traffic on the ground or aloft, the pilot should maneuver to the side, so as to keep the conflicting traffic in sight. This may involve a shallow bank turn to offset and then parallel the runway/landing area.

If the airplane was in trim for the landing approach when the go-around was commenced, it will soon require a great deal of forward elevator/stabilator pressure as the airplane accelerates away in a climb. The pilot should apply appropriate forward pressure to maintain the desired pitch attitude. Trim should be commenced immediately. The "balked landing" checklist should be reviewed as work load permits.

Flaps should be retracted before the landing gear for two reasons. First, on most airplanes, full flaps produce more drag than the extended landing gear. Secondly, the airplane will tend to settle somewhat with flap retraction, and the landing gear should be down in the event of an inadvertent, momentary touchdown.

Many multiengine airplanes have a landing gear retraction speed significantly less than the extension speed. Care should be exercised during the go-around not to exceed the retraction speed. If the pilot desires to return for a landing, it is essential to re-accomplish the entire "before landing" checklist. An interruption to a pilot's habit patterns, such as a go-around, is a classic scenario for a subsequent gear up landing.

The preceding discussion of go-arounds assumes that the maneuver was initiated from normal approach speeds or faster. If the go-around was initiated from a low airspeed, the initial pitch up to a climb attitude must be tempered with the necessity of maintaining adequate flying speed throughout the maneuver. Examples of where this applies include go-arounds initiated from the landing roundout or recovery from a bad bounce as well as a go-around initiated due to an inadvertent approach to a stall. The first priority is always to maintain control and obtain adequate flying speed. A few moments of level or near level flight may be required as the airplane accelerates up to climb speed.

REJECTED TAKEOFF

A takeoff can be rejected for the same reasons a takeoff in a single-engine airplane would be rejected. Once the decision to reject a takeoff is made, the pilot should promptly close both throttles and maintain directional control with the rudder, nosewheel steering, and brakes. Aggressive use of rudder, nosewheel steering, and brakes may be required to keep the airplane on the runway. Particularly, if an engine failure is not immediately recognized and accompanied by prompt closure of both throttles. However, the primary objective is not necessarily to stop the airplane in the shortest distance, but to maintain control of the airplane as it decelerates. In some situations, it may be preferable to continue into the overrun area under control, rather than risk directional control loss, landing gear collapse, or tire/brake failure in an attempt to stop the airplane in the shortest possible distance.

ENGINE FAILURE AFTER LIFT-OFF

A takeoff or go-around is the most critical time to suffer an engine failure. The airplane will be slow, close to the ground, and may even have landing gear and flaps extended. Altitude and time will be minimal. Until feathered, the propeller of the failed engine will be windmilling, producing a great deal of drag and yawing tendency. Airplane climb performance will be marginal or even non-existent, and obstructions may lie ahead. Add the element of surprise and the need for a plan of action before every takeoff is obvious.

With loss of an engine, it is paramount to maintain airplane control and comply with the manufacturer's recommended emergency procedures. Complete failure of one engine shortly after takeoff can be broadly categorized into one of three following scenarios.

1. **Landing gear down**. [Figure 12-11] If the engine failure occurs prior to selecting the landing gear to the UP position, close both throttles and land on the remaining runway or overrun. Depending upon how quickly the pilot reacts to the sudden yaw, the airplane may run off the side of the runway by the time action is taken. There are really no other practical options. As discussed earlier, the chances of maintaining directional control while retracting the flaps (if extended), landing gear, feathering the propeller, and accelerating are minimal. On some airplanes with a single-engine-driven hydraulic pump, failure of that engine means the only way to raise the landing gear is to allow the engine to windmill or to use a hand pump. This is not a viable alternative during takeoff.

2. **Landing gear control selected up, single-engine climb performance inadequate**. [Figure 12-12] When operating near or above the single-engine ceiling and an engine failure is experienced shortly after lift-off, a landing must be accomplished on whatever essentially lies ahead. There is also the option of continuing ahead, in a descent at V_{YSE} with the remaining engine producing power, as long as the pilot is not tempted to remain airborne beyond the airplane's performance capability. Remaining airborne, bleeding off airspeed in a futile attempt to maintain altitude is almost invariably fatal. Landing under control is paramount. The greatest hazard in a single-engine takeoff is attempting to fly when it is not within the per-

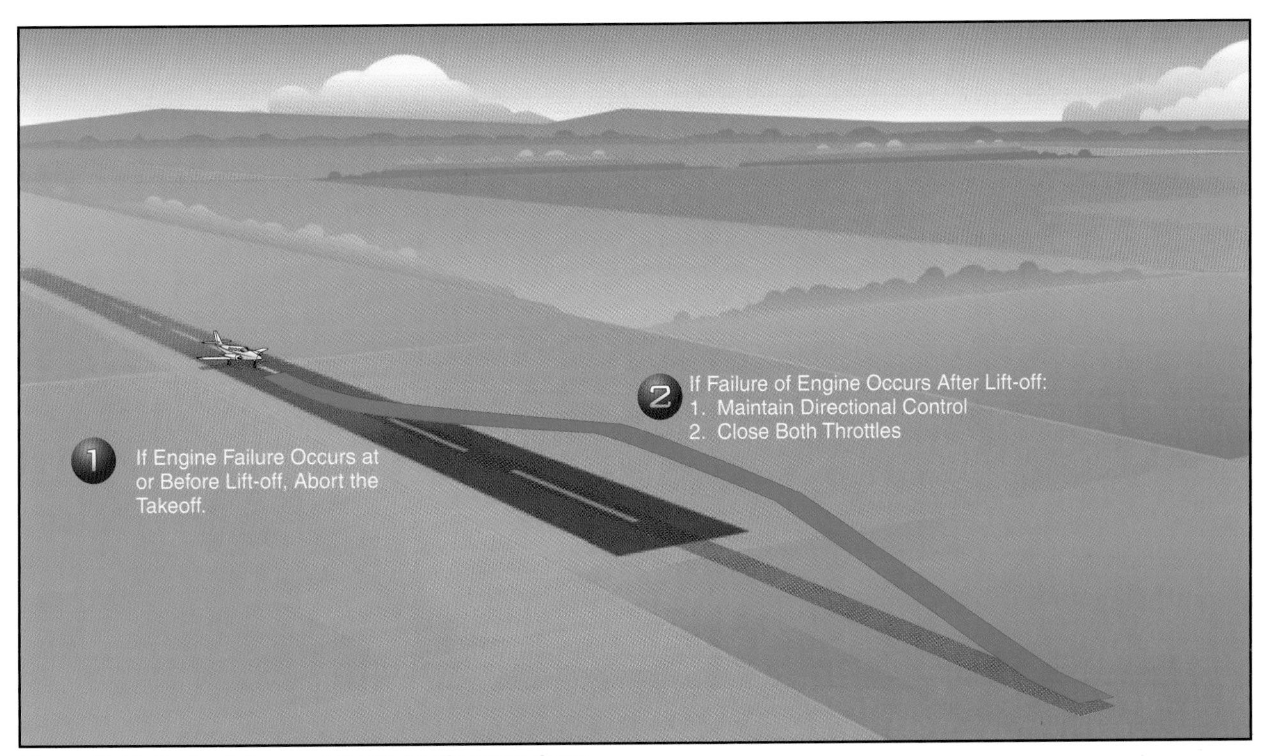

Figure 12-11. Engine failure on takeoff, landing gear down.

formance capability of the airplane to do so. An accident is inevitable.

Analysis of engine failures on takeoff reveals a very high success rate of off-airport engine inoperative landings when the airplane is landed under control. Analysis also reveals a very high fatality rate in stall-spin accidents when the pilot attempts flight beyond the performance capability of the airplane.

As mentioned previously, if the airplane's landing gear retraction mechanism is dependent upon hydraulic pressure from a certain engine-driven pump, failure of that engine can mean a loss of hundreds of feet of altitude as the pilot either windmills the engine to provide hydraulic pressure to raise the gear or raises it manually with a backup pump.

3. **Landing gear control selected up, single-engine climb performance adequate**. [Figure 12-13] If the single-engine rate of climb is adequate, the procedures for continued flight should be followed. There are four areas of concern: **control**, **configuration**, **climb**, and **checklist**.

- **CONTROL**— The first consideration following engine failure during takeoff is control of the airplane. Upon detecting an engine failure, aileron should be used to bank the airplane and rudder pressure applied, aggressively if necessary, to counteract the yaw and roll from asymmetrical thrust. The control forces, particularly on the rudder, may be high. The pitch attitude for V_{YSE} will have to be lowered from that of V_Y.

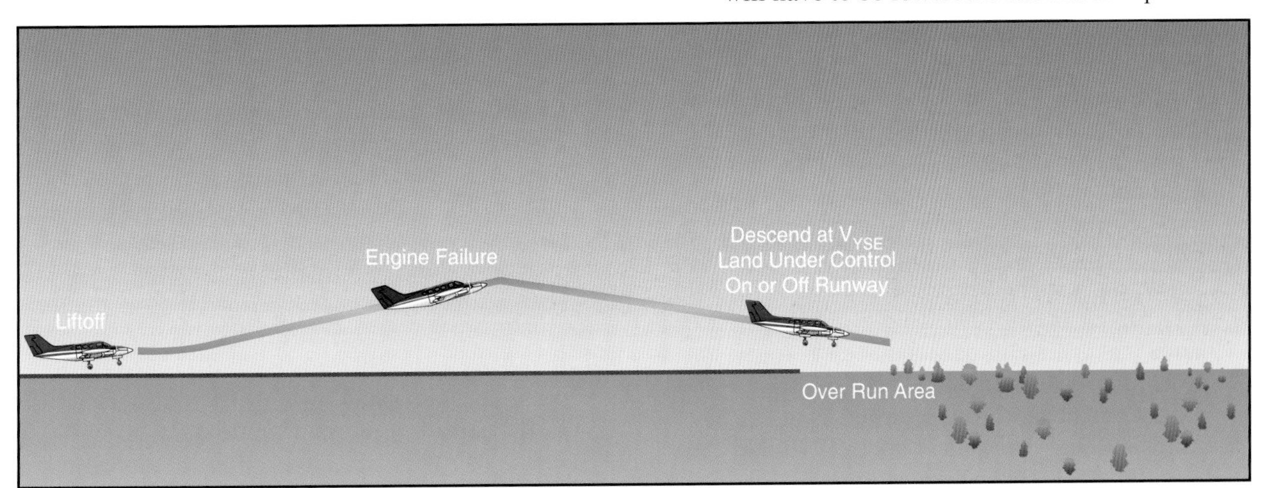

Figure 12-12. Engine failure on takeoff, inadequate climb performance.

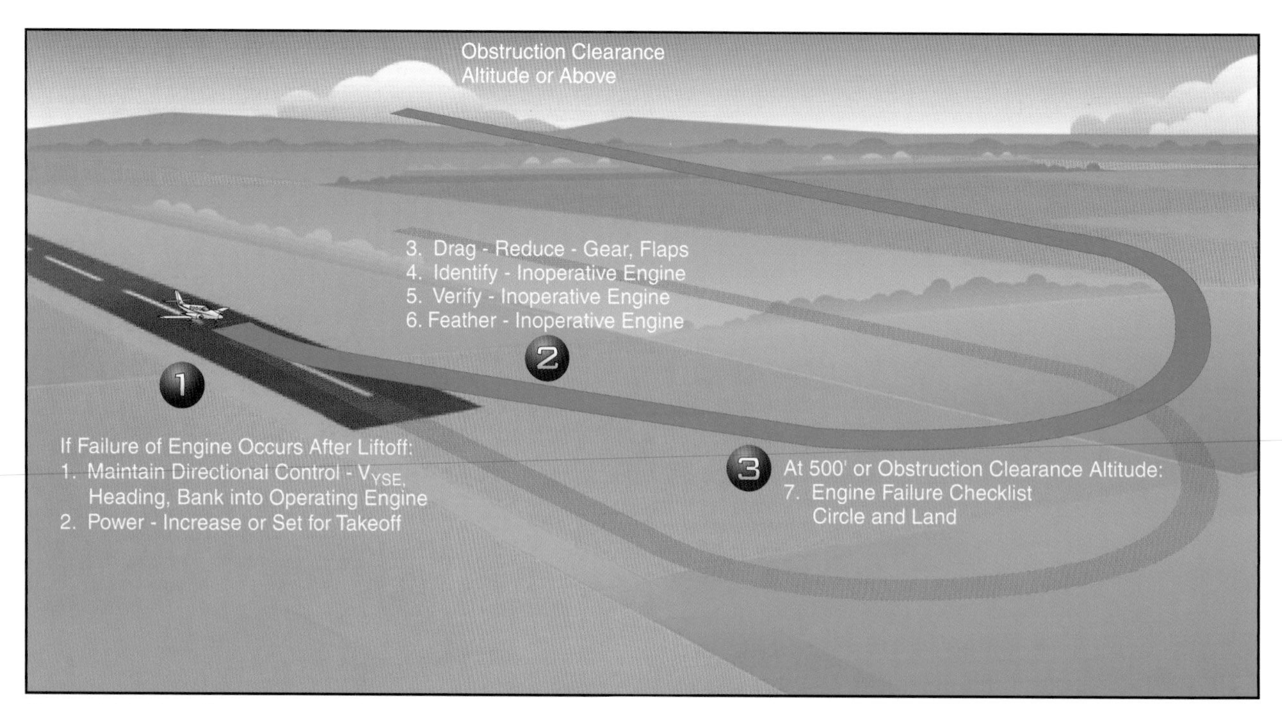

Figure 12-13. Landing gear up—adequate climb performance.

At least 5° of bank should be used, if necessary, to stop the yaw and maintain directional control. This initial bank input is held only momentarily, just long enough to establish or ensure directional control. Climb performance suffers when bank angles exceed approximately 2 or 3°, but obtaining and maintaining V_{YSE} and directional control are paramount. Trim should be adjusted to lower the control forces.

- **CONFIGURATION**—The memory items from the "engine failure after takeoff" checklist [Figure 12-14] should be promptly executed to configure the airplane for climb. The specific procedures to follow will be found in the AFM/POH and checklist for the particular airplane. Most will direct the pilot to assume V_{YSE}, set takeoff power, retract the flaps and landing gear, identify, verify, and feather the failed engine. (On some airplanes, the landing gear is to be retracted before the flaps.)

  The "identify" step is for the pilot to initially identify the failed engine. Confirmation on the engine gauges may or may not be possible, depending upon the failure mode. Identification should be primarily through the control inputs required to maintain straight flight, not the engine gauges. The "verify" step directs the pilot to retard the throttle of the engine thought to have failed. No change in performance when the suspected throttle is retarded is verification that the correct engine has been identified as failed. The corresponding propeller control should be brought fully aft to feather the engine.

ENGINE FAILURE AFTER TAKEOFF	
Airspeed	**Maintain V_{YSE}**
Mixtures	**RICH**
Propellers	**HIGH RPM**
Throttles	**FULL POWER**
Flaps	**UP**
Landing Gear	**UP**
Identify	**Determine failed engine**
Verify	**Close throttle of failed engine**
Propeller	**FEATHER**
Trim Tabs	ADJUST
Failed Engine	SECURE
As soon as practical	LAND

Bold - faced items require immediate action and are to be accomplished from memory.

Figure 12-14. Typical "engine failure after takeoff" emergency checklist.

- **CLIMB**—As soon as directional control is established and the airplane configured for climb, the bank angle should be reduced to that producing best climb performance. Without specific guidance for zero sideslip, a bank of 2° and one-third to one-half ball deflection on the slip/skid indicator is suggested. V_{YSE} is maintained with pitch control. As turning flight reduces climb performance, climb should be made straight ahead, or with shallow turns to avoid obstacles, to an altitude of at least 400 feet AGL before attempting a return to the airport.

- **CHECKLIST**—Having accomplished the memory items from the "engine failure after takeoff" checklist, the printed copy should be reviewed as time permits. The "securing failed engine" checklist [Figure 12-15] should then be accomplished. Unless the pilot suspects an engine fire, the remaining items should be accomplished deliberately and without undue haste. Airplane control should never be sacrificed to execute the remaining checklists. The priority items have already been accomplished from memory.

SECURING FAILED ENGINE	
Mixture	IDLE CUT OFF
Magnetos	OFF
Alternator	OFF
Cowl Flap	CLOSE
Boost Pump	OFF
Fuel Selector	OFF
Prop Sync	OFF
Electrical Load	Reduce
Crossfeed	Consider

Figure 12-15. Typical "securing failed engine" emergency checklist.

Other than closing the cowl flap of the failed engine, none of these items, if left undone, adversely affects airplane climb performance. There is a distinct possibility of actuating an incorrect switch or control if the procedure is rushed. The pilot should concentrate on flying the airplane and extracting maximum performance. If ATC facilities are available, an emergency should be declared.

The memory items in the "engine failure after takeoff" checklist may be redundant with the airplane's existing configuration. For example, in the third takeoff scenario, the gear and flaps were assumed to already be retracted, yet the memory items included gear and flaps. This is not an oversight. The purpose of the memory items is to either initiate the appropriate action or to confirm that a condition exists. Action on each item may not be required in all cases. The memory items also apply to more than one circumstance. In an engine failure from a go-around, for example, the landing gear and flaps would likely be extended when the failure occurred.

The three preceding takeoff scenarios all include the landing gear as a key element in the decision to land or continue. With the landing gear selector in the DOWN position, for example, continued takeoff and climb is not recommended. This situation, however, is not justification to retract the landing gear the moment the airplane lifts off the surface on takeoff as a normal procedure. The landing gear should remain selected down as long as there is usable runway or overrun available to land on. The use of wing flaps for takeoff virtually eliminates the likelihood of a single-engine climb until the flaps are retracted.

There are two time-tested memory aids the pilot may find useful in dealing with engine-out scenarios. The first, "Dead foot–dead engine" is used to assist in identifying the failed engine. Depending on the failure mode, the pilot won't be able to consistently identify the failed engine in a timely manner from the engine gauges. In maintaining directional control, however, rudder pressure will be exerted on the side (left or right) of the airplane with the operating engine. Thus, the "dead foot" is on the same side as the "dead engine." Variations on this saying include "Idle foot–idle engine" and "Working foot–working engine."

The second memory aid has to do with climb performance. The phrase "Raise the dead" is a reminder that the best climb performance is obtained with a very shallow bank, about 2° toward the operating engine. Therefore, the inoperative, or "dead" engine should be "raised" with a very slight bank.

Not all engine power losses are complete failures. Sometimes the failure mode is such that partial power may be available. If there is a performance loss when the throttle of the affected engine is retarded, the pilot should consider allowing it to run until altitude and airspeed permit safe single-engine flight, if this can be done without compromising safety. Attempts to save a malfunctioning engine can lead to a loss of the entire airplane.

ENGINE FAILURE DURING FLIGHT

Engine failures well above the ground are handled differently than those occurring at lower speeds and altitudes. Cruise airspeed allows better airplane control, and altitude may permit time for a possible diagnosis and remedy of the failure. Maintaining airplane control, however, is still paramount. Airplanes have been lost at altitude due to apparent fixation on the engine problem to the detriment of flying the airplane.

Not all engine failures or malfunctions are catastrophic in nature (catastrophic meaning a major mechanical failure that damages the engine and precludes further engine operation). Many cases of power loss are related to fuel starvation, where restoration of power may be made with the selection of another tank. An orderly inventory of gauges and switches may reveal the problem. Carburetor heat or alternate air can be selected. The affected engine may run smoothly on just one magneto or at a lower power setting. Altering the mixture may help. If fuel vapor formation is suspected, fuel boost pump operation may be used to eliminate flow and pressure fluctuations.

Although it is a natural desire among pilots to save an ailing engine with a precautionary shutdown, the engine should be left running if there is any doubt as to needing it for further safe flight. Catastrophic failure accompanied by heavy vibration, smoke, blistering paint, or large trails of oil, on the other hand, indicate a critical situation. The affected engine should be feathered and the "securing failed engine" checklist completed. The pilot should divert to the nearest suitable airport and declare an emergency with ATC for priority handling.

Fuel crossfeed is a method of getting fuel from a tank on one side of the airplane to an operating engine on the other. Crossfeed is used for extended single-engine operation. If a suitable airport is close at hand, there is no need to consider crossfeed. If prolonged flight on a single-engine is inevitable due to airport non-availability, then crossfeed allows use of fuel that would otherwise be unavailable to the operating engine. It also permits the pilot to balance the fuel consumption to avoid an out-of-balance wing heaviness.

AFM/POH procedures for crossfeed vary widely. Thorough fuel system knowledge is essential if crossfeed is to be conducted. Fuel selector positions and fuel boost pump usage for crossfeed differ greatly among multiengine airplanes. Prior to landing, crossfeed should be terminated and the operating engine returned to its main tank fuel supply.

If the airplane is above its single-engine absolute ceiling at the time of engine failure, it will slowly lose altitude. The pilot should maintain V_{YSE} to minimize the rate of altitude loss. This "drift down" rate will be greatest immediately following the failure and will decrease as the single-engine ceiling is approached. Due to performance variations caused by engine and propeller wear, turbulence, and pilot technique, the airplane may not maintain altitude even at its published single-engine ceiling. Any further rate of sink, however, would likely be modest.

An engine failure in a descent or other low power setting can be deceiving. The dramatic yaw and performance loss will be absent. At very low power settings, the pilot may not even be aware of a failure. If a failure is suspected, the pilot should advance both engine mixtures, propellers, and throttles significantly, to the takeoff settings if necessary, to correctly identify the failed engine. The power on the operative engine can always be reduced later.

ENGINE INOPERATIVE APPROACH AND LANDING

The approach and landing with one engine inoperative is essentially the same as a two-engine approach and landing. The traffic pattern should be flown at similar altitudes, airspeeds, and key positions as a two-engine approach. The differences will be the reduced power available and the fact that the remaining thrust is asymmetrical. A higher-than-normal power setting will be necessary on the operative engine.

With adequate airspeed and performance, the landing gear can still be extended on the downwind leg. In which case it should be confirmed DOWN no later than abeam the intended point of landing. Performance permitting, initial extension of wing flaps (10°, typically) and a descent from pattern altitude can also be initiated on the downwind leg. The airspeed should be no slower than V_{YSE}. The direction of the traffic pattern, and therefore the turns, is of no consequence as far as airplane controllability and performance are concerned. It is perfectly acceptable to make turns toward the failed engine.

On the base leg, if performance is adequate, the flaps may be extended to an intermediate setting (25°, typically). If the performance is inadequate, as measured by a decay in airspeed or high sink rate, delay further flap extension until closer to the runway. V_{YSE} is still the minimum airspeed to maintain.

On final approach, a normal, 3° glidepath to a landing is desirable. VASI or other vertical path lighting aids should be utilized if available. Slightly steeper approaches may be acceptable. However, a long, flat, low approach should be avoided. Large, sudden power applications or reductions should also be avoided. Maintain V_{YSE} until the landing is assured, then slow to 1.3 V_{SO} or the AFM/POH recommended speed. The final flap setting may be delayed until the landing is assured, or the airplane may be landed with partial flaps.

The airplane should remain in trim throughout. The pilot must be prepared, however, for a rudder trim change as the power of the operating engine is reduced to idle in the roundout just prior to touchdown. With drag from only one windmilling propeller, the airplane will tend to float more than on a two-engine approach. Precise airspeed control therefore is essential, especially when landing on a short, wet and/or slippery surface.

Some pilots favor resetting the rudder trim to neutral on final and compensating for yaw by holding rudder pressure for the remainder of the approach. This eliminates the rudder trim change close to the ground as the throttle is closed during the roundout for landing. This technique eliminates the need for groping for the rudder trim and manipulating it to neutral during final approach, which many pilots find to be highly distracting. AFM/POH recommendations or personal preference should be used.

Single-engine go-arounds must be avoided. As a practical matter in single-engine approaches, once the airplane is on final approach with landing gear and flaps extended, it is committed to land. If not on the intended runway, then on another runway, a taxiway, or grassy infield. The light-twin does not have the performance to climb on one engine with landing gear and flaps extended. Considerable altitude will be lost while maintaining V_{YSE} and retracting landing gear and flaps. Losses of 500 feet or more are not unusual. If the landing gear has been lowered with an alternate means of extension, retraction may not be possible, virtually negating any climb capability.

ENGINE INOPERATIVE FLIGHT PRINCIPLES

Best single-engine climb performance is obtained at V_{YSE} with maximum available power and minimum drag. After the flaps and landing gear have been retracted and the propeller of the failed engine feathered, a key element in best climb performance is minimizing sideslip.

With a single-engine airplane or a multiengine airplane with both engines operative, sideslip is eliminated when the ball of the turn and bank instrument is centered. This is a condition of **zero sideslip**, and the airplane is presenting its smallest possible profile to the relative wind. As a result, drag is at its minimum. Pilots know this as coordinated flight.

In a multiengine airplane with an inoperative engine, the centered ball is no longer the indicator of zero sideslip due to asymmetrical thrust. In fact, there is no instrument at all that will directly tell the pilot the flight conditions for zero sideslip. In the absence of a **yaw string**, minimizing sideslip is a matter of placing the airplane at a predetermined bank angle and ball position. The AFM/POH performance charts for single-engine flight were determined at zero sideslip. If this performance is even to be approximated, the zero sideslip technique **must** be utilized.

There are two different control inputs that can be used to counteract the asymmetrical thrust of a failed engine: (1) yaw from the rudder, and (2) the horizontal component of lift that results from bank with the ailerons. Used individually, neither is correct. Used together in the proper combination, zero sideslip and best climb performance are achieved.

Three different scenarios of airplane control inputs are presented below. **Neither of the first two is correct**. They are presented to illustrate the reasons for the zero sideslip approach to best climb performance.

1. Engine inoperative flight with wings level and ball centered requires large rudder input towards the operative engine. [Figure 12-16] The result is a moderate sideslip towards the inoperative engine. Climb performance will be reduced by the moderate sideslip. With wings level, V_{MC} will be significantly higher than published as there is no horizontal component of lift available to help the rudder combat asymmetrical thrust.

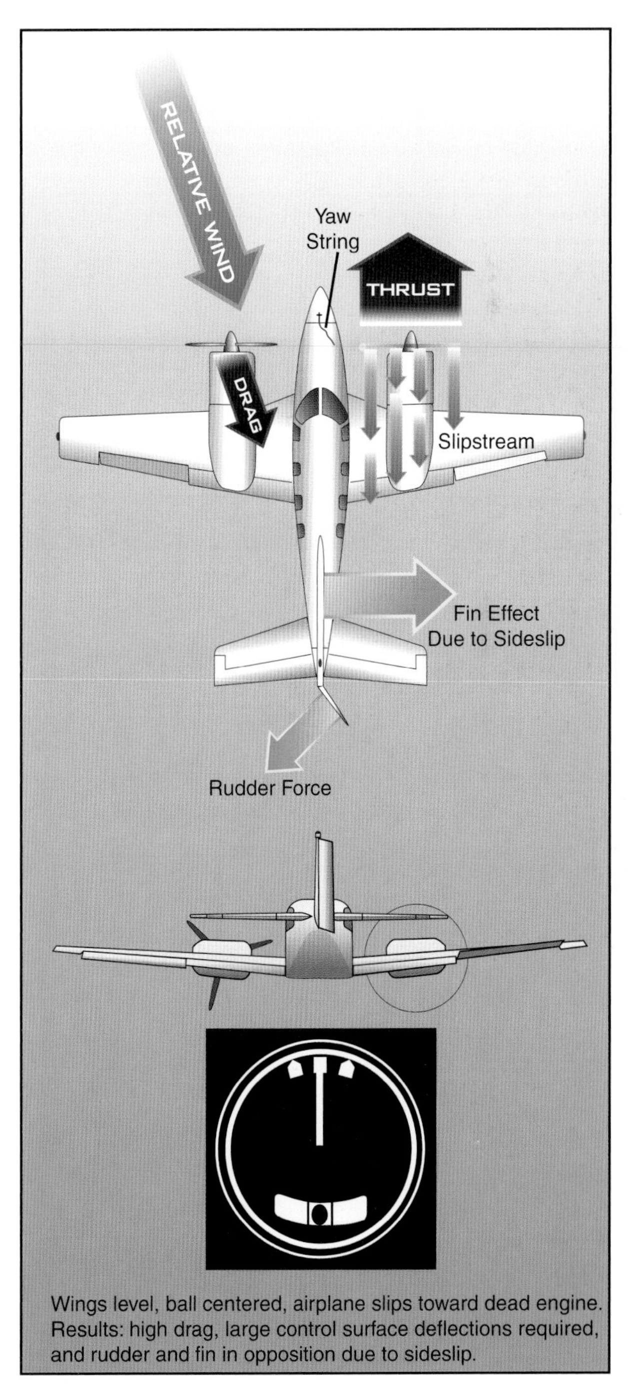

Figure 12-16. Wings level engine-out flight.

2. Engine inoperative flight using ailerons alone requires an 8 - 10° bank angle towards the operative engine. [Figure 12-17] This assumes no rudder input. The ball will be displaced well towards the operative engine. The result is a large sideslip towards the operative engine. Climb performance will be greatly reduced by the large sideslip.

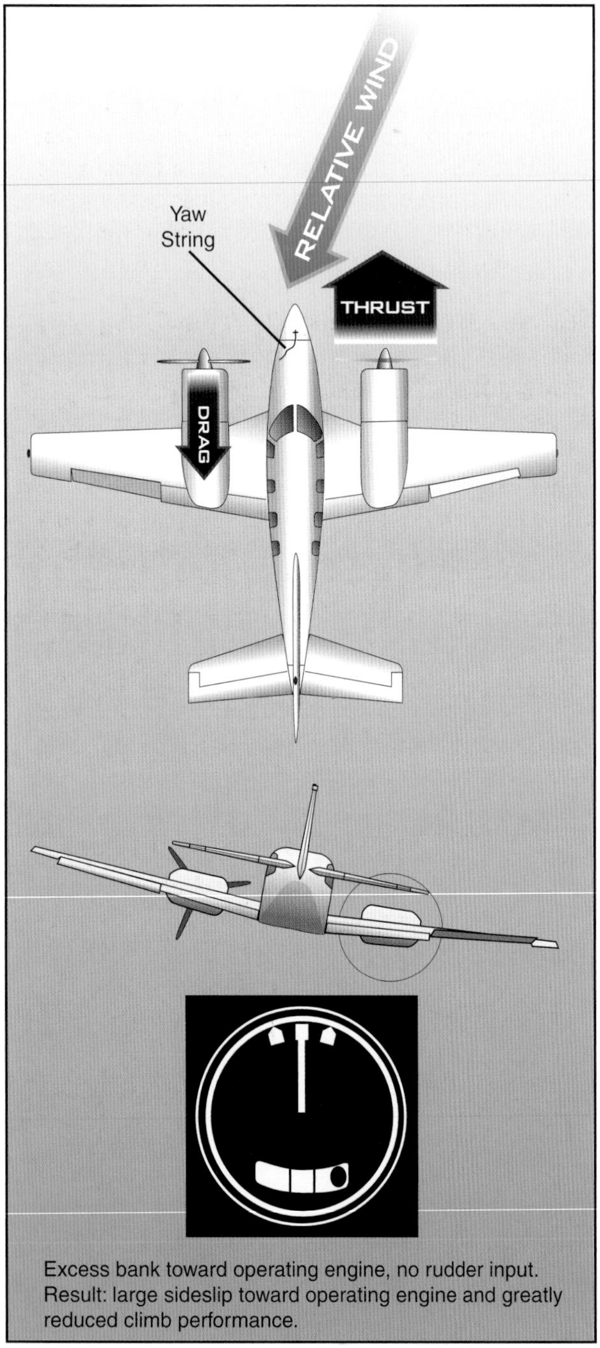

Excess bank toward operating engine, no rudder input. Result: large sideslip toward operating engine and greatly reduced climb performance.

Figure 12-17. Excessive bank engine-out flight.

3. Rudder and ailerons used together in the proper combination will result in a bank of approximately 2° towards the operative engine. The ball will be displaced approximately one-third to one-half towards the operative engine. The result is zero sideslip and maximum climb performance. [Figure 12-18] Any attitude other than zero sideslip increases drag, decreasing performance. V_{MC} under these circumstances will be higher than published, as less than the 5° bank certification limit is employed.

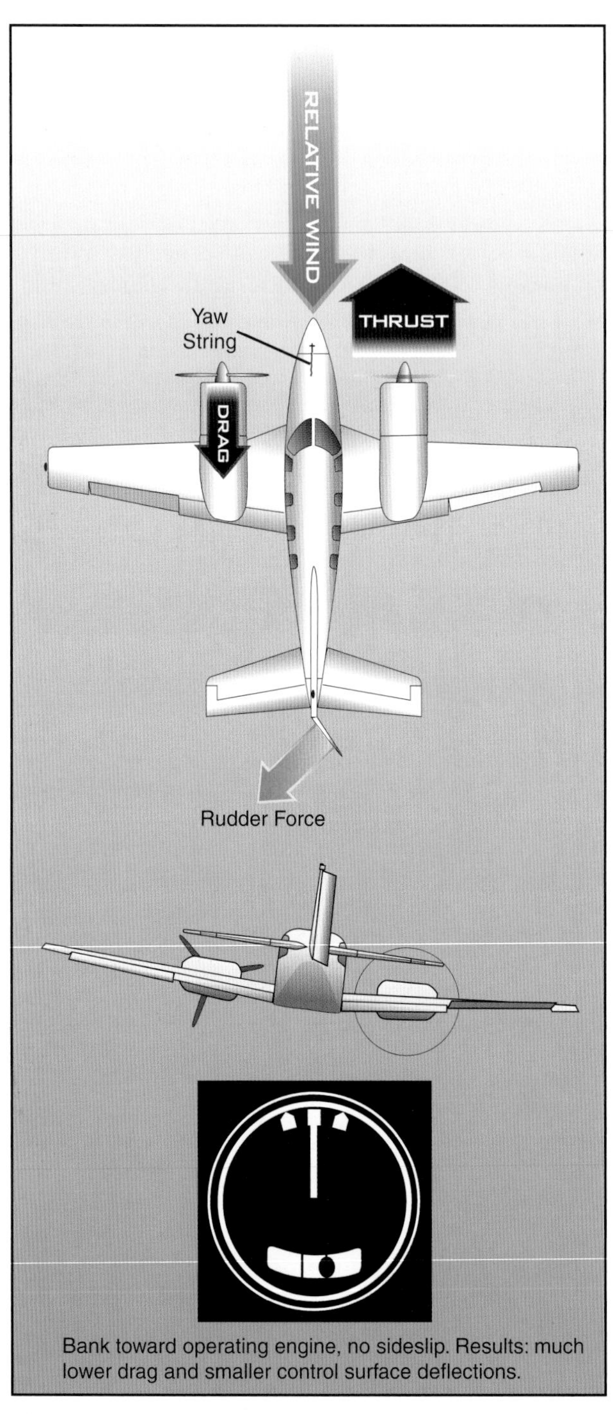

Bank toward operating engine, no sideslip. Results: much lower drag and smaller control surface deflections.

Figure 12-18. Zero sideslip engine-out flight.

The precise condition of zero sideslip (bank angle and ball position) varies slightly from model to model, and with available power and airspeed. If the airplane is not equipped with counter-rotating propellers, it will also vary slightly with the engine failed due to P-factor. The foregoing zero sideslip recommendations apply to reciprocating engine multiengine airplanes flown at V_{YSE} with the inoperative engine feathered. The zero sideslip ball position for straight flight is also the zero sideslip position for turning flight.

When bank angle is plotted against climb performance for a hypothetical twin, zero sideslip results in the best (however marginal) climb performance or the least rate of descent. Zero bank (all rudder to counteract yaw) degrades climb performance as a result of moderate sideslip. Using bank angle alone (no rudder) severely degrades climb performance as a result of a large sideslip.

The actual bank angle for zero sideslip varies among airplanes from one and one-half to two and one-half degrees. The position of the ball varies from one-third to one-half of a ball width from instrument center.

For any multiengine airplane, zero sideslip can be confirmed through the use of a yaw string. A yaw string is a piece of string or yarn approximately 18 to 36 inches in length, taped to the base of the windshield, or to the nose near the windshield, along the airplane centerline. In two-engine coordinated flight, the relative wind will cause the string to align itself with the longitudinal axis of the airplane, and it will position itself straight up the center of the windshield. This is zero sideslip. Experimentation with slips and skids will vividly display the location of the relative wind. Adequate altitude and flying speed must be maintained while accomplishing these maneuvers.

With an engine set to zero thrust (or feathered) and the airplane slowed to V_{YSE}, a climb with maximum power on the remaining engine will reveal the precise bank angle and ball deflection required for zero sideslip and best climb performance. Zero sideslip will again be indicated by the yaw string when it aligns itself vertically on the windshield. There will be very minor changes from this attitude depending upon the engine failed (with noncounter-rotating propellers), power available, airspeed and weight; but without more sensitive testing equipment, these changes are difficult to detect. The only significant difference would be the pitch attitude required to maintain V_{YSE} under different density altitude, power available, and weight conditions.

If a yaw string is attached to the airplane at the time of a V_{MC} demonstration, it will be noted that V_{MC} occurs under conditions of sideslip. V_{MC} was **not** determined under conditions of zero sideslip during aircraft certification and zero sideslip is **not** part of a V_{MC} demonstration for pilot certification.

To review, there are two different sets of bank angles used in one-engine-inoperative flight.

- To maintain directional control of a multiengine airplane suffering an engine failure at low speeds (such as climb), momentarily bank **at least** 5°, and a maximum of 10° towards the operative engine as the pitch attitude for V_{YSE} is set. This maneuver should be instinctive to the proficient multiengine pilot and take only 1 to 2 seconds to attain. It is held just long enough to assure directional control as the pitch attitude for V_{YSE} is assumed.

- To obtain the best climb performance, the airplane must be flown at V_{YSE} and zero sideslip, with the failed engine feathered and maximum available power from the operating engine. Zero sideslip is approximately **2**° of bank toward the operating engine and a one-third to one-half ball deflection, also toward the operating engine. The precise bank angle and ball position will vary somewhat with make and model and power available. If above the airplane's single-engine ceiling, this attitude and configuration will result in the minimum rate of sink.

In OEI flight at low altitudes and airspeeds such as the initial climb after takeoff, pilots must operate the airplane so as to guard against the three major accident factors: (1) loss of directional control, (2) loss of performance, and (3) loss of flying speed. All have equal potential to be lethal. Loss of flying speed will not be a factor, however, when the airplane is operated with due regard for directional control and performance.

SLOW FLIGHT

There is nothing unusual about maneuvering during slow flight in a multiengine airplane. Slow flight may be conducted in straight-and-level flight, turns, in the clean configuration, landing configuration, or at any other combination of landing gear and flaps. Pilots should closely monitor cylinder head and oil temperatures during slow flight. Some high performance multiengine airplanes tend to heat up fairly quickly under some conditions of slow flight, particularly in the landing configuration.

Simulated engine failures should not be conducted during slow flight. The airplane will be well below V_{SSE} and very close to V_{MC}. Stability, stall warning or stall avoidance devices should not be disabled while maneuvering during slow flight.

STALLS

Stall characteristics vary among multiengine airplanes just as they do with single-engine airplanes, and therefore, it is important to be familiar with them. The application of power upon stall recovery, however, has a significantly greater effect during stalls in a twin than a single-engine airplane. In the twin, an application of power blows large masses of air from the propellers directly over the wings, producing a significant amount of lift in addition to the expected thrust. The multiengine airplane, particularly at light operating weights, typically has a higher thrust-to-weight ratio, making it quicker to accelerate out of a stalled condition.

In general, stall recognition and recovery training in twins is performed similar to any high performance single-engine airplane. However, for twins, all stall maneuvers should be planned so as to be completed at least 3,000 feet AGL.

Single-engine stalls or stalls with significantly more power on one engine than the other should not be attempted due to the likelihood of a departure from controlled flight and possible spin entry. Similarly, simulated engine failures should not be performed during stall entry and recovery.

POWER-OFF STALLS (APPROACH AND LANDING)

Power-off stalls are practiced to simulate typical approach and landing scenarios. To initiate a power-off stall maneuver, the area surrounding the airplane should first be cleared for possible traffic. The airplane should then be slowed and configured for an approach and landing. A stabilized descent should be established (approximately 500 f.p.m.) and trim adjusted. The pilot should then transition smoothly from the stabilized descent attitude, to a pitch attitude that will induce a stall. Power is reduced further during this phase, and trimming should cease at speeds slower than takeoff.

When the airplane reaches a stalled condition, the recovery is accomplished by simultaneously reducing the angle of attack with coordinated use of the flight controls and smoothly applying takeoff or specified power. The flap setting should be reduced from full to approach, or as recommended by the manufacturer. Then with a positive rate of climb, the landing gear is selected up. The remaining flaps are then retracted as a climb has commenced. This recovery process should be completed with a minimum loss of altitude, appropriate to the aircraft characteristics.

The airplane should be accelerated to V_X (if simulated obstacles are present) or V_Y during recovery and climb. Considerable forward elevator/stabilator pressure will be required after the stall recovery as the airplane accelerates to V_X or V_Y. Appropriate trim input should be anticipated.

Power-off stalls may be performed with wings level, or from shallow and medium banked turns. When recovering from a stall performed from turning flight, the angle of attack should be reduced prior to leveling the wings. Flight control inputs should be coordinated.

It is usually not advisable to execute full stalls in multiengine airplanes because of their relatively high wing loading. Stall training should be limited to approaches to stalls and when a stall condition occurs. Recoveries should be initiated at the onset, or decay of control effectiveness, or when the first physical indication of the stall occurs.

POWER-ON STALLS (TAKEOFF AND DEPARTURE)

Power-on stalls are practiced to simulate typical takeoff scenarios. To initiate a power-on stall maneuver, the area surrounding the airplane should always be cleared to look for potential traffic. The airplane is slowed to the manufacturer's recommended lift-off speed. The airplane should be configured in the takeoff configuration. Trim should be adjusted for this speed. Engine power is then increased to that recommended in the AFM/POH for the practice of power-on stalls. In the absence of a recommended setting, use approximately 65 percent of maximum available power while placing the airplane in a pitch attitude that will induce a stall. Other specified (reduced) power settings may be used to simulate performance at higher gross weights and density altitudes.

When the airplane reaches a stalled condition, the recovery is made by simultaneously lowering the angle of attack with coordinated use of the flight controls and applying power as appropriate.

However, if simulating limited power available for high gross weight and density altitude situations, the power during the recovery should be limited to that specified. The recovery should be completed with a minimum loss of altitude, appropriate to aircraft characteristics.

The landing gear should be retracted when a positive rate of climb is attained, and flaps retracted, if flaps were set for takeoff. The target airspeed on recovery is V_X if (simulated) obstructions are present, or V_Y. The pilot should anticipate the need for nosedown trim as the airplane accelerates to V_X or V_Y after recovery.

Power-on stalls may be performed from straight flight or from shallow and medium banked turns. When recovering from a power-on stall performed from turning flight, the angle of attack should be reduced prior to leveling the wings, and the flight control inputs should be coordinated.

SPIN AWARENESS

No multiengine airplane is approved for spins, and their spin recovery characteristics are generally very poor. It is therefore necessary to practice spin avoidance and maintain a high awareness of situations that can result in an inadvertent spin.

In order to spin any airplane, it must first be stalled. At the stall, a yawing moment must be introduced. In a multiengine airplane, the yawing moment may be generated by rudder input or asymmetrical thrust. It follows, then, that spin awareness be at its greatest during V_{MC} demonstrations, stall practice, slow flight, or any condition of high asymmetrical thrust, particularly at low speed/high angle of attack. Single-engine stalls are **not** part of any multiengine training curriculum.

A situation that may inadvertently degrade into a spin entry is a simulated engine failure introduced at an inappropriately low speed. No engine failure should ever be introduced below safe, intentional one-engine-inoperative speed (V_{SSE}). If no V_{SSE} is published, use V_{YSE}. The "necessity" of simulating engine failures at low airspeeds is erroneous. Other than training situations, the multiengine airplane is only operated below V_{SSE} for mere seconds just after lift-off or during the last few dozen feet of altitude in preparation for landing.

For spin avoidance when practicing engine failures, the flight instructor should pay strict attention to the maintenance of proper airspeed and bank angle as the student executes the appropriate procedure. The instructor should also be particularly alert during stall and slow flight practice. Forward center-of-gravity positions result in favorable stall and spin avoidance characteristics, but do not eliminate the hazard.

When performing a V_{MC} demonstration, the instructor should also be alert for any sign of an impending stall. The student may be highly focused on the directional control aspect of the maneuver to the extent that impending stall indications go unnoticed. If a V_{MC} demonstration cannot be accomplished under existing conditions of density altitude, it may, for training purposes, be done utilizing the rudder blocking technique described in the following section.

As very few twins have ever been spin-tested (none are required to), the recommended spin recovery techniques are based only on the best information available. The departure from controlled flight may be quite abrupt and possibly disorienting. The direction of an upright spin can be confirmed from the turn needle or the symbolic airplane of the turn coordinator, if necessary. Do not rely on the ball position or other instruments.

If a spin is entered, most manufacturers recommend immediately retarding both throttles to idle, applying full rudder opposite the direction of rotation, and applying full forward elevator/stabilator pressure (with ailerons neutral). These actions should be taken as near simultaneously as possible. The controls should then be held in that position. Recovery, if possible, will take considerable altitude. The longer the delay from entry until taking corrective action, the less likely that recovery will be successful.

ENGINE INOPERATIVE—LOSS OF DIRECTIONAL CONTROL DEMONSTRATION

An engine inoperative—loss of directional control demonstration, often referred to as a "V_{MC} demonstration," is a required task on the practical test for a multiengine class rating. A thorough knowledge of the factors that affect V_{MC}, as well as its definition, is essential for multiengine pilots, and as such an essential part of that required task. V_{MC} is a speed established by the manufacturer, published in the AFM/POH, and marked on most airspeed indicators with a red radial line. The multiengine pilot must understand that V_{MC} is **not** a fixed airspeed under all conditions. V_{MC} is a fixed airspeed only for the very specific set of circumstances under which it was determined during aircraft certification. [Figure 12-19]

In reality, V_{MC} varies with a variety of factors as outlined below. The V_{MC} noted in practice and demonstration, or in actual single-engine operation, could be less or even greater than the published value, depending upon conditions and technique.

In aircraft certification, V_{MC} is the sea level calibrated airspeed at which, when the critical engine is suddenly made inoperative, it is possible to maintain control of the airplane with that engine still inoperative and then maintain straight flight at the same speed with an angle of bank of not more than 5°.

The foregoing refers to the determination of V_{MC} under "dynamic" conditions. This technique is only used by highly experienced flight test pilots during aircraft certification. It is never to be attempted outside of these circumstances.

In aircraft certification, there is also a determination of V_{MC} under "static," or steady-state conditions. If there is a difference between the dynamic and static speeds, the higher of the two is published as V_{MC}. The static determination is simply the ability to maintain straight flight at V_{MC} with a bank angle of not more than 5°. This more closely resembles the V_{MC} demonstration required in the practical test for a multiengine class rating.

The AFM/POH-published V_{MC} is determined with the "critical" engine inoperative. The critical engine is the

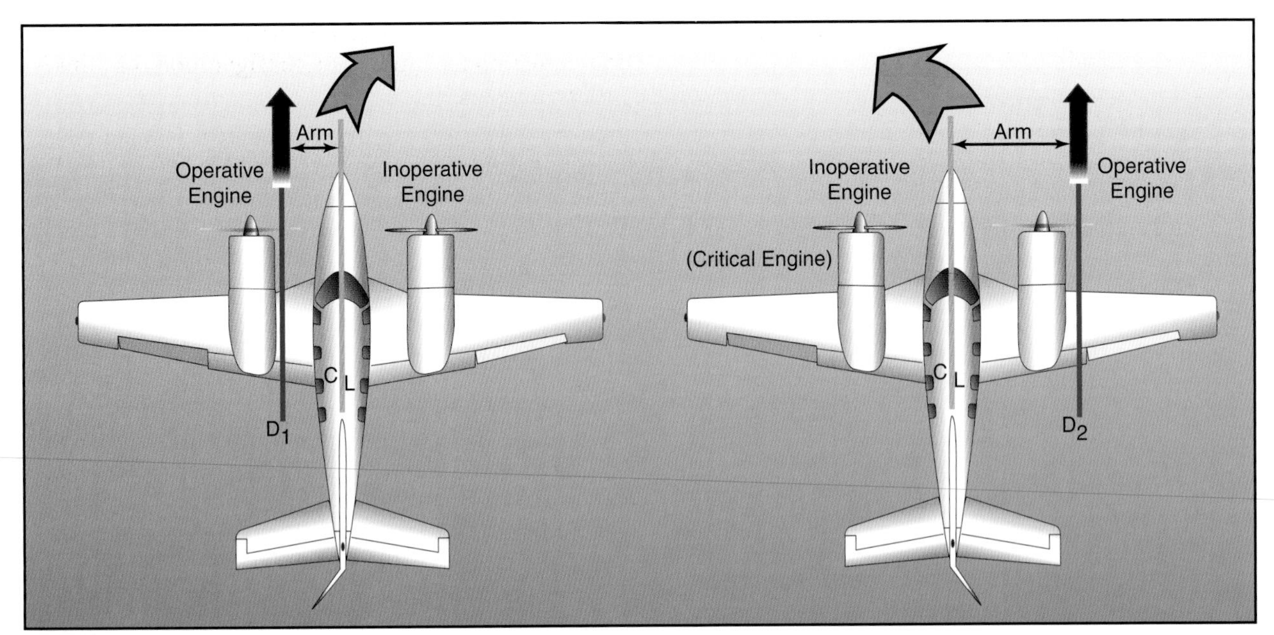

Figure 12-19. Forces created during single-engine operation.

engine whose failure has the most adverse effect on directional control. On twins with each engine rotating in conventional, clockwise rotation as viewed from the pilot's seat, the critical engine will be the left engine.

Multiengine airplanes are subject to P-factor just as single-engine airplanes are. The descending propeller blade of each engine will produce greater thrust than the ascending blade when the airplane is operated under power and at positive angles of attack. The descending propeller blade of the right engine is also a greater distance from the center of gravity, and therefore has a longer moment arm than the descending propeller blade of the left engine. As a result, failure of the left engine will result in the most asymmetrical thrust (adverse yaw) as the right engine will be providing the remaining thrust. [Figure 12-19]

Many twins are designed with a counter-rotating right engine. With this design, the degree of asymmetrical thrust is the same with either engine inoperative. No engine is more critical than the other, and a V_{MC} demonstration may be performed with either engine windmilling.

In aircraft certification, dynamic V_{MC} is determined under the following conditions.

- **Maximum available takeoff power.** V_{MC} increases as power is increased on the operating engine. With normally aspirated engines, V_{MC} is highest at takeoff power and sea level, and decreases with altitude. With turbocharged engines, takeoff power, and therefore V_{MC}, remains constant with increases in altitude up to the engine's critical altitude (the altitude where the engine can no longer maintain 100 percent power). Above the critical altitude, V_{MC} decreases just as it would with a normally aspirated engine, whose critical altitude is sea level. V_{MC} tests are conducted at a variety of altitudes. The results of those tests are then extrapolated to a single, sea level value.

- **Windmilling propeller.** V_{MC} increases with increased drag on the inoperative engine. V_{MC} is highest, therefore, when the critical engine propeller is windmilling at the low pitch, high r.p.m. blade angle. V_{MC} is determined with the critical engine propeller windmilling in the takeoff position, unless the engine is equipped with an autofeather system.

- **Most unfavorable weight and center-of-gravity position.** V_{MC} increases as the center of gravity is moved aft. The moment arm of the rudder is reduced, and therefore its effectivity is reduced, as the center of gravity is moved aft. At the same time, the moment arm of the propeller blade is increased, aggravating asymmetrical thrust. Invariably, the aft-most CG limit is the most unfavorable CG position. Currently, 14 CFR part 23 calls for V_{MC} to be determined at the most unfavorable weight. For twins certificated under CAR 3 or early 14 CFR part 23, the weight at which V_{MC} was determined was not specified. V_{MC} increases as weight is reduced. [Figure 12-20]

- **Landing gear retracted.** V_{MC} increases when the landing gear is retracted. Extended landing gear aids directional stability, which tends to decrease V_{MC}.

- **Wing flaps in the takeoff position.** For most twins, this will be 0° of flaps.

- **Cowl flaps in the takeoff position.**

- **Airplane trimmed for takeoff.**

- **Airplane airborne and the ground effect negligible.**

- **Maximum of 5° angle of bank.** V_{MC} is highly sensitive to bank angle. To prevent claims of an unrealistically low V_{MC} speed in aircraft certification, the manufacturer is permitted to use a maximum of a 5° bank angle toward the operative engine. The horizontal component of lift generated by the bank assists the rudder in counteracting the asymmetrical thrust of the operative engine. The bank angle works in the manufacturer's favor in lowering V_{MC}.

V_{MC} is reduced significantly with increases in bank angle. Conversely, V_{MC} increases significantly with decreases in bank angle. Tests have shown that V_{MC} may increase more than 3 knots for each degree of bank angle less than 5°. Loss of directional control may be experienced at speeds almost 20 knots above published V_{MC} when the wings are held level.

The 5° bank angle maximum is a regulatory limit imposed upon manufacturers in aircraft certification. The 5° bank does **not** inherently establish zero sideslip or best single-engine climb performance. Zero sideslip, and therefore best single-engine climb performance, occurs at bank angles significantly less than 5°. The determination of V_{MC} in certification is solely concerned with the minimum speed for directional control under a very specific set of circumstances, and has nothing to do with climb performance, nor is it the optimum airplane attitude or configuration for climb performance.

During dynamic V_{MC} determination in aircraft certification, cuts of the critical engine using the mixture control are performed by flight test pilots while gradually reducing the speed with each attempt. V_{MC} is the minimum speed at which directional control could be maintained within 20° of the original entry heading when a cut of the critical engine was made. During such tests, the climb angle with both engines operating was high, and the pitch attitude following the engine cut had to be quickly lowered to regain the initial speed. Pilots should never attempt to demonstrate V_{MC} with an engine cut from high power, and never intentionally fail an engine at speeds less than V_{SSE}.

The actual demonstration of V_{MC} and recovery in flight training more closely resembles static V_{MC} determination in aircraft certification. For a demonstration, the pilot should select an altitude that will allow completion of the maneuver at least 3,000 feet AGL. The following description assumes a twin with noncounter-rotating engines, where the left engine is critical.

With the landing gear retracted and the flaps set to the takeoff position, the airplane should be slowed to approximately 10 knots above V_{SSE} or V_{YSE} (whichever is higher) and trimmed for takeoff. For the remainder of the maneuver, the trim setting should not be altered. An entry heading should be selected and high r.p.m. set on both propeller controls. Power on the left engine should be throttled back to idle as the right engine power is advanced to the takeoff setting. The landing gear warning horn will sound as long as a

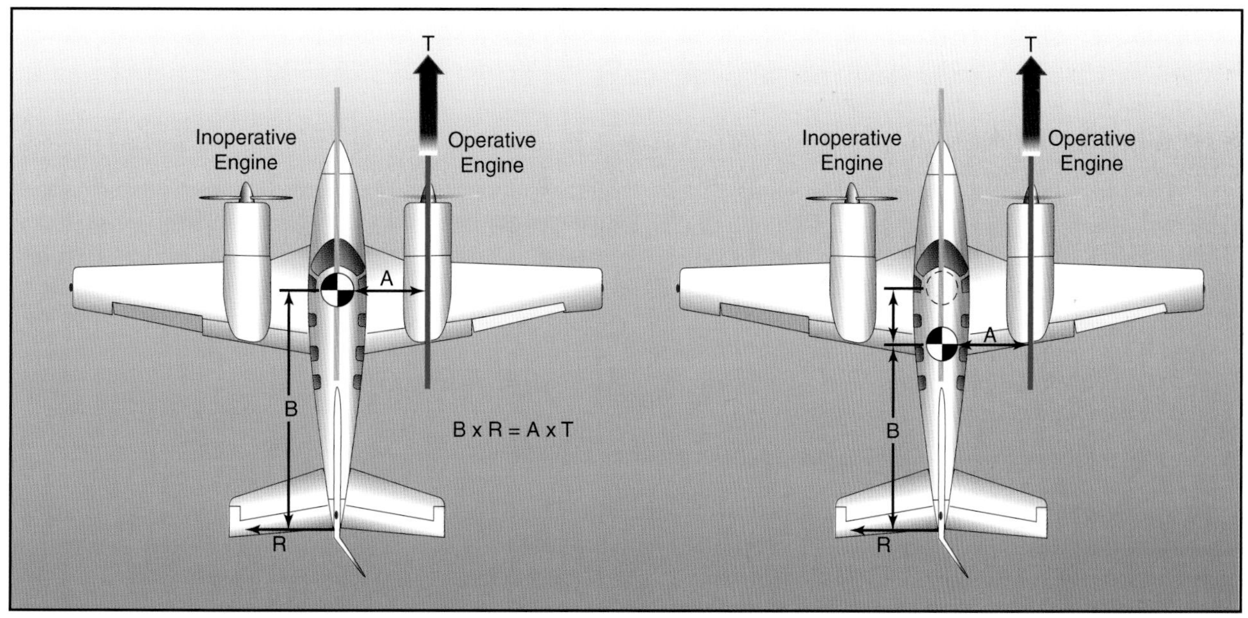

Figure 12-20. Effect of CG location on yaw.

throttle is retarded. The pilots should continue to carefully listen, however, for the stall warning horn, if so equipped, or watch for the stall warning light. The left yawing and rolling moment of the asymmetrical thrust is counteracted primarily with right rudder. A bank angle of 5° (a right bank, in this case) should also be established.

While maintaining entry heading, the pitch attitude is slowly increased to decelerate at a rate of 1 knot per second (no faster). As the airplane slows and control effectivity decays, the increasing yawing tendency should be counteracted with additional rudder pressure. Aileron displacement will also increase in order to maintain 5° of bank. An airspeed is soon reached where full right rudder travel and a 5° right bank can no longer counteract the asymmetrical thrust, and the airplane will begin to yaw uncontrollably to the left.

The moment the pilot first recognizes the uncontrollable yaw, or experiences **any** symptom associated with a stall, the operating engine throttle should be sufficiently retarded to stop the yaw as the pitch attitude is decreased. Recovery is made with a minimum loss of altitude to straight flight on the entry heading at V_{SSE} or V_{YSE}, before setting symmetrical power. The recovery should not be attempted by increasing power on the windmilling engine alone.

To keep the foregoing description simple, there were several important background details that were not covered. The rudder pressure during the demonstration can be quite high. In certification, 150 pounds of force is permitted before the limiting factor becomes rudder pressure, not rudder travel. Most twins will run out of rudder travel long before 150 pounds of pressure is required. Still, it will seem considerable.

Maintaining altitude is **not** a criterion in accomplishing this maneuver. This is a demonstration of controllability, not performance. Many airplanes will lose (or gain) altitude during the demonstration. Begin the maneuver at an altitude sufficient to allow completion by 3,000 feet AGL.

As discussed earlier, with normally aspirated engines, V_{MC} decreases with altitude. Stalling speed (V_S), however, remains the same. Except for a few models, published V_{MC} is almost always higher than V_S. At sea level, there is usually a margin of several knots between V_{MC} and V_S, but the margin decreases with altitude, and at some altitude, V_{MC} and V_S are the same. [Figure 12-21]

Should a stall occur while the airplane is under asymmetrical power, particularly high asymmetrical power, a spin entry is likely. The yawing moment induced from asymmetrical thrust is little different from that

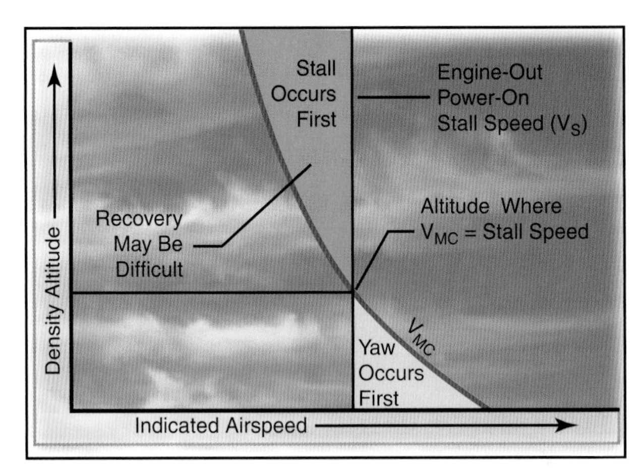

Figure 12-21. Graph depicting relationship of V_{MC} to V_S.

induced by full rudder in an intentional spin in the appropriate model of single-engine airplane. In this case, however, the airplane will depart controlled flight in the direction of the idle engine, not in the direction of the applied rudder. Twins are not required to demonstrate recoveries from spins, and their spin recovery characteristics are generally very poor.

Where V_S is encountered at or before V_{MC}, the departure from controlled flight may be quite sudden, with strong yawing and rolling tendencies to the inverted position, and a spin entry. Therefore, during a V_{MC} demonstration, if there are **any** symptoms of an impending stall such as a stall warning light or horn, airframe or elevator buffet, or rapid decay in control effectiveness, the maneuver should be terminated immediately, the angle of attack reduced as the throttle is retarded, and the airplane returned to the entry airspeed. It should be noted that if the pilots are wearing headsets, the sound of a stall warning horn will tend to be masked.

The V_{MC} demonstration only shows the earliest onset of a loss of directional control. It is not a loss of control of the airplane when performed in accordance with the foregoing procedures. A stalled condition should **never** be allowed to develop. Stalls should never be performed with asymmetrical thrust and the V_{MC} demonstration should never be allowed to degrade into a single-engine stall. A V_{MC} demonstration that is allowed to degrade into a single-engine stall with high asymmetrical thrust is very likely to result in a loss of control of the airplane.

An actual demonstration of V_{MC} may not be possible under certain conditions of density altitude, or with airplanes whose V_{MC} is equal to or less than V_S. Under those circumstances, as a training technique, a demonstration of V_{MC} may be safely conducted by artificially limiting rudder travel to simulate maximum available rudder. Limiting rudder travel should be accomplished at a speed well above V_S (approximately 20 knots).

The rudder limiting technique avoids the hazards of spinning as a result of stalling with high asymmetrical power, yet is effective in demonstrating the loss of directional control.

The V_{MC} demonstration should never be performed from a high pitch attitude with both engines operating and then reducing power on one engine. The preceding discussion should also give ample warning as to why engine failures are never to be performed at low airspeeds. An unfortunate number of airplanes and pilots have been lost from unwarranted simulated engine failures at low airspeeds that degenerated into loss of control of the airplane. V_{SSE} is the minimum airspeed at which any engine failure should be simulated.

MULTIENGINE TRAINING CONSIDERATIONS

Flight training in a multiengine airplane can be safely accomplished if both the instructor and the student are cognizant of the following factors.

- No flight should ever begin without a thorough preflight briefing of the objectives, maneuvers, expected student actions, and completion standards.
- A clear understanding must be reached as to how simulated emergencies will be introduced, and what action the student is expected to take.

The introduction, practice, and testing of emergency procedures has always been a sensitive subject. Surprising a multiengine student with an emergency without a thorough briefing beforehand has no place in flight training. Effective training must be carefully balanced with safety considerations. Simulated engine failures, for example, can very quickly become actual emergencies or lead to loss of the airplane when approached carelessly. Pulling circuit breakers can lead to a subsequent gear up landing. Stall-spin accidents in training for emergencies rival the number of stall-spin accidents from actual emergencies.

All normal, abnormal, and emergency procedures can and should be introduced and practiced in the airplane as it sits on the ground, power off. In this respect, the airplane is used as a cockpit procedures trainer (CPT), ground trainer, or simulator. The value of this training should never be underestimated. The engines do not have to be operating for real learning to occur. Upon completion of a training session, care should be taken to return items such as switches, valves, trim, fuel selectors, and circuit breakers to their normal positions.

Pilots who do not use a checklist effectively will be at a significant disadvantage in multiengine airplanes. Use of the checklist is essential to safe operation of airplanes and no flight should be conducted without one. The manufacturer's checklist or an aftermarket checklist for the specific make, model, and model year should be used. If there is a procedural discrepancy between the checklist and AFM/POH, then the AFM/POH always takes precedence.

Certain immediate action items (such as the response to an engine failure in a critical phase of flight) should be committed to memory. After they are accomplished, and as work load permits, the pilot should verify the action taken with a printed checklist.

Simulated engine failures during the takeoff ground roll should be accomplished with the mixture control. The simulated failure should be introduced at a speed no greater than 50 percent of V_{MC}. If the student does not react promptly by retarding both throttles, the instructor can always pull the other mixture.

The FAA recommends that all in-flight simulated engine failures below 3,000 feet AGL be introduced with a smooth reduction of the throttle. Thus, the engine is kept running and is available for instant use, if necessary. Throttle reduction should be smooth rather than abrupt to avoid abusing the engine and possibly causing damage. All inflight engine failures must be conducted at V_{SSE} or above.

If the engines are equipped with dynamic crankshaft counterweights, it is essential to make throttle reductions for simulated failures smoothly. Other areas leading to dynamic counterweight damage include high r.p.m. and low manifold pressure combinations, overboosting, and propeller feathering. Severe damage or repetitive abuse to counterweights will eventually lead to engine failure. Dynamic counterweights are found on larger, more complex engines—instructors should check with maintenance personnel or the engine manufacturer to determine if their engines are so equipped.

When an instructor simulates an engine failure, the student should respond with the appropriate memory items and retard the propeller control towards the FEATHER position. Assuming zero thrust will be set, the instructor should promptly move the propeller control forward and set the appropriate manifold pressure and r.p.m. It is vital that the student be kept informed of the instructor's intentions. At this point the instructor may state words to the effect, "I have the right engine; you have the left. I have set zero thrust and the right engine is simulated feathered." There should never be any ambiguity as to who is operating what systems or controls.

Following a simulated engine failure, the instructor should continue to care for the "failed" engine just as the student cares for the operative engine. If zero thrust is set to simulate a feathered propeller, the cowl flap should be closed and the mixture leaned. An occasional clearing of the engine is also desirable. If possible, avoid high power applications immediately following a prolonged cool-down at a zero-thrust power setting. The flight instructor must impress on the student multiengine pilot the critical importance of feathering the propeller in a timely manner should an actual engine failure situation be encountered. A windmilling propeller, in many cases, has given the improperly trained multiengine pilot the mistaken perception that the failed engine is still developing useful thrust, resulting in a psychological reluctance to feather, as feathering results in the cessation of propeller rotation. The flight instructor should spend ample time demonstrating the difference in the performance capabilities of the airplane with a simulated feathered propeller (zero thrust) as opposed to a windmilling propeller.

All actual propeller feathering should be performed at altitudes and positions where safe landings on established airports could be readily accomplished. Feathering and restart should be planned so as to be completed no lower than 3,000 feet AGL. At certain elevations and with many popular multiengine training airplanes, this may be above the single-engine service ceiling, and level flight will not be possible.

Repeated feathering and unfeathering is hard on the engine and airframe, and should be done only as absolutely necessary to ensure adequate training. The FAA's practical test standards for a multiengine class rating requires the feathering and unfeathering of one propeller during flight in airplanes in which it is safe to do so.

While much of this chapter has been devoted to the unique flight characteristics of the multiengine airplane with one engine inoperative, the modern, well-maintained reciprocating engine is remarkably reliable. Simulated engine failures at extremely low altitudes (such as immediately after lift-off) and/or below V_{SSE} are undesirable in view of the non-existent safety margins involved. The high risk of simulating an engine failure below 200 feet AGL does not warrant practicing such maneuvers.

For training in maneuvers that would be hazardous in flight, or for initial and recurrent qualification in an advanced multiengine airplane, a simulator training center or manufacturer's training course should be given consideration. Comprehensive training manuals and classroom instruction are available along with system training aids, audio/visuals, and flight training devices and simulators. Training under a wide variety of environmental and aircraft conditions is available through simulation. Emergency procedures that would be either dangerous or impossible to accomplish in an airplane can be done safely and effectively in a flight training device or simulator. The flight training device or simulator need not necessarily duplicate the specific make and model of airplane to be useful. Highly effective instruction can be obtained in training devices for other makes and models as well as generic training devices.

The majority of multiengine training is conducted in four to six-place airplanes at weights significantly less than maximum. Single-engine performance, particularly at low density altitudes, may be deceptively good. To experience the performance expected at higher weights, altitudes, and temperatures, the instructor should occasionally artificially limit the amount of manifold pressure available on the operative engine. Airport operations above the single-engine ceiling can also be simulated in this manner. Loading the airplane with passengers to practice emergencies at maximum takeoff weight is not appropriate.

The use of the touch-and-go landing and takeoff in flight training has always been somewhat controversial. The value of the learning experience must be weighed against the hazards of reconfiguring the airplane for takeoff in an extremely limited time as well as the loss of the follow-through ordinarily experienced in a full stop landing. Touch and goes are not recommended during initial aircraft familiarization in multiengine airplanes.

If touch and goes are to be performed at all, the student and instructor responsibilities need to be carefully briefed prior to each flight. Following touchdown, the student will ordinarily maintain directional control while keeping the left hand on the yoke and the right hand on the throttles. The instructor resets the flaps and trim and announces when the airplane has been reconfigured. The multiengine airplane needs considerably more runway to perform a touch and go than a single-engine airplane. A full stop-taxi back landing is preferable during initial familiarization. Solo touch and goes in twins are strongly discouraged.

Chapter 13

Transition to Tailwheel Airplanes

TAILWHEEL AIRPLANES

Tailwheel airplanes are often referred to as conventional gear airplanes. Due to their design and structure, tailwheel airplanes exhibit operational and handling characteristics that are different from those of tricycle gear airplanes. Tailwheel airplanes are not necessarily more difficult to takeoff, land, and/or taxi than tricycle gear airplanes; in fact under certain conditions, they may even handle with less difficulty. This chapter will focus on the operational differences that occur during ground operations, takeoffs, and landings.

LANDING GEAR

The main landing gear forms the principal support of the airplane on the ground. The tailwheel also supports the airplane, but steering and directional control are its primary functions. With the tailwheel-type airplane, the two main struts are attached to the airplane slightly ahead of the airplane's center of gravity (CG).

The rudder pedals are the primary directional controls while taxiing. Steering with the pedals may be accomplished through the forces of airflow or propeller slipstream acting on the rudder surface, or through a mechanical linkage to the steerable tailwheel. Initially, the pilot should taxi with the heels of the feet resting on the cockpit floor and the balls of the feet on the bottom of the rudder pedals. The feet should be slid up onto the brake pedals only when it is necessary to depress the brakes. This permits the simultaneous application of rudder and brake whenever needed. Some models of tailwheel airplanes are equipped with heel brakes rather than toe brakes. In either configuration the brakes are used primarily to stop the airplane at a desired point, to slow the airplane, or as an aid in making a sharp controlled turn. Whenever used, they must be applied smoothly, evenly, and cautiously at all times.

TAXIING

When beginning to taxi, the brakes should be tested immediately for proper operation. This is done by first applying power to start the airplane moving slowly forward, then retarding the throttle and simultaneously applying pressure smoothly to both brakes. If braking action is unsatisfactory, the engine should be shut down immediately.

To turn the airplane on the ground, the pilot should apply rudder in the desired direction of turn and use whatever power or brake that is necessary to control the taxi speed. The rudder should be held in the direction of the turn until just short of the point where the turn is to be stopped, then the rudder pressure released or slight opposite pressure applied as needed. While taxiing, the pilot will have to anticipate the movements of the airplane and adjust rudder pressure accordingly. Since the airplane will continue to turn slightly even as the rudder pressure is being released, the stopping of the turn must be anticipated and the rudder pedals neutralized before the desired heading is reached. In some cases, it may be necessary to apply opposite rudder to stop the turn, depending on the taxi speed.

The presence of moderate to strong headwinds and/or a strong propeller slipstream makes the use of the elevator necessary to maintain control of the pitch attitude while taxiing. This becomes apparent when considering the lifting action that may be created on the horizontal tail surfaces by either of those two factors. The elevator control should be held in the aft position (stick or yoke back) to hold the tail down.

When taxiing in a quartering headwind, the wing on the upwind side will usually tend to be lifted by the wind unless the aileron control is held in that direction (upwind aileron UP). Moving the aileron into the UP position reduces the effect of wind striking that wing, thus reducing the lifting action. This control movement will also cause the opposite aileron to be placed in the DOWN position, thus creating drag and possibly some lift on the downwind wing, further reducing the tendency of the upwind wing to rise.

When taxiing with a quartering tailwind, the elevator should be held in the full DOWN position (stick or yoke full forward), and the upwind aileron down. Since the wind is striking the airplane from behind, these control positions reduce the tendency of the wind to get under the tail and the wing possibly causing the airplane to nose over. The application of these crosswind taxi corrections also helps to minimize the weathervaning tendency and ultimately results in increased controllability.

An airplane with a tailwheel has a tendency to weathervane or turn into the wind while it is being taxied. The tendency of the airplane to weathervane is greatest while taxiing directly crosswind; consequently, directional control is somewhat difficult. Without brakes, it is almost impossible to keep the airplane from turning into any wind of considerable velocity since the airplane's rudder control capability may be inadequate to counteract the crosswind. In taxiing downwind, the tendency to weathervane is increased, due to the tailwind decreasing the effectiveness of the flight controls. This requires a more positive use of the rudder and the brakes, particularly if the wind velocity is above that of a light breeze.

Unless the field is soft, or very rough, it is best when taxiing downwind to hold the elevator control in the forward position. Even on soft fields, the elevator should be raised only as much as is absolutely necessary to maintain a safe margin of control in case there is a tendency of the airplane to nose over.

On most tailwheel-type airplanes, directional control while taxiing is facilitated by the use of a steerable tailwheel, which operates along with the rudder. The tailwheel steering mechanism remains engaged when the tailwheel is operated through an arc of about 16 to 18° each side of neutral and then automatically becomes full swiveling when turned to a greater angle. On some models the tailwheel may also be locked in place. The airplane may be pivoted within its own length, if desired, yet is fully steerable for slight turns while taxiing forward. While taxiing, the steerable tailwheel should be used for making normal turns and the pilot's feet kept off the brake pedals to avoid unnecessary wear on the brakes.

Since a tailwheel-type airplane rests on the tailwheel as well as the main landing wheels, it assumes a nose-high attitude when on the ground. In most cases this places the engine cowling high enough to restrict the pilot's vision of the area directly ahead of the airplane. Consequently, objects directly ahead of the airplane are difficult, if not impossible, to see. To observe and avoid colliding with any objects or hazardous surface conditions, the pilot should alternately turn the nose from one side to the other—that is zigzag, or make a series of short S-turns while taxiing forward. This should be done slowly, smoothly, positively, and cautiously.

NORMAL TAKEOFF ROLL

After taxiing onto the runway, the airplane should be carefully aligned with the intended takeoff direction, and the tailwheel positioned straight, or centered. In airplanes equipped with a locking device, the tailwheel should be locked in the centered position. After releasing the brakes, the throttle should be smoothly and continuously advanced to takeoff power. As the airplane starts to roll forward, the pilot should slide both feet down on the rudder pedals so that the toes or balls of the feet are on the rudder portions, not on the brake portions.

An abrupt application of power may cause the airplane to yaw sharply to the left because of the torque effects of the engine and propeller. Also, precession will be particularly noticeable during takeoff in a tailwheel-type airplane if the tail is rapidly raised from a three point to a level flight attitude. The abrupt change of attitude tilts the horizontal axis of the propeller, and the resulting precession produces a forward force on the right side (90° ahead in the direction of rotation), yawing the airplane's nose to the left. The amount of force created by this precession is directly related to the rate the propeller axis is tilted when the tail is raised. With this in mind, the throttle should always be advanced smoothly and continuously to prevent any sudden swerving.

Smooth, gradual advancement of the throttle is very important in tailwheel-type airplanes, since peculiarities in their takeoff characteristics are accentuated in proportion to how rapidly the takeoff power is applied.

As speed is gained, the elevator control will tend to assume a neutral position if the airplane is correctly trimmed. At the same time, directional control should be maintained with smooth, prompt, positive rudder corrections throughout the takeoff roll. The effects of torque and P-factor at the initial speeds tend to pull the nose to the left. The pilot must use what rudder pressure is needed to correct for these effects or for existing wind conditions to keep the nose of the airplane headed straight down the runway. The use of brakes for steering purposes should be avoided, since they will cause slower acceleration of the airplane's speed, lengthen the takeoff distance, and possibly result in severe swerving.

When the elevator trim is set for takeoff, on application of maximum allowable power, the airplane will (when sufficient speed has been attained) normally assume the correct takeoff pitch attitude on its own—the tail will rise slightly. This attitude can then be maintained by applying slight back-elevator pressure. If the elevator control is pushed forward during the takeoff roll to prematurely raise the tail, its effectiveness will rapidly build up as the speed increases, making it necessary to apply back-elevator pressure to lower the tail to the proper takeoff attitude. This erratic change in attitude will delay the takeoff and lead to directional control problems. Rudder pressure must be used promptly and smoothly to counteract yawing forces so that the airplane continues straight down the runway.

While the speed of the takeoff roll increases, more and more pressure will be felt on the flight controls, particularly the elevators and rudder. Since the tail surfaces receive the full effect of the propeller slipstream, they become effective first. As the speed continues to increase, all of the flight controls will gradually become effective enough to maneuver the airplane about its three axes. It is at this point, in the taxi to flight transition, that the airplane is being flown more than taxied. As this occurs, progressively smaller rudder deflections are needed to maintain direction.

TAKEOFF

Since a good takeoff depends on the proper takeoff attitude, it is important to know how this attitude appears and how it is attained. The ideal takeoff attitude requires only minimum pitch adjustments shortly after the airplane lifts off to attain the speed for the best rate of climb.

The tail should first be allowed to rise off the ground slightly to permit the airplane to accelerate more rapidly. At this point, the position of the nose in relation to the horizon should be noted, then elevator pressure applied as necessary to hold this attitude. The wings are kept level by applying aileron pressure as necessary.

The airplane may be allowed to fly off the ground while in normal takeoff attitude. Forcing it into the air by applying excessive back-elevator pressure would result in an excessively high pitch attitude and may delay the takeoff. As discussed earlier, excessive and rapid changes in pitch attitude result in proportionate changes in the effects of torque, making the airplane more difficult to control.

Although the airplane can be forced into the air, this is considered an unsafe practice and should be avoided under normal circumstances. If the airplane is forced to leave the ground by using too much back-elevator pressure before adequate flying speed is attained, the wing's angle of attack may be excessive, causing the airplane to settle back to the runway or even to stall. On the other hand, if sufficient back-elevator pressure is not held to maintain the correct takeoff attitude after becoming airborne, or the nose is allowed to lower excessively, the airplane may also settle back to the runway. This occurs because the angle of attack is decreased and lift is diminished to the degree where it will not support the airplane. It is important to hold the attitude constant after rotation or lift-off.

As the airplane leaves the ground, the pilot must continue to maintain straight flight, as well as holding the proper pitch attitude. During takeoffs in strong, gusty wind, it is advisable that an extra margin of speed be obtained before the airplane is allowed to leave the ground. A takeoff at the normal takeoff speed may result in a lack of positive control, or a stall, when the airplane encounters a sudden lull in strong, gusty wind, or other turbulent air currents. In this case, the pilot should hold the airplane on the ground longer to attain more speed, then make a smooth, positive rotation to leave the ground.

CROSSWIND TAKEOFF

It is important to establish and maintain the proper amount of crosswind correction prior to lift-off; that is, apply aileron pressure toward the wind to keep the upwind wing from rising and apply rudder pressure as needed to prevent weathervaning.

As the tailwheel is raised off the runway, the holding of aileron control into the wind may result in the downwind wing rising and the downwind main wheel lifting off the runway first, with the remainder of the takeoff roll being made on one main wheel. This is acceptable and is preferable to side-skipping.

If a significant crosswind exists, the main wheels should be held on the ground slightly longer than in a normal takeoff so that a smooth but definite lift-off can be made. This procedure will allow the airplane to leave the ground under more positive control so that it will definitely remain airborne while the proper amount of drift correction is being established. More importantly, it will avoid imposing excessive side loads on the landing gear and prevent possible damage that would result from the airplane settling back to the runway while drifting.

As both main wheels leave the runway, and ground friction no longer resists drifting, the airplane will be slowly carried sideways with the wind until adequate drift correction is maintained.

SHORT-FIELD TAKEOFF

Wing flaps should be lowered prior to takeoff if recommended by the manufacturer. Takeoff power should be applied smoothly and continuously, (there should be no hesitation) to accelerate the airplane as rapidly as possible. As the takeoff roll progresses, the airplane's pitch attitude and angle of attack should be adjusted to that which results in the minimum amount of drag and the quickest acceleration. The tail should be allowed to rise off the ground slightly, then held in this tail-low flight attitude until the proper lift-off or rotation airspeed is attained. For the steepest climb-out and best obstacle clearance, the airplane should be allowed to roll with its full weight on the main wheels and accelerated to the lift-off speed.

Soft-Field Takeoff

Wing flaps may be lowered prior to starting the takeoff (if recommended by the manufacturer) to provide additional lift and transfer the airplane's weight from the wheels to the wings as early as possible. The airplane should be taxied onto the takeoff surface without stopping on a soft surface. Stopping on a soft surface, such as mud or snow, might bog the airplane down. The airplane should be kept in continuous motion with sufficient power while lining up for the takeoff roll.

As the airplane is aligned with the proposed takeoff path, takeoff power is applied smoothly and as rapidly as the powerplant will accept it without faltering. The tail should be kept low to maintain the inherent positive angle of attack and to avoid any tendency of the airplane to nose over as a result of soft spots, tall grass, or deep snow.

When the airplane is held at a nose-high attitude throughout the takeoff run, the wings will, as speed increases and lift develops, progressively relieve the wheels of more and more of the airplane's weight, thereby minimizing the drag caused by surface irregularities or adhesion. If this attitude is accurately maintained, the airplane will virtually fly itself off the ground. The airplane should be allowed to accelerate to climb speed in ground effect.

Touchdown

The touchdown is the gentle settling of the airplane onto the landing surface. The roundout and touchdown should be made with the engine idling, and the airplane at minimum controllable airspeed, so that the airplane will touch down at approximately stalling speed. As the airplane settles, the proper landing attitude must be attained by applying whatever back-elevator pressure is necessary. The roundout and touchdown should be timed so that the wheels of the main landing gear and tailwheel touch down simultaneously (three-point landing). This requires proper timing, technique, and judgment of distance and altitude. [Figure 13-1]

When the wheels make contact with the ground, the elevator control should be carefully eased fully back to hold the tail down and to keep the tailwheel on the ground. This provides more positive directional control of the airplane equipped with a steerable tailwheel, and prevents any tendency for the airplane to nose over. If the tailwheel is not on the ground, easing back on the elevator control may cause the airplane to become airborne again because the change in attitude will increase the angle of attack and produce enough lift for the airplane to fly.

It is extremely important that the touchdown occur with the airplane's longitudinal axis exactly parallel to the direction the airplane is moving along the runway. Failure to accomplish this not only imposes severe side loads on the landing gear, but imparts groundlooping (swerving) tendencies. To avoid these side stresses or a ground loop, the pilot must never allow the airplane to touch down while in a crab or while drifting.

After-Landing Roll

The landing process must never be considered complete until the airplane decelerates to the normal taxi speed during the landing roll or has been brought to a complete stop when clear of the landing area. The pilot must be alert for directional control difficulties immediately upon and after touchdown due to the ground friction on the wheels. The friction creates a pivot point on which a moment arm can act. This is because the CG is behind the main wheels. [Figure 13-2]

Any difference between the direction the airplane is traveling and the direction it is headed will produce a moment about the pivot point of the wheels, and the airplane will tend to swerve. Loss of directional control may lead to an aggravated, uncontrolled, tight turn on the ground, or a ground loop. The combination of inertia acting on the CG and ground friction of the main wheels resisting it during the ground loop may cause the airplane to tip or lean enough for the outside

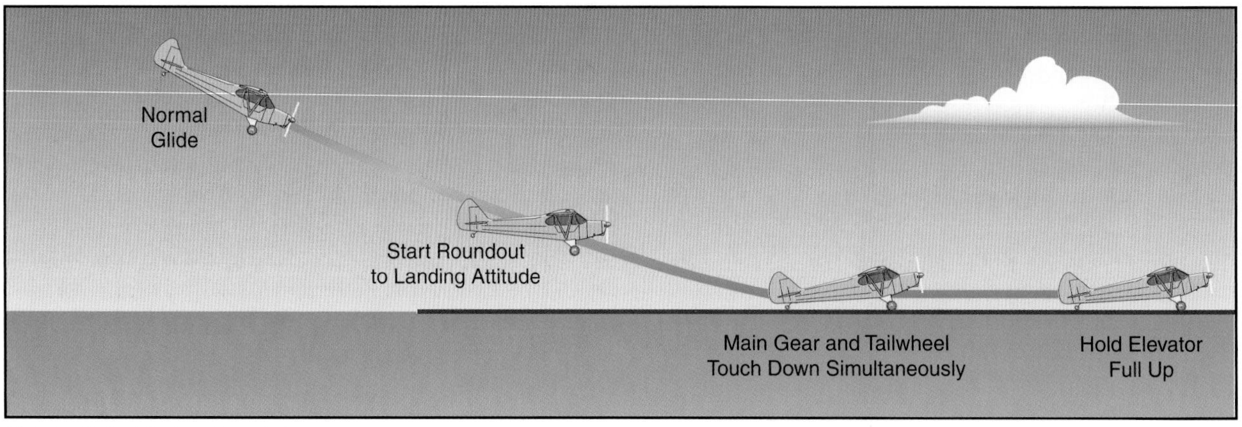

Figure 13-1. Tailwheel touchdown.

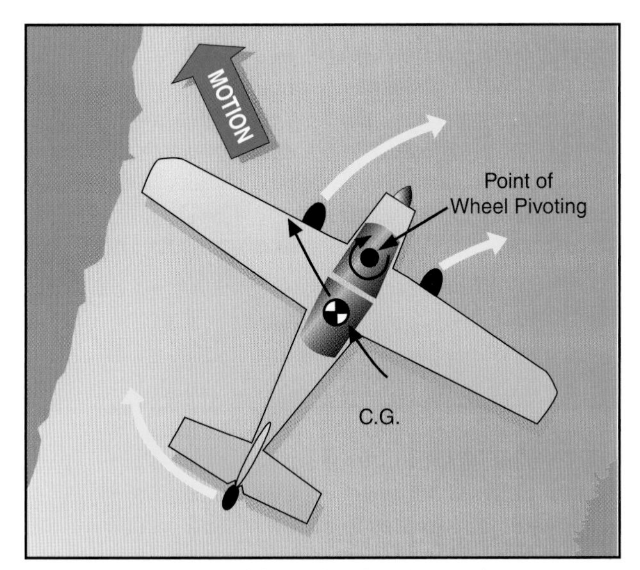

Figure 13-2. Effect of CG on directional control.

wingtip to contact the ground, and may even impose a sideward force that could collapse the landing gear. The airplane can ground loop late in the after-landing roll because rudder effectiveness decreases with the decreasing flow of air along the rudder surface as the airplane slows. As the airplane speed decreases and the tailwheel has been lowered to the ground, the steerable tailwheel provides more positive directional control.

To use the brakes, the pilot should slide the toes or feet up from the rudder pedals to the brake pedals (or apply heel pressure in airplanes equipped with heel brakes). If rudder pressure is being held at the time braking action is needed, that pressure should not be released as the feet or toes are being slid up to the brake pedals, because control may be lost before brakes can be applied. During the ground roll, the airplane's direction of movement may be changed by carefully applying pressure on one brake or uneven pressures on each brake in the desired direction. Caution must be exercised, when applying brakes to avoid overcontrolling.

If a wing starts to rise, aileron control should be applied toward that wing to lower it. The amount required will depend on speed because as the forward speed of the airplane decreases, the ailerons will become less effective.

The elevator control should be held back as far as possible and as firmly as possible, until the airplane stops. This provides more positive control with tailwheel steering, tends to shorten the after-landing roll, and prevents bouncing and skipping.

If available runway permits, the speed of the airplane should be allowed to dissipate in a normal manner by the friction and drag of the wheels on the ground. Brakes may be used if needed to help slow the airplane. After the airplane has been slowed sufficiently and has been turned onto a taxiway or clear of the landing area, it should be brought to a complete stop. Only after this is done should the pilot retract the flaps and perform other checklist items.

Crosswind Landing

If the crab method of drift correction has been used throughout the final approach and roundout, the crab must be removed before touchdown by applying rudder to align the airplane's longitudinal axis with its direction of movement. This requires timely and accurate action. Failure to accomplish this results in severe side loads being imposed on the landing gear and imparts ground looping tendencies.

If the wing-low method is used, the crosswind correction (aileron into the wind and opposite rudder) should be maintained throughout the roundout, and the touchdown made on the upwind main wheel.

During gusty or high-wind conditions, prompt adjustments must be made in the crosswind correction to assure that the airplane does not drift as it touches down.

As the forward speed decreases after initial contact, the weight of the airplane will cause the downwind main wheel to gradually settle onto the runway.

An adequate amount of power should be used to maintain the proper airspeed throughout the approach, and the throttle should be retarded to idling position after the main wheels contact the landing surface. Care must be exercised in closing the throttle before the pilot is ready for touchdown, because the sudden or premature closing of the throttle may cause a sudden increase in the descent rate that could result in a hard landing.

Crosswind After-Landing Roll

Particularly during the after-landing roll, special attention must be given to maintaining directional control by the use of rudder and tailwheel steering, while keeping the upwind wing from rising by the use of aileron. Characteristically, an airplane has a greater profile, or side area, behind the main landing gear than forward of it. [Figure 13-3] With the main wheels acting as a pivot point and the greater surface area exposed to the crosswind behind that pivot point, the airplane will tend to turn or weathervane into the wind. This weathervaning tendency is more prevalent in the tailwheel-type because the airplane's surface area behind the main landing gear is greater than in nosewheel-type airplanes.

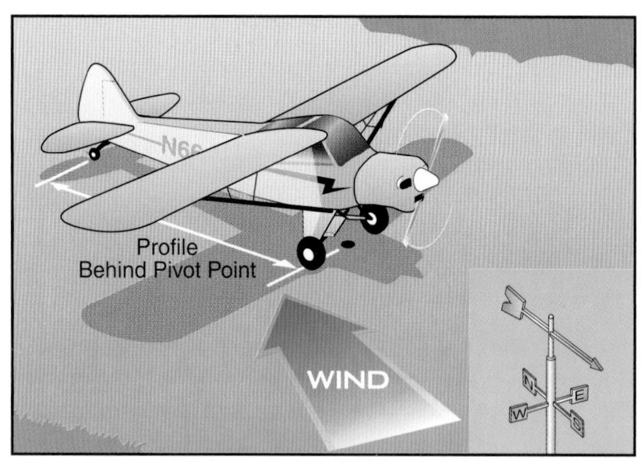

Figure 13-3. Weathervaning tendency.

Pilots should be familiar with the crosswind component of each airplane they fly, and avoid operations in wind conditions that exceed the capability of the airplane, as well as their own limitations.

While the airplane is decelerating during the after-landing roll, more aileron must be applied to keep the upwind wing from rising. Since the airplane is slowing down, there is less airflow around the ailerons and they become less effective. At the same time, the relative wind is becoming more of a crosswind and exerting a greater lifting force on the upwind wing. Consequently, when the airplane is coming to a stop, the aileron control must be held fully toward the wind.

WHEEL LANDING

Landings from power approaches in turbulence or in crosswinds should be such that the touchdown is made with the airplane in approximately level flight attitude. The touchdown should be made smoothly on the main wheels, with the tailwheel held clear of the runway. This is called a "wheel landing" and requires careful timing and control usage to prevent bouncing. These wheel landings can be best accomplished by holding the airplane in level flight attitude until the main wheels touch, then immediately but smoothly retarding the throttle, and holding sufficient forward elevator pressure to hold the main wheels on the ground. The airplane should never be forced onto the ground by excessive forward pressure.

If the touchdown is made at too high a rate of descent as the main wheels strike the landing surface, the tail is forced down by its own weight. In turn, when the tail is forced down, the wing's angle of attack increases resulting in a sudden increase in lift and the airplane may become airborne again. Then as the airplane's speed continues to decrease, the tail may again lower onto the runway. If the tail is allowed to settle too quickly, the airplane may again become airborne. This process, often called "porpoising," usually intensifies even though the pilot tries to stop it. The best corrective action is to execute a go-around procedure.

SHORT-FIELD LANDING

Upon touchdown, the airplane should be firmly held in a three-point attitude. This will provide aerodynamic braking by the wings. Immediately upon touchdown, and closing the throttle, the brakes should be applied evenly and firmly to minimize the after-landing roll. The airplane should be stopped within the shortest possible distance consistent with safety.

SOFT-FIELD LANDING

The tailwheel should touch down simultaneously with or just before the main wheels, and should then be held down by maintaining firm back-elevator pressure throughout the landing roll. This will minimize any tendency for the airplane to nose over and will provide aerodynamic braking. The use of brakes on a soft field is not needed because the soft or rough surface itself will provide sufficient reduction in the airplane's forward speed. Often it will be found that upon landing on a very soft field, the pilot will need to increase power to keep the airplane moving and from becoming stuck in the soft surface.

GROUND LOOP

A ground loop is an uncontrolled turn during ground operation that may occur while taxiing or taking off, but especially during the after-landing roll. It is not always caused by drift or weathervaning, although these things may cause the initial swerve. Careless use of the rudder, an uneven ground surface, or a soft spot that retards one main wheel of the airplane may also cause a swerve. In any case, the initial swerve tends to cause the airplane to ground loop.

Due to the characteristics of an airplane equipped with a tailwheel, the forces that cause a ground loop increase as the swerve increases. The initial swerve develops inertia and this, acting at the CG (which is located behind the main wheels), swerves the airplane even more. If allowed to develop, the force produced may become great enough to tip the airplane until one wing strikes the ground.

If the airplane touches down while drifting or in a crab, the pilot should apply aileron toward the high wing and stop the swerve with the rudder. Brakes should be used to correct for turns or swerves only when the rudder is inadequate. The pilot must exercise caution when applying corrective brake action because it is very easy to overcontrol and aggravate the situation. If brakes are used, sufficient brake should be applied on the low-wing wheel (outside of the turn) to stop the swerve. When the wings are approximately level, the new direction must be maintained until the airplane has slowed to taxi speed or has stopped.

GENERAL

The turbopropeller-powered airplane flies and handles just like any other airplane of comparable size and weight. The aerodynamics are the same. The major differences between flying a turboprop and other non-turbine-powered airplanes are found in the powerplant and systems. The powerplant is different and requires operating procedures that are unique to **gas turbine engines**. But so, too, are other systems such as the electrical system, hydraulics, environmental, flight control, rain and ice protection, and avionics. The turbopropeller-powered airplane also has the advantage of being equipped with a constant speed, full feathering and **reversing propeller**—something normally not found on piston-powered airplanes.

THE GAS TURBINE ENGINE

Both piston (reciprocating) engines and gas turbine engines are internal combustion engines. They have a similar cycle of operation that consists of induction, compression, combustion, expansion, and exhaust. In a piston engine, each of these events is a separate distinct occurrence in each cylinder. Also, in a piston engine an ignition event must occur during each cycle, in each cylinder. Unlike reciprocating engines, in gas turbine engines these phases of power occur simultaneously and continuously instead of one cycle at a time. Additionally, ignition occurs during the starting cycle and is continuous thereafter.

The basic gas turbine engine contains four sections: intake, compression, combustion, and exhaust. [Figure 14-1]

To start the engine, the compressor section is rotated by an electrical starter on small engines or an air driven starter on large engines. As compressor r.p.m. accelerates, air is brought in through the inlet duct, compressed to a high pressure, and delivered to the combustion section (combustion chambers). Fuel is then injected by a **fuel controller** through spray nozzles and ignited by igniter plugs. (Not all of the compressed air is used to support combustion. Some of the compressed air bypasses the burner section and circulates within the engine to provide internal cooling.) The fuel/air mixture in the combustion chamber is then burned in a continuous combustion process and produces a very high temperature, typically around 4,000°F, which heats

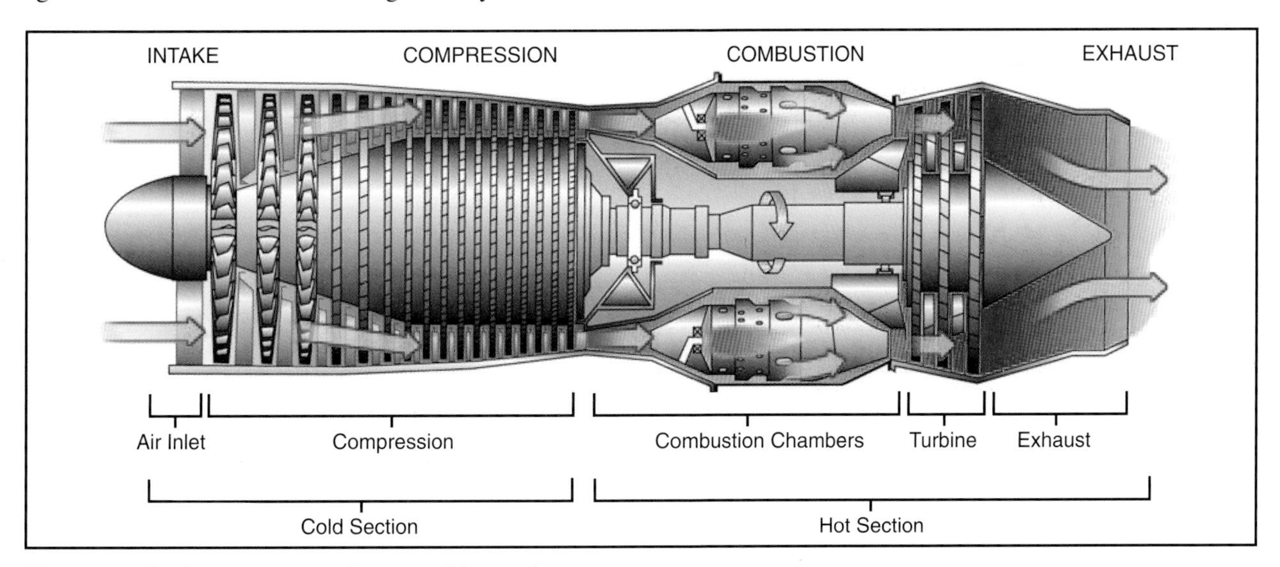

Figure 14-1. Basic components of a gas turbine engine.

the entire air mass to 1,600 – 2,400°F. The mixture of hot air and gases expands and is directed to the turbine blades forcing the turbine section to rotate, which in turn drives the compressor by means of a direct shaft. After powering the turbine section, the high velocity excess exhaust exits the tail pipe or exhaust section. Once the turbine section is powered by gases from the burner section, the starter is disengaged, and the igniters are turned off. Combustion continues until the engine is shut down by turning off the fuel supply.

High-pressure exhaust gases can be used to provide jet thrust as in a **turbojet engine**. Or, the gases can be directed through an additional turbine to drive a propeller through reduction gearing, as in a turbopropeller (turboprop) engine.

TURBOPROP ENGINES

The turbojet engine excels the reciprocating engine in top speed and altitude performance. On the other hand, the turbojet engine has limited takeoff and initial climb performance, as compared to that of a reciprocating engine. In the matter of takeoff and initial climb performance, the reciprocating engine is superior to the turbojet engine. Turbojet engines are most efficient at high speeds and high altitudes, while propellers are most efficient at slow and medium speeds (less than 400 m.p.h.). Propellers also improve takeoff and climb performance. The development of the turboprop engine was an attempt to combine in one engine the best characteristics of both the turbojet, and propeller driven reciprocating engine.

The turboprop engine offers several advantages over other types of engines such as:

- Light weight.
- Mechanical reliability due to relatively few moving parts.
- Simplicity of operation.
- Minimum vibration.
- High power per unit of weight.
- Use of propeller for takeoff and landing.

Turboprop engines are most efficient at speeds between 250 and 400 m.p.h. and altitudes between 18,000 and 30,000 feet. They also perform well at the slow speeds required for takeoff and landing, and are fuel efficient. The minimum specific fuel consumption of the turboprop engine is normally available in the altitude range of 25,000 feet up to the **tropopause**.

The power output of a piston engine is measured in horsepower and is determined primarily by r.p.m. and manifold pressure. The power of a turboprop engine, however, is measured in shaft horsepower (shp). Shaft horsepower is determined by the r.p.m. and the torque (twisting moment) applied to the propeller shaft. Since turboprop engines are gas turbine engines, some jet thrust is produced by exhaust leaving the engine. This thrust is added to the shaft horsepower to determine the total engine power, or equivalent shaft horsepower (eshp). Jet thrust usually accounts for less than 10 percent of the total engine power.

Although the turboprop engine is more complicated and heavier than a turbojet engine of equivalent size and power, it will deliver more thrust at low subsonic airspeeds. However, the advantages decrease as flight speed increases. In normal cruising speed ranges, the propulsive efficiency (output divided by input) of a turboprop decreases as speed increases.

The propeller of a typical turboprop engine is responsible for roughly 90 percent of the total thrust under sea level conditions on a standard day. The excellent performance of a turboprop during takeoff and climb is the result of the ability of the propeller to accelerate a large mass of air while the airplane is moving at a relatively low ground and flight speed. "Turboprop," however, should not be confused with "turbosupercharged" or similar terminology. All turbine engines have a similarity to normally aspirated (non-supercharged) reciprocating engines in that maximum available power decreases almost as a direct function of increased altitude.

Although power will decrease as the airplane climbs to higher altitudes, engine efficiency in terms of specific fuel consumption (expressed as pounds of fuel consumed per horsepower per hour) will be increased. Decreased specific fuel consumption plus the increased true airspeed at higher altitudes is a definite advantage of a turboprop engine.

All turbine engines, turboprop or turbojet, are defined by limiting temperatures, rotational speeds, and (in the case of turboprops) torque. Depending on the installation, the primary parameter for power setting might be temperature, torque, fuel flow or r.p.m. (either propeller r.p.m., **gas generator** (compressor) r.p.m. or both). In cold weather conditions, torque limits can be exceeded while temperature limits are still within acceptable range. While in hot weather conditions, temperature limits may be exceeded without exceeding torque limits. In any weather, the maximum power setting of a turbine engine is usually obtained with the throttles positioned somewhat *aft* of the full forward position. The transitioning pilot must understand the importance of knowing and observing limits on turbine engines. An **overtemp** or **overtorque** condition that lasts for more than a very few *seconds* can literally destroy internal engine components.

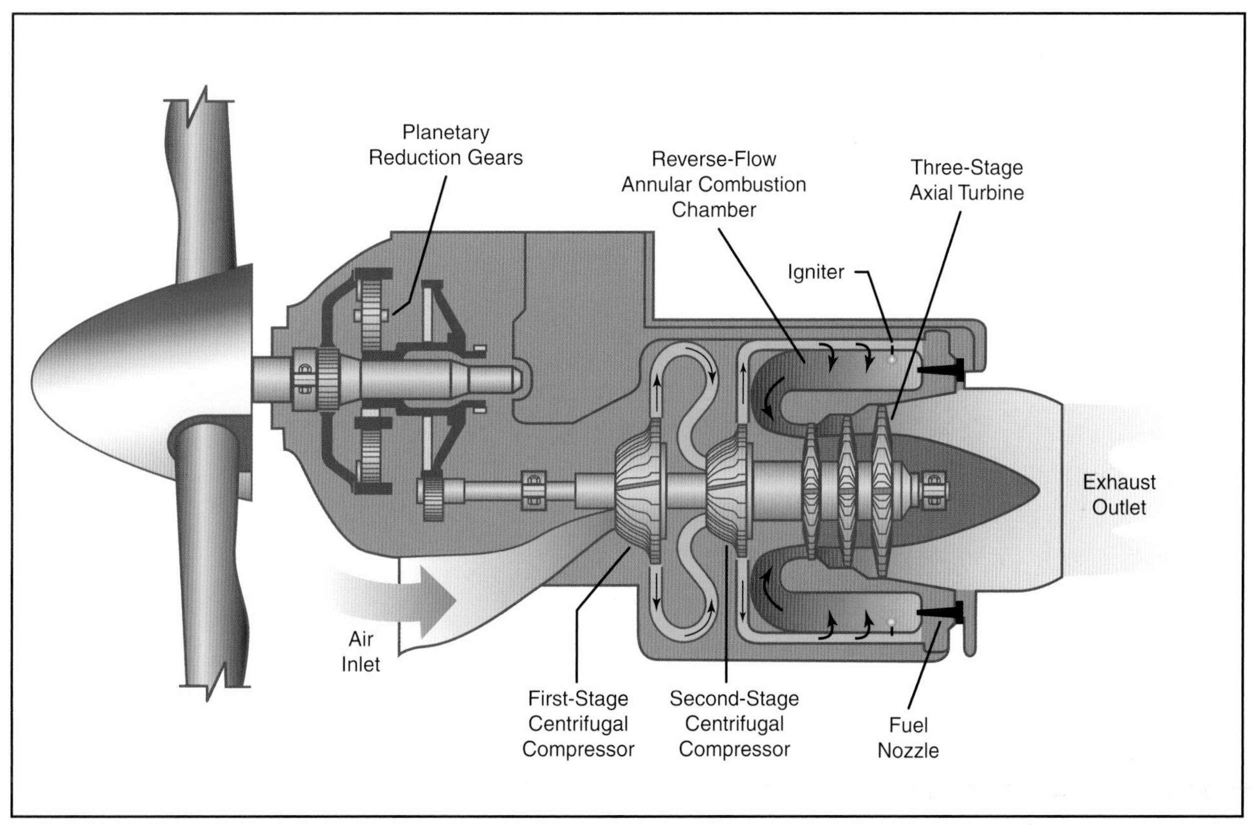

Figure 14-2. Fixed shaft turboprop engine.

TURBOPROP ENGINE TYPES

FIXED SHAFT

One type of turboprop engine is the **fixed shaft** constant speed type such as the Garrett TPE331. [Figure 14-2] In this type engine, ambient air is directed to the compressor section through the engine inlet. An acceleration/diffusion process in the two-stage compressor increases air pressure and directs it rearward to a combustor. The combustor is made up of a combustion chamber, a transition liner, and a turbine plenum. Atomized fuel is added to the air in the combustion chamber. Air also surrounds the combustion chamber to provide for cooling and insulation of the combustor.

The gas mixture is initially ignited by high-energy igniter plugs, and the expanding combustion gases flow to the turbine. The energy of the hot, high velocity gases is converted to **torque** on the main shaft by the turbine rotors. The reduction gear converts the high r.p.m.—low torque of the main shaft to low r.p.m.—high torque to drive the accessories and the propeller. The spent gases leaving the turbine are directed to the atmosphere by the exhaust pipe.

Only about 10 percent of the air which passes through the engine is actually used in the combustion process. Up to approximately 20 percent of the compressed air may be bled off for the purpose of heating, cooling, cabin pressurization, and pneumatic systems. Over half the engine power is devoted to driving the compressor, and it is the compressor which can potentially produce very high drag in the case of a failed, windmilling engine.

In the fixed shaft constant-speed engine, the engine r.p.m. may be varied within a narrow range of 96 percent to 100 percent. During ground operation, the r.p.m. may be reduced to 70 percent. In flight, the engine operates at a constant speed, which is maintained by the governing section of the propeller. Power changes are made by increasing fuel flow and propeller blade angle rather than engine speed. An increase in fuel flow causes an increase in temperature and a corresponding increase in energy available to the turbine. The turbine absorbs more energy and transmits it to the propeller in the form of torque. The increased torque forces the propeller blade angle to be increased to maintain the constant speed. Turbine temperature is a very important factor to be considered in power production. It is directly related to fuel flow and thus to the power produced. It must be limited because of strength and durability of the material in the combustion and turbine section. The control system schedules fuel flow to produce specific temperatures and to limit those temperatures so that the temperature tolerances of the combustion and turbine sections are not exceeded. The engine is designed to operate for its entire life at 100 percent. All of its components, such as compressors and turbines, are most efficient when operated at or near the r.p.m. design point.

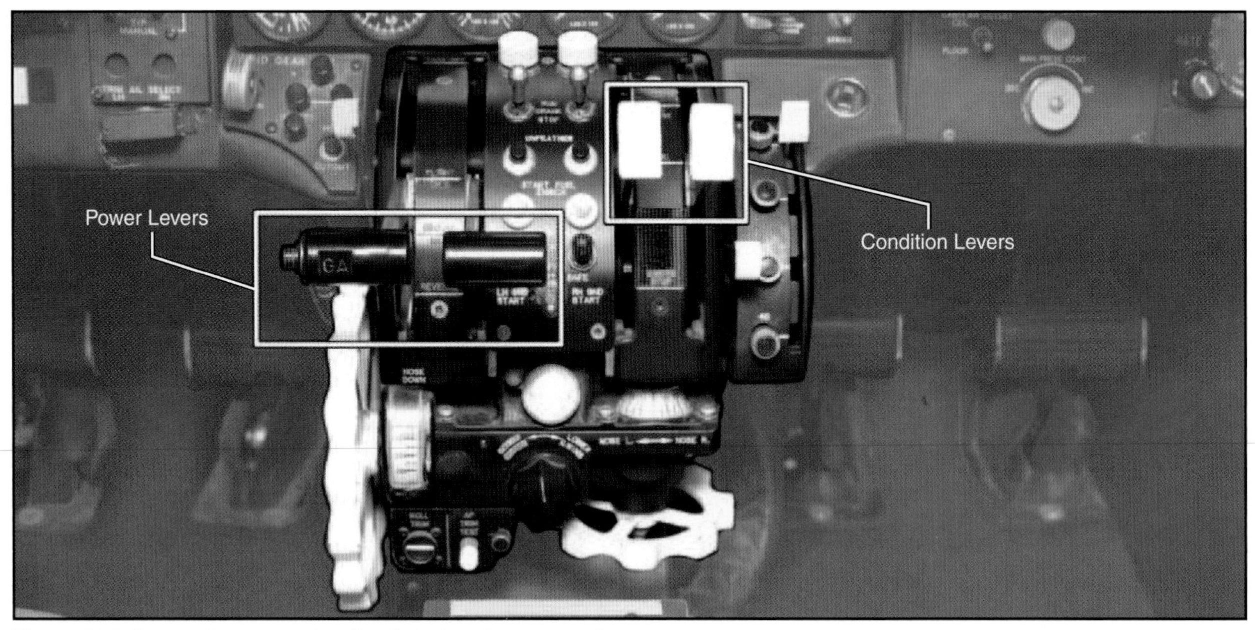

Figure 14-3. Powerplant controls—fixed shaft turboprop engine.

Powerplant (engine and propeller) control is achieved by means of a **power lever** and a **condition lever** for each engine. [Figure 14-3] There is no mixture control and/or r.p.m. lever as found on piston engine airplanes. On the fixed shaft constant-speed turboprop engine, the power lever is advanced or retarded to increase or decrease forward thrust. The power lever is also used to provide **reverse thrust**. The condition lever sets the desired engine r.p.m. within a narrow range between that appropriate for ground operations and flight.

Powerplant instrumentation in a fixed shaft turboprop engine typically consists of the following basic indicator. [Figure 14-4]

- Torque or horsepower.
- ITT – interturbine temperature.
- Fuel flow.
- RPM.

Torque developed by the turbine section is measured by a torque sensor. The torque is then reflected on a cockpit horsepower gauge calibrated in horsepower times 100. Interturbine temperature (ITT) is a measurement of the combustion gas temperature between the first and second stages of the turbine section. The gauge is calibrated in degrees Celsius. Propeller r.p.m. is reflected on a cockpit tachometer as a percentage of maximum r.p.m. Normally, a vernier indicator on the gauge dial indicates r.p.m. in 1 percent graduations as well. The fuel flow indicator indicates fuel flow rate in pounds per hour.

Propeller feathering in a fixed shaft constant-speed turboprop engine is normally accomplished with the condition lever. An engine failure in this type engine, however, will result in a serious drag condition due to the large power requirements of the compressor being absorbed by the propeller. This could create a serious airplane control problem in twin-engine airplanes unless the failure is recognized immediately and the

Figure 14-4. Powerplant instrumentation—fixed shaft turboprop engine.

affected propeller feathered. For this reason, the fixed shaft turboprop engine is equipped with **negative torque sensing** (NTS).

Negative torque sensing is a condition wherein propeller torque drives the engine and the propeller is automatically driven to high pitch to reduce drag. The function of the negative torque sensing system is to limit the torque the engine can extract from the propeller during windmilling and thereby prevent large drag forces on the airplane. The NTS system causes a movement of the propeller blades automatically toward their feathered position should the engine suddenly lose power while in flight. The NTS system is an emergency backup system in the event of sudden engine failure. It is not a substitution for the feathering device controlled by the condition lever.

SPLIT SHAFT/ FREE TURBINE ENGINE

In a **free power-turbine engine,** such as the Pratt & Whitney PT-6 engine, the propeller is driven by a separate turbine through reduction gearing. The propeller is not on the same shaft as the basic engine turbine and compressor. [Figure 14-5] Unlike the fixed shaft engine, in the split shaft engine the propeller can be feathered in flight or on the ground with the basic engine still running. The free power-turbine design allows the pilot to select a desired propeller governing r.p.m., regardless of basic engine r.p.m.

A typical free power-turbine engine has two independent counter-rotating turbines. One turbine drives the compressor, while the other drives the propeller through a reduction gearbox. The compressor in the basic engine consists of three **axial flow compressor** stages combined with a single **centrifugal compressor stage**. The axial and centrifugal stages are assembled on the same shaft, and operate as a single unit.

Inlet air enters the engine via a circular plenum near the *rear* of the engine, and flows forward through the successive compressor stages. The flow is directed outward by the centrifugal compressor stage through radial diffusers before entering the combustion chamber, where the flow direction is actually *reversed.* The gases produced by combustion are once again reversed to expand forward through each turbine stage. After leaving the turbines, the gases are collected in a peripheral exhaust scroll, and are discharged to the atmosphere through two exhaust ports near the *front* of the engine.

A pneumatic fuel control system schedules fuel flow to maintain the power set by the gas generator power lever. Except in the **beta range**, propeller speed within the governing range remains constant at any selected propeller control lever position through the action of a propeller governor.

The accessory drive at the aft end of the engine provides power to drive fuel pumps, fuel control, oil pumps, a starter/generator, and a tachometer transmitter. At this point, the speed of the drive (N_1) is the true speed of the compressor side of the engine, approximately 37,500 r.p.m.

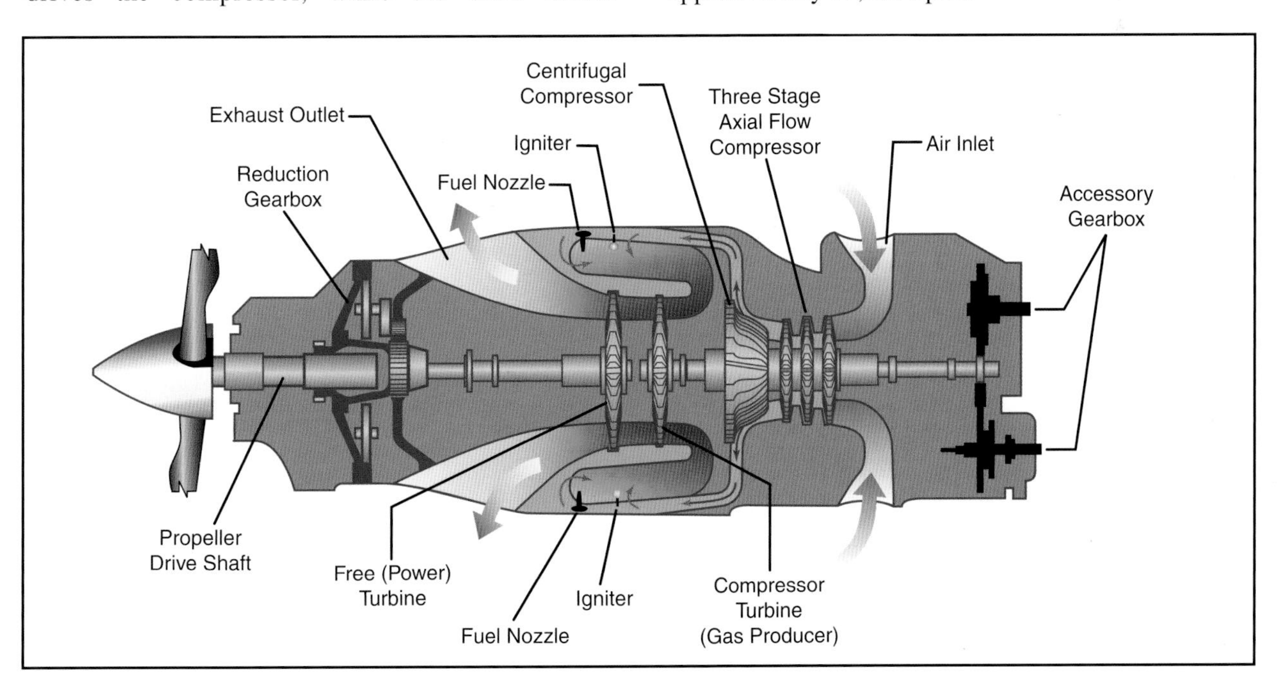

Figure 14-5. Split shaft/free turbine engine.

Figure 14-6. Powerplant controls—split shaft/free turbine engine.

Powerplant (engine and propeller) operation is achieved by three sets of controls for each engine: the power lever, propeller lever, and condition lever. [Figure 14-6] The power lever serves to control engine power in the range from idle through takeoff power. Forward or aft motion of the power lever increases or decreases gas generator r.p.m. (N_1) and thereby increases or decreases engine power. The propeller lever is operated conventionally and controls the constant-speed propellers through the primary governor. The propeller r.p.m. range is normally from 1,500 to 1,900. The condition lever controls the flow of fuel to the engine. Like the mixture lever in a piston-powered airplane, the condition lever is located at the far right of the power quadrant. But the condition lever on a turboprop engine is really just an on/off valve for delivering fuel. There are HIGH IDLE and LOW IDLE positions for ground operations, but condition levers have no metering function. Leaning is not required in turbine engines; this function is performed automatically by a dedicated **fuel control unit**.

Engine instruments in a split shaft/free turbine engine typically consist of the following basic indicators. [Figure 14-7]

- ITT (interstage turbine temperature) indicator.
- Torquemeter.
- Propeller tachometer.
- N_1 (gas generator) tachometer.
- Fuel flow indicator.
- Oil temperature/pressure indicator.

Figure 14-7. Engine instruments—split shaft/free turbine engine.

The ITT indicator gives an instantaneous reading of engine gas temperature between the compressor turbine and the power turbines. The torquemeter responds to power lever movement and gives an indication, in foot-pounds (ft/lb), of the torque being applied to the propeller. Because in the free turbine engine, the propeller is not attached physically to the shaft of the gas turbine engine, two tachometers are justified—one for the propeller and one for the gas generator. The propeller tachometer is read directly in revolutions per minute. The N_1 or gas generator is read in percent of r.p.m. In the Pratt & Whitney PT-6 engine, it is based on a figure of 37,000 r.p.m. at 100 percent. Maximum continuous gas generator is limited to 38,100 r.p.m. or 101.5 percent N_1.

The ITT indicator and torquemeter are used to set takeoff power. Climb and cruise power are established with the torquemeter and propeller tachometer while observing ITT limits. Gas generator (N_1) operation is monitored by the gas generator tachometer. Proper observation and interpretation of these instruments provide an indication of engine performance and condition.

REVERSE THRUST AND BETA RANGE OPERATIONS

The thrust that a propeller provides is a function of the angle of attack at which the air strikes the blades, and the speed at which this occurs. The angle of attack varies with the pitch angle of the propeller.

So called "flat pitch" is the blade position offering minimum resistance to rotation and no net thrust for moving the airplane. Forward pitch produces forward thrust—higher pitch angles being required at higher airplane speeds.

The "feathered" position is the highest pitch angle obtainable. [Figure 14-8] The feathered position produces no forward thrust. The propeller is generally placed in feather only in case of in-flight engine failure to minimize drag and prevent the air from using the propeller as a turbine.

In the "reverse" pitch position, the engine/propeller turns in the same direction as in the normal (forward) pitch position, but the propeller blade angle is positioned to the other side of flat pitch. [Figure 14-8] In reverse pitch, air is pushed away from the airplane rather than being drawn over it. Reverse pitch results in braking action, rather than forward thrust of the airplane. It is used for backing away from obstacles when taxiing, controlling taxi speed, or to aid in bringing the airplane to a stop during the landing roll. Reverse pitch does not mean reverse rotation of the engine. The engine delivers power just the same, no matter which side of flat pitch the propeller blades are positioned.

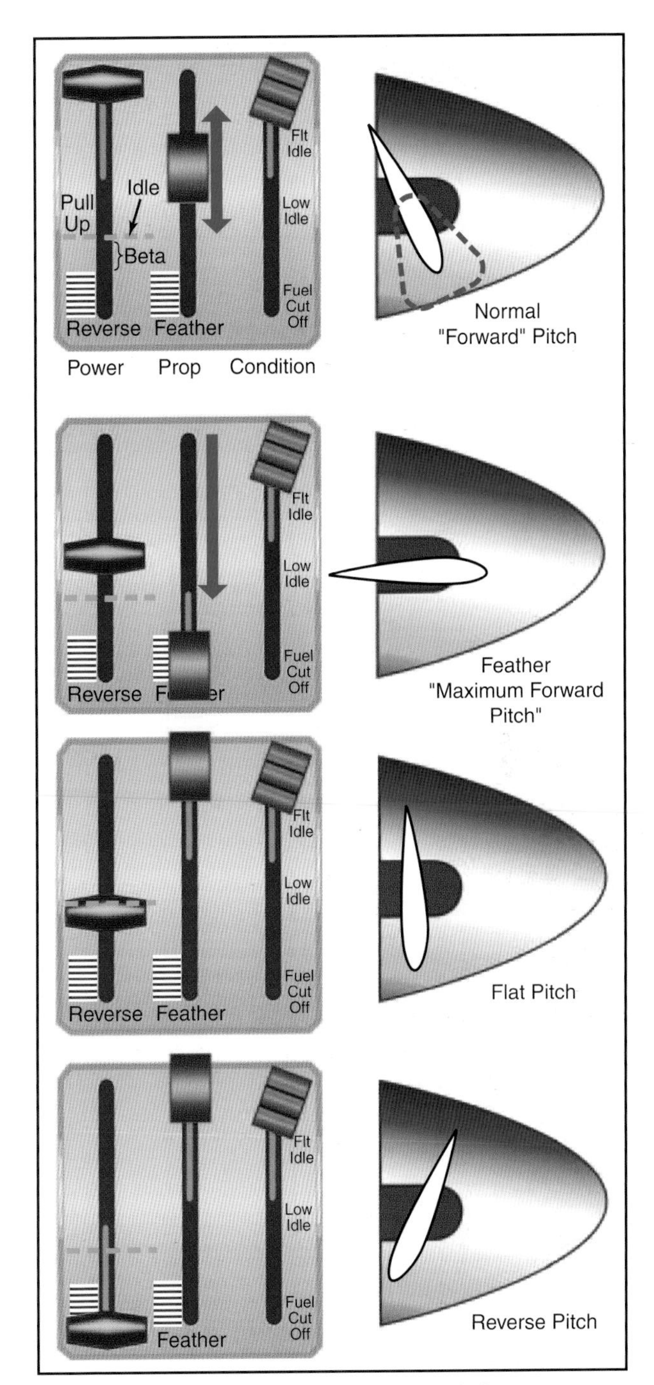

Figure 14-8. Propeller pitch angle characteristics.

With a turboprop engine, in order to obtain enough power for flight, the power lever is placed somewhere between **flight idle** (in some engines referred to as "high idle") and maximum. The power lever directs signals to a **fuel control unit** to manually select fuel. The propeller governor selects the propeller pitch needed to keep the propeller/engine on speed. This is referred to as the propeller governing or "alpha" mode of operation. When positioned aft of flight idle, however, the power lever directly controls propeller blade angle. This is known as the "beta" range of operation.

The beta range of operation consists of power lever positions from flight idle to maximum reverse.

Beginning at power lever positions just aft of flight idle, propeller blade pitch angles become progressively flatter with aft movement of the power lever until they go beyond maximum flat pitch and into negative pitch, resulting in reverse thrust. While in a fixed shaft/constant-speed engine, the engine speed remains largely unchanged as the propeller blade angles achieve their negative values. On the split shaft PT-6 engine, as the negative 5° position is reached, further aft movement of the power lever will also result in a progressive increase in engine (N_1) r.p.m. until a maximum value of about negative 11° of blade angle and 85 percent N_1 are achieved.

Operating in the beta range and/or with reverse thrust requires specific techniques and procedures depending on the particular airplane make and model. There are also specific engine parameters and limitations for operations within this area that must be adhered to. It is essential that a pilot transitioning to turboprop airplanes become knowledgeable and proficient in these areas, which are unique to turbine-engine-powered airplanes.

TURBOPROP AIRPLANE ELECTRICAL SYSTEMS

The typical turboprop airplane electrical system is a 28-volt direct current (DC) system, which receives power from one or more batteries and a **starter/generator** for each engine. The batteries may either be of the lead-acid type commonly used on piston-powered airplanes, or they may be of the nickel-cadmium (NiCad) type. The NiCad battery differs from the lead-acid type in that its output remains at relatively high power levels for longer periods of time. When the NiCad battery is depleted, however, its voltage drops off very suddenly. When this occurs, its ability to turn the compressor for engine start is greatly diminished and the possibility of engine damage due to a **hot start** increases. Therefore, it is essential to check the battery's condition before every engine start. Compared to lead-acid batteries, high-performance NiCad batteries can be recharged very quickly. But the faster the battery is recharged, the more heat it produces. Therefore, NiCad battery equipped airplanes are fitted with battery overheat annunciator lights signifying maximum safe and critical temperature thresholds.

The DC generators used in turboprop airplanes double as starter motors and are called "starter/generators." The starter/generator uses electrical power to produce mechanical torque to start the engine and then uses the engine's mechanical torque to produce electrical power after the engine is running. Some of the DC power produced is changed to 28 volt 400 cycle alternating current (AC) power for certain avionic, lighting, and indicator synchronization functions. This is accomplished by an electrical component called an **inverter**.

The distribution of DC and AC power throughout the system is accomplished through the use of **power distribution buses**. These "buses" as they are called are actually common terminals from which individual electrical circuits get their power. [Figure 14-9]

Buses are usually named for what they power (avionics bus, for example), or for where they get their power (right generator bus, battery bus). The distribution of DC and AC power is often divided into functional groups (buses) that give priority to certain equipment

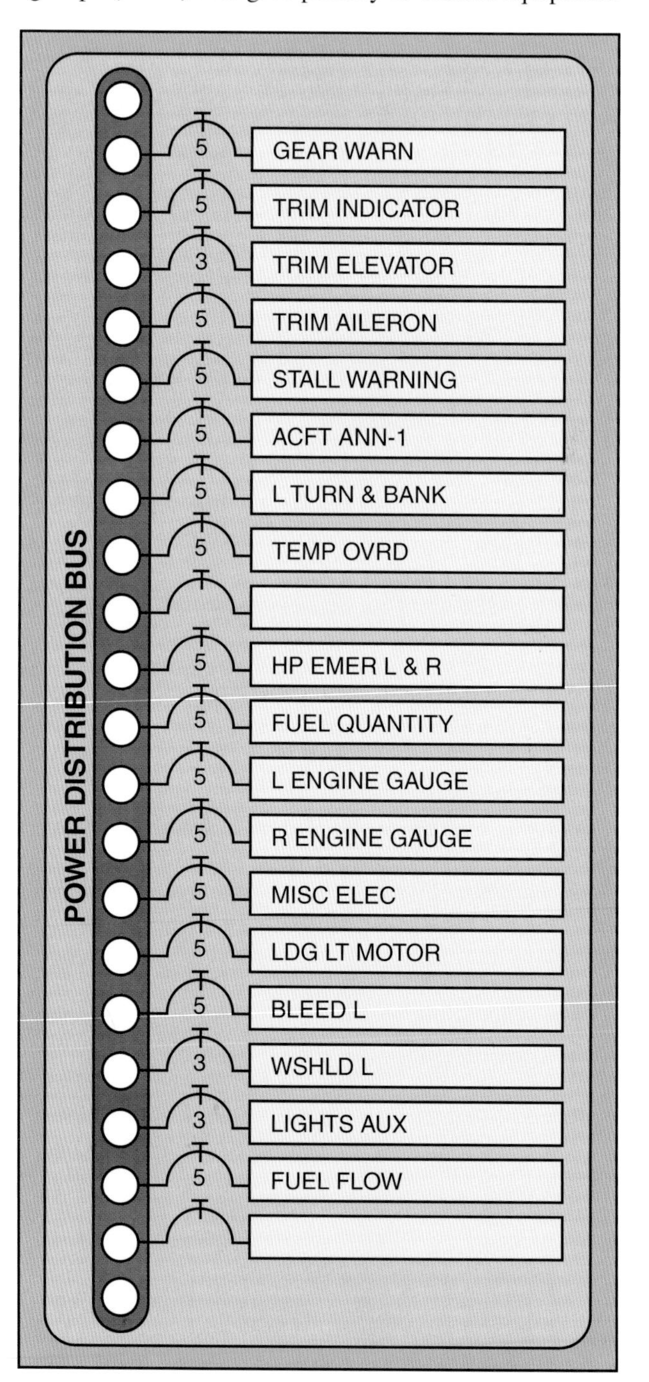

Figure 14-9. Typical individual power distribution bus.

during normal and emergency operations. Main buses serve most of the airplane's electrical equipment. Essential buses feed power to equipment having top priority. [Figure 14-10]

Multiengine turboprop airplanes normally have several power sources—a battery and at least one generator per engine. The electrical systems are usually designed so that any bus can be energized by any of the power sources. For example, a typical system might have a right and left generator buses powered normally by the right and left engine-driven generators. These buses will be connected by a normally open switch, which isolates them from each other. If one generator fails, power will be lost to its bus, but power can be restored to that bus by closing a **bus tie** switch. Closing this switch connects the buses and allows the operating generator to power both.

Power distribution buses are protected from short circuits and other malfunctions by a type of fuse called a **current limiter**. In the case of excessive current supplied by any power source, the current limiter will open the circuit and thereby isolate that power source and allow the affected bus to become separated from the system. The other buses will continue to operate normally. Individual electrical components are connected to the buses through **circuit breakers**. A circuit breaker is a device which opens an electrical circuit when an excess amount of current flows.

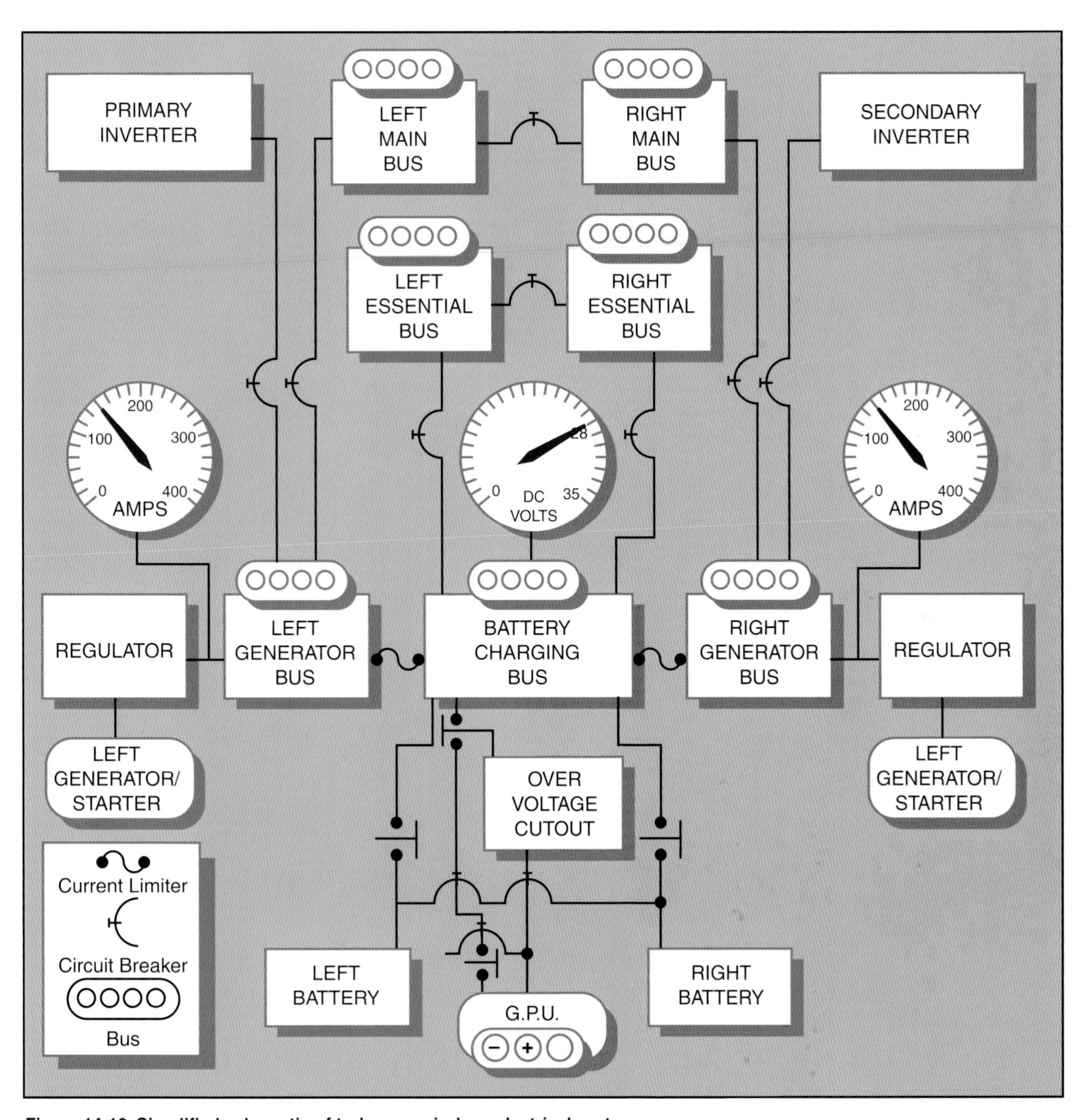

Figure 14-10. Simplified schematic of turboprop airplane electrical system.

OPERATIONAL CONSIDERATIONS

As previously stated, a turboprop airplane flies just like any other piston engine airplane of comparable size and weight. It is in the operation of the engines and airplane systems that makes the turboprop airplane different from its piston engine counterpart. Pilot errors in engine and/or systems operation are the most common cause of aircraft damage or mishap. The time of maximum vulnerability to pilot error in any gas turbine engine is during the engine start sequence.

Turbine engines are extremely heat sensitive. They cannot tolerate an overtemperature condition for more than a very few seconds without serious damage being done. Engine temperatures get hotter during starting than at any other time. Thus, turbine engines have minimum rotational speeds for introducing fuel into the combustion chambers during startup. Hypervigilant temperature and acceleration monitoring on the part of the pilot remain crucial until the engine is running at a stable speed. Successful engine starting depends on assuring the correct minimum battery voltage before initiating start, or employing a ground power unit (GPU) of adequate output.

After fuel is introduced to the combustion chamber during the start sequence, "light-off" and its associated heat rise occur very quickly. Engine temperatures may approach the maximum in a matter of 2 or 3 seconds before the engine stabilizes and temperatures fall into the normal operating range. During this time, the pilot must watch for any *tendency* of the temperatures to exceed limitations and be prepared to cut off fuel to the engine.

An engine tendency to exceed maximum starting temperature limits is termed a **hot start**. The temperature rise may be preceded by unusually high initial fuel flow, which may be the first indication the pilot has that the engine start is not proceeding normally. Serious engine damage will occur if the hot start is allowed to continue.

A condition where the engine is accelerating more slowly than normal is termed a **hung start** or **false start**. During a hung start/false start, the engine may

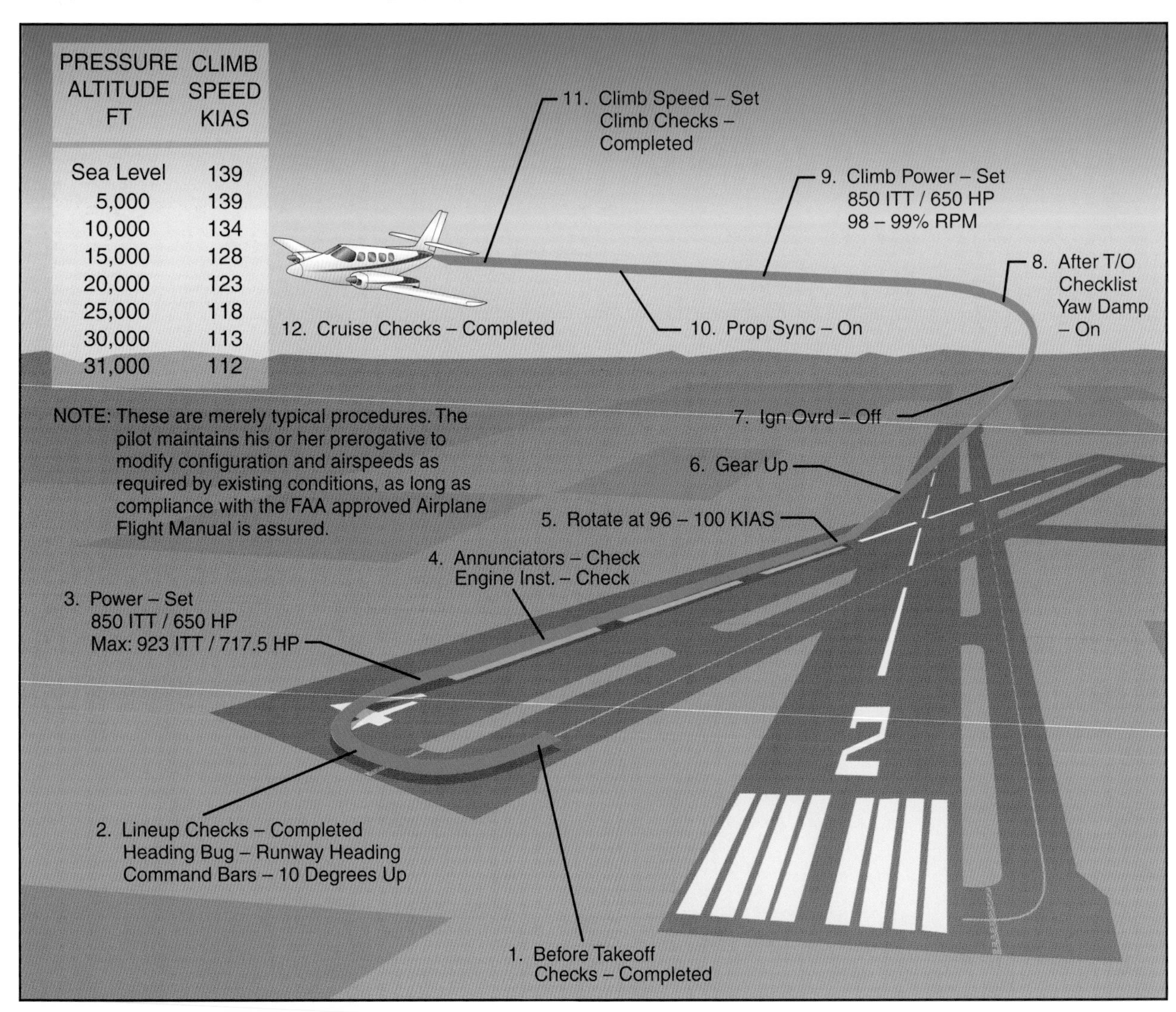

Figure 14-11. Example—typical turboprop airplane takeoff and departure profile.

stabilize at an engine r.p.m. that is not high enough for the engine to continue to run without help from the starter. This is usually the result of low battery power or the starter not turning the engine fast enough for it to start properly.

Takeoffs in turboprop airplanes are *not* made by automatically pushing the power lever full forward to the stops. Depending on conditions, takeoff power may be limited by either torque or by engine temperature. Normally, the power lever position on takeoff will be somewhat aft of full forward.

Takeoff and departure in a turboprop airplane (especially a twin-engine cabin-class airplane) should be accomplished in accordance with a standard takeoff and departure "profile" developed for the particular make and model. [Figure 14-11] The takeoff and departure profile should be in accordance with the airplane manufacturer's recommended procedures as outlined in the FAA-approved Airplane Flight Manual and/or the Pilot's Operating Handbook (AFM/POH). The increased complexity of turboprop airplanes makes the standardization of procedures a necessity for safe and efficient operation. The transitioning pilot should review the profile procedures before each takeoff to form a mental picture of the takeoff and departure process.

For any given high horsepower operation, the pilot can expect that the engine temperature will climb as altitude increases at a constant power. On a warm or hot day, maximum temperature limits may be reached at a rather low altitude, making it impossible to maintain high horsepower to higher altitudes. Also, the engine's compressor section has to work harder with decreased air density. Power capability is reduced by high-density altitude and power use may have to be modulated to keep engine temperature within limits.

In a turboprop airplane, the pilot can close the throttles(s) at any time without concern for cooling the engine too rapidly. Consequently, rapid descents with the propellers in low pitch can be dramatically steep. Like takeoffs and departures, approach and landing should be accomplished in accordance with a standard approach and landing profile. [Figure 14-12]

A **stabilized approach** is an essential part of the approach and landing process. In a stabilized approach, the airplane, depending on design and type, is placed in a stabilized descent on a glidepath ranging from 2.5 to 3.5°. The speed is stabilized at some reference from the AFM/POH—usually 1.25 to 1.30 times the stall speed in approach configuration. The descent rate is stabilized from 500 feet per minute to 700 feet per minute until the landing flare.

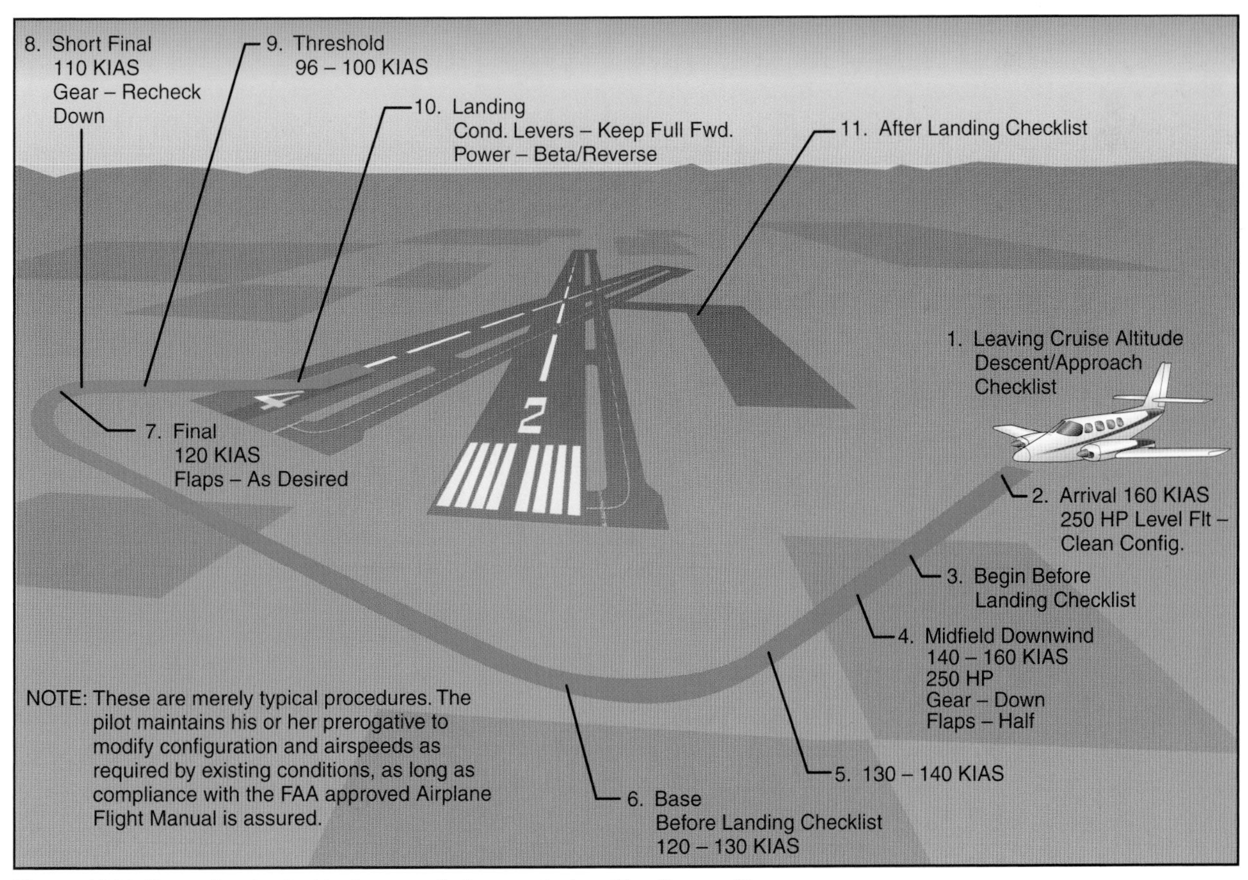

Figure 14-12. Example—typical turboprop airplane arrival and landing profile.

Landing some turboprop airplanes (as well as some piston twins) can result in a hard, premature touchdown if the engines are idled too soon. This is because large propellers spinning rapidly in low pitch create considerable drag. In such airplanes, it may be preferable to maintain power throughout the landing flare and touchdown. Once firmly on the ground, propeller beta range operation will dramatically reduce the need for braking in comparison to piston airplanes of similar weights.

TRAINING CONSIDERATIONS

The medium and high altitudes at which turboprop airplanes are flown provide an entirely different environment in terms of regulatory requirements, airspace structure, physiological requirements, and even meteorology. The pilot transitioning to turboprop airplanes, particularly those who are not familiar with operations in the high/medium altitude environment, should approach turboprop transition training with this in mind. Thorough ground training should cover all aspects of high/medium altitude flight, including the flight environment, weather, flight planning and navigation, physiological aspects of high-altitude flight, oxygen and pressurization system operation, and high-altitude emergencies.

Flight training should prepare the pilot to demonstrate a comprehensive knowledge of airplane performance, systems, emergency procedures, and operating limitations, along with a high degree of proficiency in performing all flight maneuvers and in-flight emergency procedures.

The training outline below covers the minimum information needed by pilots to operate safely at high altitudes.

a. Ground Training

- (1) The High-Altitude Flight Environment
  - (a) Airspace
  - (b) Title 14 of the Code of Federal Regulations (14 CFR) section 91.211, requirements for use of supplemental oxygen
- (2) Weather
  - (a) The atmosphere
  - (b) Winds and clear air turbulence
  - (c) Icing
- (3) Flight Planning and Navigation
  - (a) Flight planning
  - (b) Weather charts
  - (c) Navigation
  - (d) Navaids
- (4) Physiological Training
  - (a) Respiration
  - (b) Hypoxia
  - (c) Effects of prolonged oxygen use
  - (d) Decompression sickness
  - (e) Vision
  - (f) Altitude chamber (optional)
- (5) High-Altitude Systems and Components
  - (a) Oxygen and oxygen equipment
  - (b) Pressurization systems
  - (c) High-altitude components
- (6) Aerodynamics and Performance Factors
  - (a) Acceleration
  - (b) G-forces
  - (c) MACH Tuck and MACH Critical (turbojet airplanes)
- (7) Emergencies
  - (a) Decompression
  - (b) Donning of oxygen masks
  - (c) Failure of oxygen mask, or complete loss of oxygen supply/system
  - (d) In-flight fire
  - (e) Flight into severe turbulence or thunderstorms

b. Flight Training

- (1) Preflight Briefing
- (2) Preflight Planning
  - (a) Weather briefing and considerations
  - (b) Course plotting
  - (c) Airplane Flight Manual
  - (d) Flight plan
- (3) Preflight Inspection
  - (a) Functional test of oxygen system, including the verification of supply and pressure, regulator operation, oxygen flow, mask fit, and cockpit and air traffic control (ATC) communication using mask microphones
- (4) Engine Start Procedures, Runup, Takeoff, and Initial Climb
- (5) Climb to High Altitude and Normal Cruise Operations While Operating Above 25,000 Feet MSL
- (6) Emergencies
  - (a) Simulated rapid decompression, including the immediate donning of oxygen masks
  - (b) Emergency descent
- (7) Planned Descents
- (8) Shutdown Procedures
- (9) Postflight Discussion

Chapter 15

Transition to Jet Powered Airplanes

GENERAL

This chapter contains an overview of jet powered airplane operations. It is not meant to replace any portion of a formal jet airplane qualification course. Rather, the information contained in this chapter is meant to be a useful preparation for and a supplement to formal and structured jet airplane qualification training. The intent of this chapter is to provide information on the major differences a pilot will encounter when transitioning to jet powered airplanes. In order to achieve this in a logical manner, the major differences between jet powered airplanes and piston powered airplanes have been approached by addressing two distinct areas: differences in technology, or how the airplane itself differs; and differences in pilot technique, or how the pilot deals with the technological differences through the application of different techniques. If any of the information in this chapter conflicts with information contained in the FAA-approved Airplane Flight Manual for a particular airplane, the Airplane Flight Manual takes precedence.

JET ENGINE BASICS

A jet engine is a gas turbine engine. A jet engine develops thrust by accelerating a relatively small mass of air to very high velocity, as opposed to a propeller, which develops thrust by accelerating a much larger mass of air to a much slower velocity.

As stated in Chapter 14, both piston and gas turbine engines are internal combustion engines and have a similar basic cycle of operation; that is, induction, compression, combustion, expansion, and exhaust. Air is taken in and compressed, and fuel is injected and burned. The hot gases then expand and supply a surplus of power over that required for compression, and are finally exhausted. In both piston and jet engines, the efficiency of the cycle is improved by increasing the volume of air taken in and the **compression ratio**.

Part of the expansion of the burned gases takes place in the turbine section of the jet engine providing the necessary power to drive the compressor, while the remainder of the expansion takes place in the nozzle of the tail pipe in order to accelerate the gas to a high velocity jet thereby producing thrust. [Figure 15-1]

In theory, the jet engine is simpler and more directly converts thermal energy (the burning and expansion of gases) into mechanical energy (thrust). The piston or reciprocating engine, with all of its moving parts, must convert the thermal energy into mechanical energy and then finally into thrust by rotating a propeller.

One of the advantages of the jet engine over the piston engine is the jet engine's capability of producing much greater amounts of thrust horsepower at the high altitudes and high speeds. In fact, turbojet engine efficiency increases with altitude and speed.

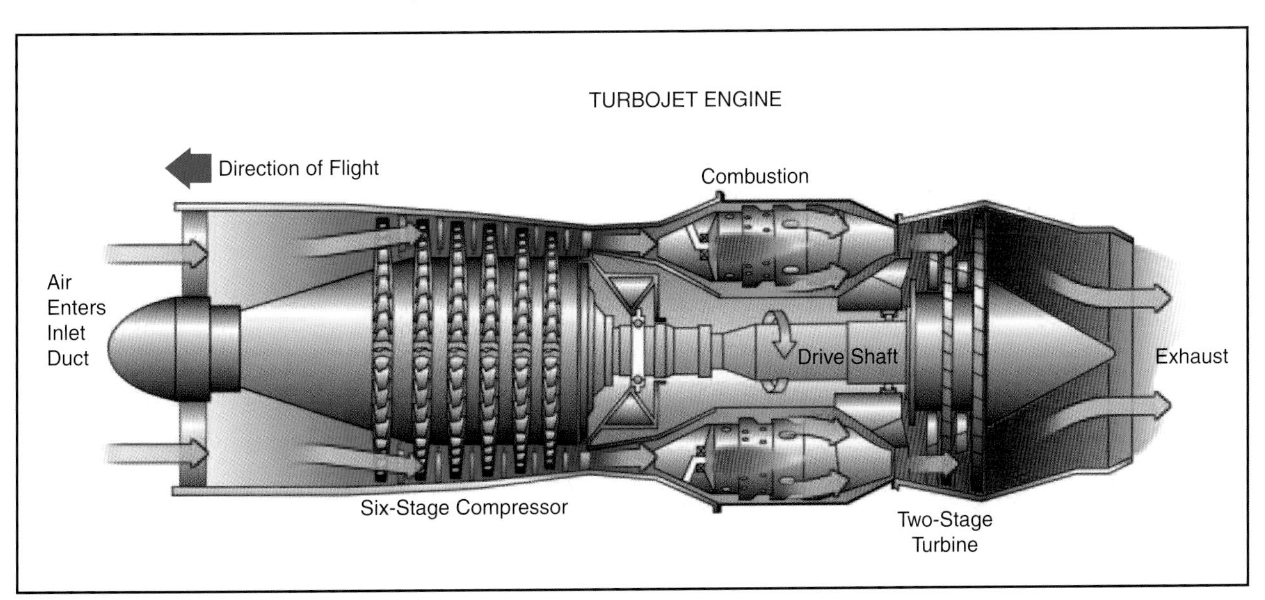

Figure 15-1. Basic turbojet engine.

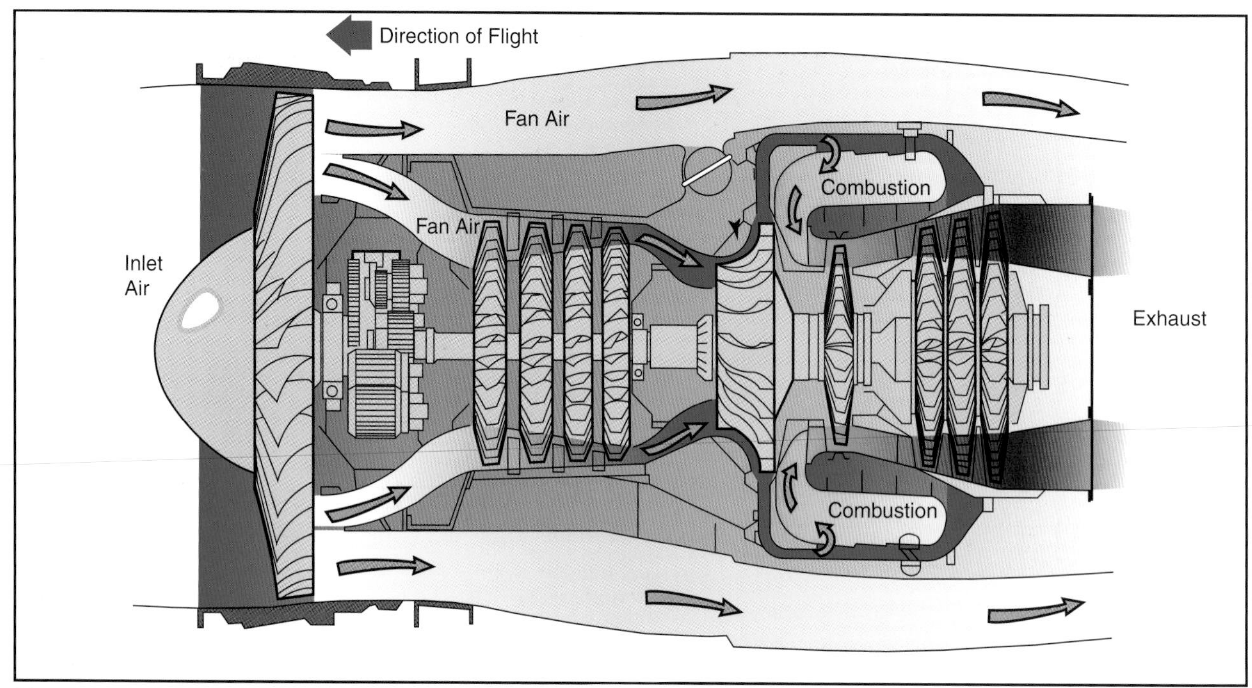

Figure 15-2. Turbofan engine.

Although the propeller driven airplane is not nearly as efficient as the jet, particularly at the higher altitudes and cruising speeds required in modern aviation, one of the few advantages the propeller driven airplane has over the jet is that maximum thrust is available almost at the start of the takeoff roll. Initial thrust output of the jet engine on takeoff is relatively lower and does not reach peak efficiency until the higher speeds. The fan-jet or **turbofan engine** was developed to help compensate for this problem and is, in effect, a compromise between the pure jet engine (turbojet) and the propeller engine.

Like other gas turbine engines, the heart of the turbofan engine is the **gas generator**—the part of the engine that produces the hot, high-velocity gases. Similar to turboprops, turbofans have a low pressure turbine section that uses most of the energy produced by the gas generator. The low pressure turbine is mounted on a concentric shaft that passes through the hollow shaft of the gas generator, connecting it to a **ducted fan** at the front of the engine. [Figure 15-2]

Air enters the engine, passes through the fan, and splits into two separate paths. Some of it flows around—bypasses—the engine core, hence its name, **bypass air**. The air drawn into the engine for the gas generator is the **core airflow**. The amount of air that bypasses the core compared to the amount drawn into the gas generator determines a turbofan's **bypass ratio**. Turbofans efficiently convert fuel into thrust because they produce low pressure energy spread over a large fan disk area. While a turbojet engine uses all of the gas generator's output to produce thrust in the form of a high-velocity exhaust gas jet, cool, low-velocity bypass air produces between 30 percent and 70 percent of the thrust produced by a turbofan engine.

The fan-jet concept increases the total thrust of the jet engine, particularly at the lower speeds and altitudes. Although efficiency at the higher altitudes is lost (turbofan engines are subject to a large lapse in thrust with increasing altitude), the turbofan engine increases acceleration, decreases the takeoff roll, improves initial climb performance, and often has the effect of decreasing specific fuel consumption.

OPERATING THE JET ENGINE

In a jet engine, thrust is determined by the amount of fuel injected into the combustion chamber. The power controls on most turbojet and turbofan powered airplanes consist of just one thrust lever for each engine, because most engine control functions are automatic. The thrust lever is linked to a fuel control and/or electronic engine computer that meters fuel flow based upon r.p.m., internal temperatures, ambient conditions, and other factors. [Figure 15-3]

In a jet engine, each major rotating section usually has a separate gauge devoted to monitoring its speed of rotation. Depending on the make and model, a jet engine may have an N_1 gauge that monitors the low pressure compressor section and/or fan speed in turbofan engines. The gas generator section may be monitored by an N_2 gauge, while **triple spool engines** may have an N_3 gauge as well. Each engine section rotates at many thousands of r.p.m. Their gauges therefore are calibrated in percent of r.p.m. rather than actual r.p.m., for ease of display and interpretation. [Figure 15-4]

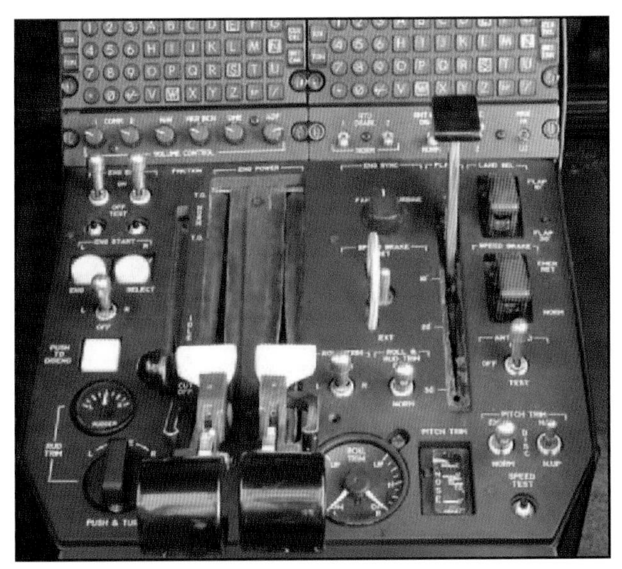

Figure 15-3. Jet engine power controls.

The temperature of turbine gases must be closely monitored by the pilot. As in any gas turbine engine, exceeding temperature limits, even for a very few seconds, may result in serious heat damage to turbine blades and other components. Depending on the make and model, gas temperatures can be measured at a number of different locations within the engine. The associated engine gauges therefore have different names according to their location. For instance:

- **Exhaust Gas Temperature** (EGT)—the temperature of the exhaust gases as they enter the tail pipe, after passing through the turbine.
- **Turbine Inlet Temperature** (TIT)—the temperature of the gases from the combustion section of the engine as they enter the first stage of the turbine. TIT is the highest temperature inside a gas turbine engine and is one of the limiting factors of the amount of power the engine can produce. TIT, however, is difficult to measure. EGT therefore, which relates to TIT, is normally the parameter measured.
- **Interstage Turbine Temperature** (ITT)—the temperature of the gases between the high pressure and low pressure turbine wheels.
- **Turbine Outlet Temperature** (TOT)—like EGT, turbine outlet temperature is taken aft of the turbine wheel(s).

JET ENGINE IGNITION

Most jet engine ignition systems consist of two igniter plugs, which are used during the ground or air starting of the engine. Once the start is completed, this ignition either automatically goes off or is turned off, and from this point on, the combustion in the engine is a continuous process.

CONTINUOUS IGNITION

An engine is sensitive to the flow characteristics of the air that enters the intake of the engine nacelle. So long as the flow of air is substantially normal, the engine will continue to run smoothly. However, particularly with rear mounted engines that are sometimes in a position to be affected by disturbed airflow from the wings, there are some abnormal flight situations that could cause a compressor stall or flameout of the engine. These abnormal flight conditions would usually be associated with abrupt pitch changes such as might be encountered in severe turbulence or a stall.

In order to avoid the possibility of engine flameout from the above conditions, or from other conditions that might cause ingestion problems such as heavy rain, ice, or possible bird strike, most jet engines are equipped with a continuous ignition system. This system can be turned on and used continuously whenever the need arises. In many jets, as an added precaution, this system is normally used during takeoffs and landings. Many jets are also equipped with an automatic ignition system that operates both igniters whenever the airplane stall warning or stick shaker is activated.

FUEL HEATERS

Because of the high altitudes and extremely cold outside air temperatures in which the jet flies, it is possible to supercool the jet fuel to the point that the small

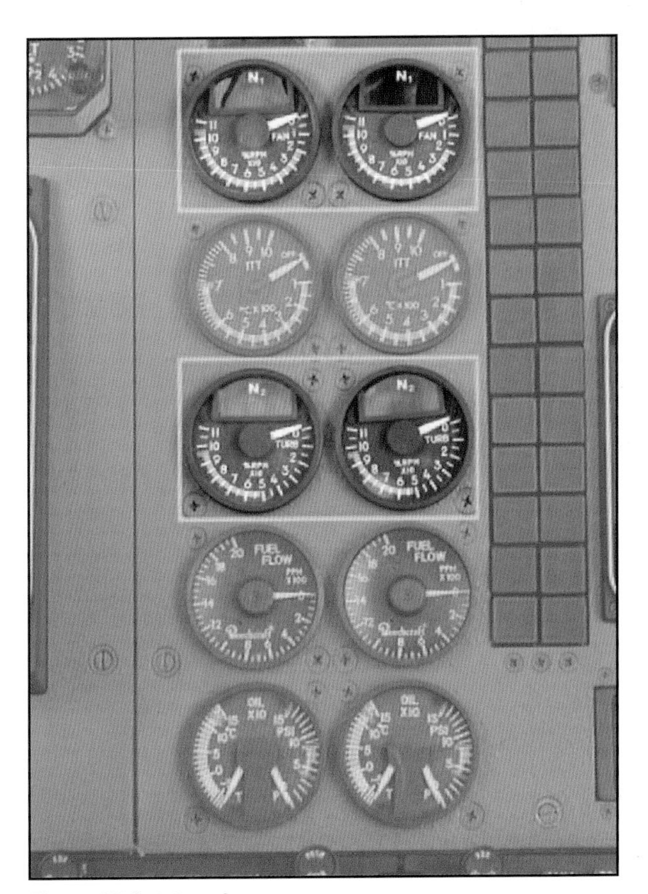

Figure 15-4. Jet engine r.p.m. gauges.

particles of water suspended in the fuel can turn to ice crystals and clog the fuel filters leading to the engine. For this reason, jet engines are normally equipped with fuel heaters. The fuel heater may be of the automatic type which constantly maintains the fuel temperature above freezing, or they may be manually controlled by the pilot from the cockpit.

SETTING POWER

On some jet airplanes, thrust is indicated by an engine pressure ratio (EPR) gauge. Engine pressure ratio can be thought of as being equivalent to the manifold pressure on the piston engine. Engine pressure ratio is the difference between turbine discharge pressure and engine inlet pressure. It is an indication of what the engine has done with the raw air scooped in. For instance, an EPR setting of 2.24 means that the discharge pressure relative to the inlet pressure is 2.24 : 1. On these airplanes, the EPR gauge is the primary reference used to establish power settings. [Figure 15-5]

Figure 15-5. EPR gauge.

Fan speed (N_1) is the primary indication of thrust on most turbofan engines. Fuel flow provides a secondary thrust indication, and cross-checking for proper fuel flow can help in spotting a faulty N_1 gauge. Turbofans also have a gas generator turbine tachometer (N_2). They are used mainly for engine starting and some system functions.

In setting power, it is usually the primary power reference (EPR or N_1) that is most critical, and will be the gauge that will first limit the forward movement of the thrust levers. However, there are occasions where the limits of either r.p.m. or temperature can be exceeded. The rule is: movement of the thrust levers must be stopped and power set at whichever the limits of EPR, r.p.m., or temperature is reached first.

THRUST TO THRUST LEVER RELATIONSHIP

In a piston engine propeller driven airplane, thrust is proportional to r.p.m., manifold pressure, and propeller blade angle, with manifold pressure being the most dominant factor. At a constant r.p.m., thrust is proportional to throttle lever position. In a jet engine, however, thrust is quite disproportional to thrust lever position. This is an important difference that the pilot transitioning into jet powered airplanes must become accustomed to.

On a jet engine, thrust is proportional to r.p.m. (mass flow) and temperature (fuel/air ratio). These are matched and a further variation of thrust results from the compressor efficiency at varying r.p.m. The jet engine is most efficient at high r.p.m., where the engine is designed to be operated most of the time. As r.p.m. increases, mass flow, temperature, and efficiency also increase. Therefore, much more thrust is produced per increment of throttle movement near the top of the range than near the bottom.

One thing that will seem different to the piston pilot transitioning into jet powered airplanes is the rather large amount of thrust lever movement between the flight idle position and full power as compared to the small amount of movement of the throttle in the piston engine. For instance, an inch of throttle movement on a piston may be worth 400 horsepower wherever the throttle may be. On a jet, an inch of thrust lever movement at a low r.p.m. may be worth only 200 pounds of thrust, but at a high r.p.m. that same inch of movement might amount to closer to 2,000 pounds of thrust. Because of this, in a situation where significantly more thrust is needed and the jet engine is at low r.p.m., it will not do much good to merely "inch the thrust lever forward." Substantial thrust lever movement is in order. This is not to say that rough or abrupt thrust lever action is standard operating procedure. If the power setting is already high, it may take only a small amount of movement. However, there are two characteristics of the jet engine that work against the normal habits of the piston engine pilot. One is the variation of thrust with r.p.m., and the other is the relatively slow acceleration of the jet engine.

VARIATION OF THRUST WITH RPM

Whereas piston engines normally operate in the range of 40 percent to 70 percent of available r.p.m., jets operate most efficiently in the 85 percent to 100 percent range, with a flight idle r.p.m. of 50 percent to 60 percent. The range from 90 percent to 100 percent in jets may produce as much thrust as the total available at 70 percent. [Figure 15-6]

SLOW ACCELERATION OF THE JET ENGINE

In a propeller driven airplane, the constant speed propeller keeps the engine turning at a constant r.p.m. within the governing range, and power is changed by varying the manifold pressure. Acceleration of the

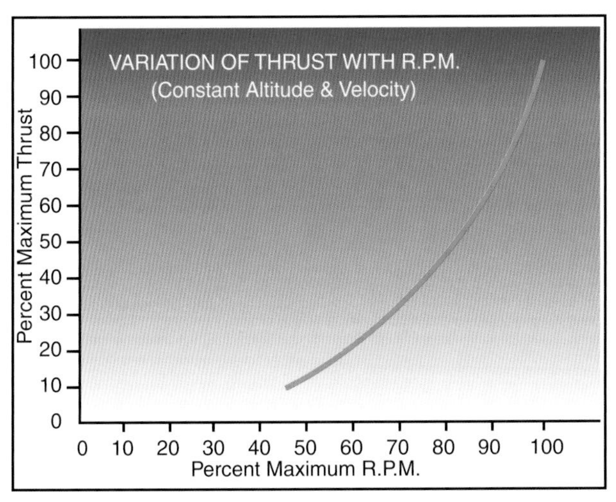

Figure 15-6. Variation of thrust with r.p.m.

piston from idle to full power is relatively rapid, somewhere on the order of 3 to 4 seconds. The acceleration on the different jet engines can vary considerably, but it is usually much slower.

Efficiency in a jet engine is highest at high r.p.m. where the compressor is working closest to its optimum conditions. At low r.p.m. the operating cycle is generally inefficient. If the engine is operating at normal approach r.p.m. and there is a sudden requirement for increased thrust, the jet engine will respond immediately and full thrust can be achieved in about 2 seconds. However, at a low r.p.m., sudden full power application will tend to overfuel the engine resulting in possible **compressor surge**, excessive turbine temperatures, **compressor stall** and/or flameout. To prevent this, various limiters such as **compressor bleed valves** are contained in the system and serve to restrict the engine until it is at an r.p.m. at which it can respond to a rapid acceleration demand without distress. This critical r.p.m. is most noticeable when the engine is at idle r.p.m. and the thrust lever is rapidly advanced to a high power position. Engine acceleration is initially very slow, but changes to very fast after about 78 percent r.p.m. is reached. [Figure 15-7]

Even though engine acceleration is nearly instantaneous after about 78 percent r.p.m., total time to accelerate from idle r.p.m. to full power may take as much as 8 seconds. For this reason, most jets are operated at a relatively high r.p.m. during the final approach to landing or at any other time that immediate power may be needed.

Jet Engine Efficiency

Maximum operating altitudes for general aviation turbojet airplanes now reach 51,000 feet. The efficiency of the jet engine at high altitudes is the primary reason for operating in the high altitude environment. The specific fuel consumption of jet engines decreases as the outside air temperature decreases for constant engine r.p.m. and true airspeed (TAS). Thus, by flying at a high altitude, the pilot is able to operate at flight levels where fuel economy is best and with the most advantageous cruise speed. For efficiency, jet airplanes are typically operated at high altitudes where cruise is usually very close to r.p.m or exhaust gas temperature limits. At high altitudes, little excess thrust may be available for maneuvering. Therefore, it is often impossible for the jet airplane to climb and turn simultaneously, and all maneuvering must be accomplished within the limits of available thrust and without sacrificing stability and controllability.

Absence of Propeller Effect

The absence of a propeller has a significant effect on the operation of jet powered airplanes that the transitioning pilot must become accustomed to. The effect is due to the absence of lift from the propeller slipstream, and the absence of propeller drag.

Absence of Propeller Slipstream

A propeller produces thrust by accelerating a large mass of air rearwards, and (especially with wing mounted engines) this air passes over a comparatively large percentage of the wing area. On a propeller driven airplane, the lift that the wing develops is the sum of the lift generated by the wing area not in the wake of the propeller (as a result of airplane speed) and the lift generated by the wing area influenced by the propeller slipstream. By increasing or decreasing the speed of the slipstream air, therefore, it is possible to increase or decrease the total lift on the wing without changing airspeed.

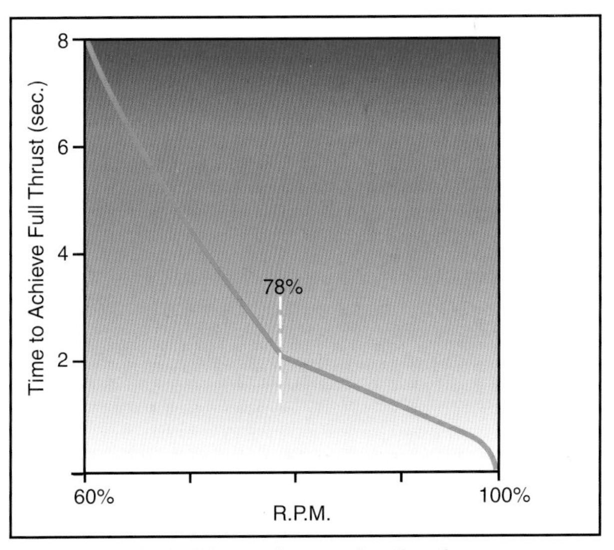

Figure 15-7. Typical Jet engine acceleration times.

For example, a propeller driven airplane that is allowed to become too low and too slow on an approach is very responsive to a quick blast of power to salvage the situation. In addition to increasing lift at a constant airspeed, stalling speed is reduced with power on. A jet engine, on the other hand, also produces thrust by accelerating a mass of air rearward, but this air does not pass over the wings. There is therefore no lift bonus at increased power at constant airspeed, and no significant lowering of power-on stall speed.

In not having propellers, the jet powered airplane is minus two assets.

- It is not possible to produce increased lift instantly by simply increasing power.
- It is not possible to lower stall speed by simply increasing power. The 10-knot margin (roughly the difference between power-off and power-on stall speed on a propeller driven airplane for a given configuration) is lost.

Add the poor acceleration response of the jet engine and it becomes apparent that there are three ways in which the jet pilot is worse off than the propeller pilot. For these reasons, there is a marked difference between the approach qualities of a piston engine airplane and a jet. In a piston engine airplane, there is some room for error. Speed is not too critical and a burst of power will salvage an increasing sink rate. In a jet, however, there is little room for error.

If an increasing sink rate develops in a jet, the pilot must remember two points in the proper sequence.

1. Increased lift can be gained only by accelerating airflow over the wings, and this can be accomplished only by accelerating the entire airplane.
2. The airplane can be accelerated, assuming altitude loss cannot be afforded, only by a rapid increase in thrust, and here, the slow acceleration of the jet engine (possibly up to 8 seconds) becomes a factor.

Salvaging an increasing sink rate on an approach in a jet can be a very difficult maneuver. The lack of ability to produce instant lift in the jet, along with the slow acceleration of the engine, necessitates a "stabilized approach" to a landing where full landing configuration, constant airspeed, controlled rate of descent, and relatively high power settings are maintained until over the threshold of the runway. This allows for almost immediate response from the engine in making minor changes in the approach speed or rate of descent and makes it possible to initiate an immediate go-around or missed approach if necessary.

ABSENCE OF PROPELLER DRAG

When the throttles are closed on a piston powered airplane, the propellers create a vast amount of drag, and airspeed is immediately decreased or altitude lost. The effect of reducing power to idle on the jet engine, however, produces no such drag effect. In fact, at an idle power setting, the jet engine still produces forward thrust. The main advantage is that the jet pilot is no longer faced with a potential drag penalty of a runaway propeller, or a reversed propeller. A disadvantage, however, is the "free wheeling" effect forward thrust at idle has on the jet. While this occasionally can be used to advantage (such as in a long descent), it is a handicap when it is necessary to lose speed quickly, such as when entering a terminal area or when in a landing flare. The lack of propeller drag, along with the aerodynamically clean airframe of the jet, are new to most pilots, and slowing the airplane down is one of the initial problems encountered by pilots transitioning into jets.

SPEED MARGINS

The typical piston powered airplane had to deal with two maximum operating speeds.

- **V_{NO}**—Maximum structural cruising speed, represented on the airspeed indicator by the upper limit of the green arc. It is, however, permissible to exceed V_{NO} and operate in the caution range (yellow arc) in certain flight conditions.
- **V_{NE}**—Never-exceed speed, represented by a red line on the airspeed indicator.

These speed margins in the piston airplanes were never of much concern during normal operations because the high drag factors and relatively low cruise power settings kept speeds well below these maximum limits.

Maximum speeds in jet airplanes are expressed differently, and always define the maximum operating speed of the airplane which is comparable to the V_{NE} of the piston airplane. These maximum speeds in a jet airplane are referred to as:

- **V_{MO}**—Maximum operating speed expressed in terms of knots.
- **M_{MO}**—Maximum operating speed expressed in terms of a decimal of Mach speed (speed of sound).

To observe both limits V_{MO} and M_{MO}, the pilot of a jet airplane needs both an airspeed indicator and a Machmeter, each with appropriate red lines. In some general aviation jet airplanes, these are combined into a single instrument that contains a pair of concentric indicators, one for the indicated airspeed and the other for indicated Mach number. Each is provided with an appropriate red line. [Figure 15-8]

Figure 15-8. Jet airspeed indicator.

A more sophisticated indicator is used on most jetliners. It looks much like a conventional airspeed indicator but has a "barber pole" that automatically moves so as to display the applicable speed limit at all times.

Because of the higher available thrust and very low drag design, the jet airplane can very easily exceed its speed margin even in cruising flight, and in fact in some airplanes in a shallow climb. The handling qualities in a jet can change drastically when the maximum operating speeds are exceeded.

High speed airplanes designed for subsonic flight are limited to some Mach number below the speed of sound to avoid the formation of shock waves that begin to develop as the airplane nears Mach 1.0. These shock waves (and the adverse effects associated with them) can occur when the airplane speed is substantially below Mach 1.0. The Mach speed at which some portion of the airflow over the wing first equals Mach 1.0 is termed the critical Mach number ($MACH_{CRIT}$). This is also the speed at which a shock wave first appears on the airplane.

There is no particular problem associated with the acceleration of the airflow up to the point where Mach 1.0 is encountered; however, a shock wave is formed at the point where the airflow suddenly returns to subsonic flow. This shock wave becomes more severe and moves aft on the wing as speed of the wing is increased, and eventually flow separation occurs behind the well-developed shock wave. [Figure 15-9]

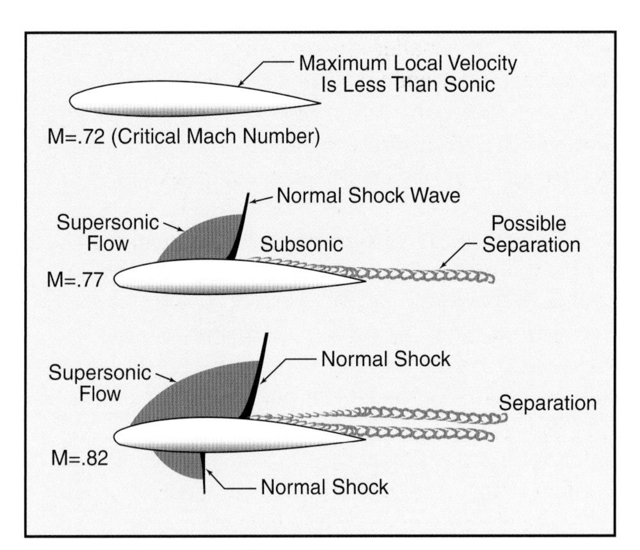

Figure 15-9. Transonic flow patterns.

If allowed to progress well beyond the M_{MO} for the airplane, this separation of air behind the shock wave can result in severe buffeting and possible loss of control or "upset."

Because of the changing center of lift of the wing resulting from the movement of the shock wave, the pilot will experience pitch change tendencies as the airplane moves through the transonic speeds up to and exceeding M_{MO}. [Figure 15-10]

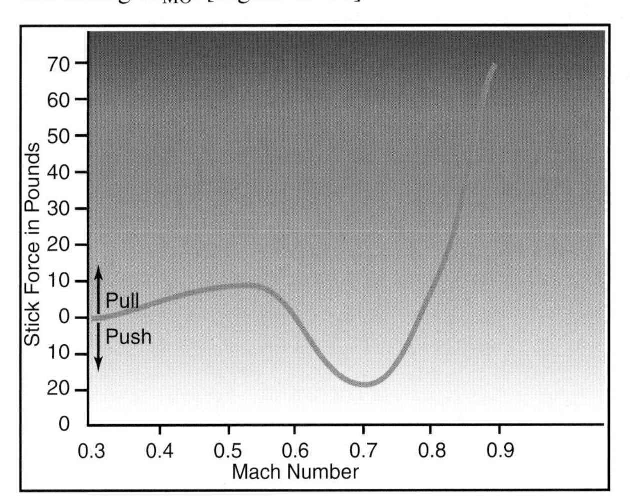

Figure 15-10. Example of Stick Forces vs. Mach Number in a typical jet airplane.

For example, as the graph in figure 15-10 illustrates, initially as speed is increased up to Mach .72 the wing develops an increasing amount of lift requiring a nose-down force or trim to maintain level flight. With increased speed and the aft movement of the shock wave, the wing's center of pressure also moves aft causing the start of a nosedown tendency or "tuck." By Mach .83 the nosedown forces are well developed to a point where a total of 70 pounds of back pressure are required to hold the nose up. If allowed to progress unchecked, **Mach tuck** may eventually occur. Although Mach tuck develops gradually, if it is allowed to progress significantly, the center of pressure can move so far rearward that there is no longer enough elevator authority available to counteract it, and the airplane could enter a steep, sometimes unrecoverable dive.

An alert pilot would have observed the high airspeed indications, experienced the onset of buffeting, and responded to aural warning devices long before encountering the extreme stick forces shown. However, in the event that corrective action is not taken and the nose allowed to drop, increasing airspeed even further, the situation could rapidly become dangerous. As the Mach speed increases beyond the airplane's M_{MO}, the effects of flow separation and turbulence behind the shock wave become more severe. Eventually, the most powerful forces causing Mach tuck are a result of the buffeting and lack of effective downwash on the horizontal stabilizer because of the disturbed airflow over the wing. This is the primary reason for the development of the T-tail configuration on some jet airplanes, which places the horizontal stabilizer as far as practical from the turbulence of the wings. Also, because of the critical aspects of high-altitude/high-Mach flight, most jet airplanes capable of operating in the Mach speed ranges are designed with some form of trim and autopilot Mach compensating device (stick puller) to alert the pilot to inadvertent excursions beyond its certificated M_{MO}.

Recovery from Overspeed Conditions

The simplest remedy for an overspeed condition is to ensure that the situation never occurs in the first place. For this reason, the pilot must be aware of all the conditions that could lead to exceeding the airplane's maximum operating speeds. Good attitude instrument flying skills and good power control are essential.

The pilot should be aware of the symptoms that will be experienced in the particular airplane as the V_{MO} or M_{MO} is being approached. These may include:

- Nosedown tendency and need for back pressure or trim.
- Mild buffeting as airflow separation begins to occur after critical Mach speed.
- Actuation of an aural warning device/stick puller at or just slightly beyond V_{MO} or M_{MO}.

The pilot's response to an overspeed condition should be to immediately slow the airplane by reducing the power to flight idle. It will also help to smoothly and easily raise the pitch attitude to help dissipate speed (in fact this is done automatically through the stick puller device when the high speed warning system is activated). The use of speed brakes can also aid in slowing the airplane. If, however, the nosedown stick forces have progressed to the extent that they are excessive, some speed brakes will tend to further aggravate the nosedown tendency. Under most conditions, this additional pitch down force is easily controllable, and since speed brakes can normally be used at any speed, they are a very real asset. A final option would be to extend the landing gear. This will create enormous drag and possibly some noseup pitch, but there is usually little risk of damage to the gear itself. The pilot transitioning into jet airplanes must be familiar with the manufacturers' recommended procedures for dealing with overspeed conditions contained in the FAA-approved Airplane Flight Manual for the particular make and model airplane.

Mach Buffet Boundaries

Thus far, only the Mach buffet that results from excessive speed has been addressed. The transitioning pilot, however, should be aware that Mach buffet is a function of the speed of the airflow over the wing—not necessarily the airspeed of the airplane. Anytime that too great a lift demand is made on the wing, whether from too fast an airspeed or from too high an angle of attack near the M_{MO}, the "high speed buffet" will occur. However, there are also occasions when the buffet can be experienced at much slower speeds known as "low speed Mach buffet."

The most likely situations that could cause the low speed buffet would be when an airplane is flown at too slow a speed for its weight and altitude causing a high angle of attack. This very high angle of attack would have the same effect of increasing airflow over the upper surface of the wing to the point that all of the same effects of the shock waves and buffet would occur as in the high speed buffet situation.

The angle of attack of the wing has the greatest effect on inducing the Mach buffet at either the high or low speed boundaries for the airplane. The conditions that increase the angle of attack, hence the speed of the airflow over the wing and chances of Mach buffet are:

- **High altitudes**—The higher the airplane flies, the thinner the air and the greater the angle of attack required to produce the lift needed to maintain level flight.
- **Heavy weights**—The heavier the airplane, the greater the lift required of the wing, and all other things being equal, the greater the angle of attack.
- **"G" loading**—An increase in the "G" loading of the wing results in the same situation as increasing the weight of the airplane. It makes

no difference whether the increase in "G" forces is caused by a turn, rough control usage, or turbulence. The effect of increasing the wing's angle of attack is the same.

An airplane's indicated airspeed decreases in relation to true airspeed as altitude increases. As the indicated airspeed decreases with altitude, it progressively merges with the low speed buffet boundary where prestall buffet occurs for the airplane at a load factor of 1.0 G. The point where the high speed Mach indicated airspeed and low speed buffet boundary indicated airspeed merge is the airplane's absolute or aerodynamic ceiling. Once an airplane has reached its aerodynamic ceiling, which is higher than the altitude stipulated in the FAA-approved Airplane Flight Manual, the airplane can neither be made to go faster without activating the design **stick puller** at Mach limit nor can it be made to go slower without activating the **stick shaker** or **stick pusher**. This critical area of the airplane's flight envelope is known as "coffin corner."

Mach buffet occurs as a result of supersonic airflow on the wing. Stall buffet occurs at angles of attack that produce airflow disturbances (burbling) over the upper surface of the wing which decreases lift. As density altitude increases, the angle of attack that is required to produce an airflow disturbance over the top of the wing is reduced until the density altitude is reached where Mach buffet and stall buffet converge (coffin corner). When this phenomenon is encountered, serious consequences may result causing loss of airplane control.

Increasing either gross weight or load factor (G factor) will increase the low speed buffet and decrease Mach buffet speeds. A typical jet airplane flying at 51,000 feet altitude at 1.0 G may encounter Mach buffet slightly above the airplane's M_{MO} (.82 Mach) and low speed buffet at .60 Mach. However, only 1.4 G (an increase of only 0.4 G) may bring on buffet at the optimum speed of .73 Mach and any change in airspeed, bank angle, or gust loading may reduce this straight-and-level flight 1.4 G protection to no protection at all. Consequently, a maximum cruising flight altitude must be selected which will allow sufficient buffet margin for necessary maneuvering and for gust conditions likely to be encountered. Therefore, it is important for pilots to be familiar with the use of charts showing cruise maneuver and buffet limits. [Figure 15-11]

The transitioning pilot must bear in mind that the maneuverability of the jet airplane is particularly critical, especially at the high altitudes. Some jet airplanes have a very narrow span between the high and low speed buffets. One airspeed that the pilot should have firmly fixed in memory is the manufacturer's recommended gust penetration speed for the particular make and model airplane. This speed is normally the speed that would give the greatest margin between the high and low speed buffets, and may be considerably higher

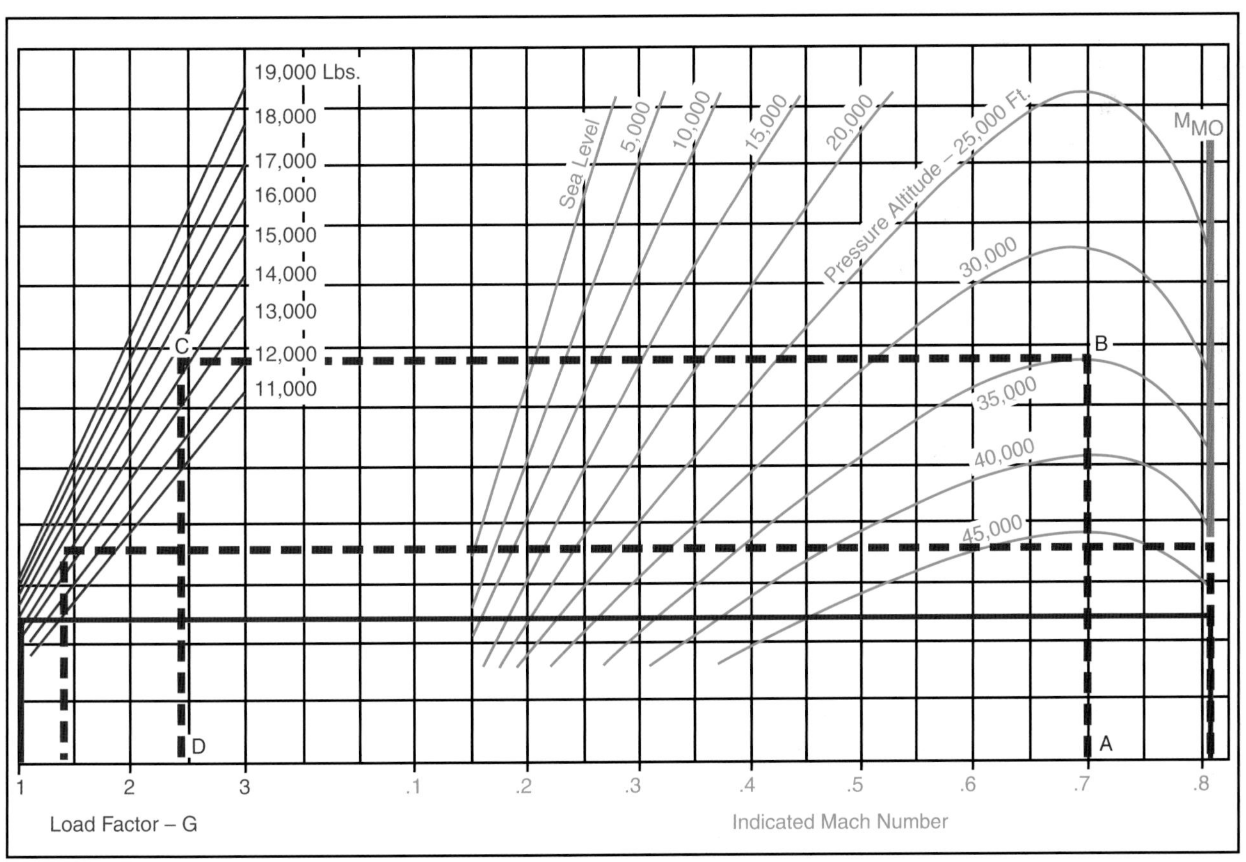

Figure 15-11. Mach buffet boundary chart.

than design maneuvering speed (V_A). This means that, unlike piston airplanes, there are times when a jet airplane should be flown in excess of V_A during encounters with turbulence. Pilots operating airplanes at high speeds must be adequately trained to operate them safely. This training cannot be complete until pilots are thoroughly educated in the critical aspects of the aerodynamic factors pertinent to Mach flight at high altitudes.

LOW SPEED FLIGHT

The jet airplane wing, designed primarily for high speed flight, has relatively poor low speed characteristics. As opposed to the normal piston powered airplane, the jet wing has less area, a lower aspect ratio (long chord/short span), and thin airfoil shape—all of which amount to less lift. The sweptwing is additionally penalized at low speeds because the effective lift, which is perpendicular to the leading edge, is always less than the airspeed of the airplane itself. In other words, the airflow on the sweptwing has the effect of persuading the wing into believing that it is flying slower than it actually is, but the wing consequently suffers a loss of lift for a given airspeed at a given angle of attack.

The first real consequence of poor lift at low speeds is a high stall speed. The second consequence of poor lift at low speeds is the manner in which lift and drag vary with speed in the lower ranges. As a jet airplane is slowed toward its minimum drag speed (V_{MD} or L/D_{MAX}), total drag increases at a much greater rate than lift, resulting in a sinking flightpath. If the pilot attempts to increase lift by increasing pitch attitude, airspeed will be further reduced resulting in a further increase in drag and sink rate as the airplane slides up the back side of the power curve. The sink rate can be arrested in one of two ways:

- Pitch attitude can be substantially reduced to reduce the angle of attack and allow the airplane to accelerate to a speed above V_{MD}, where steady flight conditions can be reestablished. This procedure, however, will invariably result in a substantial loss of altitude.

- Thrust can be increased to accelerate the airplane to a speed above V_{MD} to reestablish steady flight conditions. It should be remembered that the amount of thrust required will be quite large. The amount of thrust must be sufficient to accelerate the airplane and regain altitude lost. Also, if the airplane has slid a long way up the back side of the power required (drag) curve, drag will be very high and a very large amount of thrust will be required.

In a typical piston engine airplane, V_{MD} in the clean configuration is normally at a speed of about 1.3 V_S. [Figure 15-12] Flight below V_{MD} on a piston engine airplane is well identified and predictable. In contrast, in a jet airplane flight in the area of V_{MD} (typically 1.5 – 1.6 V_S) does not normally produce any noticeable changes in flying qualities other than a lack of speed stability—a condition where a decrease in speed leads to an increase in drag which leads to a further decrease in speed and hence a speed divergence. A pilot who is not cognizant of a developing speed divergence may find a serious sink rate developing at a constant power setting, and a pitch attitude that appears to be normal. The fact that drag increases more rapidly than lift, causing a sinking flightpath, is one of the most important aspects of jet airplane flying qualities.

STALLS

The stalling characteristics of the sweptwing jet airplane can vary considerably from those of the

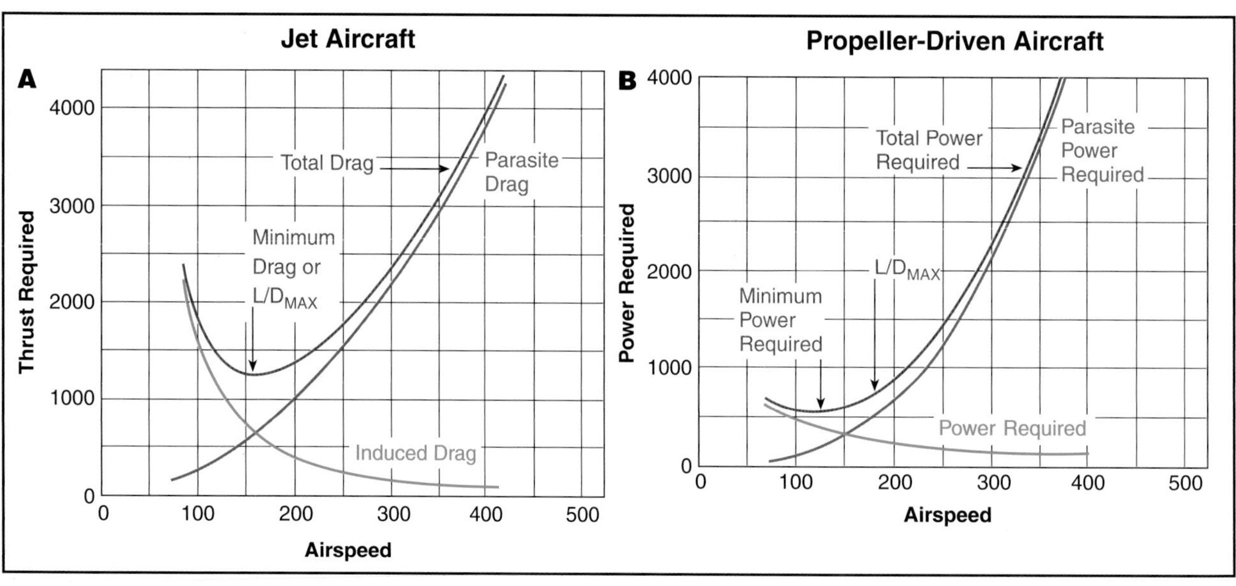

Figure 15-12. Thrust and power required curves.

normal straight wing airplane. The greatest difference that will be noticeable to the pilot is the lift developed vs. angle of attack. An increase in angle of attack of the straight wing produces a substantial and constantly increasing lift vector up to its maximum coefficient of lift, and soon thereafter flow separation (stall) occurs with a rapid deterioration of lift.

By contrast, the sweptwing produces a much more gradual buildup of lift with no well defined maximum coefficient and has the ability to fly well beyond this maximum buildup even though lift is lost. The drag curves (which are not depicted in figure 15-13) are approximately the reverse of the lift curves shown, in that a rapid increase in drag component may be expected with an increase in the angle of attack of a sweptwing airplane.

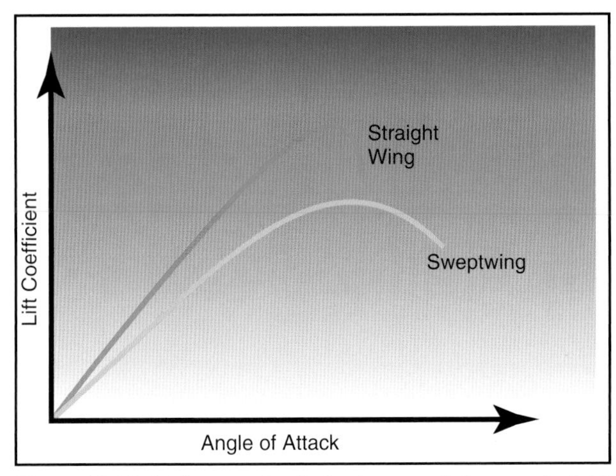

Figure 15-13. Stall vs. angle of attack—sweptwing vs. straight wing.

The differences in the stall characteristics between a conventional straight wing/low tailplane (non T-tail) airplane and a sweptwing T-tail airplane center around two main areas.

- The basic pitching tendency of the airplane at the stall.
- Tail effectiveness in stall recovery.

On a conventional straight wing/low tailplane airplane, the weight of the airplane acts downwards forward of the lift acting upwards, producing a need for a balancing force acting downwards from the tailplane. As speed is reduced by gentle up elevator deflection, the static stability of the airplane causes a nosedown tendency. This is countered by further up elevator to keep the nose coming up and the speed decreasing. As the pitch attitude increases, the low set tail is immersed in the wing wake, which is slightly turbulent, low energy air. The accompanying aerodynamic buffeting serves as a warning of impending stall. The reduced effectiveness of the tail prevents the pilot from forcing the airplane into a deeper stall. [Figure 15-14] The conventional straight wing airplane conforms to the familiar nosedown pitching tendency at the stall and gives the entire airplane a fairly pronounced nosedown pitch. At the moment of stall, the wing wake passes more or less straight rearward and passes above the tail. The tail is now immersed in high energy air where it experiences a sharp increase in positive angle of attack causing upward lift. This lift then assists the nosedown pitch and decrease in wing angle of attack essential to stall recovery.

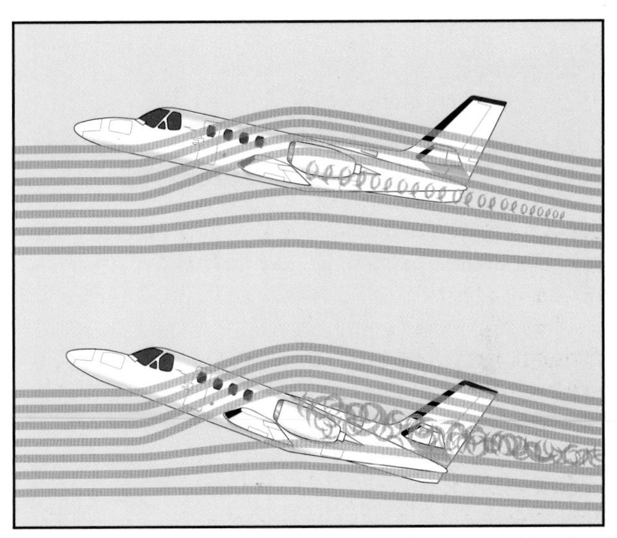

Figure 15-14. Stall progression—typical straight wing airplane.

In a sweptwing jet with a T-tail and rear fuselage mounted engines, the two qualities that are different from its straight wing low tailplane counterpart are the pitching tendency of the airplane as the stall develops and the loss of tail effectiveness at the stall. The handling qualities down to the stall are much the same as the straight wing airplane except that the high, T-tail remains clear of the wing wake and provides little or no warning in the form of a pre-stall buffet. Also, the tail is fully effective during the speed reduction towards the stall, and remains effective even after the wing has begun to stall. This enables the pilot to drive the wing into a deeper stall at a much greater angle of attack.

At the stall, two distinct things happen. After the stall, the sweptwing T-tail airplane tends to pitch up rather than down, and the T-tail is immersed in the wing wake, which is low energy turbulent air. This greatly reduces tail effectiveness and the airplane's ability to counter the noseup pitch. Also, the disturbed, relatively slow air behind the wing may sweep across the tail at such a large angle that the tail itself stalls. If this occurs, the pilot loses all pitch control and will be unable to lower the nose. The pitch up just after the stall is worsened by large reduction in lift and a large increase in drag, which causes a rapidly increasing

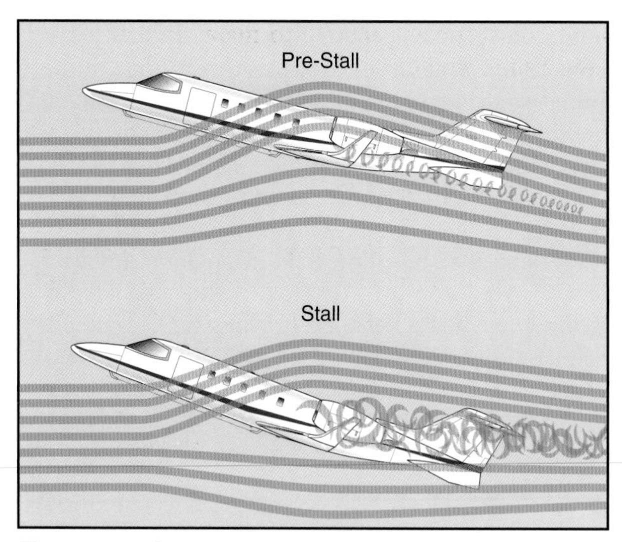

Figure 15-15. Stall progression sweptwing airplane.

descent path, thus compounding the rate of increase of the wing's angle of attack. [Figure 15-15]

The pitch up tendency after the stall is a characteristic of a swept and/or tapered wings. With these types of wings, there is a tendency for the wing to develop a strong spanwise airflow towards the wingtip when the wing is at high angles of attack. This leads to a tendency for separation of airflow, and the subsequent stall, to occur at the wingtips first. [Figure 15-16] The tip first stall, results in a shift of the center of lift of the wing in a forward direction relative to the center of gravity of the airplane, causing the nose to pitch up. Another disadvantage of a tip first stall is that it can involve the ailerons and erode roll control.

As previously stated, when flying at a speed in the area of V_{MD}, an increase in angle of attack causes drag to increase faster than lift and the airplane begins to sink. It is essential to understand that this increasing sinking tendency, at a constant pitch attitude, results in a rapid increase in angle of attack as the flightpath becomes deflected downwards. [Figure 15-17] Furthermore, once the stall has developed and a large amount of lift has been lost, the airplane will begin to sink rapidly and this will be accompanied by a corresponding rapid increase in angle of attack. This is the beginning of what is termed a **deep stall**.

As an airplane enters a deep stall, increasing drag reduces forward speed to well below normal stall speed. The sink rate may increase to many thousands of feet per minute. The airplane eventually stabilizes in a vertical descent. The angle of attack may approach

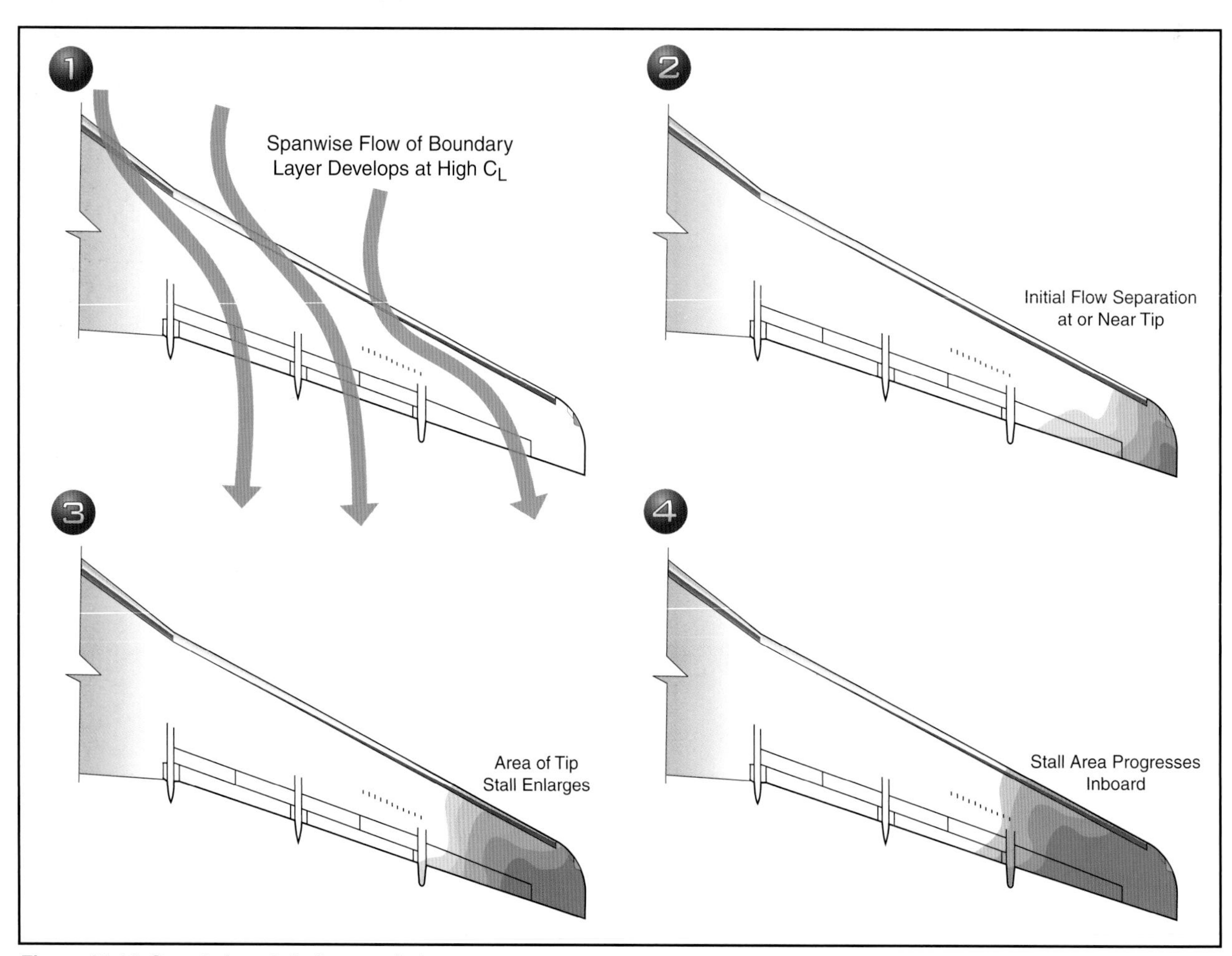

Figure 15-16. Sweptwing stall characteristics.

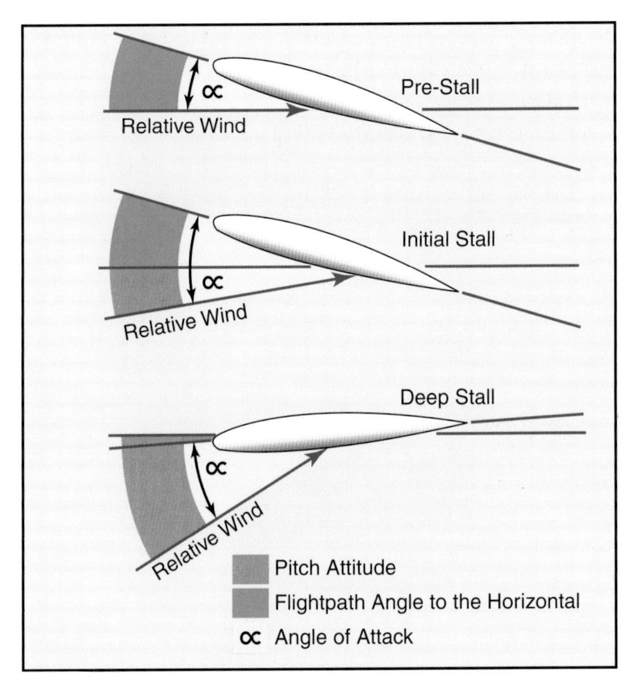

Figure 15-17. Deep stall progression.

90° and the indicated airspeed may be reduced to zero. At a 90° angle of attack, none of the airplane's control surfaces are effective. It must be emphasized that this situation can occur without an excessively nose-high pitch attitude. On some airplanes, it can occur at an apparently normal pitch attitude, and it is this quality that can mislead the pilot because it appears similar to the beginning of a normal stall recovery.

Deep stalls are virtually unrecoverable. Fortunately, they are easily avoided as long as published limitations are observed. On those airplanes susceptible to deep stalls (not all swept and/or tapered wing airplanes are), sophisticated stall warning systems such as stick shakers and stick pushers are standard equipment. A stick pusher, as its name implies, acts to automatically reduce the airplane's angle of attack before the airplane reaches a fully stalled condition.

Unless the Airplane Flight Manual procedures stipulate otherwise, a fully stalled condition in a jet airplane is to be avoided. Pilots undergoing training in jet airplanes are taught to recover at the first sign of an impending stall. Normally, this is indicated by aural stall warning devices and/or activation of the airplane's stick shaker. Stick shakers normally activate around 107 percent of the actual stall speed. At such slow speeds, very high sink rates can develop if the airplane's pitch attitude is decreased below the horizon, as is normal recovery procedure in most piston powered straight wing, light airplanes. Therefore, at the lower altitudes where plenty of engine thrust is available, the recovery technique in many sweptwing jets involves applying full available power, rolling the wings level, and holding a slightly positive pitch attitude. The amount of pitch attitude should be sufficient enough to maintain altitude or begin a slight climb.

At high altitudes, where there may be little excess thrust available to effect a recovery using power alone, it may be necessary to lower the nose below the horizon in order to accelerate away from an impending stall. This procedure may require several thousand feet or more of altitude loss to effect a recovery. Stall recovery techniques may vary considerably from airplane to airplane. The stall recovery procedures for a particular make and model airplane, as recommended by the manufacturer, are contained in the FAA-approved Airplane Flight Manual for that airplane.

Drag Devices

To the pilot transitioning into jet airplanes, going faster is seldom a problem. It is getting the airplane to slow down that seems to cause the most difficulty. This is because of the extremely clean aerodynamic design and fast momentum of the jet airplane, and also because the jet lacks the propeller drag effects that the pilot has been accustomed to. Additionally, even with the power reduced to flight idle, the jet engine still produces thrust, and deceleration of the jet airplane is a slow process. Jet airplanes have a glide performance that is double that of piston powered airplanes, and jet pilots often cannot comply with an air traffic control request to go down and slow down at the same time. Therefore, jet airplanes are equipped with drag devices such as spoilers and speed brakes.

The primary purpose of spoilers is to spoil lift. The most common type of spoiler consists of one or more rectangular plates that lie flush with the upper surface of each wing. They are installed approximately parallel to the lateral axis of the airplane and are hinged along the leading edges. When deployed, spoilers deflect up against the relative wind, which interferes with the flow of air about the wing. [Figure 15-18] This both spoils lift and increases drag. Spoilers are usually installed forward of the flaps but not in front of the ailerons so as not to interfere with roll control.

Figure 15-18. Spoilers.

Deploying spoilers results in a substantial sink rate with little decay in airspeed. Some airplanes will exhibit a noseup pitch tendency when the spoilers are deployed, which the pilot must anticipate.

When spoilers are deployed on landing, most of the wing's lift is destroyed. This action transfers the airplane's weight to the landing gear so that the wheel brakes are more effective. Another beneficial effect of deploying spoilers on landing is that they create considerable drag, adding to the overall aerodynamic braking. The real value of spoilers on landing, however, is creating the best circumstances for using wheel brakes.

The primary purpose of speed brakes is to produce drag. Speed brakes are found in many sizes, shapes, and locations on different airplanes, but they all have the same purpose—to assist in rapid deceleration. The speed brake consists of a hydraulically operated board that when deployed extends into the airstream. Deploying speed brakes results in a rapid decrease in airspeed. Typically, speed brakes can be deployed at any time during flight in order to help control airspeed, but they are most often used only when a rapid deceleration must be accomplished to slow down to landing gear and flap speeds. There is usually a certain amount of noise and buffeting associated with the use of speed brakes, along with an obvious penalty in fuel consumption. Procedures for the use of spoilers and/or speed brakes in various situations are contained in the FAA-approved Airplane Flight Manual for the particular airplane.

THRUST REVERSERS

Jet airplanes have high kinetic energy during the landing roll because of weight and speed. This energy is difficult to dissipate because a jet airplane has low drag with the nosewheel on the ground and the engines continue to produce forward thrust with the power levers at idle. While wheel brakes normally can cope, there is an obvious need for another speed retarding method. This need is satisfied by the drag provided by reverse thrust.

A **thrust reverser** is a device fitted in the engine exhaust system which effectively reverses the flow of the exhaust gases. The flow does not reverse through 180°; however, the final path of the exhaust gases is about 45° from straight ahead. This, together with the losses in the reverse flow paths, results in a net efficiency of about 50 percent. It will produce even less if the engine r.p.m. is less than maximum in reverse.

Normally, a jet engine will have one of two types of thrust reversers, either a **target reverser** or a **cascade reverser**. [Figure 15-19] Target reversers are simple clamshell doors that swivel from the stowed position at the engine tailpipe to block all of the outflow and redirect some component of the thrust forward.

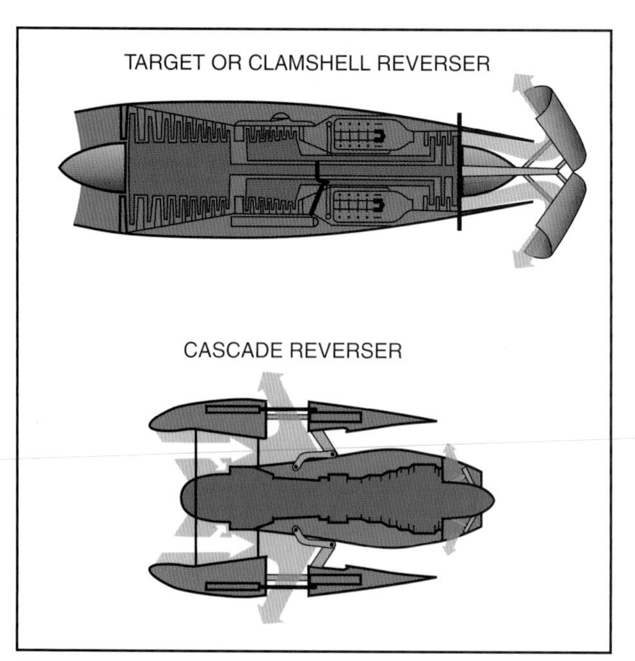

Figure 15-19. Thrust reversers.

Cascade reversers are more complex. They are normally found on turbofan engines and are often designed to reverse only the fan air portion. Blocking doors in the shroud obstructs forward fan thrust and redirects it through cascade vanes for some reverse component. Cascades are generally less effective than target reversers, particularly those that reverse only fan air, because they do not affect the engine core, which will continue to produce forward thrust.

On most installations, reverse thrust is obtained with the thrust lever at idle, by pulling up the reverse lever to a detent. Doing so positions the reversing mechanisms for operation but leaves the engine at idle r.p.m. Further upward and backward movement of the reverse lever increases engine power. Reverse is cancelled by closing the reverse lever to the idle reverse position, then dropping it fully back to the forward idle position. This last movement operates the reverser back to the forward thrust position.

Reverse thrust is much more effective at high airplane speed than at low airplane speeds, for two reasons: first, the net amount of reverse thrust increases with speed; second, the power produced is higher at higher speeds because of the increased rate of doing work. In other words, the kinetic energy of the airplane is being destroyed at a higher rate at the higher speeds. To get maximum efficiency from reverse thrust, therefore, it should be used as soon as is prudent after touchdown.

When considering the proper time to apply reverse thrust after touchdown, the pilot should remember that some airplanes tend to pitch noseup when reverse is selected on landing and this effect, particularly when combined with the noseup pitch effect from the spoilers, can cause the airplane to leave the ground again momentarily. On these types, the airplane must be firmly on the ground with the nosewheel down, before reverse is selected. Other types of airplanes have no change in pitch, and reverse idle may be selected after the main gear is down and before the nosewheel is down. Specific procedures for reverse thrust operation for a particular airplane/engine combination are contained in the FAA-approved Airplane Flight Manual for that airplane.

There is a significant difference between reverse pitch on a propeller and reverse thrust on a jet. Idle reverse on a propeller produces about 60 percent of the reverse thrust available at full power reverse and is therefore very effective at this setting when full reverse is not needed. On a jet engine, however, selecting idle reverse produces very little actual reverse thrust. In a jet airplane, the pilot must not only select reverse as soon as reasonable, but then must open up to full power reverse as soon as possible. Within Airplane Flight Manual limitations, full power reverse should be held until the pilot is certain the landing roll will be contained within the distance available.

Inadvertent deployment of thrust reversers is a very serious emergency situation. Therefore, thrust reverser systems are designed with this prospect in mind. The systems normally contain several lock systems: one to keep reversers from operating in the air, another to prevent operation with the thrust levers out of the idle detent, and/or an "auto-stow" circuit to command reverser stowage any time unwanted motion is detected. It is essential that pilots understand not only the normal procedures and limitations of thrust reverser use, but also the procedures for coping with uncommanded reverse. Those emergencies demand immediate and accurate response.

PILOT SENSATIONS IN JET FLYING

There are usually three general sensations that the pilot transitioning into jets will immediately become aware of. These are: inertial response differences, increased control sensitivity, and a much increased tempo of flight.

The varying of power settings from flight idle to full takeoff power has a much slower effect on the change of airspeed in the jet airplane. This is commonly called lead and lag, and is as much a result of the extremely clean aerodynamic design of the airplane as it is the slower response of the engine.

The lack of propeller effect is also responsible for the lower drag increment at the reduced power settings and results in other changes that the pilot will have to become accustomed to. These include the lack of effective slipstream over the lifting surfaces and control surfaces, and lack of propeller torque effect.

The aft mounted engines will cause a different reaction to power application and may result in a slightly nose-down pitching tendency with the application of power. On the other hand, power reduction will not cause the nose of the airplane to drop to the same extent the pilot is used to in a propeller airplane. Although neither of these characteristics are radical enough to cause transitioning pilots much of a problem, they must be compensated for.

Power settings required to attain a given performance are almost impossible to memorize in the jets, and the pilot who feels the necessity for having an array of power settings for all occasions will initially feel at a loss. The only way to answer the question of "how much power is needed?" is by saying, "whatever is required to get the job done." The primary reason that power settings vary so much is because of the great changes in weight as fuel is consumed during the flight. Therefore, the pilot will have to learn to use power as needed to achieve the desired performance. In time the pilot will find that the only reference to power instruments will be that required to keep from exceeding limits of maximum power settings or to synchronize r.p.m.

Proper power management is one of the initial problem areas encountered by the pilot transitioning into jet airplanes. Although smooth power applications are still the rule, the pilot will be aware that a greater physical movement of the power levers is required as compared to throttle movement in the piston engines. The pilot will also have to learn to anticipate and lead the power changes more than in the past and must keep in mind that the last 30 percent of engine r.p.m. represents the majority of the engine thrust, and below that the application of power has very little effect. In slowing the airplane, power reduction must be made sooner because there is no longer any propeller drag and the pilot should anticipate the need for drag devices.

Control sensitivity will differ between various airplanes, but in all cases, the pilot will find that they are more sensitive to any change in control displacement, particularly pitch control, than are the conventional propeller airplanes. Because of the higher speeds flown, the control surfaces are more effective and a variation of just a few degrees in pitch attitude in a jet can result in over twice the rate of altitude change that would be experienced in a slower airplane. The sensitive pitch control in jet airplanes is one of the first flight differences that the pilot will notice. Invariably the pilot will have a tendency to over-control pitch during initial training flights. The importance of accurate and smooth control cannot be overemphasized, however, and it is one of the first techniques the transitioning pilot must master.

The pilot of a sweptwing jet airplane will soon become adjusted to the fact that it is necessary and normal to fly at higher angles of attack. It is not unusual to have about 5° of noseup pitch on an approach to a landing. During an approach to a stall at constant altitude, the noseup angle may be as high as 15° to 20°. The higher deck angles (pitch angle relative to the ground) on takeoff, which may be as high as 15°, will also take some getting used to, although this is not the actual angle of attack relative to the airflow over the wing.

The greater variation of pitch attitudes flown in a jet airplane are a result of the greater thrust available and the flight characteristics of the low aspect ratio and sweptwing. Flight at the higher pitch attitudes requires a greater reliance on the flight instruments for airplane control since there is not much in the way of a useful horizon or other outside reference to be seen. Because of the high rates of climb and descent, high airspeeds, high altitudes and variety of attitudes flown, the jet airplane can only be precisely flown by applying proficient instrument flight techniques. Proficiency in attitude instrument flying, therefore, is essential to successful transition to jet airplane flying.

Most jet airplanes are equipped with a thumb operated pitch trim button on the control wheel which the pilot must become familiar with as soon as possible. The jet airplane will differ regarding pitch tendencies with the lowering of flaps, landing gear, and drag devices. With experience, the jet airplane pilot will learn to anticipate the amount of pitch change required for a particular operation. The usual method of operating the trim button is to apply several small, intermittent applications of trim in the direction desired rather than holding the trim button for longer periods of time which can lead to over-controlling.

JET AIRPLANE TAKEOFF AND CLIMB

All FAA certificated jet airplanes are certificated under Title 14 of the Code of Federal Regulations (14 CFR) part 25, which contains the airworthiness standards for transport category airplanes. The FAA certificated jet airplane is a highly sophisticated machine with proven levels of performance and guaranteed safety margins. The jet airplane's performance and safety margins can only be realized, however, if the airplane is operated in strict compliance with the procedures and limitations contained in the FAA-approved Airplane Flight Manual for the particular airplane.

The following information is generic in nature and, since most civilian jet airplanes require a minimum flight crew of two pilots, assumes a two pilot crew. If any of the following information conflicts with FAA-approved Airplane Flight Manual procedures for a particular airplane, the Airplane Flight Manual procedures take precedence. Also, if any of the following procedures differ from the FAA-approved procedures developed for use by a specific air operator and/or for use in an FAA-approved training center or pilot school curriculum, the FAA-approved procedures for that operator and/or training center/pilot school take precedence.

V-SPEEDS

The following are speeds that will affect the jet airplane's takeoff performance. The jet airplane pilot must be thoroughly familiar with each of these speeds and how they are used in the planning of the takeoff.

- **V_S**—Stall speed.
- **V_1**—Critical engine failure speed or decision speed. Engine failure below this speed should result in an aborted takeoff; above this speed the takeoff run should be continued.
- **V_R**—Speed at which the rotation of the airplane is initiated to takeoff attitude. This speed cannot be less than V_1 or less than 1.05 x V_{MCA} (minimum control speed in the air). On a single-engine takeoff, it must also allow for the acceleration to V_2 at the 35-foot height at the end of the runway.
- **V_{LO}**—The speed at which the airplane first becomes airborne. This is an engineering term used when the airplane is certificated and must meet certain requirements. If it is not listed in the Airplane Flight Manual, it is within requirements and does not have to be taken into consideration by the pilot.
- **V_2**—The takeoff safety speed which must be attained at the 35-foot height at the end of the required runway distance. This is essentially the best single-engine angle of climb speed for the airplane and should be held until clearing obstacles after takeoff, or at least 400 feet above the ground.

PRE-TAKEOFF PROCEDURES

Takeoff data, including V_1/V_R and V_2 speeds, takeoff power settings, and required field length should be computed prior to each takeoff and recorded on a takeoff data card. These data will be based on airplane weight, runway length available, runway gradient, field temperature, field barometric pressure, wind, icing conditions, and runway condition. Both pilots should separately compute the takeoff data and cross-check in the cockpit with the takeoff data card.

CAPTAIN'S BRIEFING

I will advance the thrust levers.

Follow me through on the thrust levers.

Monitor all instruments and warning lights on the takeoff roll and call out any discrepancies or malfunctions observed prior to V_1, and I will abort the takeoff. Stand by to arm thrust reversers on my command.

Give me a visual and oral signal for the following:
- 80 knots, and I will disengage nosewheel steering.
- $V_{1,}$ and I will move my hand from thrust to yoke.
- $V_{R,}$ and I will rotate.

In the event of engine failure at or after V_1, I will continue the takeoff roll to V_R, rotate and establish V_2 climb speed. I will identify the inoperative engine, and we will both verify. I will accomplish the shutdown, or have you do it on my command.

I will expect you to stand by on the appropriate emergency checklist.

I will give you a visual and oral signal for gear retraction and for power settings after the takeoff.

Our VFR emergency procedure is to............................
Our IFR emergency procedure is to.............................

Figure 15-20. Sample captain's briefing.

A captain's briefing is an essential part of cockpit resource management (CRM) procedures and should be accomplished just prior to takeoff. [Figure 15-20] The captain's briefing is an opportunity to review crew coordination procedures for takeoff, which is always the most critical portion of a flight.

The takeoff and climb-out should be accomplished in accordance with a standard takeoff and departure profile developed for the particular make and model airplane. [Figure 15-21]

TAKEOFF ROLL

The entire runway length should be available for takeoff, especially if the pre-calculated takeoff performance shows the airplane to be limited by runway length or obstacles. After taxing into position at the end of the runway, the airplane should be aligned in the center of the runway allowing equal distance on either side. The brakes should be held while the thrust levers are brought to a power setting beyond the bleed valve range (normally the vertical position) and the engines allowed to stabilized. The engine instruments should be checked for proper operation before the brakes are released or the power increased further. This procedure assures symmetrical thrust during the takeoff roll and aids in preventing overshooting the desired takeoff thrust setting. The brakes should then be released and, during the start of the takeoff roll, the thrust levers smoothly advanced to the pre-computed takeoff power setting. All final takeoff thrust adjustments should be made prior to reaching 60 knots. The final engine power adjustments are normally made by the pilot not flying. Once the thrust levers are set for takeoff power, they should not be readjusted after 60 knots. Retarding a thrust lever would only be necessary in case an engine exceeds any limitation such as ITT, fan, or turbine r.p.m.

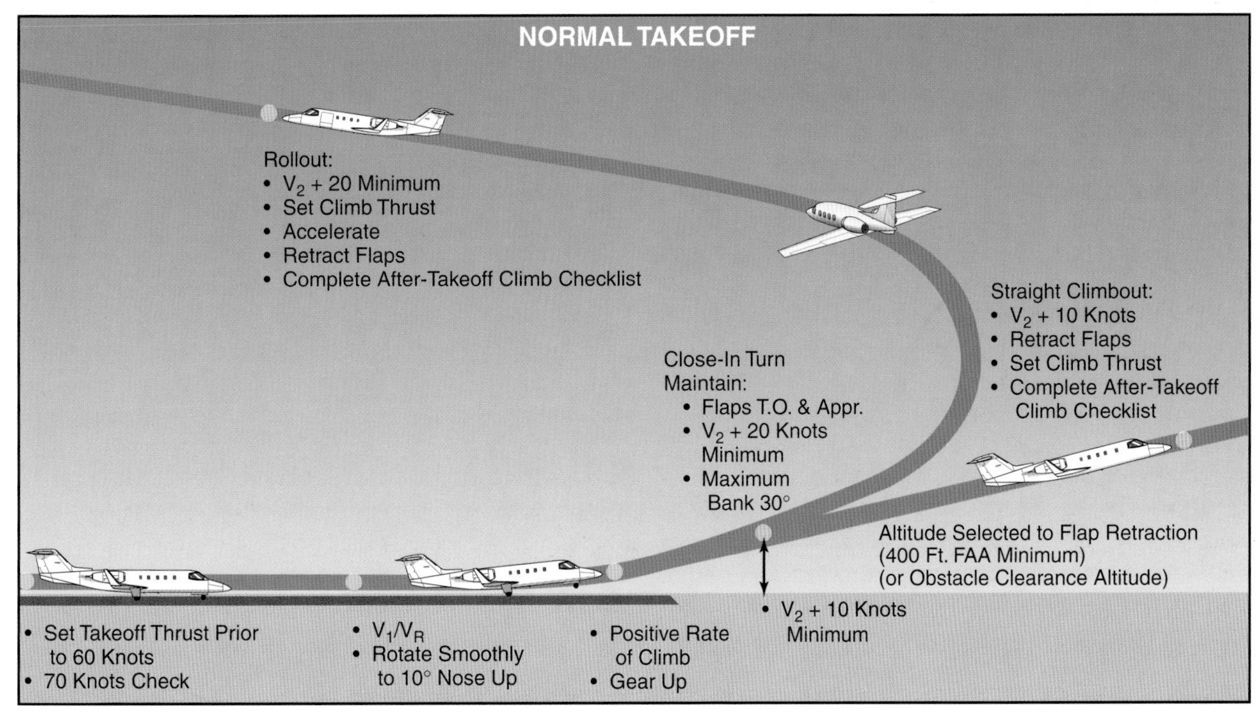

Figure 15-21. Takeoff and departure profile.

If sufficient runway length is available, a "rolling" takeoff may be made without stopping at the end of the runway. Using this procedure, as the airplane rolls onto the runway, the thrust levers should be smoothly advanced to the vertical position and the engines allowed to stabilize, and then proceed as in the static takeoff outlined above. Rolling takeoffs can also be made from the end of the runway by advancing the thrust levers from idle as the brakes are released.

During the takeoff roll, the pilot flying should concentrate on directional control of the airplane. This is made somewhat easier because there is no torque-produced yawing in a jet as there is in a propeller driven airplane. The airplane must be maintained exactly on centerline with the wings level. This will automatically aid the pilot when contending with an engine failure. If a crosswind exists, the wings should be kept level by displacing the control wheel into the crosswind. During the takeoff roll, the primary responsibility of the pilot not flying is to closely monitor the aircraft systems and to call out the proper V speeds as directed in the captain's briefing.

Slight forward pressure should be held on the control column to keep the nosewheel rolling firmly on the runway. If nosewheel steering is being utilized, the pilot flying should monitor the nosewheel steering to about 80 knots (or V_{MCG} for the particular airplane) while the pilot not flying applies the forward pressure. After reaching V_{MCG}, the pilot flying should bring his/her left hand up to the control wheel. The pilot's other hand should be on the thrust levers until at least V_1 speed is attained. Although the pilot not flying maintains a check on the engine instruments throughout the takeoff roll, the pilot flying (pilot in command) makes the decision to continue or reject a takeoff for any reason. A decision to reject a takeoff will require immediate retarding of thrust levers.

The pilot not flying should call out V_1. After passing V_1 speed on the takeoff roll, it is no longer mandatory for the pilot flying to keep a hand on the thrust levers. The point for abort has passed, and both hands may be placed on the control wheel. As the airspeed approaches V_R, the control column should be moved to a neutral position. As the pre-computed V_R speed is attained, the pilot not flying should make the appropriate callout and the pilot flying should smoothly rotate the airplane to the appropriate takeoff pitch attitude.

ROTATION AND LIFT-OFF

Rotation and lift-off in a jet airplane should be considered a maneuver unto itself. It requires planning, precision, and a fine control touch. The objective is to initiate the rotation to takeoff pitch attitude exactly at V_R so that the airplane will accelerate through V_{LOF} and attain V_2 speed at 35 feet at the end of the runway. Rotation to the proper takeoff attitude too soon may extend the takeoff roll or cause an early lift-off, which will result in a lower rate of climb, and the predicted flightpath will not be followed. A late rotation, on the other hand, will result in a longer takeoff roll, exceeding V_2 speed, and a takeoff and climb path below the predicted path.

Each airplane has its own specific takeoff pitch attitude which remains constant regardless of weight. The takeoff pitch attitude in a jet airplane is normally between 10° and 15° nose up. The rotation to takeoff pitch attitude should be made smoothly but deliberately, and at a constant rate. Depending on the particular airplane, the pilot should plan on a rate of pitch attitude increase of approximately 2.5° to 3° per second.

In training it is common for the pilot to overshoot V_R and then overshoot V_2 because the pilot not flying will call for rotation at, or just past V_R. The reaction of the pilot flying is to visually verify V_R and then rotate. The airplane then leaves the ground at or above V_2. The excess airspeed may be of little concern on a normal takeoff, but a delayed rotation can be critical when runway length or obstacle clearance is limited. It should be remembered that on some airplanes, the all-engine takeoff can be more limiting than the engine out takeoff in terms of obstacle clearance in the initial part of the climb-out. This is because of the rapidly increasing airspeed causing the achieved flightpath to fall below the engine out scheduled flightpath unless care is taken to fly the correct speeds. The transitioning pilot should remember that rotation at the right speed and rate to the right attitude will get the airplane off the ground at the right speed and within the right distance.

INITIAL CLIMB

Once the proper pitch attitude is attained, it must be maintained. The initial climb after lift-off is done at this constant pitch attitude. Takeoff power is maintained and the airspeed allowed to accelerate. Landing gear retraction should be accomplished after a positive rate of climb has been established and confirmed. Remember that in some airplanes gear retraction may temporarily increase the airplane drag while landing gear doors open. Premature gear retraction may cause the airplane to settle back towards the runway surface. Remember also that because of ground effect, the vertical speed indicator and the altimeter may not show a positive climb until the airplane is 35 to 50 feet above the runway.

The climb pitch attitude should continue to be held and the airplane allowed to accelerate to flap retraction speed. However, the flaps should not be retracted until obstruction clearance altitude or 400 feet AGL has been passed. Ground effect and landing gear drag reduction results in rapid acceleration during this phase of the takeoff and climb. Airspeed, altitude, climb rate, attitude, and heading must be monitored carefully. When the airplane settles down to a steady climb, longitudinal stick forces can be trimmed out. If a turn must be made during this phase of flight, no more than 15° to 20° of bank should be used. Because of spiral instability, and because at this point an accurate trim state on rudder and ailerons has not yet been achieved, the bank angle should be carefully monitored throughout the turn. If a power reduction must be made, pitch attitude should be reduced simultaneously and the airplane monitored carefully so as to preclude entry into an inadvertent descent. When the airplane has attained a steady climb at the appropriate en route climb speed, it can be trimmed about all axes and the autopilot engaged.

JET AIRPLANE APPROACH AND LANDING

LANDING REQUIREMENTS

The FAA landing field length requirements for jet airplanes are specified in 14 CFR part 25. It defines the minimum field length (and therefore minimum margins) that can be scheduled. The regulation describes the landing profile as the distance required from a point 50 feet above the runway threshold, through the flare to touchdown, and then stopping using the maximum stopping capability on a dry runway surface. The actual demonstrated distance is increased by 67 percent and published in the FAA-approved Airplane Flight Manual as the FAR dry runway landing distance. [Figure 15-22] For wet runways, the FAR dry runway distance is increased by an additional 15 percent. Thus the minimum dry runway field length will be 1.67 times the actual minimum air and ground distance needed and the wet runway minimum landing field length will be 1.92 times the minimum dry air and ground distance needed.

Certified landing field length requirements are computed for the stop made with speed brakes deployed and maximum wheel braking. Reverse thrust is not used in establishing the certified FAR landing distances. However, reversers should definitely be used in service.

LANDING SPEEDS

As in the takeoff planning, there are certain speeds that must be taken into consideration during any landing in a jet airplane. The speeds are as follows.

- V_{SO}—Stall speed in the landing configuration.
- V_{REF}—1.3 times the stall speed in the landing configuration.
- **Approach climb—**The speed which guarantees adequate performance in a go-around situation with an inoperative engine. The airplane's weight must be limited so that a twin-engine airplane will have a 2.1 percent climb gradient capability. (The approach climb gradient requirements for 3 and 4 engine airplanes are 2.4 percent and 2.7 percent respectively.) These criteria are based on an airplane configured with approach flaps, landing gear up, and takeoff thrust available from the operative engine(s).
- **Landing climb—**The speed which guarantees adequate performance in arresting the descent and making a go-around from the final stages of landing with the airplane in the full landing configuration and maximum takeoff power available on all engines.

The appropriate speeds should be pre-computed prior to every landing, and posted where they are visible to both pilots. The V_{REF} speed, or threshold speed, is used

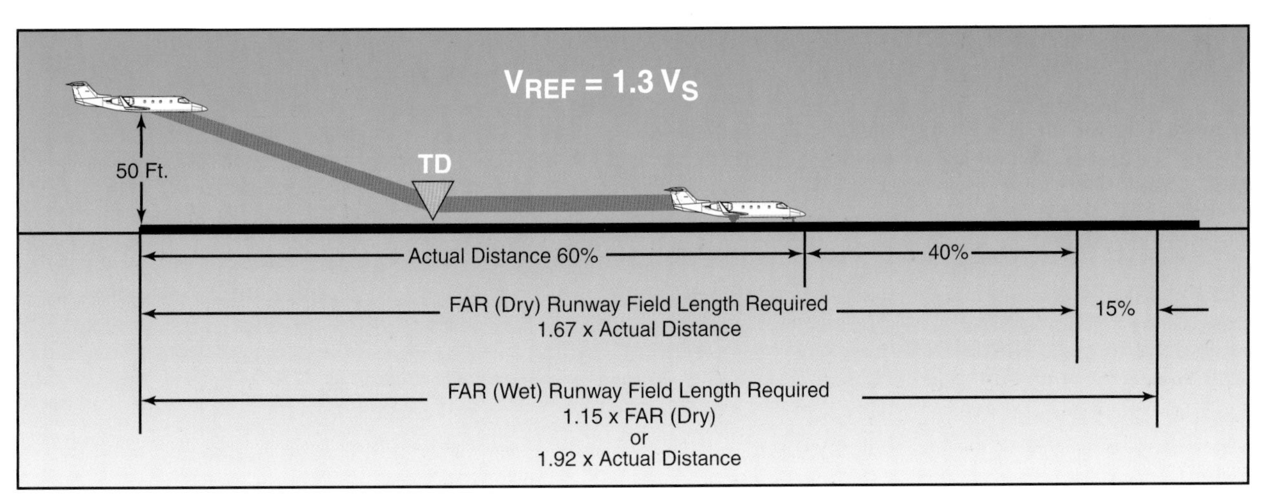

Figure 15-22. FAR landing field length required.

as a reference speed throughout the traffic pattern. For example:

Downwind leg—V_{REF} plus 20 knots.

Base leg—V_{REF} plus 10 knots.

Final approach—V_{REF} plus 5 knots.

50 feet over threshold—V_{REF}.

The approach and landing sequence in a jet airplane should be accomplished in accordance with an approach and landing profile developed for the particular airplane. [Figure 15-23]

SIGNIFICANT DIFFERENCES

A safe approach in any type of airplane culminates in a particular position, speed, and height over the runway threshold. That final flight condition is the target window at which the entire approach aims. Propeller powered airplanes are able to approach that target from wider angles, greater speed differentials, and a larger variety of glidepath angles. Jet airplanes are not as responsive to power and course corrections, so the final approach must be more stable, more deliberate, more constant, in order to reach the window accurately.

The transitioning pilot must understand that, in spite of their impressive performance capabilities, there are six ways in which a jet airplane is worse than a piston engine airplane in making an approach and in correcting errors on the approach.

- *The absence of the propeller slipstream in producing immediate extra lift at constant airspeed.* There is no such thing as salvaging a misjudged glidepath with a sudden burst of immediately available power. Added lift can only be achieved by accelerating the airframe. Not only must the pilot wait for added power but even when the engines do respond, added lift will only be available when the airframe has responded with speed.

- *The absence of the propeller slipstream in significantly lowering the power-on stall speed.* There is virtually no difference between power-on and power-off stall speed. It is not possible in a jet airplane to jam the thrust levers forward to avoid a stall.

- *Poor acceleration response in a jet engine from low r.p.m.* This characteristic requires that the approach be flown in a high drag/high power

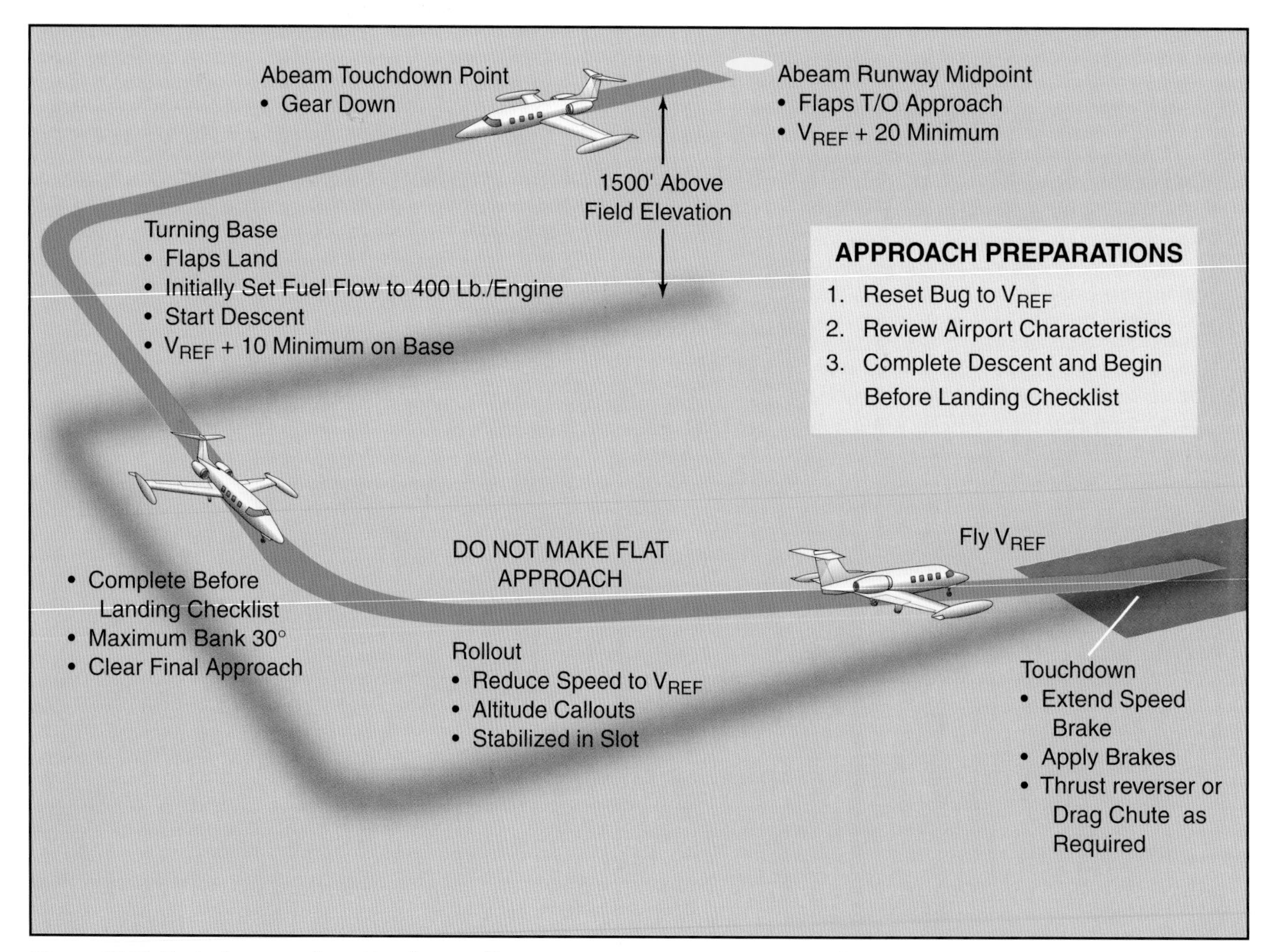

Figure 15-23. Typical approach and landing profile.

configuration so that sufficient power will be available quickly if needed.

- *The increased momentum of the jet airplane making sudden changes in the flightpath impossible.* Jet airplanes are consistently heavier than comparable sized propeller airplanes. The jet airplane, therefore, will require more indicated airspeed during the final approach due to a wing design that is optimized for higher speeds. These two factors combine to produce higher momentum for the jet airplane. Since force is required to overcome momentum for speed changes or course corrections, the jet will be far less responsive than the propeller airplane and require careful planning and stable conditions throughout the approach.

- *The lack of good speed stability being an inducement to a low speed condition.* The **drag curve** for many jet airplanes is much flatter than for propeller airplanes, so speed changes do not produce nearly as much drag change. Further, jet thrust remains nearly constant with small speed changes. The result is far less speed stability. When the speed does increase or decrease, there is little tendency for the jet airplane to re-acquire the original speed. The pilot, therefore, must remain alert to the necessity of making speed adjustments, and then make them aggressively in order to remain on speed.

- *Drag increasing faster than lift producing a high sink rate at low speeds.* Jet airplane wings typically have a large increase in drag in the approach configuration. When a sink rate does develop, the only immediate remedy is to increase pitch attitude (angle of attack). Because drag increases faster than lift, that pitch change will rapidly contribute to an even greater sink rate unless a significant amount of power is aggressively applied.

These flying characteristics of jet airplanes make a stabilized approach an absolute necessity.

THE STABILIZED APPROACH

The performance charts and the limitations contained in the FAA-approved Airplane Flight Manual are predicated on momentum values that result from programmed speeds and weights. Runway length limitations assume an exact 50-foot threshold height at an exact speed of 1.3 times V_{SO}. That "window" is critical and is a prime reason for the stabilized approach. Performance figures also assume that once through the target threshold window, the airplane will touch down in a target touchdown zone approximately 1,000 feet down the runway, after which maximum stopping capability will be used.

There are five basic elements to the stabilized approach.

- The airplane should be in the landing configuration early in the approach. The landing gear should be down, landing flaps selected, trim set, and fuel balanced. Ensuring that these tasks are completed will help keep the number of variables to a minimum during the final approach.

- The airplane should be on profile before descending below 1,000 feet. Configuration, trim, speed, and glidepath should be at or near the optimum parameters early in the approach to avoid distractions and conflicts as the airplane nears the threshold window. An optimum glidepath angle of 2.5° to 3° should be established and maintained.

- Indicated airspeed should be within 10 knots of the target airspeed. There are strong relationships between trim, speed, and power in most jet airplanes and it is important to stabilize the speed in order to minimize those other variables.

- The optimum descent rate should be 500 to 700 feet per minute. **The descent rate should not be allowed to exceed 1,000 feet per minute at any time during the approach.**

- The engine speed should be at an r.p.m. that allows best response when and if a rapid power increase is needed.

Every approach should be evaluated at 500 feet. In a typical jet airplane, this is approximately 1 minute from touchdown. If the approach is not stabilized at that height, a go-around should be initiated. (See figure 15-24 on the next page.)

APPROACH SPEED

On final approach, the airspeed is controlled with power. Any speed diversion from V_{REF} on final approach must be detected immediately and corrected. With experience the pilot will be able to detect the very first tendency of an increasing or decreasing airspeed trend, which normally can be corrected with a small adjustment in thrust. The pilot must be attentive to poor speed stability leading to a low speed condition with its attendant risk of high drag increasing the sink rate. Remember that with an increasing sink rate an apparently normal pitch attitude is no guarantee of a normal angle of attack value. If an increasing sink rate is detected, it must be countered by increasing the angle

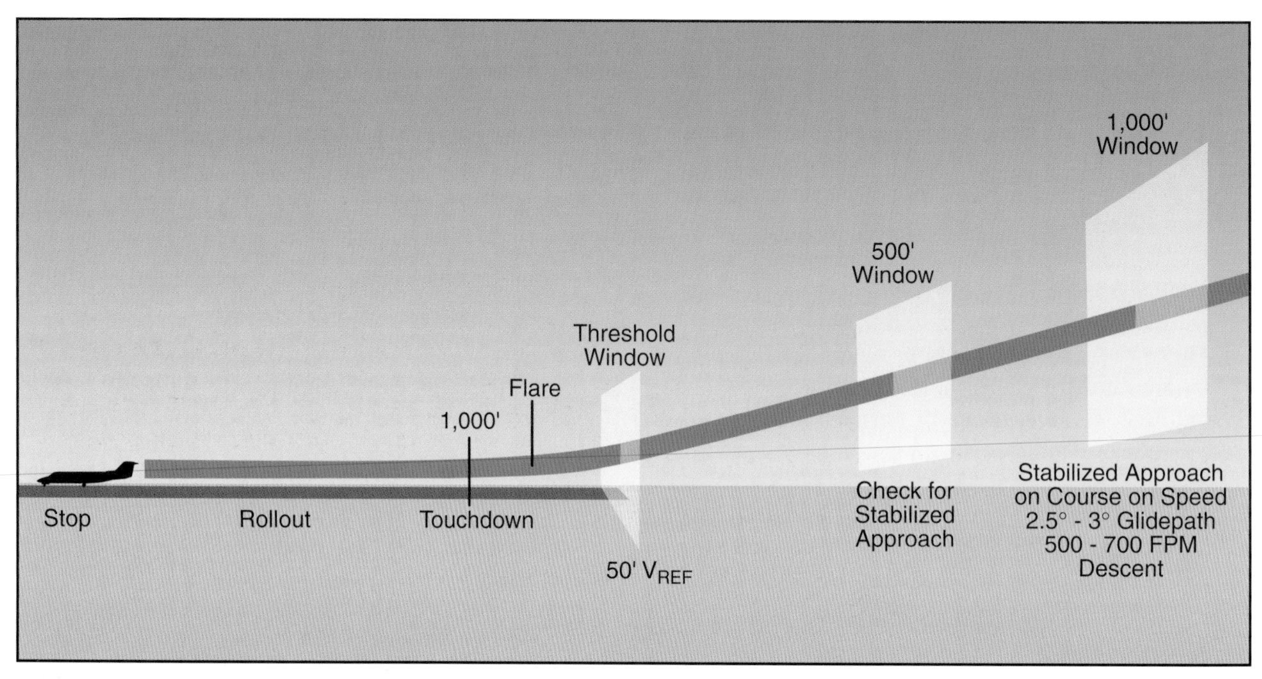

Figure 15-24. Stabilized approach.

of attack and simultaneously increasing thrust to counter the extra drag. The degree of correction required will depend on how much the sink rate needs to be reduced. For small amounts, smooth and gentle, almost anticipatory corrections will be sufficient. For large sink rates, drastic corrective measures may be required that, even if successful, would destabilize the approach.

A common error in the performance of approaches in jet airplanes is excess approach speed. Excess approach speed carried through the threshold window and onto the runway will increase the minimum stopping distance required by 20 – 30 feet per knot of excess speed for a dry runway and 40 – 50 feet for a wet runway. Worse yet, the excess speed will increase the chances of an extended flare, which will increase the distance to touchdown by approximately 250 feet for each excess knot in speed.

Proper speed control on final approach is of primary importance. The pilot must anticipate the need for speed adjustment so that only small adjustments are required. It is essential that the airplane arrive at the approach threshold window exactly on speed.

GLIDEPATH CONTROL

On final approach, at a constant airspeed, the glidepath angle and rate of descent is controlled with pitch attitude and elevator. The optimum glidepath angle is 2.5° to 3° whether or not an electronic glidepath reference is being used. On visual approaches, pilots may have a tendency to make flat approaches. A flat approach, however, will increase landing distance and should be avoided. For example, an approach angle of 2° instead of a recommended 3° will add 500 feet to landing distance.

A more common error is excessive height over the threshold. This could be the result of an unstable approach, or a stable but high approach. It also may occur during an instrument approach where the missed approach point is close to or at the runway threshold. Regardless of the cause, excessive height over the threshold will most likely result in a touchdown beyond the normal aiming point. An extra 50 feet of height over the threshold will add approximately 1,000 feet to the landing distance. It is essential that the airplane arrive at the approach threshold window exactly on altitude (50 feet above the runway).

THE FLARE

The flare reduces the approach rate of descent to a more acceptable rate for touchdown. Unlike light airplanes, a jet airplane should be flown onto the runway rather than "held off" the surface as speed dissipates. A jet airplane is aerodynamically clean even in the landing configuration, and its engines still produce residual thrust at idle r.p.m. Holding it off during the flare in a attempt to make a smooth landing will greatly increase landing distance. A *firm* landing is normal and desirable. A firm landing does not mean a hard landing, but rather a deliberate or *positive* landing.

For most airports, the airplane will pass over the end of the runway with the landing gear 30 – 45 feet above the surface, depending on the landing flap setting and the location of the touchdown zone. It will take 5 – 7 seconds from the time the airplane passes the end of the runway until touchdown. The flare is initiated by increasing the pitch attitude just enough to reduce the sink rate to 100 – 200 feet per minute when the landing gear is approximately 15 feet above the runway surface. In most jet airplanes, this will require a pitch attitude increase of only 1° to 3°. The thrust is smoothly reduced to idle as the flare progresses.

The normal speed bleed off during the time between passing the end of the runway and touchdown is 5 knots. Most of the decrease occurs during the flare when thrust is reduced. If the flare is extended (held off) while an additional speed is bled off, hundreds or even thousands of feet of runway may be used up. [Figure 15-25] The extended flare will also result in additional pitch attitude which may lead to a tail strike. **It is, therefore, essential to fly the airplane onto the runway at the target touchdown point, even if the speed is excessive.** A deliberate touchdown should be planned and practiced on every flight. A positive touchdown will help prevent an extended flare.

Pilots must learn the flare characteristics of each model of airplane they fly. The visual reference cues observed from each cockpit are different because window geometry and visibility are different. The geometric relationship between the pilot's eye and the landing gear will be different for each make and model. It is essential that the flare maneuver be initiated at the proper height—not too high and not too low.

Beginning the flare too high or reducing the thrust too early may result in the airplane floating beyond the target touchdown point or may include a rapid pitch up as the pilot attempts to prevent a high sink rate touchdown. This can lead to a tail strike. The flare that is initiated too late may result in a hard touchdown.

Proper thrust management through the flare is also important. In many jet airplanes, the engines produce a noticeable effect on pitch trim when the thrust setting is changed. A rapid change in the thrust setting requires a quick elevator response. If the thrust levers are moved to idle too quickly during the flare, the pilot must make rapid changes in pitch control. If the thrust levers are moved more slowly, the elevator input can be more easily coordinated.

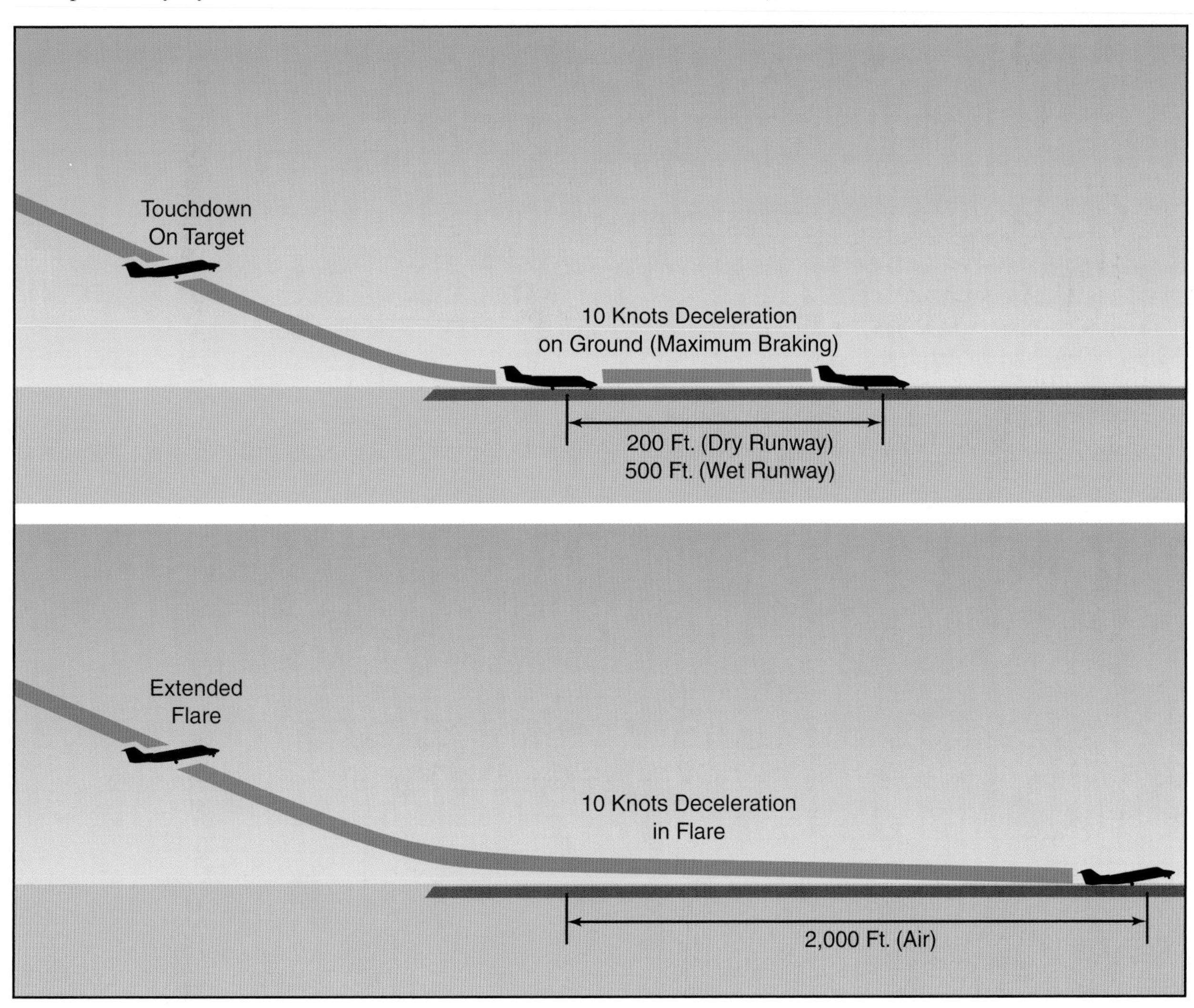

Figure 15-25. Extended flare.

TOUCHDOWN AND ROLLOUT

A proper approach and flare positions the airplane to touch down in the touchdown target zone, which is usually about 1,000 feet beyond the runway threshold. Once the main wheels have contacted the runway, the pilot must maintain directional control and initiate the stopping process. The stop must be made on the runway that remains in front of the airplane. The runway distance available to stop is longest if the touchdown was on target. The energy to be dissipated is least if there is no excess speed. The stop that begins with a touchdown that is *on the numbers* will be the easiest stop to make for any set of conditions.

At the point of touchdown, the airplane represents a very large mass that is moving at a relatively high speed. The large total energy must be dissipated by the brakes, the aerodynamic drag, and the thrust reversers. The nosewheel should be flown onto the ground immediately after touchdown because a jet airplane decelerates poorly when held in a nose-high attitude. Placing the nosewheel tire(s) on the ground will assist in maintaining directional control. Also, lowering the nose gear decreases the wing angle of attack, decreasing the lift, placing more load onto the tires, thereby increasing tire-to-ground friction. Landing distance charts for jet airplanes assume that the nosewheel is lowered onto the runway within 4 seconds of touchdown.

There are only three forces available for stopping the airplane. They are wheel braking, reverse thrust, and aerodynamic braking. Of the three, the brakes are most effective and therefore the most important stopping force for most landings. When the runway is very slippery, reverse thrust and drag may be the dominant forces. Both reverse thrust and aerodynamic drag are most effective at high speeds. Neither is affected by runway surface condition. Brakes, on the other hand, are most effective at low speed. The landing rollout distance will depend on the touchdown speed and what forces are applied and when they are applied. The pilot controls the *what and when factors*, but the maximum braking force may be limited by tire-to-ground friction.

The pilot should begin braking as soon after touchdown and wheel spin-up as possible, and to smoothly continue the braking until stopped or a safe taxi speed is reached. However, caution should be used if the airplane is not equipped with a functioning anti-skid system. In such a case, heavy braking can cause the wheels to lock and the tires to skid.

Both directional control and braking utilize tire ground friction. They share the maximum friction force the tires can provide. Increasing either will subtract from the other. Understanding tire ground friction, how runway contamination affects it, and how to use the friction available to maximum advantage is important to a jet pilot.

Spoilers should be deployed immediately after touchdown because they are most effective at high speed. Timely deployment of spoilers will increase drag by 50 to 60 percent, but more importantly, they spoil much of the lift the wing is creating, thereby causing more of the weight of the airplane to be loaded onto the wheels. The spoilers increase wheel loading by as much as 200 percent in the landing flap configuration. This increases the tire ground friction force making the maximum tire braking and cornering forces available.

Like spoilers, thrust reversers are most effective at high speeds and should be deployed quickly after touchdown. However, the pilot should not command significant reverse thrust until the nosewheel is on the ground. Otherwise, the reversers might deploy asymmetrically resulting in an uncontrollable yaw towards the side on which the most reverse thrust is being developed, in which case the pilot will need whatever nosewheel steering is available to maintain directional control.

EMERGENCY SITUATIONS

This chapter contains information on dealing with non-normal and emergency situations that may occur in flight. The key to successful management of an emergency situation, and/or preventing a non-normal situation from progressing into a true emergency, is a thorough familiarity with, and adherence to, the procedures developed by the airplane manufacturer and contained in the FAA-approved Airplane Flight Manual and/or Pilot's Operating Handbook (AFM/POH). The following guidelines are generic and are **not** meant to replace the airplane manufacturer's recommended procedures. Rather, they are meant to enhance the pilot's general knowledge in the area of non-normal and emergency operations. If any of the guidance in this chapter conflicts in any way with the manufacturer's recommended procedures for a particular make and model airplane, **the manufacturer's recommended procedures take precedence.**

EMERGENCY LANDINGS

This section contains information on emergency landing techniques in small fixed-wing airplanes. The guidelines that are presented apply to the more adverse terrain conditions for which no practical training is possible. The objective is to instill in the pilot the knowledge that almost any terrain can be considered "suitable" for a survivable crash landing if the pilot knows how to use the airplane structure for self-protection and the protection of passengers.

TYPES OF EMERGENCY LANDINGS

The different types of emergency landings are defined as follows.

- **Forced landing.** An immediate landing, on or off an airport, necessitated by the inability to continue further flight. A typical example of which is an airplane forced down by engine failure.
- **Precautionary landing.** A premeditated landing, on or off an airport, when further flight is possible but inadvisable. Examples of conditions that may call for a precautionary landing include deteriorating weather, being lost, fuel shortage, and gradually developing engine trouble.
- **Ditching.** A forced or precautionary landing on water.

A precautionary landing, generally, is less hazardous than a forced landing because the pilot has more time for terrain selection and the planning of the approach. In addition, the pilot can use power to compensate for errors in judgment or technique. The pilot should be aware that too many situations calling for a precautionary landing are allowed to develop into immediate forced landings, when the pilot uses wishful thinking instead of reason, especially when dealing with a self-inflicted predicament. The non-instrument rated pilot trapped by weather, or the pilot facing imminent fuel exhaustion who does not give any thought to the feasibility of a precautionary landing accepts an extremely hazardous alternative.

PSYCHOLOGICAL HAZARDS

There are several factors that may interfere with a pilot's ability to act promptly and properly when faced with an emergency.

- **Reluctance to accept the emergency situation.** A pilot who allows the mind to become paralyzed at the thought that the airplane will be on the ground, in a very short time, regardless of the pilot's actions or hopes, is severely handicapped in the handling of the emergency. An unconscious desire to delay the dreaded moment may lead to such errors as: failure to lower the nose to maintain flying speed, delay in the selection of the most suitable landing area within reach, and indecision in general. Desperate attempts to correct whatever went wrong, at the expense of airplane control, fall into the same category.
- **Desire to save the airplane.** The pilot who has been conditioned during training to expect to find a relatively safe landing area, whenever the flight instructor closed the throttle for a simulated forced landing, may ignore all basic rules of airmanship to avoid a touchdown in terrain where airplane damage is unavoidable. Typical consequences are: making a 180° turn back to the runway when available altitude is insufficient; stretching the glide without regard for minimum control speed in order to reach a more appealing field; accepting an approach and touchdown situation that leaves no margin for error. The desire to save the airplane, regardless of the risks involved, may be influenced by two other factors: the pilot's financial stake in the airplane and the

certainty that an undamaged airplane implies no bodily harm. There are times, however, when a pilot should be more interested in sacrificing the airplane so that the occupants can safely walk away from it.

- **Undue concern about getting hurt.** Fear is a vital part of the self-preservation mechanism. However, when fear leads to panic, we invite that which we want most to avoid. The survival records favor pilots who maintain their composure and know how to apply the general concepts and procedures that have been developed through the years. The success of an emergency landing is as much a matter of the mind as of skills.

Figure 16-1. Using vegetation to absorb energy.

BASIC SAFETY CONCEPTS

GENERAL

A pilot who is faced with an emergency landing in terrain that makes extensive airplane damage inevitable should keep in mind that the avoidance of crash injuries is largely a matter of: (1) keeping vital structure (cockpit/cabin area) relatively intact by using dispensable structure (such as wings, landing gear, and fuselage bottom) to absorb the violence of the stopping process before it affects the occupants, (2) avoiding forceful bodily contact with interior structure.

The advantage of sacrificing dispensable structure is demonstrated daily on the highways. A head-on car impact against a tree at 20 miles per hour (m.p.h.) is less hazardous for a properly restrained driver than a similar impact against the driver's door. Accident experience shows that the extent of crushable structure between the occupants and the principal point of impact on the airplane has a direct bearing on the severity of the transmitted crash forces and, therefore, on survivability.

Avoiding forcible contact with interior structure is a matter of seat and body security. Unless the occupant decelerates at the same rate as the surrounding structure, no benefit will be realized from its relative intactness. The occupant will be brought to a stop violently in the form of a secondary collision.

Dispensable airplane structure is not the only available energy absorbing medium in an emergency situation. Vegetation, trees, and even manmade structures may be used for this purpose. Cultivated fields with dense crops, such as mature corn and grain, are almost as effective in bringing an airplane to a stop with repairable damage as an emergency arresting device on a runway. [Figure 16-1] Brush and small trees provide considerable cushioning and braking effect without destroying the airplane. When dealing with natural and manmade obstacles with greater strength than the dispensable airplane structure, the pilot must plan the touchdown in such a manner that only nonessential structure is "used up" in the principal slowing down process.

The overall severity of a deceleration process is governed by speed (groundspeed) and stopping distance. The most critical of these is speed; doubling the groundspeed means quadrupling the total destructive energy, and vice versa. Even a small change in groundspeed at touchdown—be it as a result of wind or pilot technique—will affect the outcome of a controlled crash. It is important that the actual touchdown during an emergency landing be made at the lowest possible *controllable* airspeed, using all available aerodynamic devices.

Most pilots will instinctively—and correctly—look for the largest available flat and open field for an emergency landing. Actually, very little stopping distance is required if the speed can be dissipated uniformly; that is, if the deceleration forces can be spread evenly over the available distance. This concept is designed into the arresting gear of aircraft carriers that provides a nearly constant stopping force from the moment of hookup.

The typical light airplane is designed to provide protection in crash landings that expose the occupants to nine times the acceleration of gravity (9 G) in a forward direction. Assuming a uniform 9 G deceleration, at 50 m.p.h. the required stopping distance is about 9.4 feet. While at 100 m.p.h. the stopping distance is about 37.6 feet—about four times as great. [Figure 16-2] Although these figures are based on an ideal deceleration process, it is interesting to note what can be accomplished in an effectively used short stopping distance. Understanding the need for a firm but uniform deceleration process in very poor terrain enables the pilot to select touchdown conditions that will spread the breakup of dispensable structure over a short distance, thereby reducing the peak deceleration of the cockpit/cabin area.

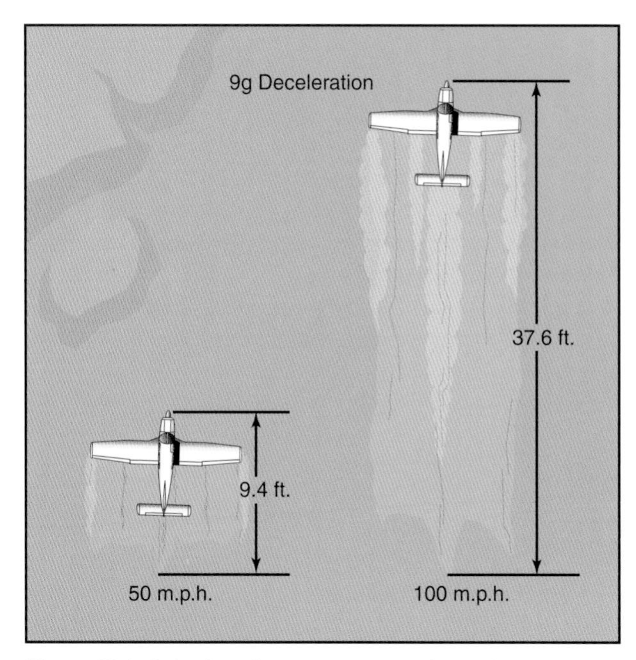

Figure 16-2. Stopping distance vs. groundspeed.

ATTITUDE AND SINK RATE CONTROL

The most critical and often the most inexcusable error that can be made in the planning and execution of an emergency landing, even in ideal terrain, is the loss of initiative over the airplane's attitude and sink rate at touchdown. When the touchdown is made on flat, open terrain, an excessive nose-low pitch attitude brings the risk of "sticking" the nose in the ground. Steep bank angles just before touchdown should also be avoided, as they increase the stalling speed and the likelihood of a wingtip strike.

Since the airplane's vertical component of velocity will be immediately reduced to zero upon ground contact, it must be kept well under control. A flat touchdown at a high sink rate (well in excess of 500 feet per minute (f.p.m.)) on a hard surface can be injurious without destroying the cockpit/cabin structure, especially during gear up landings in low-wing airplanes. A rigid bottom construction of these airplanes may preclude adequate cushioning by structural deformation. Similar impact conditions may cause structural collapse of the overhead structure in high-wing airplanes. On soft terrain, an excessive sink rate may cause digging in of the lower nose structure and severe forward deceleration.

TERRAIN SELECTION

A pilot's choice of emergency landing sites is governed by:

- The route selected during preflight planning.
- The height above the ground when the emergency occurs.
- Excess airspeed (excess airspeed can be converted into distance and/or altitude).

The only time the pilot has a very limited choice is during the low and slow portion of the takeoff. However, even under these conditions, the ability to change the impact heading only a few degrees may ensure a survivable crash.

If beyond gliding distance of a suitable open area, the pilot should judge the available terrain for its energy absorbing capability. If the emergency starts at a considerable height above the ground, the pilot should be more concerned about first selecting the desired general area than a specific spot. Terrain appearances from altitude can be very misleading and considerable altitude may be lost before the best spot can be pinpointed. For this reason, the pilot should not hesitate to discard the original plan for one that is obviously better. However, as a general rule, the pilot should not change his or her mind more than once; a well-executed crash landing in poor terrain can be less hazardous than an uncontrolled touchdown on an established field.

AIRPLANE CONFIGURATION

Since flaps improve maneuverability at slow speed, and lower the stalling speed, their use during final approach is recommended when time and circumstances permit. However, the associated increase in drag and decrease in gliding distance call for caution in the timing and the extent of their application; premature use of flap, and dissipation of altitude, may jeopardize an otherwise sound plan.

A hard and fast rule concerning the position of a retractable landing gear at touchdown cannot be given. In rugged terrain and trees, or during impacts at high sink rate, an extended gear would definitely have a protective effect on the cockpit/cabin area. However, this advantage has to be weighed against the possible side effects of a collapsing gear, such as a ruptured fuel tank. As always, the manufacturer's recommendations as outlined in the AFM/POH should be followed.

When a normal touchdown is assured, and ample stopping distance is available, a gear up landing on level, but soft terrain, or across a plowed field, may result in less airplane damage than a gear down landing. [Figure 16-3]

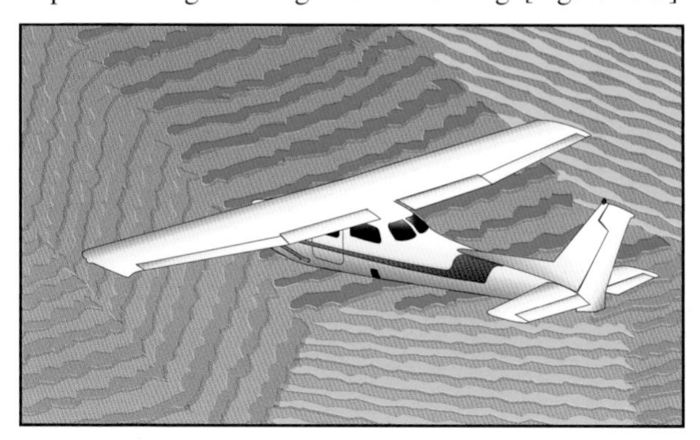
Figure 16-3. Intentional gear up landing.

Deactivation of the airplane's electrical system before touchdown reduces the likelihood of a post-crash fire. However, the battery master switch should not be turned off until the pilot no longer has any need for electrical power to operate vital airplane systems. Positive airplane control during the final part of the approach has priority over all other considerations, including airplane configuration and cockpit checks. The pilot should attempt to exploit the power available from an irregularly running engine; however, it is generally better to switch the engine and fuel off just before touchdown. This not only ensures the pilot's initiative over the situation, but a cooled down engine reduces the fire hazard considerably.

APPROACH

When the pilot has time to maneuver, the planning of the approach should be governed by three factors.

- Wind direction and velocity.
- Dimensions and slope of the chosen field.
- Obstacles in the final approach path.

These three factors are seldom compatible. When compromises have to be made, the pilot should aim for a wind/obstacle/terrain combination that permits a final approach with some margin for error in judgment or technique. A pilot who overestimates the gliding range may be tempted to stretch the glide across obstacles in the approach path. For this reason, it is sometimes better to plan the approach over an unobstructed area, regardless of wind direction. Experience shows that a collision with obstacles at the end of a ground roll, or slide, is much less hazardous than striking an obstacle at flying speed before the touchdown point is reached.

TERRAIN TYPES

Since an emergency landing on suitable terrain resembles a situation in which the pilot should be familiar through training, only the more unusual situation will be discussed.

CONFINED AREAS

The natural preference to set the airplane down on the ground should not lead to the selection of an open spot between trees or obstacles where the ground cannot be reached without making a steep descent.

Once the intended touchdown point is reached, and the remaining open and unobstructed space is very limited, it may be better to force the airplane down on the ground than to delay touchdown until it stalls (settles). An airplane decelerates faster after it is on the ground than while airborne. Thought may also be given to the desirability of ground-looping or retracting the landing gear in certain conditions.

A river or creek can be an inviting alternative in otherwise rugged terrain. The pilot should ensure that the water or creek bed can be reached without snagging the wings. The same concept applies to road landings with one additional reason for caution; manmade obstacles on either side of a road may not be visible until the final portion of the approach.

When planning the approach across a road, it should be remembered that most highways, and even rural dirt roads, are paralleled by power or telephone lines. Only a sharp lookout for the supporting structures, or poles, may provide timely warning.

TREES (FOREST)

Although a tree landing is not an attractive prospect, the following general guidelines will help to make the experience survivable.

- Use the normal landing configuration (full flaps, gear down).
- Keep the groundspeed low by heading into the wind.
- Make contact at minimum indicated airspeed, but not below stall speed, and "hang" the airplane in the tree branches in a nose-high landing attitude. Involving the underside of the fuselage and both wings in the initial tree contact provides a more even and positive cushioning effect, while preventing penetration of the windshield. [Figure 16-4]
- Avoid direct contact of the fuselage with heavy tree trunks.
- Low, closely spaced trees with wide, dense crowns (branches) close to the ground are much better than tall trees with thin tops; the latter allow too much free fall height. (A free fall from 75 feet results in an impact speed of about 40 knots, or about 4,000 f.p.m.)
- Ideally, initial tree contact should be symmetrical; that is, both wings should meet equal resistance in the tree branches. This distribution of the load helps to maintain proper airplane attitude. It may also preclude the loss of one wing, which invariably leads to a more rapid and less predictable descent to the ground.
- If heavy tree trunk contact is unavoidable once the airplane is on the ground, it is best to involve both wings simultaneously by directing the airplane between two properly spaced trees. **Do not attempt this maneuver, however, while still airborne.**

WATER (DITCHING) AND SNOW

A well-executed water landing normally involves less deceleration violence than a poor tree landing or a

Figure 16-4. Tree landing.

touchdown on extremely rough terrain. Also an airplane that is ditched at minimum speed and in a normal landing attitude will not immediately sink upon touchdown. Intact wings and fuel tanks (especially when empty) provide floatation for at least several minutes even if the cockpit may be just below the water line in a high-wing airplane.

Loss of depth perception may occur when landing on a wide expanse of smooth water, with the risk of flying into the water or stalling in from excessive altitude. To avoid this hazard, the airplane should be "dragged in" when possible. Use no more than intermediate flaps on low-wing airplanes. The water resistance of fully extended flaps may result in asymmetrical flap failure and slowing of the airplane. Keep a retractable gear up unless the AFM/POH advises otherwise.

A landing in snow should be executed like a ditching, in the same configuration and with the same regard for loss of depth perception (white out) in reduced visibility and on wide open terrain.

ENGINE FAILURE AFTER TAKEOFF (SINGLE-ENGINE)

The altitude available is, in many ways, the controlling factor in the successful accomplishment of an emergency landing. If an actual engine failure should occur immediately after takeoff and before a safe maneuvering altitude is attained, it is usually inadvisable to attempt to turn back to the field from where the takeoff was made. Instead, it is safer to immediately establish the proper glide attitude, and select a field directly ahead or slightly to either side of the takeoff path.

The decision to continue straight ahead is often difficult to make unless the problems involved in attempting to turn back are seriously considered. In the first place, the takeoff was in all probability made into the wind. To get back to the takeoff field, a downwind turn must be made. This increases the groundspeed and rushes the pilot even more in the performance of procedures and in planning the landing approach. Secondly, the airplane will be losing considerable altitude during the turn and might still be in a bank when the ground is contacted, resulting in the airplane cartwheeling (which would be a catastrophe for the occupants, as well as the airplane). After turning downwind, the apparent increase in groundspeed could mislead the pilot into attempting to prematurely slow down the airplane and cause it to stall. On the other hand, continuing straight ahead or making a slight turn allows the pilot more time to establish a safe landing attitude, and the landing can be made as slowly as possible, but more importantly, the airplane can be landed while under control.

Concerning the subject of turning back to the runway following an engine failure on takeoff, the pilot should determine the minimum altitude an attempt of such a maneuver should be made in a particular airplane. Experimentation at a safe altitude should give the pilot an approximation of height lost in a descending 180° turn at idle power. By adding a safety factor of about 25 percent, the pilot should arrive at a practical decision height. The ability to make a 180° turn does not necessarily mean that the departure runway can be reached in a power-off glide; this depends on the wind, the distance traveled during the climb, the height reached, and the glide distance of the airplane without power. The pilot should also remember that a turn back to the departure runway may in fact require more than a 180° change in direction.

Consider the following example of an airplane which has taken off and climbed to an altitude of 300 feet AGL when the engine fails. [Figure 16-5 on next page]. After a typical 4 second reaction time, the pilot elects to turn back to the runway. Using a standard rate (3° change in direction per second) turn, it will take 1 minute to turn 180°. At a glide speed of 65 knots, the radius of the turn is 2,100 feet, so at the completion of the turn, the airplane will be 4,200 feet to one side of the runway. The pilot must turn another 45° to head the airplane toward the runway. By this time the total change in direction is 225° equating to 75 seconds plus the 4 second reaction time. If the airplane in a power-off glide descends at approximately 1,000 f.p.m., it

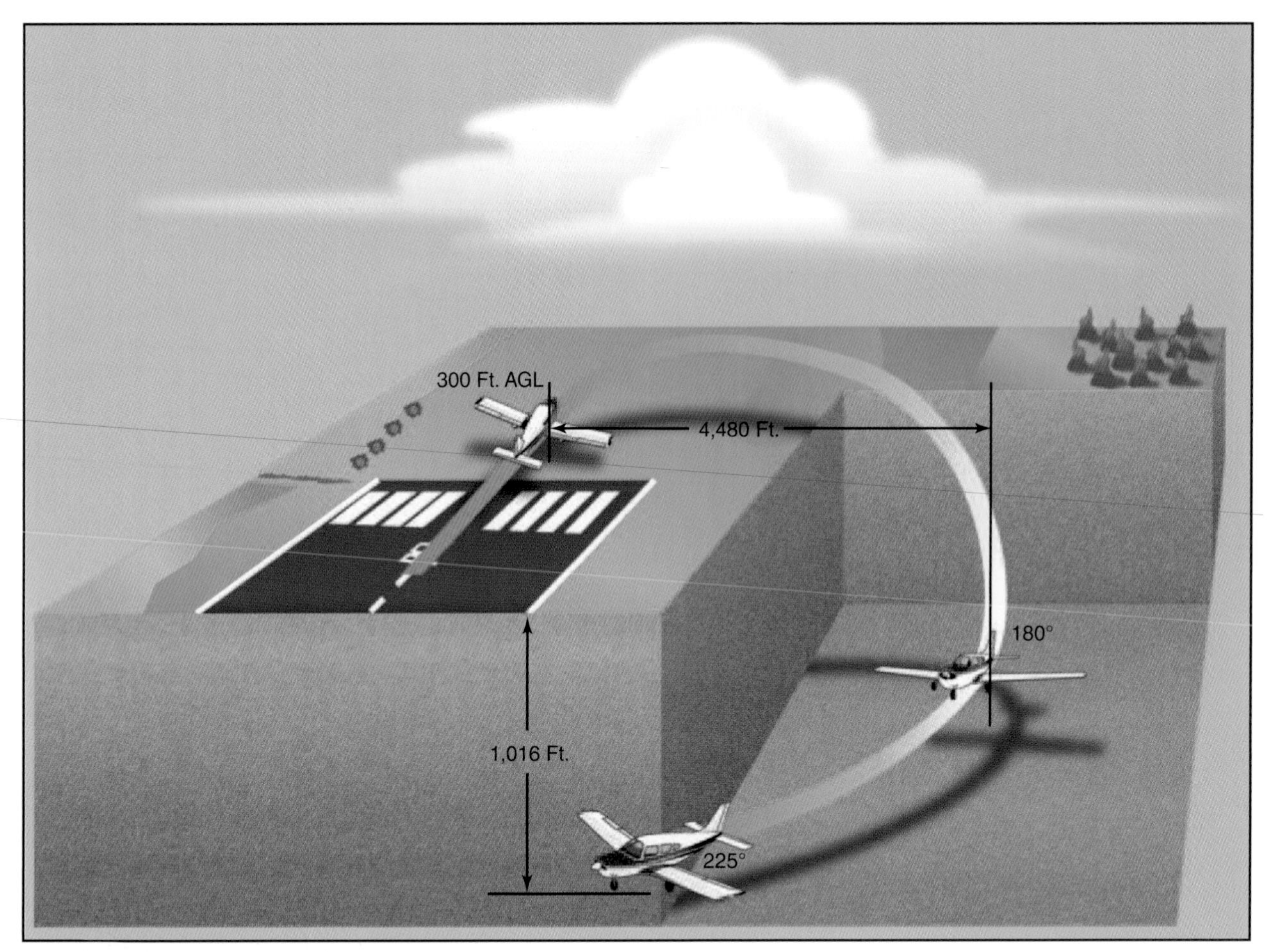

Figure 16-5. Turning back to the runway after engine failure.

will have descended 1,316, feet placing it 1,016 feet below the runway.

EMERGENCY DESCENTS

An emergency descent is a maneuver for descending as rapidly as possible to a lower altitude or to the ground for an emergency landing. [Figure 16-6] The need for this maneuver may result from an uncontrollable fire, a sudden loss of cabin pressurization, or any other situation demanding an immediate and rapid descent. The objective is to descend the airplane as soon and as rapidly as possible, within the structural limitations of the airplane. Simulated emergency descents should be made in a turn to check for other air traffic below and to look around for a possible emergency landing area. A radio call announcing descent intentions may be appropriate to alert other aircraft in the area. When initiating the descent, a bank of approximately 30 to 45° should be established to maintain positive load factors ("G" forces) on the airplane.

Emergency descent training should be performed as recommended by the manufacturer, including the configuration and airspeeds. Except when prohibited by the manufacturer, the power should be reduced to idle, and the propeller control (if equipped) should be placed in the low pitch (or high revolutions per minute (r.p.m.)) position. This will allow the propeller to act as an aerodynamic brake to help prevent an excessive airspeed buildup during the descent. The landing gear and flaps should be extended as recommended by the manufacturer. This will provide maximum drag so that the descent can be made as rapidly as possible, without excessive airspeed. The pilot should not allow the airplane's airspeed to pass the never-exceed speed (V_{NE}), the maximum landing gear extended speed (V_{LE}), or the maximum flap extended speed (V_{FE}), as applicable. In the case of an engine fire, a high airspeed descent could blow out the fire. However, the weakening of the airplane structure is a major concern and descent at low airspeed would place less stress on the airplane. If the descent is conducted in turbulent conditions, the pilot must also comply with the design maneuvering speed (V_A) limitations. The descent should be made at the maximum allowable airspeed consistent with the procedure used. This will provide increased drag and therefore the loss of altitude as quickly as possible. The recovery from an emergency descent should be initiated at a high enough altitude to ensure a safe recovery back to level flight or a precautionary landing.

When the descent is established and stabilized during training and practice, the descent should be terminated.

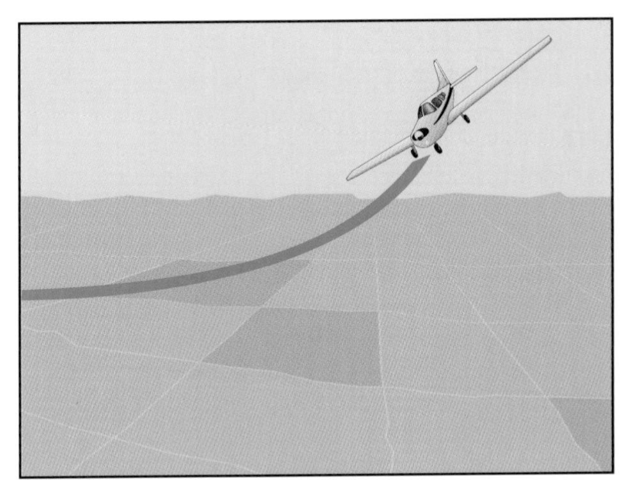

Figure 16-6. Emergency descent.

In airplanes with piston engines, prolonged practice of emergency descents should be avoided to prevent excessive cooling of the engine cylinders.

IN-FLIGHT FIRE

A fire in flight demands immediate and decisive action. The pilot therefore must be familiar with the procedures outlined to meet this emergency contained in the AFM/POH for the particular airplane. For the purposes of this handbook, in-flight fires are classified as: in-flight engine fires, electrical fires, and cabin fires.

ENGINE FIRE

An in-flight engine compartment fire is usually caused by a failure that allows a flammable substance such as fuel, oil or hydraulic fluid to come in contact with a hot surface. This may be caused by a mechanical failure of the engine itself, an engine-driven accessory, a defective induction or exhaust system, or a broken line. Engine compartment fires may also result from maintenance errors, such as improperly installed/fastened lines and/or fittings resulting in leaks.

Engine compartment fires can be indicated by smoke and/or flames coming from the engine cowling area. They can also be indicated by discoloration, bubbling, and/or melting of the engine cowling skin in cases where flames and/or smoke is not visible to the pilot. By the time a pilot becomes aware of an in-flight engine compartment fire, it usually is well developed. Unless the airplane manufacturer directs otherwise in the AFM/POH, the first step on discovering a fire should be to shut off the fuel supply to the engine by placing the mixture control in the idle cut off position and the fuel selector shutoff valve to the OFF position. The ignition switch should be left ON in order to use up the fuel that remains in the fuel lines and components between the fuel selector/shutoff valve and the engine. This procedure may starve the engine compartment of fuel and cause the fire to die naturally. If the flames are snuffed out, no attempt should be made to restart the engine.

If the engine compartment fire is oil-fed, as evidenced by thick black smoke, as opposed to a fuel-fed fire which produces bright orange flames, the pilot should consider stopping the propeller rotation by feathering or other means, such as (with constant-speed propellers) placing the pitch control lever to the minimum r.p.m. position and raising the nose to reduce airspeed until the propeller stops rotating. This procedure will stop an engine-driven oil (or hydraulic) pump from continuing to pump the flammable fluid which is feeding the fire.

Some light airplane emergency checklists direct the pilot to shut off the electrical master switch. However, the pilot should consider that unless the fire is electrical in nature, or a crash landing is imminent, deactivating the electrical system prevents the use of panel radios for transmitting distress messages and will also cause air traffic control (ATC) to lose transponder returns.

Pilots of powerless single-engine airplanes are left with no choice but to make a forced landing. Pilots of twin-engine airplanes *may* elect to continue the flight to the nearest airport. However, consideration must be given to the possibility that a wing could be seriously impaired and lead to structural failure. Even a brief but intense fire could cause dangerous structural damage. In some cases, the fire could continue to burn under the wing (or engine cowling in the case of a single-engine airplane) out of view of the pilot. Engine compartment fires which appear to have been extinguished have been known to rekindle with changes in airflow pattern and airspeed.

The pilot must be familiar with the airplane's emergency descent procedures. The pilot must bear in mind that:

- The airplane may be severely structurally damaged to the point that its ability to remain under control could be lost at any moment.
- The airplane may still be on fire and susceptible to explosion.
- The airplane is expendable and the only thing that matters is the safety of those on board.

ELECTRICAL FIRES

The initial indication of an electrical fire is usually the distinct odor of burning insulation. Once an electrical fire is detected, the pilot should attempt to identify the faulty circuit by checking circuit breakers, instruments, avionics, and lights. If the faulty circuit cannot be readily detected and isolated, and flight conditions permit, the battery master switch and alternator/generator switches should be turned off to remove the possible source of the fire. However, any materials which have been ignited may continue to burn.

If electrical power is absolutely essential for the flight, an attempt may be made to identify and isolate the faulty circuit by:

1. Turning the electrical master switch OFF.
2. Turning all individual electrical switches OFF.
3. Turning the master switch back ON.
4. Selecting electrical switches that were ON before the fire indication one at a time, permitting a short time lapse after each switch is turned on to check for signs of odor, smoke, or sparks.

This procedure, however, has the effect of recreating the original problem. The most prudent course of action is to land as soon as possible.

CABIN FIRE

Cabin fires generally result from one of three sources: (1) careless smoking on the part of the pilot and/or passengers; (2) electrical system malfunctions; (3) heating system malfunctions. A fire in the cabin presents the pilot with two immediate demands: attacking the fire, and getting the airplane safely on the ground as quickly as possible. A fire or smoke in the cabin should be controlled by identifying and shutting down the faulty system. In many cases, smoke may be removed from the cabin by opening the cabin air vents. This should be done only after the fire extinguisher (if available) is used. Then the cabin air control can be opened to purge the cabin of both smoke and fumes. If smoke increases in intensity when the cabin air vents are opened, they should be immediately closed. This indicates a possible fire in the heating system, nose compartment baggage area (if so equipped), or that the increase in airflow is feeding the fire.

On pressurized airplanes, the pressurization air system will remove smoke from the cabin; however, if the smoke is intense, it may be necessary to either depressurize at altitude, if oxygen is available for all occupants, or execute an emergency descent.

In unpressurized single-engine and light twin-engine airplanes, the pilot can attempt to expel the smoke from the cabin by opening the foul weather windows. These windows should be closed immediately if the fire becomes more intense. If the smoke is severe, the passengers and crew should use oxygen masks if available, and the pilot should initiate an immediate descent. The pilot should also be aware that on some airplanes, lowering the landing gear and/or wing flaps can aggravate a cabin smoke problem.

FLIGHT CONTROL MALFUNCTION/FAILURE

TOTAL FLAP FAILURE

The inability to extend the wing flaps will necessitate a no-flap approach and landing. In light airplanes a no-flap approach and landing is not particularly difficult or dangerous. However, there are certain factors which must be considered in the execution of this maneuver. A no-flap landing requires substantially more runway than normal. The increase in required landing distance could be as much as 50 percent.

When flying in the traffic pattern with the wing flaps retracted, the airplane must be flown in a relatively nose-high attitude to maintain altitude, as compared to flight with flaps extended. Losing altitude can be more of a problem without the benefit of the drag normally provided by flaps. A wider, longer traffic pattern may be required in order to avoid the necessity of diving to lose altitude and consequently building up excessive airspeed.

On final approach, a nose-high attitude can make it difficult to see the runway. This situation, if not anticipated, can result in serious errors in judgment of height and distance. Approaching the runway in a relatively nose-high attitude can also cause the perception that the airplane is close to a stall. This may cause the pilot to lower the nose abruptly and risk touching down on the nosewheel.

With the flaps retracted and the power reduced for landing, the airplane is slightly less stable in the pitch and roll axes. Without flaps, the airplane will tend to float considerably during roundout. The pilot should avoid the temptation to force the airplane onto the runway at an excessively high speed. Neither should the pilot flare excessively, because without flaps this might cause the tail to strike the runway.

ASYMMETRIC (SPLIT) FLAP

An asymmetric "split" flap situation is one in which one flap deploys or retracts while the other remains in position. The problem is indicated by a pronounced roll toward the wing with the least flap deflection when wing flaps are extended/retracted.

The roll encountered in a split flap situation is countered with opposite aileron. The yaw caused by the additional drag created by the extended flap will require substantial opposite rudder, resulting in a cross-control condition. Almost full aileron may be required to maintain a wings-level attitude, especially at the reduced airspeed necessary for approach and landing. The pilot therefore should not attempt to land with a crosswind from the side of the deployed flap, because the additional roll control required to counteract the crosswind may not be available.

The pilot must be aware of the difference in stall speeds between one wing and the other in a split flap situation. The wing with the retracted flap will stall considerably earlier than the wing with the deployed flap. This type of asymmetrical stall will result in an uncontrollable roll in the direction of the stalled (clean) wing. If altitude permits, a spin will result.

The approach to landing with a split flap condition should be flown at a higher than normal airspeed. The pilot should not risk an asymmetric stall and subsequent loss of control by flaring excessively. Rather, the airplane should be flown onto the runway so that the touchdown occurs at an airspeed consistent with a safe margin above flaps-up stall speed.

LOSS OF ELEVATOR CONTROL

In many airplanes, the elevator is controlled by two cables: a "down" cable and an "up" cable. Normally, a break or disconnect in only one of these cables will not result in a total loss of elevator control. In most airplanes, a failed cable results in a partial loss of pitch control. In the failure of the "up" elevator cable (the "down" elevator being intact and functional) the control yoke will move aft easily but produce no response. Forward yoke movement, however, beyond the neutral position produces a nosedown attitude. Conversely, a failure of the "down" elevator cable, forward movement of the control yoke produces no effect. The pilot will, however, have partial control of pitch attitude with aft movement.

When experiencing a loss of **up-elevator** control, the pilot can retain pitch control by:

- Applying considerable nose-up trim.
- Pushing the control yoke forward to attain and maintain desired attitude.
- Increasing forward pressure to lower the nose and relaxing forward pressure to raise the nose.
- Releasing forward pressure to flare for landing.

When experiencing a loss of **down-elevator** control, the pilot can retain pitch control by:

- Applying considerable nosedown trim.
- Pulling the control yoke aft to attain and maintain attitude.
- Releasing back pressure to lower the nose and increasing back pressure to raise the nose.
- Increasing back pressure to flare for landing.

Trim mechanisms can be useful in the event of an in-flight primary control failure. For example, if the linkage between the cockpit and the elevator fails in flight, leaving the elevator free to weathervane in the wind, the trim tab can be used to raise or lower the elevator, within limits. The trim tabs are not as effective as normal linkage control in conditions such as low airspeed, but they do have some positive effect—usually enough to bring about a safe landing.

If an elevator becomes jammed, resulting in a total loss of elevator control movement, various combinations of power and flap extension offer a limited amount of pitch control. A successful landing under these conditions, however, is problematical.

LANDING GEAR MALFUNCTION

Once the pilot has confirmed that the landing gear has in fact malfunctioned, and that one or more gear legs refuses to respond to the conventional or alternate methods of gear extension contained in the AFM/POH, there are several methods that may be useful in attempting to force the gear down. One method is to dive the airplane (in smooth air only) to V_{NE} speed (red line on the airspeed indicator) and (within the limits of safety) execute a rapid pull up. In normal category airplanes, this procedure will create a 3.8 G load on the structure, in effect making the landing gear weigh 3.8 times normal. In some cases, this may force the landing gear into the down and locked position. This procedure requires a fine control touch and good feel for the airplane. The pilot must avoid exceeding the design stress limits of the airplane while attempting to lower the landing gear. The pilot must also avoid an accelerated stall and possible loss of control while attention is directed to solving the landing gear problem.

Another method that has proven useful in some cases is to induce rapid yawing. After stabilizing at or slightly less than maneuvering speed (V_A), the pilot should alternately and aggressively apply rudder in one direction and then the other in rapid sequence. The resulting yawing action may cause the landing gear to fall into place.

If all efforts to extend the landing gear have failed, and a gear up landing is inevitable, the pilot should select an airport with crash and rescue facilities. The pilot should not hesitate to request that emergency equipment be standing by.

When selecting a landing surface, the pilot should consider that a smooth, hard-surface runway usually causes less damage than rough, unimproved grass strips. A hard surface does, however, create sparks that can ignite fuel. If the airport is so equipped, the pilot can request that the runway surface be foamed. The pilot should consider burning off excess fuel. This will reduce landing speed and fire potential.

If the landing gear malfunction is limited to one main landing gear leg, the pilot should consume as much fuel from that side of the airplane as practicable, thereby reducing the weight of the wing on that side. The reduced weight makes it possible to delay the unsupported wing from contacting the surface during the landing roll until the last possible moment. Reduced impact speeds result in less damage.

If only one landing gear leg fails to extend, the pilot has the option of landing on the available gear legs, or landing with all the gear legs retracted. Landing on only one main gear usually causes the airplane to veer strongly in the direction of the faulty gear leg after touchdown. If the landing runway is narrow, and/or ditches and obstacles line the runway edge, maximum directional control after touchdown is a necessity. In this situation, a landing with all three gear retracted may be the safest course of action.

If the pilot elects to land with one main gear retracted (and the other main gear and nose gear down and locked), the landing should be made in a nose-high attitude with the wings level. As airspeed decays, the pilot should apply whatever aileron control is necessary to keep the unsupported wing airborne as long as possible. [Figure 16-7] Once the wing contacts the surface, the pilot can anticipate a strong yaw in that direction. The pilot must be prepared to use full opposite rudder and aggressive braking to maintain some degree of directional control.

Figure 16-7. Landing with one main gear retracted.

When landing with a retracted nosewheel (and the main gear extended and locked) the pilot should hold the nose off the ground until *almost* full up-elevator has been applied. [Figure 16-8] The pilot should then release back pressure in such a manner that the nose settles slowly to the surface. Applying and holding *full* up-elevator will result in the nose abruptly dropping to the surface as airspeed decays, possibly resulting in burrowing and/or additional damage. Brake pressure should not be applied during the landing roll unless absolutely necessary to avoid a collision with obstacles.

Figure 16-8. Landing with nosewheel retracted.

If the landing must be made with only the nose gear extended, the initial contact should be made on the aft fuselage structure with a nose-high attitude. This procedure will help prevent porpoising and/or wheelbarrowing. The pilot should then allow the nosewheel to gradually touch down, using nosewheel steering as necessary for directional control.

SYSTEMS MALFUNCTIONS

ELECTRICAL SYSTEM

The loss of electrical power can deprive the pilot of numerous critical systems, and therefore should not be taken lightly even in day/VFR conditions. Most in-flight failures of the electrical system are located in the generator or alternator. Once the generator or alternator system goes off line, the electrical source in a typical light airplane is a battery. If a warning light or ammeter indicates the probability of an alternator or generator failure in an airplane with only one generating system, however, the pilot may have very little time available from the battery.

The rating of the airplane battery provides a clue to how long it *may* last. With batteries, the higher the amperage load, the less the usable total amperage. Thus a 25-amp hour battery could produce 5 amps per hour for 5 hours, but if the load were increased to 10 amps, it might last only 2 hours. A 40-amp load might discharge the battery fully in about 10 or 15 minutes. Much depends on the battery condition at the time of the system failure. If the battery has been in service for a few years, its power may be reduced substantially because of internal resistance. Or if the system failure was not detected immediately, much of the stored energy may have already been used. It is essential, therefore, that the pilot immediately shed non-essential loads when the generating source fails. [Figure 16-9] The pilot should then plan to land at the nearest suitable airport.

What constitutes an "emergency" load following a generating system failure cannot be predetermined, because the actual circumstances will always be somewhat different—for example, whether the flight is VFR or IFR, conducted in day or at night, in clouds or in the clear. Distance to nearest suitable airport can also be a factor.

Electrical Loads for Light Single	Number of Units	Total Amperes
A. Continuous Load		
Pitot Heating (Operating)	1	3.30
Wingtip Lights	4	3.00
Heater Igniter	1	1-20
**Navigation Receivers	1-4	1-2 each
**Communications Receivers	1-2	1-2 each
Fuel Indicator	1	0.40
Instrument Lights (overhead)	2	0.60
Engine Indicator	1	0.30
Compass Light	1	0.20
Landing Gear Indicator	1	0.17
Flap Indicator	1	0.17
B. Intermittent Load		
Starter	1	100.00
Landing Lights	2	17.80
Heater Blower Motor	1	14.00
Flap Motor	1	13.00
Landing Gear Motor	1	10.00
Cigarette Lighter	1	7.50
Transceiver (keyed)	1	5-7
Fuel Boost Pump	1	2.00
Cowl Flap Motor	1	1.00
Stall Warning Horn	1	1.50

** Amperage for radios varies with equipment. In general, the more recent the model, the less amperage required.
NOTE: Panel and indicator lights usually draw less than one amp.

Figure 16-9. Electrical load for light single.

The pilot should remember that the electrically powered (or electrically selected) landing gear and flaps will not function properly on the power left in a partially depleted battery. Landing gear and flap motors use up power at rates much greater than most other types of electrical equipment. The result of selecting these motors on a partially depleted battery may well result in an immediate total loss of electrical power.

If the pilot should experience a complete in-flight loss of electrical power, the following steps should be taken:

- Shed all but the most necessary electrically-driven equipment.
- Understand that any loss of electrical power is critical in a small airplane—notify ATC of the situation immediately. Request radar vectors for a landing at the nearest suitable airport.
- If landing gear or flaps are electrically controlled or operated, plan the arrival well ahead of time. Expect to make a no-flap landing, and anticipate a manual landing gear extension.

PITOT-STATIC SYSTEM

The source of the pressure for operating the airspeed indicator, the vertical speed indicator, and the altimeter is the pitot-static system. The major components of the pitot-static system are the impact pressure chamber and lines, and the static pressure chamber and lines, each of which are subject to total or partial blockage by ice, dirt, and/or other foreign matter. Blockage of the pitot-static system will adversely affect instrument operation. [Figure 16-10 on next page]

Partial static system blockage is insidious in that it may go unrecognized until a critical phase of flight. During takeoff, climb, and level-off at cruise altitude the altimeter, airspeed indicator, and vertical speed indicator may operate normally. No indication of malfunction may be present until the airplane begins a descent.

If the static reference system is severely restricted, but not entirely blocked, as the airplane descends, the static reference pressure at the instruments begins to lag behind the actual outside air pressure. While descending, the altimeter may indicate that the airplane is higher than actual because the obstruction slows the airflow from the static port to the altimeter. The vertical speed indicator confirms the altimeter's information regarding rate of change, because the reference pressure is not changing at the same rate as the outside air pressure. The airspeed indicator, unable to tell whether it is experiencing more airspeed pitot pressure or less static reference pressure, indicates a higher airspeed than actual. To the pilot, the instruments indicate that the airplane is too high, too fast, and descending at a rate much less than desired.

If the pilot levels off and then begins a climb, the altitude indication may still lag. The vertical speed indicator will indicate that the airplane is not climbing as fast as actual. The indicated airspeed, however, may begin to decrease at an alarming rate. The least amount of pitch-up attitude may cause the airspeed needle to indicate dangerously near stall speed.

Managing a static system malfunction requires that the pilot know and understand the airplane's pitot-static system. If a system malfunction is suspected, the pilot should confirm it by opening the alternate static source. This should be done while the airplane is climbing or descending. If the instrument needles move significantly when this is done, a static pressure problem exists and the alternate source should be used during the remainder of the flight.

ABNORMAL ENGINE INSTRUMENT INDICATIONS

The AFM/POH for the specific airplane contains information that should be followed in the event of any

Effect of Blocked Pitot/Static Sources on Airspeed, Altimeter and Vertical Speed Indications	 Indicated Airspeed	 Indicated Altitude	 Indicated Vertical Speed
Pitot Source Blocked	Increases with altitude gain; decreases with altitude loss.	Unaffected	Unaffected
One Static Source Blocked	Inaccurate while sideslipping; very sensitive in turbulence.		
Both Static Sources Blocked	Decreases with altitude gain; increases with altitude loss.	Does not change with actual gain or loss of altitude.	Does not change with actual variations in vertical speed.
Both Static and Pitot Sources Blocked	All indications remain constant, regardless of actual changes in airspeed, altitude and vertical speed.		

Figure 16-10. Effects of blocked pitot-static sources.

abnormal engine instrument indications. The table on the next page offers generic information on some of the more commonly experienced in-flight abnormal engine instrument indications, their possible causes, and corrective actions. [Table 1]

DOOR OPENING IN FLIGHT

In most instances, the occurrence of an inadvertent door opening is not of great concern to the safety of a flight, but rather, the pilot's reaction at the moment the incident happens. A door opening in flight may be accompanied by a sudden loud noise, sustained noise level and possible vibration or buffeting. If a pilot allows himself or herself to become distracted to the point where attention is focused on the open door rather than maintaining control of the airplane, loss of control may result, even though disruption of airflow by the door is minimal.

In the event of an inadvertent door opening in flight or on takeoff, the pilot should adhere to the following.

- Concentrate on flying the airplane. Particularly in light single- and twin-engine airplanes; a cabin door that opens in flight seldom if ever compromises the airplane's ability to fly. There may be some handling effects such as roll and/or yaw, but in most instances these can be easily overcome.
- If the door opens after lift-off, do not rush to land. Climb to normal traffic pattern altitude, fly a normal traffic pattern, and make a normal landing.
- Do not release the seat belt and shoulder harness in an attempt to reach the door. Leave the door alone. Land as soon as practicable, and close the door once safely on the ground.
- Remember that most doors will not stay wide open. They will usually bang open, then settle partly closed. A slip towards the door may cause it to open wider; a slip away from the door may push it closed.
- Do not panic. Try to ignore the unfamiliar noise and vibration. Also, do not rush. Attempting to get the airplane on the ground as quickly as possible may result in steep turns at low altitude.
- Complete all items on the landing checklist.
- Remember that accidents are almost never caused by an open door. Rather, an open door accident is caused by the pilot's distraction or failure to maintain control of the airplane.

INADVERTENT VFR FLIGHT INTO IMC

GENERAL

It is beyond the scope of this handbook to incorporate a course of training in basic attitude instrument flying. This information is contained in FAA-H-8083-15, *Instrument Flying Handbook*. Certain pilot certificates and/or associated ratings require training in instrument flying and a demonstration of specific instrument flying tasks on the practical test.

MALFUNCTION	PROBABLE CAUSE	CORRECTIVE ACTION
Loss of r.p.m. during cruise flight (non-altitude engines)	Carburetor or induction icing or air filter clogging	Apply carburetor heat. If dirty filter is suspected and non-filtered air is available, switch selector to unfiltered position.
Loss of manifold pressure during cruise flight	Same as above Turbocharger failure	Same as above. Possible exhaust leak. Shut down engine or use lowest practicable power setting. Land as soon as possible.
Gain of manifold pressure during cruise flight	Throttle has opened, propeller control has decreased r.p.m., or improper method of power reduction	Readjust throttle and tighten friction lock. Reduce manifold pressure prior to reducing r.p.m.
High oil temperature	Oil congealed in cooler Inadequate engine cooling Detonation or preignition Forth coming internal engine faiure Defective thermostatic oil cooler control	Reduce power. Land. Preheat engine. Reduce power. Increase airspeed. Observe cylinder head temperatures for high reading. Reduce manifold pressure. Enrich mixture. Land as soon as possible or feather propeller and stop engine. Land as soon as possible. Consult maintenance personnel.
Low oil temperature	Engine not warmed up to operating temperature	Warm engine in prescribed manner.
High oil pressure	Cold oil Possible internal plugging	Same as above. Reduce power. Land as soon as possible.
Low oil pressure	Broken pressure relief valve Insufficient oil Burned out bearings	Land as soon as possible or feather propeller and stop engine. Same as above. Same as above.
Fluctuating oil pressure	Low oil supply, loose oil lines, defective pressure relief valve	Same as above.
High cylinder head temperature	Improper cowl flap adjustment Insufficient airspeed for cooling Improper mixture adjustment Detonation or preignition	Adjust cowl flaps. Increase airspeed. Adjust mixture. Reduce power, enrich mixture, increase cooling airflow.
Low cylinder head temperature	Excessive cowl flap opening Excessively rich mixture Exteneded glides without clearing engine	Adjust cowl flaps. Adjust mixture control. Clear engine long enough to keep temperatures at minimum range.
Ammeter indicating discharge	Alternator or generator failure	Shed unnecessary electrical load. Land as soon as practicable.
Load meter indicating zero	Same as above	Same as above.
Surging r.p.m. and overspeeding	Defective propeller Defective engine Defective propeller governor Defective tachometer Improper mixture setting	Adjust propeller r.p.m. Consult maintenance. Adjust propeller control. Attempt to restore normal operation. Consult maintenance. Readjust mixture for smooth operation.
Loss of airspeed in cruise flight with manifold pressure and r.p.m. constant	Possible loss of one or more cylinders	Land as soon as possible.
Rough running engine	Improper mixture control setting Defective ignition or valves Detonation or preignition Induction air leak Plugged fuel nozzle (Fuel injection) Excessive fuel pressure or fuel flow	Adjust mixture for smooth operation. Consult maintenance personnel. Reduce power, enrich mixture, open cowl flaps to reduce cylinder head temp. Land as soon as practicable. Reduce power. Consult maintenance. Same as above. Lean mixture control.
Loss of fuel pressure	Engine driven pump failure No fuel	Turn on boost tanks. Switch tanks, turn on fuel.

Table 1.

Pilots and flight instructors should refer to FAA-H-8083-15 for guidance in the performance of these tasks, and to the appropriate practical test standards for information on the standards to which these required tasks must be performed for the particular certificate level and/or rating. The pilot should remember, however, that unless these tasks are practiced on a continuing and regular basis, skill erosion begins almost immediately. In a very short time, the pilot's *assumed* level of confidence will be much higher than the performance he or she will actually be able to demonstrate should the need arise.

Accident statistics show that the pilot who has not been trained in attitude instrument flying, or one whose instrument skills have eroded, will lose control of the airplane in about 10 minutes once forced to rely solely on instrument reference. The purpose of this section is to provide guidance on practical emergency measures to maintain airplane control for a limited period of time in the event a VFR pilot encounters IMC conditions. The main goal is *not* precision instrument flying; rather, it is to help the VFR pilot keep the airplane under adequate control until suitable visual references are regained.

The first steps necessary for surviving an encounter with instrument meteorological conditions (IMC) by a VFR pilot are:

- Recognition and acceptance of the seriousness of the situation and the need for immediate remedial action.
- Maintaining control of the airplane.
- Obtaining the appropriate assistance in getting the airplane safely on the ground.

RECOGNITION

A VFR pilot is in IMC conditions anytime he or she is unable to maintain airplane attitude control by reference to the natural horizon, regardless of the circumstances or the prevailing weather conditions. Additionally, the VFR pilot is, in effect, in IMC anytime he or she is inadvertently, or intentionally for an indeterminate period of time, unable to navigate or establish geographical position by visual reference to landmarks on the surface. These situations must be accepted by the pilot involved as a genuine emergency, requiring appropriate action.

The pilot must understand that unless he or she is trained, qualified, and current in the control of an airplane solely by reference to flight instruments, he or she will be unable to do so for any length of time. Many hours of VFR flying using the attitude indicator as a reference for airplane control may lull a pilot into a false sense of security based on an overestimation of his or her personal ability to control the airplane solely by instrument reference. In VFR conditions, even though the pilot thinks he or she is controlling the airplane by instrument reference, the pilot also receives an overview of the natural horizon and may subconsciously rely on it more than the cockpit attitude indicator. If the natural horizon were to suddenly disappear, the untrained instrument pilot would be subject to vertigo, spatial disorientation, and inevitable control loss.

MAINTAINING AIRPLANE CONTROL

Once the pilot recognizes and accepts the situation, he or she must understand that the only way to control the airplane safely is by using and trusting the flight instruments. Attempts to control the airplane *partially* by reference to flight instruments while searching outside the cockpit for visual confirmation of the information provided by those instruments will result in inadequate airplane control. This may be followed by spatial disorientation and complete control loss.

The most important point to be stressed is that the pilot **must not panic**. The task at hand may seem overwhelming, and the situation may be compounded by extreme apprehension. The pilot therefore must make a conscious effort to relax.

The pilot must understand the most important concern—in fact the only concern at this point—is to keep the wings level. An uncontrolled turn or bank usually leads to difficulty in achieving the objectives of any desired flight condition. The pilot will find that good bank control has the effect of making pitch control much easier.

The pilot should remember that a person cannot feel control pressures with a tight grip on the controls. Relaxing and learning to "control with the eyes and the brain" instead of only the muscles, usually takes considerable conscious effort.

The pilot must believe what the flight instruments show about the airplane's attitude regardless of what the natural senses tell. The vestibular sense (motion sensing by the inner ear) can and will confuse the pilot. Because of inertia, the sensory areas of the inner ear cannot detect slight changes in airplane attitude, nor can they accurately sense attitude changes which occur at a uniform rate over a period of time. On the other hand, *false* sensations are often generated, leading the pilot to believe the attitude of the airplane *has* changed when, in fact, it has not. These false sensations result in the pilot experiencing spatial disorientation.

ATTITUDE CONTROL

An airplane is, by design, an inherently stable platform and, except in turbulent air, will maintain approximately straight-and-level flight if properly trimmed and left alone. It is designed to maintain a state of equilibrium in pitch, roll, and yaw. The pilot must be aware, however, that a change about one axis will affect the stability of the others. The typical light airplane exhibits a good deal of stability in the yaw axis, slightly less in the pitch axis, and even lesser still in the roll axis. The key to emergency airplane attitude control, therefore, is to:

- Trim the airplane with the elevator trim so that it will maintain hands-off level flight at cruise airspeed.
- Resist the tendency to over control the airplane. Fly the attitude indicator with fingertip control. No attitude changes should be made unless the flight instruments indicate a definite need for a change.
- Make all attitude changes smooth and small, yet with positive pressure. Remember that a small change as indicated on the horizon bar corresponds to a proportionately much larger change in actual airplane attitude.
- Make use of any available aid in attitude control such as autopilot or wing leveler.

The primary instrument for attitude control is the attitude indicator. [Figure 16-11] Once the airplane is trimmed so that it will maintain hands-off level flight at cruise airspeed, that airspeed need not vary until the airplane must be slowed for landing. All turns, climbs and descents can and should be made at this airspeed. Straight flight is maintained by keeping the wings level using "fingertip pressure" on the control wheel. Any pitch attitude change should be made by using no more than one bar width up or down.

Figure 16-11. Attitude indicator.

TURNS

Turns are perhaps the most potentially dangerous maneuver for the untrained instrument pilot for two reasons.

- The normal tendency of the pilot to over control, leading to steep banks and the possibility of a "graveyard spiral."
- The inability of the pilot to cope with the instability resulting from the turn.

When a turn must be made, the pilot must anticipate and cope with the relative instability of the roll axis. The smallest practical bank angle should be used—in any case no more than 10° bank angle. [Figure 16-12] A shallow bank will take very little vertical lift from the wings resulting in little if any deviation in altitude. It may be helpful to turn a few degrees and then return to level flight, if a large change in heading must be made. Repeat the process until the desired heading is reached. This process may relieve the progressive overbanking that often results from prolonged turns.

Figure 16-12. Level turn.

CLIMBS

If a climb is necessary, the pilot should raise the miniature airplane on the attitude indicator no more than one bar width and apply power. [Figure 16-13] The pilot should not attempt to attain a specific climb speed but accept whatever speed results. The objective is to deviate as little as possible from level flight attitude in order to disturb the airplane's equilibrium as little as possible. If the airplane is properly trimmed, it will assume a nose-up attitude on its own commensurate with the amount of power applied. Torque and P-factor will cause the airplane to have a

Figure 16-13. Level climb.

tendency to bank and turn to the left. This must be anticipated and compensated for. If the initial power application results in an inadequate rate of climb, power should be increased in increments of 100 r.p.m. or 1 inch of manifold pressure until the desired rate of climb is attained. Maximum available power is seldom necessary. The more power used the more the airplane will want to bank and turn to the left. Resuming level flight is accomplished by first decreasing pitch attitude to level on the attitude indicator using slow but deliberate pressure, allowing airspeed to increase to near cruise value, and then decreasing power.

DESCENTS

Descents are very much the opposite of the climb procedure if the airplane is properly trimmed for hands-off straight-and-level flight. In this configuration, the airplane requires a certain amount of thrust to maintain altitude. The pitch attitude is controlling the airspeed. The engine power, therefore, (translated into thrust by the propeller) is maintaining the selected altitude. Following a power reduction, however slight, there will be an almost imperceptible decrease in airspeed. However, even a slight change in speed results in less down load on the tail, whereupon the designed nose heaviness of the airplane causes it to pitch down just enough to maintain the airspeed for which it was trimmed. The airplane will then descend at a rate directly proportionate to the amount of thrust that has been removed. Power reductions should be made in increments of 100 r.p.m. or 1 inch of manifold pressure and the resulting rate of descent should never exceed 500 f.p.m. The wings should be held level on the attitude indicator, and the pitch attitude should not exceed one bar width below level. [Figure 16-14]

Figure 16-14. Level descent.

COMBINED MANEUVERS

Combined maneuvers, such as climbing or descending turns should be avoided if at all possible by an untrained instrument pilot already under the stress of an emergency situation. Combining maneuvers will only compound the problems encountered in individual maneuvers and increase the risk of control loss. Remember that the objective is to maintain airplane control by deviating as little as possible from straight-and-level flight attitude and thereby maintaining as much of the airplane's natural equilibrium as possible.

When being assisted by air traffic controllers from the ground, the pilot may detect a sense of urgency as he or she is being directed to change heading and/or altitude. This sense of urgency reflects a normal concern for safety on the part of the controller. But the pilot must not let this prompt him or her to attempt a maneuver that could result in loss of control.

TRANSITION TO VISUAL FLIGHT

One of the most difficult tasks a trained and qualified instrument pilot must contend with is the transition from instrument to visual flight prior to landing. For the untrained instrument pilot, these difficulties are magnified.

The difficulties center around acclimatization and orientation. On an instrument approach the trained instrument pilot must prepare in advance for the transition to visual flight. The pilot must have a mental picture of what he or she expects to see once the transition to visual flight is made and quickly acclimatize to the new environment. Geographical orientation must also begin before the transition as the pilot must visualize where the airplane will be in relation to the airport/runway when the transition occurs so that the approach and landing may be completed by visual reference to the ground.

In an ideal situation the transition to visual flight is made with ample time, at a sufficient altitude above terrain, and to visibility conditions sufficient to accommodate acclimatization and geographical orientation. This, however, is not always the case. The untrained instrument pilot may find the visibility still limited, the terrain completely unfamiliar, and altitude above terrain such that a "normal" airport traffic pattern and landing approach is not possible. Additionally, the pilot will most likely be under considerable self-induced psychological pressure to get the airplane on the ground. The pilot must take this into account and, if possible, allow time to become acclimatized and geographically oriented before attempting an approach and landing, even if it means flying straight and level for a time or circling the airport. This is especially true at night.

100

W 10 x

100-HOUR INSPECTION— An inspection, identical in scope to an annual inspection. Must be conducted every 100 hours of flight on aircraft of under 12,500 pounds that are used for hire.

ABSOLUTE ALTITUDE— The vertical distance of an airplane above the terrain, or above ground level (AGL).

ABSOLUTE CEILING— The altitude at which a climb is no longer possible.

ACCELERATE-GO DISTANCE— The distance required to accelerate to V_1 with all engines at takeoff power, experience an engine failure at V_1 and continue the takeoff on the remaining engine(s). The runway required includes the distance required to climb to 35 feet by which time V_2 speed must be attained.

ACCELERATE-STOP DISTANCE—The distance required to accelerate to V_1 with all engines at takeoff power, experience an engine failure at V_1, and abort the takeoff and bring the airplane to a stop using braking action only (use of thrust reversing is not considered).

ACCELERATION—Force involved in overcoming inertia, and which may be defined as a change in velocity per unit of time.

ACCESSORIES—Components that are used with an engine, but are not a part of the engine itself. Units such as magnetos, carburetors, generators, and fuel pumps are commonly installed engine accessories.

ADJUSTABLE STABILIZER— A stabilizer that can be adjusted in flight to trim the airplane, thereby allowing the airplane to fly hands-off at any given airspeed.

ADVERSE YAW—A condition of flight in which the nose of an airplane tends to yaw toward the outside of the turn. This is caused by the higher induced drag on the outside wing, which is also producing more lift. Induced drag is a by-product of the lift associated with the outside wing.

AERODYNAMIC CEILING— The point (altitude) at which, as the indicated airspeed decreases with altitude, it progressively merges with the low speed buffet boundary where prestall buffet occurs for the airplane at a load factor of 1.0 G.

AERODYNAMICS—The science of the action of air on an object, and with the motion of air on other gases. Aerodynamics deals with the production of lift by the aircraft, the relative wind, and the atmosphere.

AILERONS—Primary flight control surfaces mounted on the trailing edge of an airplane wing, near the tip. Ailerons control roll about the longitudinal axis.

AIR START—The act or instance of starting an aircraft's engine while in flight, especially a jet engine after flameout.

AIRCRAFT LOGBOOKS— Journals containing a record of total operating time, repairs, alterations or inspections performed, and all Airworthiness Directive (AD) notes complied with. A maintenance logbook should be kept for the airframe, each engine, and each propeller.

AIRFOIL—An airfoil is any surface, such as a wing, propeller, rudder, or even a trim tab, which provides aerodynamic force when it interacts with a moving stream of air.

AIRMANSHIP SKILLS—The skills of coordination, timing, control touch, and speed sense in addition to the motor skills required to fly an aircraft.

AIRMANSHIP— A sound acquaintance with the principles of flight, the ability to operate an airplane with competence and precision both on the ground and in the air, and the exercise of sound judgment that results in optimal operational safety and efficiency.

AIRPLANE FLIGHT MANUAL (AFM)—A document developed by the airplane manufacturer and approved by the Federal Aviation Administration (FAA). It is specific to a particular make and model airplane by serial number and it contains operating procedures and limitations.

AIRPLANE OWNER/ INFORMATION MANUAL—A document developed by the airplane manufacturer containing general information about the make and model of an airplane. The airplane owner's manual is not FAA-approved and is not specific to a particular serial numbered airplane. This manual is not kept current, and therefore cannot be substituted for the AFM/POH.

AIRPORT/FACILITY DIRECTORY— A publication designed primarily as a pilot's operational manual containing all airports, seaplane bases, and heliports open to the public including communications data, navigational facilities, and certain special notices and procedures. This publication is issued in seven volumes according to geographical area.

AIRWORTHINESS—A condition in which the aircraft conforms to its type certificated design including supplemental type certificates, and field approved alterations. The aircraft must also be in a condition for safe flight as determined by annual, 100 hour, preflight and any other required inspections.

AIRWORTHINESS CERTIFICATE—
A certificate issued by the FAA to all aircraft that have been proven to meet the minimum standards set down by the Code of Federal Regulations.

AIRWORTHINESS DIRECTIVE—A regulatory notice sent out by the FAA to the registered owner of an aircraft informing the owner of a condition that prevents the aircraft from continuing to meet its conditions for airworthiness. Airworthiness Directives (AD notes) must be complied with within the required time limit, and the fact of compliance, the date of compliance, and the method of compliance must be recorded in the aircraft's maintenance records.

ALPHA MODE OF OPERATION—The operation of a turboprop engine that includes all of the flight operations, from takeoff to landing. Alpha operation is typically between 95 percent to 100 percent of the engine operating speed.

ALTERNATE AIR—A device which opens, either automatically or manually, to allow induction airflow to continue should the primary induction air opening become blocked.

ALTERNATE STATIC SOURCE—
A manual port that when opened allows the pitot static instruments to sense static pressure from an alternate location should the primary static port become blocked.

ALTERNATOR/GENERATOR—A device that uses engine power to generate electrical power.

ALTIMETER—A flight instrument that indicates altitude by sensing pressure changes.

ALTITUDE (AGL)—The actual height above ground level (AGL) at which the aircraft is flying.

ALTITUDE (MSL)—The actual height above mean sea level (MSL) at which the aircraft is flying.

ALTITUDE CHAMBER—A device that simulates high altitude conditions by reducing the interior pressure. The occupants will suffer from the same physiological conditions as flight at high altitude in an unpressurized aircraft.

ALTITUDE ENGINE—
A reciprocating aircraft engine having a rated takeoff power that is producible from sea level to an established higher altitude.

ANGLE OF ATTACK—The acute angle between the chord line of the airfoil and the direction of the relative wind.

ANGLE OF INCIDENCE—
The angle formed by the chord line of the wing and a line parallel to the longitudinal axis of the airplane.

ANNUAL INSPECTION—
A complete inspection of an aircraft and engine, required by the Code of Federal Regulations, to be accomplished every 12 calendar months on all certificated aircraft. Only an A&P technician holding an Inspection Authorization can conduct an annual inspection.

ANTI-ICING—The prevention of the formation of ice on a surface. Ice may be prevented by using heat or by covering the surface with a chemical that prevents water from reaching the surface. Anti-icing should not be confused with deicing, which is the removal of ice after it has formed on the surface.

ATTITUDE INDICATOR—
An instrument which uses an artificial horizon and miniature airplane to depict the position of the airplane in relation to the true horizon. The attitude indicator senses roll as well as pitch, which is the up and down movement of the airplane's nose.

ATTITUDE— The position of an aircraft as determined by the relationship of its axes and a reference, usually the earth's horizon.

AUTOKINESIS—This is caused by staring at a single point of light against a dark background for more than a few seconds. After a few moments, the light appears to move on its own.

AUTOPILOT—An automatic flight control system which keeps an aircraft in level flight or on a set course. Automatic pilots can be directed by the pilot, or they may be coupled to a radio navigation signal.

AXES OF AN AIRCRAFT—Three imaginary lines that pass through an aircraft's center of gravity. The axes can be considered as imaginary axles around which the aircraft turns. The three axes pass through the center of gravity at 90° angles to each other. The axis from nose to tail is the longitudinal axis, the axis that passes from wingtip to wingtip is the lateral axis, and the axis that passes vertically through the center of gravity is the vertical axis.

AXIAL FLOW COMPRESSOR—
A type of compressor used in a turbine engine in which the airflow through the compressor is essentially linear. An axial-flow compressor is made up of several stages of alternate rotors and stators. The compressor ratio is determined by the decrease in area of the succeeding stages.

BACK SIDE OF THE POWER CURVE— Flight regime in which flight at a higher airspeed requires a lower power setting and a lower airspeed requires a higher power setting in order to maintain altitude.

BALKED LANDING—
A go-around.

BALLAST—Removable or permanently installed weight in an aircraft used to bring the center of gravity into the allowable range.

BALLOON—The result of a too aggressive flare during landing causing the aircraft to climb.

BASIC EMPTY WEIGHT (GAMA)—Basic empty weight includes the standard empty weight plus optional and special equipment that has been installed.

BEST ANGLE OF CLIMB (V_X)—The speed at which the aircraft will produce the most gain in altitude in a given distance.

BEST GLIDE—The airspeed in which the aircraft glides the furthest for the least altitude lost when in non-powered flight.

BEST RATE OF CLIMB (V_Y)—The speed at which the aircraft will produce the most gain in altitude in the least amount of time.

BLADE FACE—The flat portion of a propeller blade, resembling the bottom portion of an airfoil.

BLEED AIR—Compressed air tapped from the compressor stages of a turbine engine by use of ducts and tubing. Bleed air can be used for deice, anti-ice, cabin pressurization, heating, and cooling systems.

BLEED VALVE—In a turbine engine, a flapper valve, a popoff valve, or a bleed band designed to bleed off a portion of the compressor air to the atmosphere. Used to maintain blade angle of attack and provide stall-free engine acceleration and deceleration.

BOOST PUMP—An electrically driven fuel pump, usually of the centrifugal type, located in one of the fuel tanks. It is used to provide fuel to the engine for starting and providing fuel pressure in the event of failure of the engine driven pump. It also pressurizes the fuel lines to prevent vapor lock.

BUFFETING—The beating of an aerodynamic structure or surface by unsteady flow, gusts, etc.; the irregular shaking or oscillation of a vehicle component owing to turbulent air or separated flow.

BUS BAR—An electrical power distribution point to which several circuits may be connected. It is often a solid metal strip having a number of terminals installed on it.

BUS TIE—A switch that connects two or more bus bars. It is usually used when one generator fails and power is lost to its bus. By closing the switch, the operating generator powers both busses.

BYPASS AIR—The part of a turbofan's induction air that bypasses the engine core.

BYPASS RATIO—The ratio of the mass airflow in pounds per second through the fan section of a turbofan engine to the mass airflow that passes through the gas generator portion of the engine. Or, the ratio between fan mass airflow (lb/sec.) and core engine mass airflow (lb/sec.).

CABIN PRESSURIZATION—A condition where pressurized air is forced into the cabin simulating pressure conditions at a much lower altitude and increasing the aircraft occupants comfort.

CALIBRATED AIRSPEED (CAS)—Indicated airspeed corrected for installation error and instrument error. Although manufacturers attempt to keep airspeed errors to a minimum, it is not possible to eliminate all errors throughout the airspeed operating range. At certain airspeeds and with certain flap settings, the installation and instrument errors may total several knots. This error is generally greatest at low airspeeds. In the cruising and higher airspeed ranges, indicated airspeed and calibrated airspeed are approximately the same. Refer to the airspeed calibration chart to correct for possible airspeed errors.

CAMBERED—The camber of an airfoil is the characteristic curve of its upper and lower surfaces. The upper camber is more pronounced, while the lower camber is comparatively flat. This causes the velocity of the airflow immediately above the wing to be much higher than that below the wing.

CARBURETOR ICE— Ice that forms inside the carburetor due to the temperature drop caused by the vaporization of the fuel. Induction system icing is an operational hazard because it can cut off the flow of the fuel/air charge or vary the fuel/air ratio.

CARBURETOR—1. Pressure: A hydromechanical device employing a closed feed system from the fuel pump to the discharge nozzle. It meters fuel through fixed jets according to the mass airflow through the throttle body and discharges it under a positive pressure. Pressure carburetors are distinctly different from float-type carburetors, as they do not incorporate a vented float chamber or suction pickup from a discharge nozzle located in the venturi tube. 2. Float-type: Consists essentially of a main air passage through which the engine draws its supply of air, a mechanism to control the quantity of fuel discharged in relation to the flow of air, and a means of regulating the quantity of fuel/air mixture delivered to the engine cylinders.

CASCADE REVERSER—A thrust reverser normally found on turbofan engines in which a blocker door and a series of cascade vanes are used to redirect exhaust gases in a forward direction.

CENTER OF GRAVITY (CG)—The point at which an airplane would balance if it were possible to suspend it at that point. It is the mass center of the airplane, or the theoretical point at which the entire weight of the airplane is assumed to be concentrated. It may be expressed in inches from the reference datum, or in percent of mean aerodynamic chord (MAC). The location depends on the distribution of weight in the airplane.

CENTER-OF-GRAVITY LIMITS—The specified forward and aft points within which the CG must be located during flight. These limits are indicated on pertinent airplane specifications.

CENTER-OF-GRAVITY RANGE—The distance between the forward and aft CG limits indicated on pertinent airplane specifications.

CENTRIFUGAL FLOW COMPRESSOR—An impeller-shaped device that receives air at its center and slings air outward at high velocity into a diffuser for increased pressure. Also referred to as a radial outflow compressor.

CHORD LINE—An imaginary straight line drawn through an airfoil from the leading edge to the trailing edge.

CIRCUIT BREAKER—A circuit-protecting device that opens the circuit in case of excess current flow. A circuit breakers differs from a fuse in that it can be reset without having to be replaced.

CLEAR AIR TURBULENCE—Turbulence not associated with any visible moisture.

CLIMB GRADIENT—The ratio between distance traveled and altitude gained.

COCKPIT RESOURCE MANAGEMENT—Techniques designed to reduce pilot errors and manage errors that do occur utilizing cockpit human resources. The assumption is that errors are going to happen in a complex system with error-prone humans.

COEFFICIENT OF LIFT—See LIFT COEFFICIENT.

COFFIN CORNER—The flight regime where any increase in airspeed will induce high speed mach buffet and any decrease in airspeed will induce low speed mach buffet.

COMBUSTION CHAMBER—The section of the engine into which fuel is injected and burned.

COMMON TRAFFIC ADVISORY FREQUENCY—The common frequency used by airport traffic to announce position reports in the vicinity of the airport.

COMPLEX AIRCRAFT—An aircraft with retractable landing gear, flaps, and a controllable-pitch propeller, or is turbine powered.

COMPRESSION RATIO—1. In a reciprocating engine, the ratio of the volume of an engine cylinder with the piston at the bottom center to the volume with the piston at top center. 2. In a turbine engine, the ratio of the pressure of the air at the discharge to the pressure of air at the inlet.

COMPRESSOR BLEED AIR—See BLEED AIR.

COMPRESSOR BLEED VALVES—See BLEED VALVE.

COMPRESSOR SECTION—The section of a turbine engine that increases the pressure and density of the air flowing through the engine.

COMPRESSOR STALL—In gas turbine engines, a condition in an axial-flow compressor in which one or more stages of rotor blades fail to pass air smoothly to the succeeding stages. A stall condition is caused by a pressure ratio that is incompatible with the engine r.p.m. Compressor stall will be indicated by a rise in exhaust temperature or r.p.m. fluctuation, and if allowed to continue, may result in flameout and physical damage to the engine.

COMPRESSOR SURGE—A severe compressor stall across the entire compressor that can result in severe damage if not quickly corrected. This condition occurs with a complete stoppage of airflow or a reversal of airflow.

CONDITION LEVER—In a turbine engine, a powerplant control that controls the flow of fuel to the engine. The condition lever sets the desired engine r.p.m. within a narrow range between that appropriate for ground and flight operations.

CONFIGURATION—This is a general term, which normally refers to the position of the landing gear and flaps.

CONSTANT SPEED PROPELLER—A controllable-pitch propeller whose pitch is automatically varied in flight by a governor to maintain a constant r.p.m. in spite of varying air loads.

CONTROL TOUCH—The ability to sense the action of the airplane and its probable actions in the immediate future, with regard to attitude and speed variations, by sensing and evaluation of varying pressures and resistance of the control surfaces transmitted through the cockpit flight controls.

CONTROLLABILITY—A measure of the response of an aircraft relative to the pilot's flight control inputs.

CONTROLLABLE PITCH PROPELLER—A propeller in which the blade angle can be changed during flight by a control in the cockpit.

CONVENTIONAL LANDING GEAR—Landing gear employing a third rear-mounted wheel. These airplanes are also sometimes referred to as tailwheel airplanes.

COORDINATED FLIGHT—Application of all appropriate flight and power controls to prevent slipping or skidding in any flight condition.

COORDINATION—The ability to use the hands and feet together subconsciously and in the proper relationship to produce desired results in the airplane.

CORE AIRFLOW—Air drawn into the engine for the gas generator.

COWL FLAPS—Devices arranged around certain air-cooled engine cowlings which may be opened or closed to regulate the flow of air around the engine.

CRAB—A flight condition in which the nose of the airplane is pointed into the wind a sufficient amount to counteract a crosswind and maintain a desired track over the ground.

CRAZING—Small fractures in aircraft windshields and windows caused from being exposed to the ultraviolet rays of the sun and temperature extremes.

CRITICAL ALTITUDE—The maximum altitude under standard atmospheric conditions at which a turbocharged engine can produce its rated horsepower.

CRITICAL ANGLE OF ATTACK—The angle of attack at which a wing stalls regardless of airspeed, flight attitude, or weight.

CRITICAL ENGINE—The engine whose failure has the most adverse effect on directional control.

CROSS CONTROLLED—A condition where aileron deflection is in the opposite direction of rudder deflection.

CROSSFEED—A system that allows either engine on a twin-engine airplane to draw fuel from any fuel tank.

CROSSWIND COMPONENT—The wind component, measured in knots, at 90° to the longitudinal axis of the runway.

CURRENT LIMITER—A device that limits the generator output to a level within that rated by the generator manufacturer.

DATUM (REFERENCE DATUM)—An imaginary vertical plane or line from which all measurements of moment arm are taken. The datum is established by the manufacturer. Once the datum has been selected, all moment arms and the location of CG range are measured from this point.

DECOMPRESSION SICKNESS—A condition where the low pressure at high altitudes allows bubbles of nitrogen to form in the blood and joints causing severe pain. Also known as the bends.

DEICER BOOTS—Inflatable rubber boots attached to the leading edge of an airfoil. They can be sequentially inflated and deflated to break away ice that has formed over their surface.

DEICING—Removing ice after it has formed.

DELAMINATION—The separation of layers.

DENSITY ALTITUDE—This altitude is pressure altitude corrected for variations from standard temperature. When conditions are standard, pressure altitude and density altitude are the same. If the temperature is above standard, the density altitude is higher than pressure altitude. If the temperature is below standard, the density altitude is lower than pressure altitude. This is an important altitude because it is directly related to the airplane's performance.

DESIGNATED PILOT EXAMINER (DPE)—An individual designated by the FAA to administer practical tests to pilot applicants.

DETONATION—The sudden release of heat energy from fuel in an aircraft engine caused by the fuel-air mixture reaching its critical pressure and temperature. Detonation occurs as a violent explosion rather than a smooth burning process.

DEWPOINT—The temperature at which air can hold no more water.

DIFFERENTIAL AILERONS—Control surface rigged such that the aileron moving up moves a greater distance than the aileron moving down. The up aileron produces extra parasite drag to compensate for the additional induced drag caused by the down aileron. This balancing of the drag forces helps minimize adverse yaw.

DIFFUSION—Reducing the velocity of air causing the pressure to increase.

DIRECTIONAL STABILITY—Stability about the vertical axis of an aircraft, whereby an aircraft tends to return, on its own, to flight aligned with the relative wind when disturbed from that equilibrium state. The vertical tail is the primary contributor to directional stability, causing an airplane in flight to align with the relative wind.

DITCHING—Emergency landing in water.

DOWNWASH—Air deflected perpendicular to the motion of the airfoil.

DRAG—An aerodynamic force on a body acting parallel and opposite to the relative wind. The resistance of the atmosphere to the relative motion of an aircraft. Drag opposes thrust and limits the speed of the airplane.

DRAG CURVE—A visual representation of the amount of drag of an aircraft at various airspeeds.

DRIFT ANGLE—Angle between heading and track.

DUCTED-FAN ENGINE—An engine-propeller combination that has the propeller enclosed in a radial shroud. Enclosing the propeller improves the efficiency of the propeller.

DUTCH ROLL—A combination of rolling and yawing oscillations that normally occurs when the dihedral effects of an aircraft are more powerful than the directional stability. Usually dynamically stable but objectionable in an airplane because of the oscillatory nature.

DYNAMIC HYDROPLANING—A condition that exists when landing on a surface with standing water deeper than the tread depth of the tires. When the brakes are applied, there is a possibility that the brake will lock up and the tire will ride on the surface of the water, much like a water ski. When the tires are hydroplaning, directional control and braking action are virtually impossible. An effective anti-skid system can minimize the effects of hydroplaning.

DYNAMIC STABILITY—
The property of an aircraft that causes it, when disturbed from straight-and-level flight, to develop forces or moments that restore the original condition of straight and level.

ELECTRICAL BUS—
See BUS BAR.

ELECTROHYDRAULIC—
Hydraulic control which is electrically actuated.

ELEVATOR—
The horizontal, movable primary control surface in the tail section, or empennage, of an airplane. The elevator is hinged to the trailing edge of the fixed horizontal stabilizer.

EMERGENCY LOCATOR TRANSMITTER—A small, self-contained radio transmitter that will automatically, upon the impact of a crash, transmit an emergency signal on 121.5, 243.0, or 406.0 MHz.

EMPENNAGE—The section of the airplane that consists of the vertical stabilizer, the horizontal stabilizer, and the associated control surfaces.

ENGINE PRESSURE RATIO (EPR)—The ratio of turbine discharge pressure divided by compressor inlet pressure that is used as an indication of the amount of thrust being developed by a turbine engine.

ENVIRONMENTAL SYSTEMS—
In an aircraft, the systems, including the supplemental oxygen systems, air conditioning systems, heaters, and pressurization systems, which make it possible for an occupant to function at high altitude.

EQUILIBRIUM—A condition that exists within a body when the sum of the moments of all of the forces acting on the body is equal to zero. In aerodynamics, equilibrium is when all opposing forces acting on an aircraft are balanced (steady, unaccelerated flight conditions).

EQUIVALENT SHAFT HORSEPOWER (ESHP)—
A measurement of the total horsepower of a turboprop engine, including that provided by jet thrust.

EXHAUST GAS TEMPERATURE (EGT)—The temperature of the exhaust gases as they leave the cylinders of a reciprocating engine or the turbine section of a turbine engine.

EXHAUST MANIFOLD—The part of the engine that collects exhaust gases leaving the cylinders.

EXHAUST—The rear opening of a turbine engine exhaust duct. The nozzle acts as an orifice, the size of which determines the density and velocity of the gases as they emerge from the engine.

FALSE HORIZON—An optical illusion where the pilot confuses a row of lights along a road or other straight line as the horizon.

FALSE START—
See HUNG START.

FEATHERING PROPELLER (FEATHERED)—A controllable pitch propeller with a pitch range sufficient to allow the blades to be turned parallel to the line of flight to reduce drag and prevent further damage to an engine that has been shut down after a malfunction.

FIXATION—
A psychological condition where the pilot fixes attention on a single source of information and ignores all other sources.

FIXED SHAFT TURBOPROP ENGINE—A turboprop engine where the gas producer spool is directly connected to the output shaft.

FIXED-PITCH PROPELLERS—
Propellers with fixed blade angles. Fixed-pitch propellers are designed as climb propellers, cruise propellers, or standard propellers.

FLAPS—Hinged portion of the trailing edge between the ailerons and fuselage. In some aircraft, ailerons and flaps are interconnected to produce full-span "flaperons." In either case, flaps change the lift and drag on the wing.

FLAT PITCH—
A propeller configuration when the blade chord is aligned with the direction of rotation.

FLICKER VERTIGO—
A disorientating condition caused from flickering light off the blades of the propeller.

FLIGHT DIRECTOR—An automatic flight control system in which the commands needed to fly the airplane are electronically computed and displayed on a flight instrument. The commands are followed by the human pilot with manual control inputs or, in the case of an autopilot system, sent to servos that move the flight controls.

FLIGHT IDLE—Engine speed, usually in the 70-80 percent range, for minimum flight thrust.

FLOATING—A condition when landing where the airplane does not settle to the runway due to excessive airspeed.

FORCE (F)—The energy applied to an object that attempts to cause the object to change its direction, speed, or motion. In aerodynamics, it is expressed as F, T (thrust), L (lift), W (weight), or D (drag), usually in pounds.

FORM DRAG—The part of parasite drag on a body resulting from the integrated effect of the static pressure acting normal to its surface resolved in the drag direction.

FORWARD SLIP—A slip in which the airplane's direction of motion continues the same as before the slip was begun. In a forward slip, the airplane's longitudinal axis is at an angle to its flightpath.

FREE POWER TURBINE ENGINE—A turboprop engine where the gas producer spool is on a separate shaft from the output shaft. The free power turbine spins independently of the gas producer and drives the output shaft.

FRICTION DRAG—The part of parasitic drag on a body resulting from viscous shearing stresses over its wetted surface.

FRISE-TYPE AILERON—Aileron having the nose portion projecting ahead of the hinge line. When the trailing edge of the aileron moves up, the nose projects below the wing's lower surface and produces some parasite drag, decreasing the amount of adverse yaw.

FUEL CONTROL UNIT—
The fuel-metering device used on a turbine engine that meters the proper quantity of fuel to be fed into the burners of the engine. It integrates the parameters of inlet air temperature, compressor speed, compressor discharge pressure, and exhaust gas temperature with the position of the cockpit power control lever.

FUEL EFFICIENCY—Defined as the amount of fuel used to produce a specific thrust or horsepower divided by the total potential power contained in the same amount of fuel.

FUEL HEATERS—A radiator-like device which has fuel passing through the core. A heat exchange occurs to keep the fuel temperature above the freezing point of water so that entrained water does not form ice crystals, which could block fuel flow.

FUEL INJECTION—
A fuel metering system used on some aircraft reciprocating engines in which a constant flow of fuel is fed to injection nozzles in the heads of all cylinders just outside of the intake valve. It differs from sequential fuel injection in which a timed charge of high-pressure fuel is sprayed directly into the combustion chamber of the cylinder.

FUEL LOAD—The expendable part of the load of the airplane. It includes only usable fuel, not fuel required to fill the lines or that which remains trapped in the tank sumps.

FUEL TANK SUMP—A sampling port in the lowest part of the fuel tank that the pilot can utilize to check for contaminants in the fuel.

FUSELAGE—The section of the airplane that consists of the cabin and/or cockpit, containing seats for the occupants and the controls for the airplane.

GAS GENERATOR—The basic power producing portion of a gas turbine engine and excluding such sections as the inlet duct, the fan section, free power turbines, and tailpipe. Each manufacturer designates what is included as the gas generator, but generally consists of the compressor, diffuser, combustor, and turbine.

GAS TURBINE ENGINE—A form of heat engine in which burning fuel adds energy to compressed air and accelerates the air through the remainder of the engine. Some of the energy is extracted to turn the air compressor, and the remainder accelerates the air to produce thrust. Some of this energy can be converted into torque to drive a propeller or a system of rotors for a helicopter.

GLIDE RATIO—The ratio between distance traveled and altitude lost during non-powered flight.

GLIDEPATH—The path of an aircraft relative to the ground while approaching a landing.

GLOBAL POSITION SYSTEM (GPS)—A satellite-based radio positioning, navigation, and time-transfer system.

GO-AROUND—
Terminating a landing approach.

GOVERNING RANGE—The range of pitch a propeller governor can control during flight.

GOVERNOR—A control which limits the maximum rotational speed of a device.

GROSS WEIGHT—
The total weight of a fully loaded aircraft including the fuel, oil, crew, passengers, and cargo.

GROUND ADJUSTABLE TRIM TAB—A metal trim tab on a control surface that is not adjustable in flight. Bent in one direction or another while on the ground to apply trim forces to the control surface.

GROUND EFFECT—A condition of improved performance encountered when an airplane is operating very close to the ground. When an airplane's wing is under the influence of ground effect, there is a reduction in upwash, downwash, and wingtip vortices. As a result of the reduced wingtip vortices, induced drag is reduced.

GROUND IDLE—Gas turbine engine speed usually 60-70 percent of the maximum r.p.m. range, used as a minimum thrust setting for ground operations.

GROUND LOOP—A sharp, uncontrolled change of direction of an airplane on the ground.

GROUND POWER UNIT (GPU)—
A type of small gas turbine whose purpose is to provide electrical power, and/or air pressure for starting aircraft engines. A ground unit is connected to the aircraft when needed. Similar to an aircraft-installed auxiliary power unit.

GROUNDSPEED (GS)—The actual speed of the airplane over the ground. It is true airspeed adjusted for wind. Groundspeed decreases with a headwind, and increases with a tailwind.

GROUND TRACK—The aircraft's path over the ground when in flight.

GUST PENETRATION SPEED—The speed that gives the greatest margin between the high and low mach speed buffets.

GYROSCOPIC PRECESSION—An inherent quality of rotating bodies, which causes an applied force to be manifested 90° in the direction of rotation from the point where the force is applied.

HAND PROPPING—Starting an engine by rotating the propeller by hand.

HEADING—The direction in which the nose of the aircraft is pointing during flight.

HEADING BUG—A marker on the heading indicator that can be rotated to a specific heading for reference purposes, or to command an autopilot to fly that heading.

HEADING INDICATOR—An instrument which senses airplane movement and displays heading based on a 360° azimuth, with the final zero omitted. The heading indicator, also called a directional gyro, is fundamentally a mechanical instrument designed to facilitate the use of the magnetic compass. The heading indicator is not affected by the forces that make the magnetic compass difficult to interpret.

HEADWIND COMPONENT—The component of atmospheric winds that acts opposite to the aircraft's flightpath.

HIGH PERFORMANCE AIRCRAFT—An aircraft with an engine of more than 200 horsepower.

HORIZON—The line of sight boundary between the earth and the sky.

HORSEPOWER—The term, originated by inventor James Watt, means the amount of work a horse could do in one second. One horsepower equals 550 foot-pounds per second, or 33,000 foot-pounds per minute.

HOT START—In gas turbine engines, a start which occurs with normal engine rotation, but exhaust temperature exceeds prescribed limits. This is usually caused by an excessively rich mixture in the combustor. The fuel to the engine must be terminated immediately to prevent engine damage.

HUNG START—In gas turbine engines, a condition of normal light off but with r.p.m. remaining at some low value rather than increasing to the normal idle r.p.m. This is often the result of insufficient power to the engine from the starter. In the event of a hung start, the engine should be shut down.

HYDRAULICS—The branch of science that deals with the transmission of power by incompressible fluids under pressure.

HYDROPLANING—A condition that exists when landing on a surface with standing water deeper than the tread depth of the tires. When the brakes are applied, there is a possibility that the brake will lock up and the tire will ride on the surface of the water, much like a water ski. When the tires are hydroplaning, directional control and braking action are virtually impossible. An effective anti-skid system can minimize the effects of hydroplaning.

HYPOXIA—A lack of sufficient oxygen reaching the body tissues.

IFR (INSTRUMENT FLIGHT RULES)—Rules that govern the procedure for conducting flight in weather conditions below VFR weather minimums. The term "IFR" also is used to define weather conditions and the type of flight plan under which an aircraft is operating.

IGNITER PLUGS—The electrical device used to provide the spark for starting combustion in a turbine engine. Some igniters resemble spark plugs, while others, called glow plugs, have a coil of resistance wire that glows red hot when electrical current flows through the coil.

IMPACT ICE—Ice that forms on the wings and control surfaces or on the carburetor heat valve, the walls of the air scoop, or the carburetor units during flight. Impact ice collecting on the metering elements of the carburetor may upset fuel metering or stop carburetor fuel flow.

INCLINOMETER—An instrument consisting of a curved glass tube, housing a glass ball, and damped with a fluid similar to kerosene. It may be used to indicate inclination, as a level, or, as used in the turn indicators, to show the relationship between gravity and centrifugal force in a turn.

INDICATED AIRSPEED (IAS)—The direct instrument reading obtained from the airspeed indicator, uncorrected for variations in atmospheric density, installation error, or instrument error. Manufacturers use this airspeed as the basis for determining airplane performance. Takeoff, landing, and stall speeds listed in the AFM or POH are indicated airspeeds and do not normally vary with altitude or temperature.

INDICATED ALTITUDE—The altitude read directly from the altimeter (uncorrected) when it is set to the current altimeter setting.

INDUCED DRAG—That part of total drag which is created by the production of lift. Induced drag increases with a decrease in airspeed.

INDUCTION MANIFOLD—The part of the engine that distributes intake air to the cylinders.

INERTIA—The opposition which a body offers to a change of motion.

INITIAL CLIMB—This stage of the climb begins when the airplane leaves the ground, and a pitch attitude has been established to climb away from the takeoff area.

INTEGRAL FUEL TANK—
A portion of the aircraft structure, usually a wing, which is sealed off and used as a fuel tank. When a wing is used as an integral fuel tank, it is called a "wet wing."

INTERCOOLER—A device used to reduce the temperature of the compressed air before it enters the fuel metering device. The resulting cooler air has a higher density, which permits the engine to be operated with a higher power setting.

INTERNAL COMBUSTION ENGINES—An engine that produces power as a result of expanding hot gases from the combustion of fuel and air within the engine itself. A steam engine where coal is burned to heat up water inside the engine is an example of an external combustion engine.

INTERSTAGE TURBINE TEMPERATURE (ITT)—The temperature of the gases between the high pressure and low pressure turbines.

INVERTER—An electrical device that changes DC to AC power.

ISA (INTERNATIONAL STANDARD ATMOSPHERE)—
Standard atmospheric conditions consisting of a temperature of 59°F (15°C), and a barometric pressure of 29.92 in. Hg. (1013.2 mb) at sea level. ISA values can be calculated for various altitudes using a standard lapse rate of approximately 2°C per 1,000 feet.

JET POWERED AIRPLANE—An aircraft powered by a turbojet or turbofan engine.

KINESTHESIA—The sensing of movements by feel.

LATERAL AXIS—An imaginary line passing through the center of gravity of an airplane and extending across the airplane from wingtip to wingtip.

LATERAL STABILITY (ROLLING)—The stability about the longitudinal axis of an aircraft. Rolling stability or the ability of an airplane to return to level flight due to a disturbance that causes one of the wings to drop.

LEAD-ACID BATTERY—
A commonly used secondary cell having lead as its negative plate and lead peroxide as its positive plate. Sulfuric acid and water serve as the electrolyte.

LEADING EDGE DEVICES—
High lift devices which are found on the leading edge of the airfoil. The most common types are fixed slots, movable slats, and leading edge flaps.

LEADING EDGE—The part of an airfoil that meets the airflow first.

LEADING EDGE FLAP—
A portion of the leading edge of an airplane wing that folds downward to increase the camber, lift, and drag of the wing. The leading-edge flaps are extended for takeoffs and landings to increase the amount of aerodynamic lift that is produced at any given airspeed.

LICENSED EMPTY WEIGHT—
The empty weight that consists of the airframe, engine(s), unusable fuel, and undrainable oil plus standard and optional equipment as specified in the equipment list. Some manufacturers used this term prior to GAMA standardization.

LIFT—One of the four main forces acting on an aircraft. On a fixed-wing aircraft, an upward force created by the effect of airflow as it passes over and under the wing.

LIFT COEFFICIENT— A coefficient representing the lift of a given airfoil. Lift coefficient is obtained by dividing the lift by the free-stream dynamic pressure and the representative area under consideration.

LIFT/DRAG RATIO—
The efficiency of an airfoil section. It is the ratio of the coefficient of lift to the coefficient of drag for any given angle of attack.

LIFT-OFF—The act of becoming airborne as a result of the wings lifting the airplane off the ground, or the pilot rotating the nose up, increasing the angle of attack to start a climb.

LIMIT LOAD FACTOR—Amount of stress, or load factor, that an aircraft can withstand before structural damage or failure occurs.

LOAD FACTOR—The ratio of the load supported by the airplane's wings to the actual weight of the aircraft and its contents. Also referred to as G-loading.

LONGITUDINAL AXIS—
An imaginary line through an aircraft from nose to tail, passing through its center of gravity. The longitudinal axis is also called the roll axis of the aircraft. Movement of the ailerons rotates an airplane about its longitudinal axis.

LONGITUDINAL STABILITY (PITCHING)—Stability about the lateral axis. A desirable characteristic of an airplane whereby it tends to return to its trimmed angle of attack after displacement.

MACH—Speed relative to the speed of sound. Mach 1 is the speed of sound.

MACH BUFFET—
Airflow separation behind a shock-wave pressure barrier caused by airflow over flight surfaces exceeding the speed of sound.

MACH COMPENSATING DEVICE—A device to alert the pilot of inadvertent excursions beyond its certified maximum operating speed.

MACH CRITICAL—The MACH speed at which some portion of the airflow over the wing first equals MACH 1.0. This is also the speed at which a shock wave first appears on the airplane.

MACH TUCK—A condition that can occur when operating a swept-wing airplane in the transonic speed range. A shock wave could form in the root portion of the wing and cause the air behind it to separate. This shock-induced separation causes the center of pressure to move aft. This, combined with the increasing amount of nose down force at higher speeds to maintain left flight, causes the nose to "tuck." If not corrected, the airplane could enter a steep, sometimes unrecoverable dive.

MAGNETIC COMPASS—A device for determining direction measured from magnetic north.

MAIN GEAR—The wheels of an aircraft's landing gear that supports the major part of the aircraft's weight.

MANEUVERABILITY—Ability of an aircraft to change directions along a flightpath and withstand the stresses imposed upon it.

MANEUVERING SPEED (V_A) — The maximum speed where full, abrupt control movement can be used without overstressing the airframe.

MANIFOLD PRESSURE (MP)— The absolute pressure of the fuel/air mixture within the intake manifold, usually indicated in inches of mercury.

MAXIMUM ALLOWABLE TAKEOFF POWER—The maximum power an engine is allowed to develop for a limited period of time; usually about one minute.

MAXIMUM LANDING WEIGHT—The greatest weight that an airplane normally is allowed to have at landing.

MAXIMUM RAMP WEIGHT— The total weight of a loaded aircraft, including all fuel. It is greater than the takeoff weight due to the fuel that will be burned during the taxi and runup operations. Ramp weight may also be referred to as taxi weight.

MAXIMUM TAKEOFF WEIGHT—The maximum allowable weight for takeoff.

MAXIMUM WEIGHT— The maximum authorized weight of the aircraft and all of its equipment as specified in the Type Certificate Data Sheets (TCDS) for the aircraft.

MAXIMUM ZERO FUEL WEIGHT (GAMA)—The maximum weight, exclusive of usable fuel.

MINIMUM CONTROLLABLE AIRSPEED—An airspeed at which any further increase in angle of attack, increase in load factor, or reduction in power, would result in an immediate stall.

MINIMUM DRAG SPEED (L/D_{MAX})—The point on the total drag curve where the lift-to-drag ratio is the greatest. At this speed, total drag is minimized.

MIXTURE—The ratio of fuel to air entering the engine's cylinders.

M_{MO}—Maximum operating speed expressed in terms of a decimal of mach speed.

MOMENT ARM—The distance from a datum to the applied force.

MOMENT INDEX (OR INDEX)— A moment divided by a constant such as 100, 1,000, or 10,000. The purpose of using a moment index is to simplify weight and balance computations of airplanes where heavy items and long arms result in large, unmanageable numbers.

MOMENT—The product of the weight of an item multiplied by its arm. Moments are expressed in pound-inches (lb-in). Total moment is the weight of the airplane multiplied by the distance between the datum and the CG.

MOVABLE SLAT—A movable auxiliary airfoil on the leading edge of a wing. It is closed in normal flight but extends at high angles of attack. This allows air to continue flowing over the top of the wing and delays airflow separation.

MUSHING—A flight condition caused by slow speed where the control surfaces are marginally effective.

N_1, N_2, N_3—Spool speed expressed in percent rpm. N_1 on a turboprop is the gas producer speed. N_1 on a turbofan or turbojet engine is the fan speed or low pressure spool speed. N_2 is the high pressure spool speed on engine with 2 spools and medium pressure spool on engines with 3 spools with N_3 being the high pressure spool.

NACELLE— A streamlined enclosure on an aircraft in which an engine is mounted. On multiengine propeller-driven airplanes, the nacelle is normally mounted on the leading edge of the wing.

NEGATIVE STATIC STABILITY—The initial tendency of an aircraft to continue away from the original state of equilibrium after being disturbed.

NEGATIVE TORQUE SENSING (NTS)— A system in a turboprop engine that prevents the engine from being driven by the propeller. The NTS increases the blade angle when the propellers try to drive the engine.

NEUTRAL STATIC STABILITY—The initial tendency of an aircraft to remain in a new condition after its equilibrium has been disturbed.

NICKEL-CADMIUM BATTERY (NICAD)— A battery made up of alkaline secondary cells. The positive plates are nickel hydroxide, the negative plates are cadmium hydroxide, and potassium hydroxide is used as the electrolyte.

NORMAL CATEGORY— An airplane that has a seating configuration, excluding pilot seats, of nine or less, a maximum certificated takeoff weight of 12,500 pounds or less, and intended for nonacrobatic operation.

NORMALIZING (TURBONORMALIZING)—A turbocharger that maintains sea level pressure in the induction manifold at altitude.

OCTANE—The rating system of aviation gasoline with regard to its antidetonating qualities.

OVERBOOST—A condition in which a reciprocating engine has exceeded the maximum manifold pressure allowed by the manufacturer. Can cause damage to engine components.

OVERSPEED—A condition in which an engine has produced more r.p.m. than the manufacturer recommends, or a condition in which the actual engine speed is higher than the desired engine speed as set on the propeller control.

OVERTEMP—A condition in which a device has reached a temperature above that approved by the manufacturer or any exhaust temperature that exceeds the maximum allowable for a given operating condition or time limit. Can cause internal damage to an engine.

OVERTORQUE—A condition in which an engine has produced more torque (power) than the manufacturer recommends, or a condition in a turboprop or turboshaft engine where the engine power has exceeded the maximum allowable for a given operating condition or time limit. Can cause internal damage to an engine.

PARASITE DRAG—That part of total drag created by the design or shape of airplane parts. Parasite drag increases with an increase in airspeed.

PAYLOAD (GAMA)—The weight of occupants, cargo, and baggage.

P-FACTOR—A tendency for an aircraft to yaw to the left due to the descending propeller blade on the right producing more thrust than the ascending blade on the left. This occurs when the aircraft's longitudinal axis is in a climbing attitude in relation to the relative wind. The P-factor would be to the right if the aircraft had a counterclockwise rotating propeller.

PILOT'S OPERATING HANDBOOK (POH)—A document developed by the airplane manufacturer and contains the FAA-approved Airplane Flight Manual (AFM) information.

PISTON ENGINE—A reciprocating engine.

PITCH—The rotation of an airplane about its lateral axis, or on a propeller, the blade angle as measured from plane of rotation.

PIVOTAL ALTITUDE—A specific altitude at which, when an airplane turns at a given groundspeed, a projecting of the sighting reference line to a selected point on the ground will appear to pivot on that point.

PNEUMATIC SYSTEMS—The power system in an aircraft used for operating such items as landing gear, brakes, and wing flaps with compressed air as the operating fluid.

PORPOISING—Oscillating around the lateral axis of the aircraft during landing.

POSITION LIGHTS—Lights on an aircraft consisting of a red light on the left wing, a green light on the right wing, and a white light on the tail. CFRs require that these lights be displayed in flight from sunset to sunrise.

POSITIVE STATIC STABILITY—The initial tendency to return to a state of equilibrium when disturbed from that state.

POWER DISTRIBUTION BUS—See BUS BAR.

POWER LEVER—The cockpit lever connected to the fuel control unit for scheduling fuel flow to the combustion chambers of a turbine engine.

POWER—Implies work rate or units of work per unit of time, and as such, it is a function of the speed at which the force is developed. The term "power required" is generally associated with reciprocating engines.

POWERPLANT—A complete engine and propeller combination with accessories.

PRACTICAL SLIP LIMIT—The maximum slip an aircraft is capable of performing due to rudder travel limits.

PRECESSION—The tilting or turning of a gyro in response to deflective forces causing slow drifting and erroneous indications in gyroscopic instruments.

PREIGNITION—Ignition occurring in the cylinder before the time of normal ignition. Preignition is often caused by a local hot spot in the combustion chamber igniting the fuel/air mixture.

PRESSURE ALTITUDE—The altitude indicated when the altimeter setting window (barometric scale) is adjusted to 29.92. This is the altitude above the standard datum plane, which is a theoretical plane where air pressure (corrected to 15°C) equals 29.92 in. Hg. Pressure altitude is used to compute density altitude, true altitude, true airspeed, and other performance data.

PROFILE DRAG—The total of the skin friction drag and form drag for a two-dimensional airfoil section.

PROPELLER BLADE ANGLE—The angle between the propeller chord and the propeller plane of rotation.

PROPELLER LEVER—The control on a free power turbine turboprop that controls propeller speed and the selection for propeller feathering.

PROPELLER SLIPSTREAM—The volume of air accelerated behind a propeller producing thrust.

PROPELLER SYNCHRONIZATION—A condition in which all of the propellers have their pitch automatically adjusted to maintain a constant r.p.m. among all of the engines of a multiengine aircraft.

PROPELLER—A device for propelling an aircraft that, when rotated, produces by its action on the air, a thrust approximately perpendicular to its plane of rotation. It includes the control components normally supplied by its manufacturer.

RAMP WEIGHT—The total weight of the aircraft while on the ramp. It differs from takeoff weight by the weight of the fuel that will be consumed in taxiing to the point of takeoff.

RATE OF TURN—The rate in degrees/second of a turn.

RECIPROCATING ENGINE—An engine that converts the heat energy from burning fuel into the reciprocating movement of the pistons. This movement is converted into a rotary motion by the connecting rods and crankshaft.

REDUCTION GEAR—The gear arrangement in an aircraft engine that allows the engine to turn at a faster speed than the propeller.

REGION OF REVERSE COMMAND—Flight regime in which flight at a higher airspeed requires a lower power setting and a lower airspeed requires a higher power setting in order to maintain altitude.

REGISTRATION CERTIFICATE—A State and Federal certificate that documents aircraft ownership.

RELATIVE WIND—The direction of the airflow with respect to the wing. If a wing moves forward horizontally, the relative wind moves backward horizontally. Relative wind is parallel to and opposite the flightpath of the airplane.

REVERSE THRUST—A condition where jet thrust is directed forward during landing to increase the rate of deceleration.

REVERSING PROPELLER—A propeller system with a pitch change mechanism that includes full reversing capability. When the pilot moves the throttle controls to reverse, the blade angle changes to a pitch angle and produces a reverse thrust, which slows the airplane down during a landing.

ROLL—The motion of the aircraft about the longitudinal axis. It is controlled by the ailerons.

ROUNDOUT (FLARE)—A pitch-up during landing approach to reduce rate of descent and forward speed prior to touchdown.

RUDDER—The movable primary control surface mounted on the trailing edge of the vertical fin of an airplane. Movement of the rudder rotates the airplane about its vertical axis.

RUDDERVATOR—A pair of control surfaces on the tail of an aircraft arranged in the form of a V. These surfaces, when moved together by the control wheel, serve as elevators, and when moved differentially by the rudder pedals, serve as a rudder.

RUNWAY CENTERLINE LIGHTS—Runway centerline lights are installed on some precision approach runways to facilitate landing under adverse visibility conditions. They are located along the runway centerline and are spaced at 50-foot intervals. When viewed from the landing threshold, the runway centerline lights are white until the last 3,000 feet of the runway. The white lights begin to alternate with red for the next 2,000 feet, and for the last 1,000 feet of the runway, all centerline lights are red.

RUNWAY CENTERLINE MARKINGS—The runway centerline identifies the center of the runway and provides alignment guidance during takeoff and landings. The centerline consists of a line of uniformly spaced stripes and gaps.

RUNWAY EDGE LIGHTS—Runway edge lights are used to outline the edges of runways during periods of darkness or restricted visibility conditions. These light systems are classified according to the intensity or brightness they are capable of producing: they are the High Intensity Runway Lights (HIRL), Medium Intensity Runway Lights (MIRL), and the Low Intensity Runway Lights (LIRL). The HIRL and MIRL systems have variable intensity controls, whereas the LIRLs normally have one intensity setting.

RUNWAY END IDENTIFIER LIGHTS (REIL)—One component of the runway lighting system. These lights are installed at many airfields to provide rapid and positive identification of the approach end of a particular runway.

RUNWAY INCURSION—Any occurrence at an airport involving an aircraft, vehicle, person, or object on the ground that creates a collision hazard or results in loss of separation with an aircraft taking off, intending to takeoff, landing, or intending to land.

RUNWAY THRESHOLD MARKINGS—Runway threshold markings come in two configurations. They either consist of eight longitudinal stripes of uniform dimensions disposed symmetrically about the runway centerline, or the number of stripes is related to the runway width. A threshold marking helps identify the beginning of the runway that is available for landing. In some instances, the landing threshold may be displaced.

SAFETY (SQUAT) SWITCH—An electrical switch mounted on one of the landing gear struts. It is used to sense when the weight of the aircraft is on the wheels.

SCAN—A procedure used by the pilot to visually identify all resources of information in flight.

SEA LEVEL—A reference height used to determine standard atmospheric conditions and altitude measurements.

SEGMENTED CIRCLE—A visual ground based structure to provide traffic pattern information.

SERVICE CEILING—The maximum density altitude where the best rate-of-climb airspeed will produce a 100 feet-per-minute climb at maximum weight while in a clean configuration with maximum continuous power.

SERVO TAB—An auxiliary control mounted on a primary control surface, which automatically moves in the direction opposite the primary control to provide an aerodynamic assist in the movement of the control.

SHAFT HORSE POWER (SHP)—Turboshaft engines are rated in shaft horsepower and calculated by use of a dynamometer device. Shaft horsepower is exhaust thrust converted to a rotating shaft.

SHOCK WAVES—A compression wave formed when a body moves through the air at a speed greater than the speed of sound.

SIDESLIP—A slip in which the airplane's longitudinal axis remains parallel to the original flightpath, but the airplane no longer flies straight ahead. Instead, the horizontal component of wing lift forces the airplane to move sideways toward the low wing.

SINGLE ENGINE ABSOLUTE CEILING—The altitude that a twin-engine airplane can no longer climb with one engine inoperative.

SINGLE ENGINE SERVICE CEILING—The altitude that a twin-engine airplane can no longer climb at a rate greater then 50 f.p.m. with one engine inoperative.

SKID—A condition where the tail of the airplane follows a path outside the path of the nose during a turn.

SLIP—An intentional maneuver to decrease airspeed or increase rate of descent, and to compensate for a crosswind on landing. A slip can also be unintentional when the pilot fails to maintain the aircraft in coordinated flight.

SPECIFIC FUEL CONSUMPTION—Number of pounds of fuel consumed in 1 hour to produce 1 HP.

SPEED—The distance traveled in a given time.

SPEED BRAKES—A control system that extends from the airplane structure into the airstream to produce drag and slow the airplane.

SPEED INSTABILITY—A condition in the region of reverse command where a disturbance that causes the airspeed to decrease causes total drag to increase, which in turn, causes the airspeed to decrease further.

SPEED SENSE—The ability to sense instantly and react to any reasonable variation of airspeed.

SPIN—An aggravated stall that results in what is termed an "autorotation" wherein the airplane follows a downward corkscrew path. As the airplane rotates around the vertical axis, the rising wing is less stalled than the descending wing creating a rolling, yawing, and pitching motion.

SPIRAL INSTABILITY—A condition that exists when the static directional stability of the airplane is very strong as compared to the effect of its dihedral in maintaining lateral equilibrium.

SPIRALING SLIPSTREAM—The slipstream of a propeller-driven airplane rotates around the airplane. This slipstream strikes the left side of the vertical fin, causing the airplane to yaw slightly. Vertical stabilizer offset is sometimes used by aircraft designers to counteract this tendency.

SPLIT SHAFT TURBINE ENGINE—See FREE POWER TURBINE ENGINE.

SPOILERS—High-drag devices that can be raised into the air flowing over an airfoil, reducing lift and increasing drag. Spoilers are used for roll control on some aircraft. Deploying spoilers on both wings at the same time allows the aircraft to descend without gaining speed. Spoilers are also used to shorten the ground roll after landing.

SPOOL—A shaft in a turbine engine which drives one or more compressors with the power derived from one or more turbines.

STABILATOR—A single-piece horizontal tail surface on an airplane that pivots around a central hinge point. A stabilator serves the purposes of both the horizontal stabilizer and the elevator.

STABILITY—The inherent quality of an airplane to correct for conditions that may disturb its equilibrium, and to return or to continue on the original flightpath. It is primarily an airplane design characteristic.

STABILIZED APPROACH—A landing approach in which the pilot establishes and maintains a constant angle glidepath towards a predetermined point on the landing runway. It is based on the pilot's judgment of certain visual cues, and depends on the maintenance of a constant final descent airspeed and configuration.

STALL—A rapid decrease in lift caused by the separation of airflow from the wing's surface brought on by exceeding the critical angle of attack. A stall can occur at any pitch attitude or airspeed.

STALL STRIPS—A spoiler attached to the inboard leading edge of some wings to cause the center section of the wing to stall before the tips. This assures lateral control throughout the stall.

STANDARD ATMOSPHERE—
At sea level, the standard atmosphere consists of a barometric pressure of 29.92 inches of mercury (in. Hg.) or 1013.2 millibars, and a temperature of 15°C (59°F). Pressure and temperature normally decrease as altitude increases. The standard lapse rate in the lower atmosphere for each 1,000 feet of altitude is approximately 1 in. Hg. and 2°C (3.5°F). For example, the standard pressure and temperature at 3,000 feet mean sea level (MSL) is 26.92 in. Hg. (29.92 - 3) and 9°C (15°C - 6°C).

STANDARD DAY—
See STANDARD ATMOSPHERE.

STANDARD EMPTY WEIGHT (GAMA)—This weight consists of the airframe, engines, and all items of operating equipment that have fixed locations and are permanently installed in the airplane; including fixed ballast, hydraulic fluid, unusable fuel, and full engine oil.

STANDARD WEIGHTS—These have been established for numerous items involved in weight and balance computations. These weights should not be used if actual weights are available.

STANDARD-RATE TURN—A turn at the rate of 3° per second which enables the airplane to complete a 360° turn in 2 minutes.

STARTER/GENERATOR—
A combined unit used on turbine engines. The device acts as a starter for rotating the engine, and after running, internal circuits are shifted to convert the device into a generator.

STATIC STABILITY—The initial tendency an aircraft displays when disturbed from a state of equilibrium.

STATION—A location in the airplane that is identified by a number designating its distance in inches from the datum. The datum is, therefore, identified as station zero. An item located at station +50 would have an arm of 50 inches.

STICK PULLER—A device that applies aft pressure on the control column when the airplane is approaching the maximum operating speed.

STICK PUSHER—A device that applies an abrupt and large forward force on the control column when the airplane is nearing an angle of attack where a stall could occur.

STICK SHAKER—An artificial stall warning device that vibrates the control column.

STRESS RISERS—
A scratch, groove, rivet hole, forging defect or other structural discontinuity that causes a concentration of stress.

SUBSONIC—Speed below the speed of sound.

SUPERCHARGER—An engine- or exhaust-driven air compressor used to provide additional pressure to the induction air so the engine can produce additional power.

SUPERSONIC—Speed above the speed of sound.

SUPPLEMENTAL TYPE CERTIFICATE (STC)—
A certificate authorizing an alteration to an airframe, engine, or component that has been granted an Approved Type Certificate.

SWEPT WING—A wing planform in which the tips of the wing are farther back than the wing root.

TAILWHEEL AIRCRAFT—
SEE CONVENTIONAL LANDING GEAR.

TAKEOFF ROLL (GROUND ROLL)—The total distance required for an aircraft to become airborne.

TARGET REVERSER—A thrust reverser in a jet engine in which clamshell doors swivel from the stowed position at the engine tailpipe to block all of the outflow and redirect some component of the thrust forward.

TAXIWAY LIGHTS—
Omnidirectional lights that outline the edges of the taxiway and are blue in color.

TAXIWAY TURNOFF LIGHTS—
Flush lights which emit a steady green color.

TETRAHEDRON—
A large, triangular-shaped, kite-like object installed near the runway. Tetrahedrons are mounted on a pivot and are free to swing with the wind to show the pilot the direction of the wind as an aid in takeoffs and landings.

THROTTLE—The valve in a carburetor or fuel control unit that determines the amount of fuel-air mixture that is fed to the engine.

THRUST LINE—An imaginary line passing through the center of the propeller hub, perpendicular to the plane of the propeller rotation.

THRUST REVERSERS—Devices which redirect the flow of jet exhaust to reverse the direction of thrust.

THRUST—The force which imparts a change in the velocity of a mass. This force is measured in pounds but has no element of time or rate. The term, thrust required, is generally associated with jet engines. A forward force which propels the airplane through the air.

TIMING—The application of muscular coordination at the proper instant to make flight, and all maneuvers incident thereto, a constant smooth process.

TIRE CORD—Woven metal wire laminated into the tire to provide extra strength. A tire showing any cord must be replaced prior to any further flight.

TORQUE METER—An indicator used on some large reciprocating engines or on turboprop engines to indicate the amount of torque the engine is producing.

TORQUE SENSOR—
See TORQUE METER.

TORQUE—**1.** A resistance to turning or twisting. **2.** Forces that produce a twisting or rotating motion. **3.** In an airplane, the tendency of the aircraft to turn (roll) in the opposite direction of rotation of the engine and propeller.

TOTAL DRAG—The sum of the parasite and induced drag.

TOUCHDOWN ZONE LIGHTS—
Two rows of transverse light bars disposed symmetrically about the runway centerline in the runway touchdown zone.

TRACK—The actual path made over the ground in flight.

TRAILING EDGE—The portion of the airfoil where the airflow over the upper surface rejoins the lower surface airflow.

TRANSITION LINER—
The portion of the combustor that directs the gases into the turbine plenum.

TRANSONIC—At the speed of sound.

TRANSPONDER—The airborne portion of the secondary surveillance radar system. The transponder emits a reply when queried by a radar facility.

TRICYCLE GEAR—Landing gear employing a third wheel located on the nose of the aircraft.

TRIM TAB—A small auxiliary hinged portion of a movable control surface that can be adjusted during flight to a position resulting in a balance of control forces.

TRIPLE SPOOL ENGINE—
Usually a turbofan engine design where the fan is the N_1 compressor, followed by the N_2 intermediate compressor, and the N_3 high pressure compressor, all of which rotate on separate shafts at different speeds.

TROPOPAUSE—The boundary layer between the troposphere and the mesosphere which acts as a lid to confine most of the water vapor, and the associated weather, to the troposphere.

TROPOSPHERE—The layer of the atmosphere extending from the surface to a height of 20,000 to 60,000 feet depending on latitude.

TRUE AIRSPEED (TAS)—
Calibrated airspeed corrected for altitude and nonstandard temperature. Because air density decreases with an increase in altitude, an airplane has to be flown faster at higher altitudes to cause the same pressure difference between pitot impact pressure and static pressure. Therefore, for a given calibrated airspeed, true airspeed increases as altitude increases; or for a given true airspeed, calibrated airspeed decreases as altitude increases.

TRUE ALTITUDE—The vertical distance of the airplane above sea level—the actual altitude. It is often expressed as feet above mean sea level (MSL). Airport, terrain, and obstacle elevations on aeronautical charts are true altitudes.

T-TAIL—An aircraft with the horizontal stabilizer mounted on the top of the vertical stabilizer, forming a T.

TURBINE BLADES—The portion of the turbine assembly that absorbs the energy of the expanding gases and converts it into rotational energy.

TURBINE OUTLET TEMPERATURE (TOT)—
The temperature of the gases as they exit the turbine section.

TURBINE PLENUM—The portion of the combustor where the gases are collected to be evenly distributed to the turbine blades.

TURBINE ROTORS—The portion of the turbine assembly that mounts to the shaft and holds the turbine blades in place.

TURBINE SECTION—The section of the engine that converts high pressure high temperature gas into rotational energy.

TURBOCHARGER—
An air compressor driven by exhaust gases, which increases the pressure of the air going into the engine through the carburetor or fuel injection system.

TURBOFAN ENGINE—A turbojet engine in which additional propulsive thrust is gained by extending a portion of the compressor or turbine blades outside the inner engine case. The extended blades propel bypass air along the engine axis but between the inner and outer casing. The air is not combusted but does provide additional thrust.

TURBOJET ENGINE—A jet engine incorporating a turbine-driven air compressor to take in and compress air for the combustion of fuel, the gases of combustion being used both to rotate the turbine and create a thrust producing jet.

TURBOPROP ENGINE—A turbine engine that drives a propeller through a reduction gearing arrangement. Most of the energy in the exhaust gases is converted into torque, rather than its acceleration being used to propel the aircraft.

TURBULENCE—An occurrence in which a flow of fluid is unsteady.

TURN COORDINATOR—A rate gyro that senses both roll and yaw due to the gimbal being canted. Has largely replaced the turn-and-slip indicator in modern aircraft.

TURN-AND-SLIP INDICATOR—
A flight instrument consisting of a rate gyro to indicate the rate of yaw and a curved glass inclinometer to indicate the relationship between gravity and centrifugal force. The turn-and-slip indicator indicates the relationship between angle of bank and rate of yaw. Also called a turn-and-bank indicator.

TURNING ERROR—One of the errors inherent in a magnetic compass caused by the dip compensating weight. It shows up only on turns to or from northerly headings in the Northern Hemisphere and southerly headings in the Southern Hemisphere. Turning error causes the compass to lead turns to the north or south and lag turns away from the north or south.

ULTIMATE LOAD FACTOR—
In stress analysis, the load that causes physical breakdown in an aircraft or aircraft component during a strength test, or the load that according to computations, should cause such a breakdown.

UNFEATHERING ACCUMULATOR—Tanks that hold oil under pressure which can be used to unfeather a propeller.

UNICOM—
A nongovernment air/ground radio communication station which may provide airport information at public use airports where there is no tower or FSS.

UNUSABLE FUEL—Fuel that cannot be consumed by the engine. This fuel is considered part of the empty weight of the aircraft.

USEFUL LOAD—The weight of the pilot, copilot, passengers, baggage, usable fuel, and drainable oil. It is the basic empty weight subtracted from the maximum allowable gross weight. This term applies to general aviation aircraft only.

UTILITY CATEGORY—
An airplane that has a seating configuration, excluding pilot seats, of nine or less, a maximum certificated takeoff weight of 12,500 pounds or less, and intended for limited acrobatic operation.

V-BARS—The flight director displays on the attitude indicator that provide control guidance to the pilot.

V-SPEEDS—Designated speeds for a specific flight condition.

VAPOR LOCK—A condition in which air enters the fuel system and it may be difficult, or impossible, to restart the engine. Vapor lock may occur as a result of running a fuel tank completely dry, allowing air to enter the fuel system. On fuel-injected engines, the fuel may become so hot it vaporizes in the fuel line, not allowing fuel to reach the cylinders.

V_A—The design maneuvering speed. This is the "rough air" speed and the maximum speed for abrupt maneuvers. If during flight, rough air or severe turbulence is encountered, reduce the airspeed to maneuvering speed or less to minimize stress on the airplane structure. It is important to consider weight when referencing this speed. For example, V_A may be 100 knots when an airplane is heavily loaded, but only 90 knots when the load is light.

VECTOR—A force vector is a graphic representation of a force and shows both the magnitude and direction of the force.

VELOCITY—The speed or rate of movement in a certain direction.

VERTICAL AXIS—An imaginary line passing vertically through the center of gravity of an aircraft. The vertical axis is called the z-axis or the yaw axis.

VERTICAL CARD COMPASS—
A magnetic compass that consists of an azimuth on a vertical card, resembling a heading indicator with a fixed miniature airplane to accurately present the heading of the aircraft. The design uses eddy current damping to minimize lead and lag during turns.

VERTICAL SPEED INDICATOR (VSI)—
An instrument that uses static pressure to display a rate of climb or descent in feet per minute. The VSI can also sometimes be called a vertical velocity indicator (VVI).

VERTICAL STABILITY—Stability about an aircraft's vertical axis. Also called yawing or directional stability.

V_{FE}—The maximum speed with the flaps extended. The upper limit of the white arc.

V_{FO}—The maximum speed that the flaps can be extended or retracted.

VFR TERMINAL AREA CHARTS (1:250,000)—
Depict Class B airspace which provides for the control or segregation of all the aircraft within the Class B airspace. The chart depicts topographic information and aeronautical information which includes visual and radio aids to navigation, airports, controlled airspace, restricted areas, obstructions, and related data.

V-G DIAGRAM—A chart that relates velocity to load factor. It is valid only for a specific weight, configuration, and altitude and shows the maximum amount of positive or negative lift the airplane is capable of generating at a given speed. Also shows the safe load factor limits and the load factor that the aircraft can sustain at various speeds.

VISUAL APPROACH SLOPE INDICATOR (VASI)—
The most common visual glidepath system in use. The VASI provides obstruction clearance within 10° of the extended runway centerline, and to 4 nautical miles (NM) from the runway threshold.

VISUAL FLIGHT RULES (VFR)—
Code of Federal Regulations that govern the procedures for conducting flight under visual conditions.

V_{LE}—Landing gear extended speed. The maximum speed at which an airplane can be safely flown with the landing gear extended.

V_{LOF}—Lift-off speed. The speed at which the aircraft departs the runway during takeoff.

V_{LO}—Landing gear operating speed. The maximum speed for extending or retracting the landing gear if using an airplane equipped with retractable landing gear.

V_{MC}—Minimum control airspeed. This is the minimum flight speed at which a twin-engine airplane can be satisfactorily controlled when an engine suddenly becomes inoperative and the remaining engine is at takeoff power.

V_{MD}—Minimum drag speed.

V_{MO}—Maximum operating speed expressed in knots.

V_{NE}—Never-exceed speed. Operating above this speed is prohibited since it may result in damage or structural failure. The red line on the airspeed indicator.

V_{NO}—Maximum structural cruising speed. Do not exceed this speed except in smooth air. The upper limit of the green arc.

V_P—Minimum dynamic hydroplaning speed. The minimum speed required to start dynamic hydroplaning.

V_R—Rotation speed. The speed that the pilot begins rotating the aircraft prior to lift-off.

V_{S0}—Stalling speed or the minimum steady flight speed in the landing configuration. In small airplanes, this is the power-off stall speed at the maximum landing weight in the landing configuration (gear and flaps down). The lower limit of the white arc.

V_{S1}—Stalling speed or the minimum steady flight speed obtained in a specified configuration. For most airplanes, this is the power-off stall speed at the maximum takeoff weight in the clean configuration (gear up, if retractable, and flaps up). The lower limit of the green arc.

V_{SSE}—Safe, intentional one-engine inoperative speed. The minimum speed to intentionally render the critical engine inoperative.

V-TAIL—A design which utilizes two slanted tail surfaces to perform the same functions as the surfaces of a conventional elevator and rudder configuration. The fixed surfaces act as both horizontal and vertical stabilizers.

V_X—Best angle-of-climb speed. The airspeed at which an airplane gains the greatest amount of altitude in a given distance. It is used during a short-field takeoff to clear an obstacle.

V_{XSE}—Best angle of climb speed with one engine inoperative. The airspeed at which an airplane gains the greatest amount of altitude in a given distance in a light, twin-engine airplane following an engine failure.

V_Y—Best rate-of-climb speed. This airspeed provides the most altitude gain in a given period of time.

V_{YSE}—Best rate-of-climb speed with one engine inoperative. This airspeed provides the most altitude gain in a given period of time in a light, twin-engine airplane following an engine failure.

WAKE TURBULENCE—Wingtip vortices that are created when an airplane generates lift. When an airplane generates lift, air spills over the wingtips from the high pressure areas below the wings to the low pressure areas above them. This flow causes rapidly rotating whirlpools of air called wingtip vortices or wake turbulence.

WASTE GATE—A controllable valve in the tailpipe of an aircraft reciprocating engine equipped with a turbocharger. The valve is controlled to vary the amount of exhaust gases forced through the turbocharger turbine.

WEATHERVANE—The tendency of the aircraft to turn into the relative wind.

WEIGHT—A measure of the heaviness of an object. The force by which a body is attracted toward the center of the Earth (or another celestial body) by gravity. Weight is equal to the mass of the body times the local value of gravitational acceleration. One of the four main forces acting on an aircraft. Equivalent to the actual weight of the aircraft. It acts downward through the aircraft's center of gravity toward the center of the Earth. Weight opposes lift.

WEIGHT AND BALANCE—The aircraft is said to be in weight and balance when the gross weight of the aircraft is under the max gross weight, and the center of gravity is within limits and will remain in limits for the duration of the flight.

WHEELBARROWING—A condition caused when forward yoke or stick pressure during takeoff or landing causes the aircraft to ride on the nosewheel alone.

WIND CORRECTION ANGLE—Correction applied to the course to establish a heading so that track will coincide with course.

WIND DIRECTION INDICATORS—Indicators that include a wind sock, wind tee, or tetrahedron. Visual reference will determine wind direction and runway in use.

WIND SHEAR—A sudden, drastic shift in windspeed, direction, or both that may occur in the horizontal or vertical plane.

WINDMILLING—When the air moving through a propeller creates the rotational energy.

WINDSOCK—A truncated cloth cone open at both ends and mounted on a freewheeling pivot that indicates the direction from which the wind is blowing.

WING—Airfoil attached to each side of the fuselage and are the main lifting surfaces that support the airplane in flight.

WING AREA—The total surface of the wing (square feet), which includes control surfaces and may include wing area covered by the fuselage (main body of the airplane), and engine nacelles.

WING SPAN—
The maximum distance from wingtip to wingtip.

WINGTIP VORTICES—
The rapidly rotating air that spills over an airplane's wings during flight. The intensity of the turbulence depends on the airplane's weight, speed, and configuration. It is also referred to as wake turbulence. Vortices from heavy aircraft may be extremely hazardous to small aircraft.

WING TWIST—A design feature incorporated into some wings to improve aileron control effectiveness at high angles of attack during an approach to a stall.

YAW—Rotation about the vertical axis of an aircraft.

YAW STRING—A string on the nose or windshield of an aircraft in view of the pilot that indicates any slipping or skidding of the aircraft.

ZERO FUEL WEIGHT—
The weight of the aircraft to include all useful load except fuel.

ZERO SIDESLIP—A maneuver in a twin-engine airplane with one engine inoperative that involves a small amount of bank and slightly uncoordinated flight to align the fuselage with the direction of travel and minimize drag.

ZERO THRUST (SIMULATED FEATHER)—
An engine configuration with a low power setting that simulates a propeller feathered condition.

Index

D

E

F

G

I

J

K

L

M

N

O

P

R

S

T

U

V

W

Y